Fundamentals of Signal Processing in Metric Spaces with Lattice Properties

ALGEBRAIC APPROACH

Fundamentals of Signal Processing in Metric Spaces with Lattice Properties

ALGEBRAIC APPROACH

Andrey Popoff

CRC Press
Taylor & Francis Group
Boca Raton London New York

CRC Press is an imprint of the
Taylor & Francis Group, an **informa** business

CRC Press
Taylor & Francis Group
6000 Broken Sound Parkway NW, Suite 300
Boca Raton, FL 33487-2742

© 2018 by Andrey Popoff
CRC Press is an imprint of Taylor & Francis Group, an Informa business

No claim to original U.S. Government works

Printed on acid-free paper

International Standard Book Number-13: 978-1-138-09938-8 (Hardback)

This book contains information obtained from authentic and highly regarded sources. Reasonable efforts have been made to publish reliable data and information, but the author and publisher cannot assume responsibility for the validity of all materials or the consequences of their use. The authors and publishers have attempted to trace the copyright holders of all material reproduced in this publication and apologize to copyright holders if permission to publish in this form has not been obtained. If any copyright material has not been acknowledged please write and let us know so we may rectify in any future reprint.

Except as permitted under U.S. Copyright Law, no part of this book may be reprinted, reproduced, transmitted, or utilized in any form by any electronic, mechanical, or other means, now known or hereafter invented, including photocopying, microfilming, and recording, or in any information storage or retrieval system, without written permission from the publishers.

For permission to photocopy or use material electronically from this work, please access www.copyright.com (http://www.copyright.com/) or contact the Copyright Clearance Center, Inc. (CCC), 222 Rosewood Drive, Danvers, MA 01923, 978-750-8400. CCC is a not-for-profit organization that provides licenses and registration for a variety of users. For organizations that have been granted a photocopy license by the CCC, a separate system of payment has been arranged.

Trademark Notice: Product or corporate names may be trademarks or registered trademarks, and are used only for identification and explanation without intent to infringe.

Library of Congress Cataloging-in-Publication Data

Names: Popoff, Andrey, author.
Title: Fundamentals of signal processing in metric spaces with lattice
properties : algebraic approach / Andrey Popoff.
Description: Boca Raton : CRC Press, Taylor & Francis Group, 2017. | Includes
bibliographical references and index.
Identifiers: LCCN 2017029313| ISBN 9781138099388 (hardback : acid-free
paper) | ISBN 9781315104119 (ebook)
Subjects: LCSH: Signal processing--Mathematics. | Metric spaces.
Classification: LCC TK5102.9 P655 2017 | DDC 621.382/2--dc23
LC record available at https://lccn.loc.gov/2017029313

Visit the Taylor & Francis Web site at
http://www.taylorandfrancis.com

and the CRC Press Web site at
http://www.crcpress.com

Contents

List of Figures   ix

Preface   xiii

Introduction   xix

List of Abbreviations   xxvii

Notation System   xxix

1 General Ideas of Natural Science, Signal Theory, and Information Theory   1
    1.1 General System of Notions and Principles of Modern Research Methodology   2
    1.2 Information Theory and Natural Sciences   5
    1.3 Overcoming Logical Difficulties   10

2 Information Carrier Space Built upon Generalized Boolean Algebra with a Measure   25
    2.1 Information Carrier Space   26
    2.2 Geometrical Properties of Metric Space Built upon Generalized Boolean Algebra   29
       2.2.1 Main Relationships between Elements of Metric Space Built upon Generalized Boolean Algebra with a Measure   29
       2.2.2 Notion of Line in Metric Space Built upon Generalized Boolean Algebra with a Measure   32
       2.2.3 Notions of Sheet and Plane in Metric Space Built upon Generalized Boolean Algebra with a Measure   41
       2.2.4 Axiomatic System of Metric Space Built upon Generalized Boolean Algebra with a Measure   46
       2.2.5 Metric and Trigonometrical Relationships in Space Built upon Generalized Boolean Algebra with a Measure   51
       2.2.6 Properties of Metric Space with Normalized Metric Built upon Generalized Boolean Algebra with a Measure   54
    2.3 Informational Properties of Information Carrier Space   59

3 Informational Characteristics and Properties of Stochastic Processes 75
3.1 Normalized Function of Statistical Interrelationship of Stochastic Processes 76
3.2 Normalized Measure of Statistical Interrelationship of Stochastic Processes 82
3.3 Probabilistic Measure of Statistical Interrelationship of Stochastic Processes 98
3.4 Information Distribution Density of Stochastic Processes 101
3.5 Informational Characteristics and Properties of Stochastic Processes 108

4 Signal Spaces with Lattice Properties 123
4.1 Physical and Informational Signal Spaces 123
4.2 Homomorphic Mappings in Signal Space Built upon Generalized Boolean Algebra 133
    4.2.1 Homomorphic Mappings of Continuous Signal Into a Finite (Countable) Sample Set in Signal Space Built upon Generalized Boolean Algebra with a Measure 133
    4.2.2 Theorems on Isomorphism in Signal Space Built upon Generalized Boolean Algebra with a Measure 143
4.3 Features of Signal Interaction in Signal Spaces with Various Algebraic Properties 146
    4.3.1 Informational Paradox of Additive Signal Interaction in Linear Signal Space: Notion of Ideal Signal Interaction ... 147
    4.3.2 Informational Relationships Characterizing Signal Interaction in Signal Spaces with Various Algebraic Properties 156
4.4 Metric and Informational Relationships between Signals Interacting in Signal Space 163

5 Communication Channel Capacity 173
5.1 Information Quantity Carried by Discrete and Continuous Signals 174
    5.1.1 Information Quantity Carried by Binary Signals 174
    5.1.2 Information Quantity Carried by m-ary Signals 179
    5.1.3 Information Quantity Carried by Continuous Signals 190
5.2 Capacity of Noiseless Communication Channels 196
    5.2.1 Discrete Noiseless Channel Capacity 199
    5.2.2 Continuous Noiseless Channel Capacity 200
    5.2.3 Evaluation of Noiseless Channel Capacity 201
        5.2.3.1 Evaluation of Capacity of Noiseless Channel Matched with Stochastic Stationary Signal with Uniform Information Distribution Density 201
        5.2.3.2 Evaluation of Capacity of Noiseless Channel Matched with Stochastic Stationary Signal with Laplacian Information Distribution Density 204

6 Quality Indices of Signal Processing in Metric Spaces with L-group Properties — 207

- 6.1 Formulation of Main Signal Processing Problems — 208
- 6.2 Quality Indices of Signal Filtering in Metric Spaces with L-group Properties — 210
- 6.3 Quality Indices of Unknown Nonrandom Parameter Estimation — 218
- 6.4 Quality Indices of Classification and Detection of Deterministic Signals — 229
  - 6.4.1 Quality Indices of Classification of Deterministic Signals in Metric Spaces with L-group Properties — 229
  - 6.4.2 Quality Indices of Deterministic Signal Detection in Metric Spaces with L-group Properties — 237
- 6.5 Capacity of Communication Channels Operating in Presence of Interference (Noise) — 242
  - 6.5.1 Capacity of Continuous Communication Channels Operating in Presence of Interference (Noise) in Metric Spaces with L-group Properties — 244
  - 6.5.2 Capacity of Discrete Communication Channels Operating in Presence of Interference (Noise) in Metric Spaces with L-group Properties — 247
- 6.6 Quality Indices of Resolution-Detection in Metric Spaces with L-group Properties — 250
- 6.7 Quality Indices of Resolution-Estimation in Metric Spaces with Lattice Properties — 259

7 Synthesis and Analysis of Signal Processing Algorithms — 267

- 7.1 Signal Spaces with Lattice Properties — 269
- 7.2 Estimation of Unknown Nonrandom Parameter in Sample Space with Lattice Properties — 271
  - 7.2.1 Efficiency of Estimator $\hat{\lambda}_{n,\wedge}$ in Sample Space with Lattice Properties with Respect to Estimator $\hat{\lambda}_{n,+}$ in Linear Sample Space — 276
  - 7.2.2 Quality Indices of Estimators in Metric Sample Space — 282
- 7.3 Extraction of Stochastic Signal in Metric Space with Lattice Properties — 287
  - 7.3.1 Synthesis of Optimal Algorithm of Stochastic Signal Extraction in Metric Space with L-group Properties — 288
  - 7.3.2 Analysis of Optimal Algorithm of Signal Extraction in Metric Space with L-group Properties — 294
  - 7.3.3 Possibilities of Further Processing of Narrowband Stochastic Signal Extracted on Basis of Optimal Filtering Algorithm in Metric Space with L-group Properties — 302
- 7.4 Signal Detection in Metric Space with Lattice Properties — 304
  - 7.4.1 Deterministic Signal Detection in Presence of Interference (Noise) in Metric Space with Lattice Properties — 305

- 7.4.2 Detection of Harmonic Signal with Unknown Nonrandom Amplitude and Initial Phase with Joint Estimation of Time of Signal Arrival (Ending) in Presence of Interference (Noise) in Metric Space with L-group Properties 311
- 7.4.3 Detection of Harmonic Signal with Unknown Nonrandom Amplitude and Initial Phase with Joint Estimation of Amplitude, Initial Phase, and Time of Signal Arrival (Ending) in Presence of Interference (Noise) in Metric Space with L-group Properties 324
- 7.4.4 Features of Detection of Linear Frequency Modulated Signal with Unknown Nonrandom Amplitude and Initial Phase with Joint Estimation of Time of Signal Arrival (Ending) in Presence of Interference (Noise) in Metric Space with L-group Properties 342
- 7.5 Classification of Deterministic Signals in Metric Space with Lattice Properties 352
- 7.6 Resolution of Radio Frequency Pulses in Metric Space with Lattice Properties 358
- 7.7 Methods of Mapping of Signal Spaces into Signal Space with Lattice Properties 378
  - 7.7.1 Method of Mapping of Linear Signal Space into Signal Space with Lattice Properties 379
  - 7.7.2 Method of Mapping of Signal Space with Semigroup Properties into Signal Space with Lattice Properties 383

Conclusion   385

Bibliography   389

Index   403

List of Figures

I.1   Dependences $Q(m_1^2/m_2)$ of some normalized signal processing quality index Q on the ratio m_1^2/m_2 xxii

I.2   Dependences $Q(q^2)$ of some normalized signal processing quality index Q on the signal-to-noise ratio q^2 xxii

2.2.1   Hexahedron built upon a set of all subsets of element $A + B$.   30

2.2.2   Tetrahedron built upon elements $\mathbf{O}, A, B, C$ 30

2.2.3   Simplex $Sx(A_i + A_j + A_k)$ in metric space Ω   55

2.3.1   Information quantities between two sets A and B   61

2.3.2   Elements of set $A = \bigcup_\alpha A_\alpha$ situated on n-dimensional sphere $Sp(\mathbf{O}, R)$ 62

3.2.1   Metric relationships between samples of signals in metric signal space $(\mathbf{\Gamma}, \mu)$ 95

3.2.2   Metric relationships between samples of signals in metric space $(\mathbf{\Gamma}, \mu)$ elucidating Theorem 3.2.14   97

3.5.1   IDDs $i_\xi(t_j, t), i_\xi(t_k, t)$ of samples $\xi(t_j), \xi(t_k)$ of stochastic process $\xi(t)$ 109

3.5.2   IDD $i_\xi(t_j, t)$ and mutual IDD $i_{\xi\eta}(t_k^0, t)$ of samples $\xi(t_j), \eta(t_k')$ of stochastic processes $\xi(t), \eta(t')$ 119

4.2.1   Graphs of dependences of ratio $I_\Delta(X')/I(X)$ on discretization interval Δt   142

4.3.1   Dependences $I_{bx}(I_{ax})$ of quantity of mutual information I_{bx} on quantity of mutual information I_{ax}   160

5.1.1   NFSI $\psi_u(\tau)$ and normalized ACF $r_u(\tau)$ of stochastic process .   175

5.1.2   IDD $i_u(\tau)$ of stochastic process   176

5.1.3   NFSI of multipositional sequence   181

5.1.4   IDD of multipositional sequence   181

6.2.1   Metric relationships between signals elucidating Theorem 6.2.1   213

6.2.2   Dependences of NMSIs $\nu_{s\hat{s}}(q), \nu_{sx}(q)$ on signal-to-noise ratio $q^2 = S_0/N_0$ 217

6.2.3   Dependences of metrics $\mu_{s\hat{s}}(q), \mu_{sx}(q)$ on signal-to-noise ratio $q^2 = S_0/N_0$ 217

6.3.1   CDFs of estimation (measurement) results $X_{\vee,i}, X_{\wedge,i}$ and estimators $\hat{\lambda}_{n,\wedge}, \hat{\lambda}_{n,\vee}$ 221

6.3.2 Graphs of metrics $\mu(\hat{\lambda}_{n,\wedge}, \lambda)$ and $\mu'(\hat{\lambda}_{n,\vee}, \lambda)$ depending on parameter λ 226
6.3.3 Dependences $q\{\hat{\lambda}_{n,+}\}$ and $q\{\hat{\lambda}_{n,\wedge}\}$ on size of samples n 228
6.4.1 Dependences $\nu_{s\hat{s}}(q^2)|_{\perp}$ and $\nu_{s\hat{s}}(q^2)|_{-}$ 235
6.5.1 Generalized structural scheme of communication channel functioning in presence of interference (noise) 243
6.7.1 PDFs of estimators $\hat{s}_{\wedge}(t)|_{H_{11}, H_{01}}$ and $\hat{s}_{\vee}(t)|_{H_{11}, H_{01}}$ 263
6.7.2 Signal $\hat{s}_{\wedge}(t)$ in output of filter forming estimator (6.7.8) 265

7.2.1 PDF $p_{\hat{\lambda}n,\wedge}(z)$ of estimator $\hat{\lambda}_{n,\wedge}$ 276
7.2.2 Upper $\Delta_{\wedge}(n)$ and lower $\Delta_{+}(n)$ bounds of quality indices $q\{\hat{\lambda}_{n,\wedge}\}$ and $q\{\hat{\lambda}_{n,+}\}$ 286
7.3.1 Block diagram of processing unit realizing general algorithm $Ext[s(t)]$ 293
7.3.2 Realization $s^*(t)$ of useful signal $s(t)$ acting in input of filtering unit and possible realization $w^*(t)$ of process $w(t)$ in output of adder 296
7.3.3 Generalized block diagram of signal processing unit 303
7.4.1 Block diagram of deterministic signal detection unit 308
7.4.2 Signals $z(t)$ and $u(t)$ in outputs of correlation integral computing unit and strobing circuit 310
7.4.3 Estimator $\hat{\theta}$ of unknown nonrandom parameter θ of useful signal $s(t)$ in output of decision gate 310
7.4.4 Block diagram of processing unit that realizes harmonic signal detection with joint estimation of time of signal arrival (ending) 319
7.4.5 Useful signal $s(t)$ and realizations $w^*(t)$, $v^*(t)$ of signals in outputs of adder and median filter 320
7.4.6 Useful signal $s(t)$ and realization $v^*(t)$ of signal in output of median filter $v(t)$ 320
7.4.7 Useful signal $s(t)$, realization $v^*(t)$ of signal in output of median filter $v(t)$, and δ-pulse determining time position of estimator $\hat{t}_1$ 321
7.4.8 Useful signal $s(t)$, realization $E_v^*(t)$ of envelope $E_v(t)$ of signal $v(t)$, and δ-pulse determining time position of estimator $\hat{t}_1$ .. 321
7.4.9 Block diagram of unit processing observations defined by Equations: (a) (7.4.11a); (b) (7.4.11b) 323
7.4.10 Block diagram of processing unit that realizes harmonic signal detection with joint estimation of amplitude, initial phase, and time of signal arrival (ending) 336
7.4.11 Useful signal $s(t)$ and realization $w^*(t)$ of signal $w(t)$ in output of adder 337
7.4.12 Useful signal $s(t)$ and realization $v^*(t)$ of signal $v(t)$ in output of median filter 337
7.4.13 Useful signal $s(t)$ and realization $W^*(t)$ of signal $W(t)$ in output of adder 338

List of Figures

7.4.14	Useful signal $s(t)$, realization $V^*(t)$ of signal $V(t)$, and realization $E_V^*(t)$ of its envelope	338
7.4.15	Block diagram of processing unit that realizes LFM signal detection with joint estimation of time of signal arrival (ending) . .	349
7.4.16	Useful signal $s(t)$ and realization $w^*(t)$ of signal $w(t)$ in output of adder	350
7.4.17	Useful signal $s(t)$ and realization $v^*(t)$ of signal $v(t)$ in output of median filter	350
7.4.18	Useful signal $s(t)$ and realization $v^*(t)$ of signal $v(t)$ in output of median filter, and δ-pulse determining time position of estimator $\hat{t}_1$	351
7.4.19	Useful signal $s(t)$ and realization $E_v^*(t)$ of envelope $E_v(t)$ of signal $v(t)$	351
7.5.1	Block diagram of deterministic signals classification unit	355
7.5.2	Signals in outputs of correlation integral computation circuit $z_i(t)$ and strobing circuit $u_i(t)$	357
7.6.1	Block diagram of signal resolution unit	366
7.6.2	Signal $w(t)$ in input of limiter in absence of interference (noise)	368
7.6.3	Normalized time-frequency mismatching function $\rho(\delta\tau, \delta F)$ of filter	370
7.6.4	Cut projections of normalized time-frequency mismatching function	370
7.6.5	Realization $w^*(t)$ of signal $w(t)$ including signal response and residual overshoots	373
7.6.6	Realization $w^*(t)$ of stochastic process $w(t)$ in input of limiter and residual overshoots	376
7.6.7	Realization $v^*(t)$ of stochastic process $v(t)$ in output of median filter	376
7.6.8	Normalized time-frequency mismatching function of filter in presence of strong interference	377
7.7.1	Block diagram of mapping unit	379
7.7.2	Directional field patterns $F_A(\theta)$ and $F_B(\theta)$ of antennas A and B	380
C.1	Suggested scheme of interrelations between information theory, signal processing theory, and algebraic structures	387

Preface

Electronics is one of the main high-end technological sectors of the economics providing development and manufacture of civil, military, and double-purpose production, whose level defines the technological, economical, and informational security of the leading countries of the world. Electronics serves as a catalyst and a locomotive of scientific and technological progress, promoting the stable growth of the various industry branches, and world economics as a whole. The majority of electronic systems, means, and sets are the units of information (signal) transmitting, receiving, and processing.

As in other branches of science, the progress in both information theory and signal processing theory is directly related to understanding and investigating their fundamental principles. While it is acceptable in the early stages of the development of a theory to use some approximations, over the course of time it is necessary to have a closed theory which is able to predict unknown earlier phenomena and facts. This book was conceived as an introduction into the field of signal processing in non-Euclidean spaces with special properties based on measure of information quantity.

Successful research in the 21st century is impossible without comprehensive study of a specific discipline. Specialists often have poor understanding of their colleagues working in adjacent branches of science. This is not surprising because a severance of scientific disciplines ignores the interrelations existing between various areas of science. To a degree, detriment caused by a narrow specialization is compensated by popular scientific literature acquainting the reader with a wider range of the phenomena and also offset by the works pretending to cover some larger subject matter domain of research.

The research subject of this book includes the methodology of constructing the unified mathematical fundamentals of both information theory and signal processing theory, the methods of synthesis of signal processing algorithms under prior uncertainty conditions, and also the methods of evaluating their efficiency.

While this book does not constitute a transdisciplinary approach, it starts with generalized methodology based on natural sciences to create new research concepts with application to information and signal processing theories.

Two principal problems will be investigated. The first involves unified mathematical fundamentals of information theory and signal processing theory. Its solution is provided by definition of an information quantity measure connected in a unique way to the notion of signal space. The second problem is the need to increase signal processing efficiency under parametric and nonparametric prior uncertainty conditions. The resolution of that problem rests on the first problem solution using signal spaces with various algebraic properties such as groups, lattices, and generalized Boolean algebras.

This book differs from traditional monographs in (1) algebraic structure of signal spaces, (2) measures of information quantities, (3) metrics in signal spaces, (4) common signal processing problems, and (5) methods for solving such problems (methods for overcoming prior uncertainty).

This book is intended for professors, researchers, post-graduates under-graduate students, and specialists in signal processing and information theories, electronics, radiophysics, telecommunications, various engineering disciplines including radio-engineering, and information technology. It presents alternative approaches to constructing signal processing theory and information theory based upon Boolean algebra and lattice theory. The book may be useful for mathematicians and physicists interested in applied problems in their areas. The material contained in the book differs from the traditional approaches of classical information theory and signal processing theory and may interest pre-graduate students specializing in the directions: "radiophysics", "telecommunications", "electronic systems", "system analysis and control", "automation and control", "robotics", "electronics and communication engineering", "control systems engineering", "electronics technology", "information security", and others.

Signal processing theory is currently one of the most active areas of research and constitutes a "proving ground" in which mathematical ideas and methods find their realization in quite physical analogues. Moreover, the analysis of IT tendencies in the 21st century, and perspective transfer towards quantum and optic systems of information storage, transmission, and processing allow us to claim that in the nearest future the parallel development of signal processing theory and information theory will be realized on the basis of their stronger interpenetration of each other. Also, the author would like to express his confidence that these topics could awaken interest of young researchers, who will test their own strengths in this direction.

The possible directions of future applications of new ideas and technologies based on signal processing in spaces with lattice properties described in the book could be extremely wide. Such directions could cover, for example, search for extraterrestrial intelligence (SETI) program, electromagnetic compatibility problems, and military applications intended to provide steady operation of electronic systems under severe jam conditions.

Models of real signals discussed in this book are based on stochastic processes. Nevertheless, the use of generalized Boolean algebra with a measure allows us to expand the results upon the signal models in the form of stochastic fields that constitutes the subject of independent discussing. Traditional aspects of classical signal processing and information theories are not covered in the book because they have been presented widely in the existing literature. Readers of this book should understand the basics of set theory, abstract algebra, mathematical analysis, and probability theory. The final two chapters require knowledge of mathematical statistics or statistical radiophysics (radioengineering).

The book is arranged in the following way. Chapter 1 is introductory in nature and describes general methodology questions of classical signal theory and information theory. Brief analysis of general concepts, notions, and ideas of natural science, signal theory, and information theory is carried out. The relations between information theory and natural sciences are briefly discussed. Theoretical difficulties within foundations of both signal and information theories are shown. Overcoming these obstacles is discussed.

Chapter 2 formulates an approach to constructing a signal space on the basis of generalized Boolean algebra with a measure. Axiomatic system of metric space built upon generalized Boolean algebra with a measure is formulated. Chapter 2 establishes metric and trigonometrical relationships in space built upon generalized Boolean algebra with a measure and considers informational properties of metric space built upon generalized Boolean algebra with a measure. The properties are introduced by the axiom of a measure of a binary operation of generalized Boolean algebra.

Chapter 3 introduces probabilistic characteristics of stochastic processes which are invariant with respect to groups of their mappings. The interconnection between introduced probabilistic characteristics and metric relations between the instantaneous values (the samples) of stochastic processes in metric space is shown. Informational characteristics of stochastic processes are introduced. Chapter 3 establishes the necessary condition according to which a stochastic process possesses the ability to carry information. The mapping that allows considering an arbitrary stochastic process as subalgebra of generalized Boolean algebra with a measure is introduced. Informational properties of stochastic processes are introduced on the base of the axiom of a measure of a binary operation of generalized Boolean algebra. The main results are formulated in the form of corresponding theorems.

Chapter 4 together with Chapter 7 occupies the central place in the book and deals with the notions of informational and physical signal spaces. On the basis of probabilistic and informational characteristics of the signals and their elements, that are introduced in the previous chapter, Chapter 4 considers the characteristics and the properties of informational signal space built upon generalized Boolean algebra with a measure. At the same time, the separate signal carrying information is considered as a subalgebra of generalized Boolean algebra with a measure. We state that a measure on Boolean algebra accomplishes a twofold function: firstly, it is a measure of information quantity and secondly it induces a metric in signal space. The interconnection between introduced measure and logarithmic measure of information quantity is shown. Some homomorphic mappings in informational signal space are considered. Particularly for this signal space, the sampling theorem is formulated. Theorems on isomorphisms are established for informational signal space. The informational paradox of additive signal interaction in linear signal space is considered. Informational relations, taking place under signal interaction in signal spaces with various algebraic properties, are established. It is shown that from the standpoint of providing minimum losses of information contained in the signals, one should carry out their processing in signal spaces with lattice properties.

Chapter 5 establishes the relationships determining the quantities of information carried by discrete and continuous signals. The upper bound of information quantity, which can be transmitted by discrete random sequence, is determined by a number of symbols of the sequence only, and does not depend on code base (alphabet size). This fact is stipulated by the fundamental property of information, i.e., by its ability to exist exclusively within a statistical collection of structural elements of its carrier (the signal). On the basis of introduced information quantity measure, relationships are established to determine the capacities of discrete and continuous noiseless communication channels. Boundedness of capacity of discrete and continuous noiseless channels is proved. Chapter 5 concludes with examples of evaluating the capacity of noiseless channel matched with stochastic stationary signal characterized by certain informational properties.

Chapter 6 considers quality indices of signal processing in metric spaces with L-group properties (i.e., with both group and lattice properties). These indices are based on metric relationships between the instantaneous values (the samples) of the signals determined in Chapter 3. The obtained quality indices correspond to the main problems of signal processing, i.e., signal detection; signal filtering; signal classification; signal parameter estimation; and signal resolution. Chapter 6 provides brief comparative analysis of the obtained relationships, determining the differences of signal processing qualities in the spaces with mentioned properties. Potential quality indices of signal processing in metric spaces with lattice properties are characterized by the invariance property with respect to parametric and nonparametric prior uncertainty conditions. The relationships determining capacity of communication channel operating in the presence of interference (noise) in metric spaces with L-group properties are obtained. All the main signal processing problems are considered from the standpoint of estimating the signals and/or their parameters.

Chapter 7 deals with synthesis of algorithms and units of signal processing in metric spaces with lattice properties, so that the developed approaches allow operating with minimum necessary prior data concerning characteristics and properties of interacting useful and interference signals. This means that no prior data concerning probabilistic distribution of useful signals and interference are supposed to be present. Second, a priori, the kinds of useful signal (signals) are assumed to be known, i.e., we know that they are either deterministic (quasi-deterministic) or stochastic. The quality indices of synthesized signal processing algorithms and units are obtained. Algorithms and units of signal processing in metric spaces with lattice properties are characterized by the invariance property with respect to parametric and nonparametric prior uncertainty conditions. Chapter 7 concludes with methods of mapping of signal spaces with group (semigroup) properties into signal spaces with lattice properties.

The first, second, and seventh chapters are relatively independent and may be read separately. Before reading the third and the fourth chapters it is desirable to get acquainted with main content of the first and second chapters. Understanding Chapters 4 through 6 to a great extent relies on the main ideas stated in the third chapter. Proofs provided for an exigent and strict reader may be ignored at least on first reading.

There is a triple numeration of formulas in the book: the first number corresponds to the chapter number, the second one denotes the section number, the third one corresponds to the equation number within the current section, for example, (2.3.4) indicates the 4th formula in Section 2.3 of Chapter 2. The similar numeration system is used for axioms, theorems, lemmas, corollaries, examples, and definitions. Proofs of theorems and lemmas are denoted by □ symbols. Endings of examples are indicated by the ▽ symbols. Outlines generalizing the obtained results are placed at the end of the corresponding sections if necessary. In general, the chapters are not summarized.

While translating the book, the author attempted to provide "isomorphism" of the monograph perception within principal ideas, approaches, methods, and main statements formulated in the form of mathematical relationships along with key notions.

As remarked by Johannes Kepler, "... If there is no essential strictness in terms, elucidations, proofs and inference, then a book will not be a mathematical one. If the strictness is provided, the book reading becomes very tiresome ...".[1]

The author's intention in writing this book was to provide readers with concise comprehensible content while ensuring adequate coverage of signal processing and information theories and their applications to metric signal spaces with lattice properties. The author recognizes the complexity of the subject matter and the fast pace of related research. He welcomes all comments by readers and can be reached (at andoff@rambler.ru).

The author would like to extend his sincere appreciation, thanks, and gratitude to Victor Astapenya, Ph.D.; Alexander Geleseff, D.Sc.; Vladimir Horoshko, D.Sc.; Sergey Rodionov, Ph.D.; Vladimir Rudakov, D.Sc.; and Victor Seletkov, D.Sc. for attention, support, and versatile help that contributed greatly to expediting the writing of this book and improving its content.

The author would like to express his frank acknowledgment to Allerton Press, Inc. and also to its Senior Vice President Ruben de Semprun for granted permissions that allow using the material published by the author in *Radioelectronics and Communications System*.

The author would like to acknowledge understanding, patience, and support provided within CRC Press by its staff and the assistance from Nora Konopka, Michele Dimont, Kyra Lindholm, and unknown copy editor(s).

Finally, the author would like to express his thanks to all LaTeX developers whose tremendous efforts greatly lighten the author's burden.

ANDREY POPOFF

---

[1] J. Kepler. Astronomia Nova, Prague, 1609.

Introduction

At the frontier of the 21st century, the amounts of information obtained and processed in all fields of human activity, from oceanic depths to remote parts of cosmic space, increase exponentially every year. It is impossible to satisfy growing needs of humanity in transmitting, receiving, and processing of all sorts of information without continuous improvement of acoustic, optic, electronic systems and signal processing methods. The last two tendencies provide the development of information theory, signal processing theory, and synthesis foundations of such systems.

The subject of information theory includes the analysis of qualitative and quantitative relationships that take place through transmitting, receiving, and processing of information contained in both messages and signals, so that the useful signal $s(t)$ is considered a one-to-one function of a transmitted message $m(t)$: $s(t) = M[c(t), m(t)]$ (where M is the modulating one-to-one function and $c(t)$ is a signal carrier; $m(t) = M^{-1}[c(t), s(t)]$).

The subject of signal processing theory includes the analysis of probabilistic-statistic models of the signals interacting properly with each other and statistical inference in specific aspects of the process of extracting information contained in signals. Information theory and signal processing theory continue to develop in various directions.

Thus, one should refer the following related directions to information theory in its classical formulation:

1. Analysis of stochastic signals and messages, including the questions on information quantity evaluation
2. Analysis of informational characteristics of message sources and communication channels in the presence and absence of noise (interference)
3. Development of message encoding/decoding methods and means
4. Development of mathematical foundations of information theory
5. Development of communication system optimization methodology
6. Development of all other problems whose formulations includes the notion of information

One can trace the development of information theory within the aforementioned six directions in References [1–7], [8–24], [25–49], [50–70], [71–84], [85–98], respectively.

We distinguish several interrelated directions, along which the signal processing theory developed:

1. Analysis of probabilistic-statistic characteristics of stochastic signals, interference (noise), their transformations and their influence on the operation of electronic means and systems

2. Optimization and analysis of useful signals set used to solve the main signal processing problems in electronic means and systems of different functionality

3. Synthesis of signal processing algorithms and units while solving the main signal processing problems: signal detection (including multiple-alternative detection), classification of signals, signal extraction, signal filtering, signal parameter estimation, resolution of the signals and recognition of the signals

4. Development of mathematical foundations of signal processing theory

The main ideas and approaches to solving the signal processing theory problems in their historical retrospective within the aforementioned four directions are included in References [99–117], [118–124], [125–142], [143–167], correspondingly.

It should be noted that this bibliography does not include purely mathematical works, especially in probability theory and mathematical statistics, which had an impact on the course of the development of both information theory and signal processing theory.

The analysis of the state of the art of signal processing theory and information theory reveals a contradiction between relative proximity of subject matter domains of these theories and their rather weak interactions at the levels of the key notions and used methods.

Information quantity measure was introduced in information theory, ignoring the notion of signal space. On the other hand, the approaches to the synthesis of signal processing algorithms and units along with the approaches to their efficiency evaluation within the framework of signal processing theory were developed without use of information theory methods and categories.

At first, the material of this book was developed exclusively within the mathematical foundations of signal processing theory in non-Euclidean spaces. However it became impossible to write some acceptable variant of signal processing fundamentals without referring to the basics of information theory. Besides, some aspects of classical information theory need more serious refinement of principal approaches than conceptual basics of the direction Lewis Franks named the *signal theory*. Dissatisfaction with both theories' state of the art, when information carriers (the signals) and the principles of their processing exist per se, while information transferred by them and the approaches intended to describe a wide range of informational processes are apart from information carriers made the author to focus his effort on unifying the mathematical foundations of both signal processing theory and information theory.

Most of the signal processing theory problems (along with the problems of statistical radiophysics, statistical radioengineering, and automatic control theory) make no sense in the absence of noise and interference and in general cases, without taking into account the random character of the received signals. An interaction between useful and interference (noise) signals usually is described through the superposition principle; however, it has neither fundamental nor universal character. This means that while constructing fundamentals of information theory and signal processing theory, there is no need to confine the space of information material carriers (signal space) artificially by the properties which are inherent exclusively to linear spaces.

Meanwhile, in most publications, signal processing problems are formulated within the framework of additive commutative groups of linear spaces, where interaction between useful and interference (noise) signals is described by a binary operation of addition. Rather seldom these problems are considered in terminology of ring binary operations, i.e., in additive commutative groups with introduced multiplication operation connected with addition by distributive laws. In this case, interaction between useful and interference (noise) signals is described by multiplication and addition operations in the presence of multiplicative and additive noise (interference), respectively.

Capabilities of signal processing in linear space are confined within potential quality indices of optimal systems demonstrated by the results of classical signal processing theory. Besides, modern electronic systems and means of different functionalities operate under prior uncertainty conditions adversely affecting quality indices of signal processing.

Under prior uncertainty conditions, the efficiency of signal processing algorithms and units can be evaluated upon some i-th distribution family of interference (noise) $\mathbf{D}_i[\{a_{i,k}\}]$, where $\{a_{i,k}\}$ is a set of shape parameters of this distribution family; $i, k \in \mathbf{N}$, $\mathbf{N}$ is the set of natural numbers, on the basis of the dependences $Q(m_1^2/m_2)$, $Q(q^2)$ of some normalized signal processing quality index on the ratio m_1^2/m_2 between squared average m_1^2 and the second order moment m_2 of interference (noise) envelope [156], and also on signal-to-noise (signal-to-interference) ratio q^2, respectively. By normalized signal processing quality index Q we will mean any signal processing quality index that takes its values in the interval $[0, 1]$, so that 1 and 0 correspond to the best and the worst values of Q, respectively. For instance, it could be conditional probability of correct detection; correlation coefficient between useful signal and its estimator, etc. By the envelope $E_x(t)$ of stochastic process $x(t)$ (particularly of interference) we mean the function: $E_x(t) = \sqrt{x^2(t) + x_\mathcal{H}^2(t)}$, where $x_\mathcal{H}(t) = -\frac{1}{\pi} \int\limits_{-\infty}^{\infty} \frac{x(\tau)}{\tau-t} d\tau$ (as a principal value) is the Hilbert transform of initial stochastic process $x(t)$. Dependences $Q(m_1^2/m_2)$ of such normalized signal processing quality index Q on the ratio m_1^2/m_2 (q^2=const) for an arbitrary i-family of interference (noise) distribution $\mathbf{D}_i[\{a_{i,k}\}]$ may look like the curves 1, 2, and 3 shown in Fig. I.1. Optimal Bayesian decisions [143, 146, 150, 152] and the decisions obtained on the basis of robust methods [157, 166, 168] and on the basis of nonparametric statistics methods [150, 169–172], at a qualitative level, as a rule, are characterized by the curves 1, 2, and 3, respectively. Figure I.1 conveys generalized behaviors of some groups of signal processing algorithms operating in nonparametric prior uncertainty conditions, when (1) distribution family of interference is

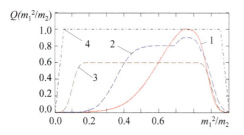 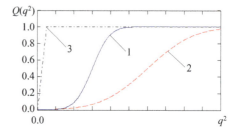

FIGURE I.1 Dependences $Q(m_1^2/m_2)$ of some normalized signal processing quality index Q on the ratio m_1^2/m_2 that characterize (1) optimal Bayesian decisions; decisions, obtained on the basis of (2) robust methods; (3) nonparametric statistics methods; (4) desirable dependence

FIGURE I.2 Dependences $Q(q^2)$ of some normalized signal processing quality index Q on the signal-to-noise ratio q^2 that characterize (1) the best case of signal receiving; (2) the worst case of signal receiving; (3) desirable dependence

known, but concrete type of distribution is unknown (i.e., m_1^2/m_2 is unknown); (2) the worst case, when even distribution family of interference is unknown. Conditions when $m_1^2/m_2 \in]0, 0.7]$, $m_1^2/m_2 \in]0.7, 0.8]$, $m_1^2/m_2 \in]0.8, 1.0[$ correspond to interference of pulse, intermediate, and harmonic kind, respectively. At the same time, while solving the signal processing problems under prior uncertainty conditions, it is desirable for all, or at least several interference (noise) distribution families, to obtain the dependence in the form of the curve 4 shown in Fig. I.1: $Q(m_1^2/m_2) \to 1(m_1^2/m_2 - \varepsilon) - 1(m_1^2/m_2 - 1 + \varepsilon)$, where $1(*)$ is Heaviside step function and ε is an indefinitely small positive number.

The curve 4 in Fig. I.1 is desirable, since quality index Q is equal to 1 in the whole interval $]0, 1[$ of the ratio m_1^2/m_2, i.e., signal processing algorithm is absolutely robust ($Q=1$) with respect to nonparametric prior uncertainty conditions.

Figure I.2 illustrates generalized behaviors of signal processing algorithms operating in parametric prior uncertainty conditions, when (1) interference distribution is known and energetic/spectral characteristics of useful and interference signals are known (curve 1) and (2) interference distribution is unknown and/or energetic/spectral characteristics of useful and/or interference signals are unknown (curve 2).

Figure I.2 shows dependences $Q(q^2)$ of some normalized signal processing quality index Q on the signal-to-noise ratio q^2 that characterize: (1) the best case of signal receiving, (2) the worst case of signal receiving; (3) desirable dependence. The curve 3 in Fig. I.2 is desirable, since quality index Q is equal to 1 in the whole interval $]0, \infty[$ of q^2, i.e., signal processing algorithm is absolutely robust ($Q=1$) with respect to parametric prior uncertainty conditions.

Thus, depending on interference (noise) distribution and also on characteristics of a useful signal, optimal variants of a solution of some signal processing problem provide the location of functional dependence $Q(q^2)$ within the interval between the curves 1 and 2 in Fig. I.2, which, at a qualitative level, characterize the most

Introduction xxiii

and the least desired cases of useful signal receiving (the best and the worst cases, respectively). However, while solving the signal processing problems under prior uncertainty conditions, for all interference (noise) distribution families, the most desired dependence $Q(q^2)$ undoubtedly is the curve 3 in Fig. I.2: $Q(q^2) \to 1(q^2 - \varepsilon)$, where, similarly, $1(*)$ is Heaviside step function and ε is indefinitely small positive number. Besides, at a theoretical level, the maximum possible proximity between the functions $Q(q^2)$ and $1(q^2)$ should be indicated.

Providing constantly growing requirements for the real signal processing quality indices under prior uncertainty conditions on the basis of known approaches formulated, for instance, in [150, 152, 157, 166], appears to be problematic; thus, other ideas should be used in this case. This book is devoted to a great large-scale problem, i.e., providing signal processing quality indices Q in the form of dependencies 4 and 3 shown in Figs. I.1 and I.2, correspondingly.

The basic concepts for signal processing theory and information theory are *signal space* and *information quantity*, respectively. The development of the notions of information quantity and signal space leads to important methodological conclusions concerning interrelation between informational theory (within its syntactical aspects) and signal processing theory. But this leads to other questions. For example, what is the interrelation between, on the one hand, set theory (ST), mathematical analysis (MA), probability theory (PT), and mathematical statistics (MS); and on the other hand, between information theory (IT) and signal processing theory (SPT)? Earlier it seemed normal to formulate axiomatic grounds of probability theory on the basis of set theory by introducing a specific measure, i.e., the probabilistic one and the grounds of signal theory (SgT) on the basis of mathematical analysis (function spaces of special kind, i.e., linear spaces with scalar product). In a classical scheme, the following one: the connection ST $\to$ PT $\to$ IT was traced separately, the relation PT $\to$ MS $\to$ SPT was observed rather independently, and absolutely apart there existed the link MA $\to$ SgT. Here and below, arrows indicate how one or another theory is related to its mathematical foundations.

The "set" interpretation of probability theory has its own weak position noted by many mathematicians, including its author A.N. Kolmogoroff. Since the second part of the 20th century, we know an approach, according to which Boolean algebra with a measure (BA) forms an adequate mathematical model of the notion called an "event set"; thus, the interrelation is established: BA $\to$ PT [173–176]. Correspondingly, in abstract algebra (AA) within the framework of lattice theory (LT), Boolean algebras are considered to be further development of algebraic structures with special properties, namely, the lattices. Instead of traditional schemes of theories' interrelations, in this work, to develop the relationship between information theory and signal processing theory, we use the following:

$$\text{LT} \to \text{BA} \to \{\text{PT} \to \{\text{IT} \leftrightarrow \text{SPT} \leftarrow \{\text{MS} + (\text{SgT} \leftarrow \{\text{AA} + \text{MA}\})\}\}\}.$$

Undoubtedly, this scheme is simplified, because known interrelations between abstract algebra, geometry, topology, mathematical analysis, and also adjacent parts of modern mathematics are not shown.

The choice of Boolean algebra as mathematical apparatus for foundations of both signal processing theory and syntactical aspects of information theory is not an arbitrary one. Boolean algebra, considered as a set of the elements (in this case, each signal is a set of the elements of its instantaneous values) possessing the certain properties, is intended for signal space description. A measure defined upon it is intended for the quantitative description of informational relationships between the signals and their elements, i.e., the instantaneous values (the samples).

In this work, special attention is concentrated on interrelation between information theory and signal processing theory (IT $\leftrightarrow$ SPT). Unfortunately, such a direct and tangible interrelation between classical variants of these theories is not observed. The answer to the principal question of signal processing theory—how, in fact, one should process the results of interaction of useful and interference (noise) signals, is given not with information theory, but by means of applying the special parts of mathematical statistics, i.e., statistical hypothesis testing and estimation theory. While investigating such an interrelation (IT $\leftrightarrow$ SPT), it is important that information theory and signal processing theory could answer interrelated questions below.

So, information theory with application to signal processing theory has to resolve certain issues:

1.1. Information quantity measure, its properties and its relation to signal space as a set of material carriers of information with special properties.

1.2. The relation between the notion of signal space in information theory and the notion of signal space in signal processing theory.

1.3. Main informational relationships between the signals in signal space that is the category of information theory.

1.4. Interrelation between potential quality indices (confining the efficiency) of signal processing in signal space which is the category of information theory, and the main informational relationships between the signals.

1.5. Algebraic properties of signal spaces, where the best signal processing quality indices may be obtained by providing minimum losses of information contained in useful signals.

1.6. Informational characteristics of communication channels built upon signal spaces with the properties mentioned in Item 1.5.

Signal processing theory with application to information theory has the following issues:

2.1. Main informational relations between the signals in signal space that is the category of signal processing theory.

2.2. Informational interrelation between main signal processing problems.

2.3. Interrelation between potential quality indices (confining the efficiency) of signal processing in signal space with certain algebraic properties and main informational relationships between the signals.

2.4. The synthesis of algorithms and units of optimal signal processing in signal space with special algebraic properties, taking into consideration Item 1.5 (the synthesis problem).

2.5. Quality indices of signal processing algorithms and units in signal space with special algebraic properties, taking into consideration Item 1.5 (the analysis problem).

Apparently, it is possible to resolve the above issues only by constructing signal processing theory while providing the unity at the level of its basics with information theory (at least within the framework of syntactical aspects), thus providing their theoretical compatibility and harmonious associativity.

List of Abbreviations

Notion	Abbreviation
Abstract algebra	AA
Autocorrelation function	ACF
Characteristic function	CF
Cumulative distribution function	CDF
Decision gate	DG
Envelope computation unit	ECU
Estimator formation unit	EFU
Generalized Boolean algebra	GBA
Hyperspectral density	HSD
Information distribution density	IDD
Information theory	IT
Lattice theory	LT
Least modules method	LMM
Least squares method	LSM
Linear frequency modulated signal	LFM signal
Matched filtering unit	MFU
Mathematical analysis	MA
Mathematical statistics	MS
Median filter	MF
Mutual information distribution density	mutual IDD
Mutual normalized function of statistical interrelationship	mutual NFSI
Normalized function of statistical interrelationship	NFSI
Normalized measure of statistical interrelationship	NMSI
Overall quantity of information	o.q.i.
Power spectral density	PSD
Probabilistic measure of statistical interrelationship	PMSI
Probability density function	PDF
Probability theory	PT
Radio frequency	RF
Relative quantity of information	r.q.i.
Set theory	ST
Signal detection unit	SDU
Signal extraction unit	SEU
Signal processing theory	SPT
Signal theory	SgT
White Gaussian noise	WGN

Notation System

Notion	Notation
ABSTRACT ALGEBRA AND UNIVERSAL ALGEBRA	
Additive semigroup	$\mathbf{SG}(+)$
Boolean algebra	$\mathbf{B}'$
Boolean lattice	$\mathbf{BL}'$
Boolean ring	$\mathbf{BR}'$
Generalized Boolean algebra	$\mathbf{B}$
Generalized Boolean lattice	$\mathbf{BL}$
Generalized Boolean ring	$\mathbf{BR}$
Group	$\mathbf{G}$
Lattice	$\mathbf{L}$
Measure on generalized Boolean algebra	m
Metric upon generalized Boolean algebra	$\rho(A,B) = m(A \Delta B)$
Multiplicative semigroup	$\mathbf{SG}(\cdot)$
Null element	$\mathbf{O}$
Ring	$\mathbf{R}(+,\cdot)$
Semigroup	$\mathbf{SG}$
Signature of universal algebra $\mathbf{A}$	$T_\mathbf{A}$
Unit element	$\mathbf{I}$
Universal algebra	$\mathbf{A} = (X, T_\mathbf{A})$
PROBABILITY THEORY AND MATHEMATICAL STATISTICS	
Characteristic function	$\Phi_w(u)$
Coefficient of statistical interrelation	$\psi_{\xi\eta}$
Cumulative distribution function	$F_{\xi\eta}(x,y),\ \ F_\xi(x)$
Estimator of parameter λ	$\hat{\lambda}$
Hyperspectral density	$\sigma_\xi(\omega)$
Linear sample space	$\mathcal{LS}(\mathcal{X}, \mathcal{B}_\mathcal{X}; +)$
Mathematical expectation	$\mathbf{M}(*)$
Matrix of probabilities of transition	$\mathbf{\Pi}_n$
Metric between samples $\xi_t,\ \eta_t$	$\mu(\xi_t, \eta_t)$
Metric between stochastic processes $\xi(t),\ \eta(t)$	$\rho_{\xi\eta}$
Mutual normalized function of statistical interrelationship	$\psi_{\xi\eta}(t_j, t'_k)$
Negative part of stochastic process $v(t)$	$v_-(t)$
Normalized correlation function	$r_\xi(t_1, t_2),\ r(\tau)$
Normalized function of statistical interrelationship	$\psi_\xi(t_j, t_k),\ \psi(\tau)$
Normalized measure of statistical interrelationship	$\nu(\xi_t, \eta_t),\ \nu(a_t, b_t)$
Normalized variance function	$\theta_\xi(t_1, t_2)$
Positive part of stochastic process $v(t)$	$v_+(t)$
Power spectral density	$S(\omega)$

Continued on next page

Notion	Notation
Probabilistic measure of statistical interrelationship	$\nu_\mathbf{P}(\xi_t, \eta_t),\ \nu_\mathbf{P}(a_t, b_t)$
Probability	$\mathbf{P}$
Probability density function	$p_{\xi\eta}(x,y),\ p_\xi(x)$
Sample space with lattice properties	$\mathcal{L}(\mathcal{Y}, \mathcal{B}_\mathcal{Y}; \vee, \wedge)$
Sample space with L-group properties	$\mathcal{L}(\mathcal{X}, \mathcal{B}_\mathcal{X}; +, \vee, \wedge)$
Stochastic process	$\xi(t), \eta(t), \ldots$
SIGNAL PROCESSING THEORY	
Carrier frequency	f_0
Conditional probability of correct detection	D
Conditional probability of false alarm	F
Cross-correlation coefficient between two signals	r_{ik}
Detection of signal $s(t)$	$Det[s(t)]$
Domain of definition of signal $s(t)$	T_s
Dynamic error of signal filtering	δ_{d0}
Dynamic error of signal smoothing	$\delta_{d,\mathrm{sm}}$
Energy falling at one bit of information	E_b
Energy of signal $s(t)$	E_s
Envelope of signal $v(t)$	$E_v(t)$
Estimation of time parameter t_1	$Est[t_1]$
Estimator of the signal $s(t)$	$\hat{s}(t)$
Estimator of time parameter t_1	$\hat{t}_1$
Extraction of the signal $s(t)$	$Ext[s(t)]$
Frequency shift	Δ_F
Intermediate processing of signal $s(t)$	$IP[s(t)]$
Intermediate smoothing of signal $s(t)$	$ISm[s(t)]$
Linear signal space	$\mathcal{LS}(+)$
Matched filtering	$MF[s(t)]$
Metric signal space	$(\mathbf{\Gamma}, \mu)$
Mismatching function	$w(\boldsymbol{\lambda}', \boldsymbol{\lambda})$
Modulating function	$M[*,*]$
Noise power spectral density	N_0
Normalized mismatching function	$\rho(\boldsymbol{\lambda}', \boldsymbol{\lambda})$
Normalized time-frequency mismatching function	$\rho(\delta\tau, \delta F)$
Number of periods of harmonic signal $s(t)$	N_s
Observation interval	T_obs
Period of signal carrier	T_0
Physical signal space	$\mathbf{\Gamma}$
Primary filtering	$PF[s(t)]$
Probability of correct formation of signal estimator $\hat{s}(t)$	$P(C_\mathrm{c})$
Probability of error formation of signal estimator $\hat{s}(t)$	$P(C_\mathrm{e})$
Realization of signal $x(t)$	$x^*(t)$
Relative frequency shift	δ_F
Relative time delay	δ_τ
Resolution of signals $s_1(t), s_2(t)$	$Res[s_{1,2}(t)]$
Signal space with L-group properties	$\mathcal{L}(+, \vee, \wedge), \mathbf{\Gamma}(+, \vee, \wedge)$
Signal space with lattice properties	$\mathcal{L}(\vee, \wedge), \mathbf{\Gamma}(\vee, \wedge)$
Signals	$a(t), b(t), \ldots$

Continued on next page

Notion	Notation
Smoothing of signal $s(t)$	$Sm[s(t)]$
Time delay	Δ_τ
Time of arrival of signal	t_0
INFORMATION THEORY	
Channel capacity	C
Information distribution density	$i_\xi(t_\alpha, t)$, $i_\xi(\tau)$
Information losses of first genus	I'_L
Information losses of second genus	I''_L
Informational signal space	Ω
Mutual information distribution density	$i_{\xi\eta}(t'_k, t)$, $i_{\eta\xi}(t_j, t')$
Overall quantity of information	$I(A)$, $I[\xi(t)]$
Quantity of absolute information	I_A
Quantity of mutual information	$I_{A \cdot B}$
Quantity of overall information	I_{A+B}
Quantity of particular relative information	I_{A-B}
Quantity of relative information	$I_{A \Delta B}$
Relative quantity of information	$I_\Delta(A)$, $I_\Delta[\xi(t)]$
GEOMETRY	
Barycenter of n-dimensional simplex	$bc[Sx(A)]$
Curvature of set of elements A	$c(A)$
Line	l
Plane	α_{ABC}
Sheet	L_{AB}
Simplex	$Sx(A)$
Sphere	Sp
Triangle	$\triangle ABC$
GENERAL NOTATIONS	
Cardinality of set $U = \{u_i\}$	$\operatorname{Card} U$
Dirac delta function	$\delta(x)$
End of example	$\triangledown$
End of proof	$\square$
Euclidean space (n-dimensional)	$\boldsymbol{R}^n$
Fourier transform	$\mathcal{F}[*]$
Heaviside step function	$1(\tau)$
Hilbert space	$\mathcal{HS}$
Hilbert transform	$\mathcal{H}[*]$
Indexed set	$\{A_t\}_{t \in T}$
Set of natural numbers	$\boldsymbol{N}$

1

General Ideas of Natural Science, Signal Theory, and Information Theory

All scientific research has both subject and methodological content. The last is connected with critical reconsideration of existing conceptual apparatus and approaches for interpretation of phenomena of interest.

A researcher working with real world physical objects eventually must choose a methodological basis to describe researched phenomena. This basis determines a proper mathematical apparatus.

The choice of correct methodological basis is the key to success and often the reason more useful information than expected is obtained. Mathematical principles and laws contain a lot of hidden information accumulated during the ages. Mathematical ideas could give much more than expected before. That is why a person engaged in some fundamental mathematical research may not foresee all the possible applications in natural science, sometimes creating precedent when "the most principal thing has been said by someone, who does not understand it". For example, Johannes Kepler possessed all the necessary information to formulate universal gravitation law, but he did not do so, etc.

It is clear that any mathematical model developed on the basis of a specific methodology may describe the real phenomenon studied with a certain amount of accuracy. But it may happen that the mathematical model used until now does not satisfy the requirements of completeness, adequacy, and internal consistency anymore. In such a case, a researcher has two choices. The first is to explain all known facts, all new facts, and appearing paradoxes within the Procrustean bed of an old theory. The second option is to devise a new mathematical model to explain all the available facts.

The modern methodologies of all natural sciences involve its base, i.e., methodological basis (a certain mathematical apparatus), general system of notions and principles, and also particular system of notions and principles of a given scientific direction. These components driving natural science research are interconnected within their disciplines and to other branches of science.

We first consider general system of notions and principles of modern research methodology within natural sciences and then discuss a particular system of notions of specific research directions, in this case signal processing theory and information theory. We also explain the relationships of these subjects to various branches of the natural sciences.

Phenomena and processes related to information and signal processing theories are essential constituent parts of an entire physical world. The foundations of these theories rely on the same fundamental notions and the same research techniques utilized by all branches of science. The first step is to find out the content of general system of notions used in research methodology of natural sciences, without going into details of its structure, which is well stated in the proper literature. Next, the researcher should analyze the suitability of the systems of notions that are used in classical information theory and signal processing theory. Finally, one should determine how both theories converge within the framework of research methodology of natural sciences.

1.1 General System of Notions and Principles of Modern Research Methodology

Since ancient times, philosophers tried to understand natural phenomena, considering them within the unity of preservation and change of parts and the properties of the whole; within the unity of order and disorder, regular and random; and within the relation between diversity and duplication, discreteness and continuity. As natural science continued to develop, new concepts explained the relationships between symmetry versus asymmetry and linearity versus nonlinearity.

Science progressed on the basis of *deterministic principles* for most of its history. However, scientific facts revealed in the last few centuries demonstrated the need for a *probabilistic approach*, i.e., the rejection of the unique description of the considered phenomena. Thermodynamics became the first branch of physics in which probabilistic methods were tested. The subject was later named statistical thermodynamics. Scientists of the 19th century conceived the wave properties of light as streams of discrete particles. By the 20th century, the use of the probabilistic approach to describe these phenomena stimulated further review of fundamental ideas and resulted in the creation of quantum mechanics. Applying probabilistic methods to classical physics led to the development of statistical physics and its first parts, i.e., statistical thermodynamics and quantum statistics. The transition from discreteness to continuity and from deterministic to probabilistic methods proved productive in studies of natural phenomena.

We know that the discrete versus continuous and deterministic versus probabilistic properties of matter should not be in opposition because they are closely linked and effective if used in combination. Determining interrelations of various properties of materials is not a simple endeavor. The degree of success is determined by the depth of investigation of studied objects.

Symmetry in the natural world dates back to antiquity; its extraordinary versatility was not known until the 19th century. Most natural phenomena display symmetry and that fact played an important role in scientific advancement [177], [178], [179], [180].

Studies of symmetry led to the discovery of *invariance principles* confining the diversity of the objects' structure. On the contrary, asymmetry leads to the objects' diversity grounded in a given structural basis. Commonality of properties constitutes symmetry; distinct properties reveal asymmetry.

Symmetry is almost impossible to define exactly. In every single phenomenon, symmetry inevitably takes the form corresponding to it. Examples of symmetry include the metrics of poetry and music, coloring in painting, word arrangements in literature, constellations in astronomy. Symmetry reveals itself within limitations of physical processes passing. These constraints are described by laws of preservation of energy, mass, momentum, electrical charge, etc. The creation of relativistic quantum theory evoked the discovery of a new type of symmetry. This is symmetry of nature's laws with respect to simultaneous transformations of charge conjugation, parity transformation, and time reversal designated CPT-symmetry. Symmetry underlies elementary particle classification in chemistry and allowed Dmitri Mendeleev to devise his periodic table of elements. Gregor Mendel applied the idea of symmetry to characterize hereditary factors. In abstract algebra and topology, symmetry appears in isomorphic and homomorphic mappings respectively.

One more unusual qualitative measure of investigative complexity is *nonlinearity*. One of the first attempts to solve the wave equation for a pendulum demonstrated that its deflections were not always negligibly small and one should use more exact expressions to determine the deflecting force that creates a nonlinear appearance. Even small nonlinear additions qualitatively change a situation; a sum of several solutions may not satisfy an equation. There is no principle of superposition. One cannot obtain a general solution from a group of specific solutions; other approaches may be required.

Nonlinearity appears within the majority of real systems [181], [182], [183]. In investigation of nonlinearity, difficulties taking place are stipulated by the fact that the world of nonlinear phenomena, which require the special models for their description, is much wider than "linear world". There are a lot of important nonlinear equations which should be studied. To top it all off, the majority of such equations cannot be solved analytically.

Nonlinearity is a feature describing the interactions of components within a physical structure; it is also a universal property of various types of matter that provide the diversity of material forms found throughout the world.

Among the most important ideas upon which all our knowledge systems are built is the notion of *space*. By the notion of space modern geometry means the set of some geometrical objects that are connected to each other by certain relations. Depending on the character of these ties, we differ Euclidean and non-Euclidean types of spaces.

The discoveries of nonEuclidean geometry and group theory represented turning points in the history of mathematics. Since then, a new era of mathematical development has begun and various types of geometries beyond the Euclidean started to appear [184], [185], [186]. New algebraic systems that had no classical analogues advance mathematics beyond the basic arithmetic of real numbers.

The greatest challenges at present are intensive research of cosmic space and studies of the behaviors of elementary articles. In this connection, we now want to understand the structure of the universe and the geometry and algebra of intra-atomic space. In the opinion of Paul Dirac:

> The modern physical developments have required a mathematics that continually shifts its foundation and gets more abstract. Non-Euclidean geometry and noncommutative algebra, which at one time were considered to be purely fictions of the mind and pastimes of logical thinkers, have now been found to be very necessary for the description of general facts of the physical world [187].

Modern science has no definitive answers to these questions. At present, we can express only the most general understanding. The main elements of cosmic space are objects called *straight lines*. They are trajectories of light wave movements or trajectories of movements of particles bearing light energy, i.e., photons. The gravity field lines surrounding all masses of matter are considered rectilinear. Trajectories of material particles (cosmic rays) moving freely throughout the universe are rectilinear.

All these straight lines analyzed on Earth's scale are considered identical but that conclusion may not be correct. We have no reason yet to speak about the geometry of the universe. We can speak only about the geometries of light rays and gravitation fields, etc. It is quite possible that these geometries can be absolutely different, and the issue becomes even more complicated because the concepts of general relativity theory, electromagnetic waves, and gravity fields are dependent on each other.

The violation of the rectilinearity of light waves within gravity fields was established theoretically and confirmed by observations. Light rays passing a heavy body, for example, near the Sun, are distorted. The geometry of light rays in space is complicated because huge masses of matter are distributed nonuniformly throughout the universe.

General relativity theory revealed the interdependence of gravity field space, electromagnetic field space, and time. These objects define four-dimensional space whose laws have been explained by modern physicists, astronomers, and mathematicians.

At present, we can say only that the properties of these objects are not described by Euclidean geometry.

The geometry of the intra-atomic world is more ambiguous. In cosmic space, we can indicate straight lines in a certain sense, but it is impossible to do this with atomic nuclei. We have little to say about the geometry of intra-atomic space but we can certainly say there is no Euclidean geometry there.

Although the word *information* has served as a catch-all for a long time, its use in the middle of the 20th century evolved to describe a specific concept that plays a critical role in all fields of science. Application of the information approach expanded greatly since then and Claude Shannon reminded scientists working in social and humanitarian disciplines of the need to keep their houses in first class order [188]. At present, there are a lot of definitions of this notion. The definition choice sufficiently depends on researchers' directions, goals, techniques, and available technologies of research in every individual case.

The definition of information has been covered widely in scientific literature for decades but no single definition has appeared. The main reason is a persistent development of information theory and other sciences that actively use a theoretical-informational approach in their methodology. Expansion of the notion of information continues to impact its definition and reveal new features and properties. The alternative is either to develop a new definition of the notion based on a simple generalization of new data, or to give qualitatively new definition of the notion of information, which would be broad enough to not require revision whenever new revelations appear.

The analysis of the development of information theory allows the selection of the most universal and sufficient features. The notion of information in its most general form was described [189] as a reflected diversity and an interconnection between reflection and diversity. Nevertheless, whether information is considered a philosophical category or not, it occupies an important place in modern science and a return to the past, when natural sciences could operate without this notion, is impossible. The role of information in natural science has expanded steadily. The 1970 statement by Russian scientists A.I. Berg and B.V. Biryukov that, "Information, probably, will take the first place in the 21st century in a world of scientific and practically efficient notions" was certainly prophetic [190].

No concept or approach can be constructed without using a combination of key notions. In this book, the key notions are probability and probabilistic approaches, symmetry and invariance principles, space and its geometry, nonlinearity, information and its measure, discrete versus continuous, and deterministic versus probabilistic. The main key notions of this approach are signal space and a measure of information quantity. All necessary key notions regarding the suggested approach are considered in their relation to research methodology used in natural sciences.

1.2 Information Theory and Natural Sciences

Analytical apparatus of information theory was created after the edifice of mathematical statistics had already been built. The central problem of mathematical statistics is the development of methods that allow us to extract from data storage the most vital information about the phenomena of interest. It is no wonder that the first steps in studying information as a science were taken by Ronald A. Fischer who is considered the founder of modern mathematical statistics. Apparently, he was the first mathematician who understood this notion needs more accurate definition and he introduced the notion of sufficient statistics, i.e., an extract from observable data that contains all information about distribution parameters.

Fischer's measure of information quantity contained in data concerning an unknown parameter is well known to statisticians. This measure is the first use of the notion of "information quantity" introduced mainly for the needs of estimation theory.

After the appearance of the works of Claude Shannon [51] and Norbert Wiener [85], interest in information theory and its utility in the fields of physics, biology, psychology, and other hard and soft science fields increased. Solomon Kullback connects it also with Wiener's statement, that in practice of statistics, his (Wiener's) definition of information could be used instead of Fischer's [89]. We should note Leonard Savage's remark that, "The ideas of Shannon and Wiener, though concerned with probability, seem rather far from statistics. It is, therefore, something of an accident that the term 'information' coined by them should be not altogether inappropriate in statistics".

The main thesis of Wiener's book titled *Cybernetics, or Control and Communication in the Animal and the Machine* [85] was the similarity of control and communications processes in machines, living organisms, and societies. These processes encompass transmission, storage, and processing of information (signals carrying messages).

One of the brightest examples are the processes of genetic information transmission. Genetic information transmission plays a vital role in all forms of life. About 2 million species of flora and fauna inhabit the Earth. Transmission of genetic information determines the development of all organisms from single cells to their adult forms. Transmitted genetic data governs species structures and individual features for both present and future generations. All this information is preserved within a small volume of elementary cell nucleus and is transmitted through intricate ways to all the other cells originated from a given one by cell fission; this information is preserved also during the process of further reproduction of next generations of similar species.

Every field of natural science and technology relies on information transmission, receiving, and transformation. Visible light reports a lot of live creatures up to 90% of data concerning the surrounding world, electromagnetic waves and fluxes of particles carry an imprint of the processes from the universe remote parts. All living organisms depend on information on their relationships with each other, with other living things, and with inanimate objects. Physicists, philosophers, and all others who studied all aspects of our existence and our world depended on the availability of *information* and *information quantity* long before Shannon and Wiener defined those terms.

When information theory developed in the works of mathematicians, it dropped out of sight of physics and other nature sciences, and some scientists opined that information was an intangible notion that had nothing to do with energy transfer and other physical phenomena. The reasons for such misunderstandings could be explained by some peculiarities of the origin of information theory, its further development and application. Mathematical communication theory appearance was stimulated by the achievements of electronics which in those times was based on classical electrodynamics with its inherent ideas of continuity and absolute simultaneous measurability of all the parameters of physical objects.

As communication technologies evolved, information theory was used to solve the problems of both communication channel optimization and encoding method optimization, which transformed into a vast independent part of mathematics, having

1.2 Information Theory and Natural Sciences

moved into the background the problems of physical constraints in communication systems.

Nevertheless, the analogy between Shannon – Wiener's *entropy* and Boltzmann's *thermodynamic entropy* attracted a lot of research attention. Physical ideas in information theory never seemed to be strange to outstanding representatives of its mathematical direction. The use of high frequency intervals of the electromagnetic spectrum and miniaturization of information processing systems precipitated the development of electronics that approach the limit at which quantum mechanical regularities and related constraints become essential.

By the end of the 20th century, quantum computing was accepted as a new branch of science. Using the laws of quantum physics allowed us to create new types of high capacity supercomputers. We know that quantum systems can provide high computation speed and that messages transmitted via quantum communication channels cannot be copied or intercepted [191], [192]. Russian scientist Alexander Holevo said, "Regardless of the fact of how soon such projects could be realized, quantum information theory is a new direction giving a key to understanding the Nature's fundamental regularities that were until now out of the field of researchers' vision" [97].

What is the specificity of a physical approach to the problems of information theory? Does the need for radical revision of Shannon's theory really exist or does the answer lie in application of physical equations to describe the properties of communication channels? How should one connect the inferences of statistical information theory with general constraints imposed by Nature's laws on communication channels? How can one establish the limits of fundamental realizability of various information transmitting and processing systems? We know that under a given signal power in the absence of interference (noise), channel capacity is always a finite quantity [92].

Measurement is one of the fundamental problems of exact sciences. More generally, the problem is extracting information during detection, extraction, estimation, classification, resolution, recognition, and other informational procedures. They are widely used in physics (within relativity theory and quantum theory), mathematical statistics (within statistical hypotheses testing and estimation theory), and other disciplines. One should not consider the measurement process simply a comparison of measured parameters with measurement units. One should take into account the conditions under which measurements are made.

Measurement process is a specific case of the informational process; it is the extraction of information about an object or its parameters.

In fact, the discussion on some theoretical physics questions has a theoretical-informational character, at least, those which are connected with the causality principle. It is enough to notice that the notion of the signal is used in deduction of the laws of special theory of relativity.

Regarding the close connection of information theory and measurement (or estimation), we recall Wiener's remark [193]: "...One single voltage measured to an accuracy of one part in ten trillion could convey all the information in the *Encyclo-*

pedia Britannica, if the circuit noise did not hold us to an accuracy of measurement of perhaps one part in ten thousand".

Another important issue is the specificity of measurement processes that has no analogues in classical physics. According to quantum theory, probabilistic prediction of measurement results of signal receiving cannot be determined only by a signal state; a list of measurable parameters should be specified. Exact measurement of some parameters usually fails to include reliable estimates regarding other parameters. For example, while aspiring to measure velocity of an elementary particle in quantum physics or a target in radiolocation or hydrolocation, the ability to obtain information about an object's position in space is limited.

The essential peculiarity of some measurement processes, especially those based upon indirect methods of measurements, is nonlinearity of a measurement space (or a sample space). An appearing problem of data processing optimization methods is not trivial, and in general has not been resolved satisfactorily as of this writing.

The development of electronic systems that serve various functions and are characterized by the use of the microwave and optic ranges of the electromagnetic spectrum requires study of the influences of physical phenomena on the processes of signal transmitting, receiving, and processing.

The application of physical methods to information theory involves reverse process: the use of information to solve some key problems of theoretical physics.

An information interchange is an example of a process developing from the past into the future. One can say, time "has a direction". A listener cannot understand a compact disk spinning in the reverse direction. Reverse pulling of film strips was used widely to create surrealistic effects in the early phases of cinematography development. The world may look absurd when the courses of events are reversed.

The laws of classical mechanics discovered by Isaac Newton are reversible. In the equations time can figure with either positive or negative signs. Thus, time can be considered reversible or irreversible. Time direction plays an important role in research on life processes, meteorology, thermodynamics, quantum physics, and other scientific areas.

The concept of an obvious irreversibility of time has found the most clear-cut formulation in thermodynamics in the form of the Second Law, according to which a quantitative characteristic called *entropy* never decreases. Originally, thermodynamics was used to study the properties of the gases, i.e., large ensembles of particles that are in persistent movement and interaction with each other. We can obtain only partial data about such ensembles. Although Newton's laws are applicable to every single particle, it is impossible to observe each of them, or to distinguish one particle from another. Their characteristics can not be determined exactly; they can be studied only on the base of probabilistic relationships.

One can obtain certain data concerning macroscopic properties of this ensemble of particles, for instance, concerning a number of degrees of freedom or dimension, pressure, volume, temperature, energy. Some properties can be represented by statistical distributions, for instance, particles' velocity distribution. Also one can observe some microscopic movements, but it is impossible to obtain complete data about every particle.

1.2 Information Theory and Natural Sciences

An important place among probabilistic systems is occupied by information interchange systems. Both types of systems can utilize the same statistical methods to describe their behaviors. Information sources do not contain complete data. We can explain the properties of an ensemble or encoding system even though some constraints are imposed on their messages or signals. A priori, we cannot know the details of a message source or determine which message will be transmitted at the next moment. We cannot anticipate coming events or obtain complete information from signals.

In statistical thermodynamics, *entropy* is the function of probabilities of the states of the particles that constitute a gas. In statistical communication theory, information quantity is the same function of message source states. In both cases, we have some ensemble. In the case of a gas, the ensemble is the totality of particles whose states (energies) are distributed according to some function of probability. In communication theory, totality of messages or message source states are also described by some function of probability. The relation between information and entropy is revealed in the following formula of Wiener and Shannon:

$$H = -\sum_i p_i \log p_i,$$

where p_i is the probability of the i-th state of message source.

The amazing connection between two areas of knowledge can be used in two ways. One can follow the exact mathematical methods, taking into account the appropriateness of their use or follow a less formal method. A lot has been written about the entropy of social and economic systems and their applications in various areas of research suffering from "methodological hunger." As noted by Edward Cherry [88]:

> It is the kind of sweeping generality which people will clutch like a straw. Some part of these interpretations has indeed been valid and useful, but the concept of entropy is one of considerable difficulty and of a deceptively apparent simplicity. It is essentially a mathematical concept and the rules of its application are clearly laid down.

Unfortunately, the ease of transfer of thermodynamics ideas to information theory and other sciences already led to several unsubstantiated and speculative ideas. Shannon clearly sensed that and in a critical article [188] he wrote:

> Starting as a technical tool for the communication engineer, it (information theory) has received an extraordinary amount of publicity in the popular as well as the scientific press. ... As a consequence, it has perhaps been ballooned to an importance beyond its actual accomplishments. ... What can be done to inject a note of moderation in this situation? In the first place, workers in other fields should realize that the basic results of the subject are aimed in a very specific direction... Secondly, we must keep our own house in first class order.

Of course, keeping "our own house in first class order" in the context of information theory means neither dust removal nor furniture transposition. Now it is realized by the addition of outhouse extensions to the edifice of informational theory.

Among them a researcher can select unified information theory [54], semantic information theory [86], [91], quantum information theory [97], algebraic information theory [95], dynamic information theory [98], combinatorial methods [59] and topological approaches [87] to the definition of information. We cannot claim that architectural harmony exists among the various choices. We can only hope that information theory in the foreseeable future will be built upon an integrated load-carrying frame.

1.3 Overcoming Logical Difficulties

Modern teachings about space are logical, consistent, and branched, and rest on a synthesis of physical, mathematical, and philosophical ideas. In antiquity, the first expression of mathematical abstraction of real space was developed and is now known as Euclidean geometry. This geometry reflects the space relationships of our daily experience with great accuracy, and has been used for more than 2,000 years. Until recently it was the only space theory that represented lucidity, harmony, and entirety for all natural sciences.

During the development of natural sciences and philosophy of New Time, teaching about space became inseparably linked with principles of mechanics developed by Isaac Newton and Galileo Galilei. Newton put forth the notions of absolute space and relative space into the base of his mechanics. Within contemporary interpretation, Newton's theses on classical mechanics [194] state:

1. Space exists independently of anything in the world.
2. Space contains all the objects of nature and gives the place to all of its phenomena, but does not experience their influence upon itself.
3. Space is everywhere and the same with respect to its properties. All its points are equitable and the same — space is isotropic.
4. In all times space is invariably the same.
5. Space is ranging along all the directions unrestrictedly and has infinite volume.
6. Space has three dimensions.
7. Space is described by Euclidean geometry.

Newton's understanding was considered the only true one for almost two centuries despite Gottfried Leibniz' opposing position [195]. Leibniz detected a boundedness of Newton's overview of space and considered the separation of space from matter to be erroneous. He wrote, "What is the space without a material object? ...Unhesitatingly I will answer that I do not know anything about this." Leibniz

1.3 Overcoming Logical Difficulties

considered space as a property of the world of material objects without which it could not exist. This interpretation of the notion of space, while logical, did not correspond to the majority opinion about the unity and universality of Euclidean geometry. Although Leibniz's idea seemed incontestable and obvious, the scientific community would not even discuss the possible existence of a geometry other than Euclidean.

The New Time's teaching about space developed two opposite concepts: (a) space is a receptacle of all the material objects and (b) space is a property of a set of material objects. Concept (a) is connected with Newton and (b) is attributed to Leibniz.

In a surprising development in geometry since the discoveries of Nikolai Lobachevsky and János Bolyai, Bernhard Riemann pursued differential-metrical research and Felix Klein studied a group-theoretical approach. These developments in geometry found acceptance in philosophy and the natural sciences.

According to Riemann's ideas about constructing geometry [184], [196], one should define a variety of the elements characterized by their coordinates in some absolute Euclidean space. He also proposed defining a differential quadratic form that determines linear elements, i.e., follows the law of measurement of distances between infinitely close elements of the variety.

Conversely, according Klein [184], [197], a mathematician constructing a geometry should define some variety of the elements and the group of mappings allowing to transform these elements into each other. In this case, he sees the problem of geometry as studying relationships between these elements which are invariable under all the mappings of this group.

Two opposing philosophical concepts (Newton's concept of absolute space and Leibniz's treatment of space as a property of material objects) developed within the space geometry construction theories of Riemann and Klein respectively.

The development of the idea of space in signal theory repeats the concept of space in physics and mathematics. Vladimir Kotelnikov was one of the first scientists to utilize the concept of *signal space* to investigate noise immunity in communication systems [127], thus, he laid the foundation of signal theory. Lewis Franks' book titled *Signal Theory* [122] is the first source in this field.

According to Franks, mathematical analysis provides the base of signal theory. It relies on the notion of signal space corresponding to the function space in terms of mathematical analysis. The two types of function spaces are metric and linear. The linear spaces are further classified as normed linear spaces and linear spaces with scalar product. The linear spaces with scalar product (more simply, Euclidean spaces) serve as the basis of modern signal theory.

This approach to the construction of signal space includes a number of disadvantages. For example, a stochastic signal carrying a certain formation quantity and a deterministic signal having known parameters (but not carrying information) are represented by an element of linear space with scalar product (by a vector). Thus, regardless of their informational properties, all signals are represented equally in signal space.

Is it logical to ask how to describe signals in terms of linear space with scalar product taking into consideration their informational properties? If we use the modern view of the notion of space in physics, the question can be formulated in the following form. Do any signals that carry or do not carry information, deterministic or stochastic, wideband or narrowband, generate exact Euclidean spaces and no others? Existing signal theory indicates the answer is affirmative. On the basis of the sampling theorem, the transformation progresses from continuous signals to discrete signals, i.e., into the form of a vector of the signal samples. Further, the techniques and methods of linear algebra are used to solve the signal processing problems.

Another theoretical problem lies in the use of linear space with scalar product to describe real signals. This type of description imposes hard constraints on signal properties and signal processing capabilities.

We take as an example a linear space S defined by the signal set $\{s_i(t)\}$ with scalar product with certain probabilistic properties: all the elements $\{s_i(t)\}$ of the space S are represented by signal sample vectors. The signals are assumed to be stationary Gaussian stochastic processes. We also let $G = \{g_{ij}\}$ be a group of linear isomorphisms (bijection linear mappings) of signals $\{s_i\}$ into each other in space S.

Whatever linear mapping g_{ij} over the elements s_i of the space S produced, there is no place in it for nonGaussian stochastic processes and their nonlinear transformations. Thus, linearity property essentially confines the diversity of signal space by the probabilistic properties of the signals (in this case, the Gaussian stochastic signals form only signal space) and the type of signal processing (linear signal transforms only). Note that $s_i + s_j$ is also a Gaussian stochastic process (closure property of a group). For this (and only this) example, the probabilistic properties of the signals are confined by their stationarity and Gaussian distribution and all possible signal processing methods are confined by linear signal processing. If no constraints are imposed on probabilistic properties of stochastic processes and the group of isomorphisms $G = \{g_{ij}\}$, then S is a linear space with nonGaussian stochastic processes and $G = \{g_{ij}\}$ is a group of nonlinear isomorphisms.

The next theoretical difficulty lies in relation between the continuous signal $s(t)$ and the result of its discretization. The sample vector $s=[s_j]=[s(t_j)]$ represents the elements of distinct spaces with different properties. The first is the element of Hilbert space $\mathcal{HS}$. The second element belongs to finite-dimensional Euclidean space $\mathbf{R}^n$. These two spaces are not isomorphic to each other.

The problem arises because the signal sample vector $s = [s_j]$ and its single sample $s(t_j)$ are represented absolutely equally in the space $\mathbf{R}^n$ by a point (vector). Thus, both the single instantaneous value $s(t_j)$ of the signal and the signal $s(t)$ as a whole, are represented equally by the least element of the space $\mathbf{R}^n$ (by a point).

Theoretical basis of transfer from the Hilbert space into finite-dimensional Euclidean space is *sampling theorem* that embodies an attempt to achieve the harmony between continuous and discrete entities. Its inferences and relationships have paramount theoretical and applied importance. They are used to solve various problems of signal processing theory and allow the providing of continuous and discrete signal processing from the same standpoint. Nevertheless, using the sampling theorem as an exact statement with respect to real signals and imposing technological

1.3 Overcoming Logical Difficulties

methods of continuous signal processing without information losses will create numerous problems.

One theoretical disadvantage of the sampling theorem is its orientation toward using deterministic signals with known parameters that cannot carry information. Nikolai Zheleznov [198] suggested applying the sampling theorem interpretation to stochastic signals. This idea considers the signals as nonstationary stochastic processes with some power spectral densities. Zheleznov also proposed using an idea concerning the boundedness of correlation interval and its smallness as against the signal duration. The correlation interval would be considered equal to the sampling interval. That is, the utility of using the sampling theorem with stochastic signals lies in the requirement for lack of correlation of neighbor samples during the transformation of a continuous signal to a discrete one.

The main feature of Zheleznov's variant of sampling theorem is the statement that the sequence of discrete samples in time domain can provide only an approximation of the initial signal. Meanwhile, we know that classical formulation of the sampling theorem claims absolute accuracy of signal representation.

Unfortunately this interpretation also presents disadvantages. The correlation concept used here describes only linear statistical relations and thus limits the use of this interpretation on nonGaussian random processes (signals). Even if one assumes the processed signal is completely deterministic and is described by some analytic function (even a very complicated one), the use of the sampling theorem will create theoretical and practical difficulties.

First, a real deterministic signal has a finite duration T. In the frequency domain, the signal has an unbounded spectrum. However, due to the properties of real signal sources and the boundedness of real channel passband, one can consider the signal spectrum to be bounded by some limited frequency F. The spectrum is usually defined on the basis of an energetic criterion, i.e., it is bounded within the frequency interval from 0 to F where most of the signal energy concentrates.

This spectrum boundedness leads to a loss of some information. As a result, restoration of a bounded signal by its samples in time domain based on the sampling theorem constrained by limits on the signal spectrum can be only approximate. Errors arise also from the constraints on the finite number of samples lying within time interval T (equal to $2FT$ according to the theorem). These errors appear to be due to neglecting the infinite number of expansion functions corresponding to samples outside the interval T.

Second, the restoration procedure causes another error arising from the impossibility of creating pulses of infinitesimal duration and transmitting them through real communication channels. The maximum output signal corresponding to the reaction of an ideal low-pass filter on delta-function action has a delay time that tends to infinity. For a finite time T, every sample function and their sums that are copies of initial continuous signals will be formed only approximately. Less T means more rough approximation.

Nevertheless, some formulations of the sample theorem are free of these disadvantages, as will be shown in Chapter 4.

The first steps in developing the notion of *information quantity* were undertaken by Ralph Hartley, an American communication specialist [50]. He suggested characterizing the information quantity of a message consisting of N symbols of the alphabet containing q symbols by a quantity equal to logarithm of a total number of messages from an ensemble (by *logarithmic measure*):

$$H = N \log q.$$

Hartley considered messages consisting of both discrete symbols and continuous waveforms. He stated that, "The sender is unable to control the form of the function with complete accuracy", thus the messages of the last type include infinite quantity of information. Hartley considered information quantity that could be transmitted within the frequency band F for the time T was proportional to the product $FT \log S$, where S is the number of "distinguishable intensities of a signal". Hartley understood that using *measure of information quantity* as a convenient approach to a number of practical problems was imperfect because it did not account for the statistical structures of messages.

If we assume a correspondence between the symbols of some alphabet a_i (or states of signal waveforms) and probabilities p_i, Hartley's measure determining the information quantity in a message consisting of k symbols takes the form:

$$H = -\sum_{i=1}^{k} p_i \log p_i. \tag{1.3.1}$$

A number of authors obtained this equation by various methods. Shannon [51] and Wiener [85] based their work on a statistical approach to information transmission.

However, statistical approaches vary. In the approach to information quantity above, we dealt with average values, not information transmitted by a single symbol. Thus, Equation (1.3.1) represents an average value. It can be overwritten as:

$$H = -\overline{\log p_i}.$$

Specialists in probability theory and mathematical statistics may call H a mathematical expectation of a logarithm of probability of symbols' appearance. As well as a variance, H represents a measure of scattering of the values of a random variable (or a stochastic process) with respect to the mean, i.e., a measure of statistical "rarity" of message symbols. Note that for the most common random variable (stochastic process) models, this measure is proportional to some monotonically increasing function of variance. Cherry [88] noted, "In a sense, it is a pity that the mathematical concepts stemming from Hartley have been called 'information' at all". Equation (1.3.1) for entropy defines only one aspect of information: statistical rarity of a single instantaneous value (symbol) appearance in continuous or discrete message.

In attempts to extend the notion of entropy upon the continuous random variables, another difficulty appears. Let us determine the entropy of a continuous random variable ξ with probability density function (PDF) $p(x)$ starting from entropy of discrete (quantized) random variable per Equation (1.3.1). If the quantization

1.3 Overcoming Logical Difficulties

step Δx of the random variable ξ is small enough against its range, the probability that random variable will take its values within the i-th quantization interval will be approximately equal to:

$$p_i = p(x_i)\Delta x,$$

where $p(x_i)$ is PDF value at the point x_i. Substituting the value p_i into Equation (1.3.1), we have:

$$H(x) = \lim_{\Delta x \to 0}\{-\sum_i p(x_i)\Delta x \log[p(x_i)\Delta x]\} =$$

$$= \lim_{\Delta x \to 0}\{-\sum_i [p(x_i)\log p(x_i)]\Delta x - \log \Delta x \sum_i p(x_i)\Delta x\}.$$

Taking into account normalization property of PDF $p(x)$:

$$\lim_{\Delta x \to 0} \sum_i p(x_i)\Delta x = 1,$$

and the limit transfer

$$\lim_{\Delta x \to 0}\{-\sum_i [p(x_i)\log p(x_i)]\Delta x\} = -\int_{-\infty}^{\infty} p(x)\log p(x)dx,$$

we obtain:

$$H(x) = -\int_{-\infty}^{\infty} p(x)\log p(x)dx - \lim_{\Delta x \to 0}(\log \Delta x). \tag{1.3.2}$$

As shown by Equation (1.3.2), while transferring to continuous random variable entropy, the last tends to infinity. Therefore, continuous random variables do not allow introducing a finite absolute quantitative measure of information. Definition of *differential entropy* $H(x)$ is realized by rejection of the infinite term $\lim_{\Delta x \to 0}(\log \Delta x)$ of the Equation (1.3.2):

$$H(x) = -\int_{-\infty}^{\infty} p(x)\log p(x)dx. \tag{1.3.3}$$

Thus, to define continuous random variable entropy, Shannon's approach [51] lies in a limit passage from discrete random variables to continuous ones while ignoring infinity, rejecting a continuous random variable entropy devoid of sense, and replacing it by differential entropy. Shannon's preference for discrete communication channels does not look logical based on recent developments in communications technology. Shannon's information quantity during the transition to continuous noiseless channels is coupled too closely with thermodynamic entropy so problems arising from both concepts have the same characteristics caused by the same reason.

The additive noise that limits the signal receiving accuracy according to the classical information theory imparts a finite unambiguous sense to information quantity.

An analogous situation occurs in determining the capacity of continuous communication channel with additive Gaussian noise [51]:

$$C = F \log(1 + \frac{P}{N}), \qquad (1.3.4)$$

where F is a channel bandwidth, P is the signal power, N is the noise power.

Analysis of this expression leads to a conclusion about the possibility of transmitting any information quantity per second by indefinitely weak signals in the absence of noise. This result is based on the assumption that at low levels of interference (noise), one can distinguish two indefinitely close to each other signals with any reliability, causing an unlimited increase in channel capacity while decreasing noise power to zero. This assumption seems absurd from a theoretical view because Nature's fundamental laws limit measurement accuracy and they are insuperable by any technological methods and means.

Despite attempts to eliminate infinity during a transition from discrete random variable entropy to continuous random variable entropy, Shannon's theory can cause the same problem arising from defining continuous channel capacity based on differential entropy. The indispensable condition is a presence of noise in channels, because information quantity transmitted by a signal per time unit tends to infinity in the absence of noise.

Another difficulty with classical information theory arises when differential entropy (1.3.3) compared with Equation (1.3.1) is not preserved under bijection mappings of stochastic signals; this situation can produce paradoxical results. For example, the process $y(t)$ obtained from initial signal $x(t)$ by its amplification k times ($k > 1$) possesses greater differential entropy than the original signal $x(t)$ in the input of the amplifier. This, of course, does not mean that the signal $y(t) = kx(t)$ carries more information than the original one $x(t)$. Note an important circumstance. Shannon's theory excludes the notion of *quantity of absolute information* generally contained in a signal.

The question of the quantity of information contained in a signal (stochastic process) $x(t)$ has no place in Shannon's theory. In this theory, the notion of information quantity makes sense only with respect to a pair of signals. In that case, the appropriate question is: how much information does the signal $y(t)$ contain with respect to the signal $x(t)$? If the signals are Gaussian with correlation coefficient ρ_{xy}, *quantity of mutual information* $I[x(t), y(t)]$ contained in the signal $y(t)$ with respect to the signal $x(t)$, assuming their linear relation $y(t) = kx(t)$, is equal to infinity:

$$I[x(t), y(t)] = \int_{-\infty}^{\infty} \int_{-\infty}^{\infty} p(x,y) \log \frac{p(x,y)}{p(x)p(y)} dx dy = -\log\sqrt{1 - \rho_{xy}^2} = \infty.$$

Evidently the answer to the question about the quantity of mutual information and the quantity of absolute information is too weak. The question about the quantity of absolute information can be formulated more generally and neutrally.

1.3 Overcoming Logical Difficulties

Let $y(t)$ be the result of a nonlinear (in general case) one-to-one transformation of Gaussian stochastic signal $x(t)$:

$$y(t) = f[x(t)]. \tag{1.3.5}$$

The question is: will the information contained in signals $x(t)$ and $y(t)$ be the same if their probabilistic-statistical characteristics differ? In the case above, their probability density functions, autocorrelation functions, and power spectral densities could differ. For example, if the result of nonlinear transformation (1.3.5) of quasi-white Gaussian stochastic process $x(t)$ with power spectral density width F_x is the stochastic process $y(t)$ with power spectral density width F_y, $F_y > F_x$, does the resulting stochastic process $y(t)$ carry more information than its original in the input of a transformer? It is hard to believe, that classical information theory can provide a perspicuous answer to the question.

The so-called cryptographic encoding paradox relates to this question and can be explained. According to Shannon [72], information quantities obtained from two independent sources, should be added. The cryptographic encoder must provide statistical independence between the input $x(t)$ and the output $y(t)$ signals:

$$p(x,y) = p(x)p(y).$$

where $p(x,y)$ and $p(x)$, $p(y)$ are joint and univariate probability density functions, respectively.

We can consider the signals at the input $x(t)$ and the output $y(t)$ of a cryptographic encoder as independent message sources. The general quantity of information I obtained from each of them (I_x, I_y) should equal their sum: $I = I_x + I_y$. However, under any one-to-one transformation under Equation (1.3.5), the identity: $I = I_x = I_y$ must hold inasmuch as both $x(t)$ and $y(t)$ carry the same information.

The information theory conclusion that Gaussian noises possessing the most interference effect (maximum entropy) among all types of noises with limited average power is connected closely with *differential entropy noninvariance property* [Equation (1.3.3)] with respect to a group of signal mappings. This statement contradicts known results, for example, of signal detection theory and estimation theory in which the examples of interference (noise) exert stronger influence on signal processing systems than Gaussian interference (noise).

Shannon's *measure of information quantity*, like Hartley's measure, cannot pretend to cover all factors determining "uncertainty of outcome" in an arbitrary sense. For example, these measures do not account for time aspects of signals. Entropy (1.3.1) is defined by the probabilities p_i of various outcomes. It does not depend on the nature of outcomes — whether they are close or distant [199]. The uncertainty degree will be the same for two discrete random variables ξ and η characterized by identical probability distributions $p_\xi(x)$ and $p_\eta(y)$:

$$p_\xi(x) = \sum_i p_i^\xi \cdot \delta(x - x_i), \; p_\eta(y) = \sum_i p_i^\eta \cdot \delta(y - y_i), \; p_i^\xi = p_i^\eta,$$

for equipotent countable sets of the values $\{x_i\} = x_1, x_2, \ldots, x_n$, $\{y_i\} =$

$y_1, y_2, \ldots, y_n$, although the average absolute deviations of random variables ξ and η can differ:

$$\int_{-\infty}^{\infty} |x - m_\xi| p_\xi(x) dx << \int_{-\infty}^{\infty} |y - m_\eta| p_\eta(y) dy,$$

where $m_\xi = \int_{-\infty}^{\infty} x p_\xi(x) dx$, $m_\eta = \int_{-\infty}^{\infty} y p_\eta(y) dy$.

Applying these concepts to the messages (signals) means that under Shannon's approach, one should consider so-called conditional entropy to account for the statistical relationships between single fragments of the messages.

If the chaos is interpreted as the absence of statistical coupling between time series of events, then Shannon's entropy (1.3.1) is an uncertainty measure in timeless space or in space with a time disorder. Jonathan Swift described the situation in *Gulliver's Travels*. The protagonist visits Laputian Academy of Lagado and encounters a wonder machine with which "the most ignorant person, at a reasonable charge, and with a little bodily labour, might write books in philosophy, poetry, politics, laws, mathematics, and theology, without the least assistance from genius of study." The sentences in such books are formed by random combinations of "particles, nouns, and verbs, and other parts of speech." Every press of the machine's handle produces a new phrase with the help of "all the words of their language, in their several moods, tenses, and declensions, but without any order."

Émile Borel described an experiment involving a monkey and a typewriter. A monkey randomly pressing the keys could create "texts" with maximal information content according Shannon. It seems appropriate here to repeat Ilya Prigogine's statement about the "impossibility of surrounding world description without constructive role of time" [200]. We contend that neglecting the time component (or statistical relations between the instantaneous values of the signals) is unsatisfactory when constructing the foundations of signal theory and information theory.

The paradox of subjective perception of a message by sender and addressee is specific to statistical information theory. The message M represents a deterministic set of the elements (signs, symbols, etc.) to the sender. Thus, from the standpoint of the sender, the message contains the quantity of information $I(M)$ equal to zero. For the addressee, the same message M^* is a probabilistic-statistical totality of the elements. Therefore, from the standpoint of the addressee, this message contains the quantity of information $I(M^*)$ that does not equal to zero. A paradoxical situation occurs when the sender knowingly sends a message that contains a quantity of information equal to zero while the receiver is sure the message contains real content.

Three considerations for constructing the foundations of information theory are:

1. Accepting the sender's view that the message is completely known and its elements form a deterministic totality

2. Considering the view of the addressee — the message is unknown and its elements form a probabilistic-statistical totality

3. As an ideal observer, attempting to unify the views of the sender and the addressee

The authors of the sampling theory accept the view of the sender who knows the content of the message sent and they prefer to work with deterministic functions. The creators of statistical information theory considered the view of the addressee indisputable. Researchers in semantic information theory tried to improve the situation by treating the message as an invariant of information [91], [201]. From this view, the quantity of semantic information transmitted by a message has to be the same for both the sender and the addressee.

Analogously, quantity of syntactical information $I(M)$ contained in deterministic (for the sender) message M must be equal to the quantity of syntactical information $I(M^*)$ contained in the received message M^* that represents for the addressee the probabilistic-statistical totality of the elements: $I(M) = I(M^*)$.

This circumstance demands appropriate elucidation of information theory to ensure that a measure of information quantity combine both probabilistic and deterministic approaches.

The author sees a resolution to this predicament using an approach that does not require identification of measure of information quantity and physical entropy. Its essence lies in representation of the signals carrying information by a physical system with a set of probabilistic states, and not by abstract random variables and a set of numbers.

The informational structure of a random function (stochastic signal) represents an internal organization of the system (the signal) that determines robust relations between its elements. The totality of the elements of informational structure of the signal is the set of the elements of metric signal space where the metric between any two elements is invariant with respect to a group of signal mappings.

In summary, these considerations can be divided into two groups. The first group concerns *signal space concept*, its current state (Group 1 below) and its suggested development (Group 1A below). The second group covers issues related to a measure of information quantity, its current state (Group 2 below) and its suggested development (Group 2A below).

Group 1:

1. Linear spaces with scalar product (Euclidean spaces) serve as the basis of the modern variant of signal theory construction.

2. In classical signal theory, any signal, whether stochastic (carrying a certain quantity of information) or deterministic (containing no information), is represented by an element of linear space with scalar product (a vector). Regardless of their informational properties, all signals are represented equally in the signal space.

3. Using linear space with scalar product to describe the real signals imposes strong constraints on signal properties and signal processing. The probabilistic properties of signals are described, mainly, by Gaussian distribution, and optimal processing of such signals is confined within linear

processing. In the case of nonGaussian inputs, optimal signal processing algorithms require the use of nonlinear processing methods.

4. The least element of classical signal space is a signal represented by a point (or vector) in Euclidean space. Quite similarly a single element of the signal (instantaneous value of the signal) is described. To maintain logic, an instantaneous value of a signal must be the least element of signal space. This circumstance influences the profundity of constructing the signal space concept, possibilities of investigating the real signals properties, and also possibilities of optimal signal processing.

5. The concept of signal space is related closely to formulating sampling theorems and provides dialectical unity of continuous and discrete forms of signal representation. The use of existing variant of sampling theorem for passing from Hilbert to Euclidean space meets both theoretical and practical difficulties.

Based on items 1 through 5, we can formulate a requirement list for our signal space concept.

Group 1A:

1. Signal space has no absolute character; it is not a receptacle of material objects (signals). Signal space is formed by a set of the signals interacting with each other.

2. To construct the geometry of signal space, the set of the space elements (the set of the signals) should be defined. Also one should define a group of morphisms to allow mapping these elements into each other. The geometry of signal space requires studying relationships of the signals that remain invariable through all mappings of a given group.

3. The signal space concept must be based on metric spaces and not confined exclusively to linear spaces with scalar product.

4. The signal space concept must describe interactions of signals with arbitrary probabilistic and informational properties.

5. The signal space concept must consider time aspect of signals.

6. Transition from continuous form of signal representation to discrete form with the help of sampling theorem or its analogue must ensure the signals belong to the same space. The least element of metric signal space (the point of metric space) must be a single instantaneous value (a signal sample), not a signal as a whole.

7. The signal space concept must be in accordance with a measure of information quantity carried by the signals.

Adduce the considerations concerning classical measure of information quantity that could be united into the second group.

Group 2:

1. Introducing entropy as a measure of information quantity was advocated by communication engineering of the past, and was not based on the needs of information theory. Entropy reflects a measure of fortuitousness of events (random variable) or a measure of statistical rarity. From the standpoint of analyzing the characteristics of random variable, entropy is a measure of spread in the values of random variable around its mean value.

   A measure of information quantity of a statistical totality of elements (a message or a signal) must reflect the diversity of all the elements, not simply a single element of this totality (a sample of either a message or a signal). Information contained in a totality of the elements is based on the diversity of statistical relationships of the elements, not by the diversity of the possible values of the elements. Information is not a characteristic of a single random variable considered outside statistical totality. The statement that separately taken random variable contains a quantity of information is nonsensical. Information can be used with respect to statistical totality as a whole or with respect to its single elements. With application to stochastic processes (signals), the measure of information quantity must include consideration of time statistical relationships between single instantaneous values (samples) of stochastic process.

2. Another theoretical inconvenience of classical information theory is the inability to construct a signal space using entropy as a measure of information quantity. The situation when information exists outside a signal space that should have been formed by signals acting as material carriers of information (when information exists per se, and the signal space exists per se) is absurd.

   The absence of a measure of information quantity built on relationships of the elements of statistical totality (the signal) does not allow uniting within the framework of the unified signal space the sets of single elements (samples) of the signals and the sets of the signals.

3. Classical information theory is slanted toward stochastic processes in the form of discrete random sequences (random sequences with domain of definition and range of values in the form of countable sets). Entropy introduced for this purpose depends on the average statistical rarity of the values of discrete random variable. Attempts to extend entropy to continuous random variables encounter the difficulty that entropy of continuous random variables is an infinitely large quantity due to an unbounded increasing statistical rarity of random variable values.

   Trying to overcome infinity by rejecting an infinite term and thus forming the notion of differential entropy may appear similar to entropy of a discrete random variable but the action is baseless. The result is an incorrect inference of infinitely large channel capacity in the absence of noise.

4. Statistical information theory is one-sided: it does not allow using the possibilities of nonprobabilistic approaches to describe the deterministic totalities of the elements. The orientation of classical information theory toward random sequences is clear. It is impossible for two observers to compare information quantity of the same message, when one of them exactly knows the message content, but another does not know it. This situation does not permit using the quantity of absolute information contained in a message (signal) independently of subjective factors that is invariant with respect to a group of mappings of the messages (signals).

In summary, we can formulate a list of requirements to suggested *measure of information quantity*. Measure of information quantity must:

Group 2A:

1. Be a characteristic of a single element of statistical totality and statistical totality as a whole.
2. Be a measure of structural diversity of this totality.
3. Take into account statistical relationships between the elements of totality: time statistical relationships of stochastic processes and space-time statistical relationships of stochastic fields.
4. Evaluate information quantity within deterministic and probabilistic-statistical totalities of the elements, thus combining the possibilities of probabilistic and nonprobabilistic approaches.
5. Conform to the notion of signal space.
6. Encompass stochastic processes (signals) of all kinds utilizing domain of definition and range of values as countable sets and continuum, or both.
7. Be invariant of a group of signal mappings in signal space.

The last consideration is formulated as the *main axiom of signal processing theory*.

Axiom 1.3.1. *The main axiom of signal processing theory.* Information quantity $I[y]$ contained in the signal y in the output of arbitrary processing unit $f[*]$ does not exceed the information quantity $I[x]$ contained in the initial signal x before processing:

$$I[y] \leq I[x]; \quad y = f[x], \qquad (1.3.6)$$

so that the inequality (1.3.6) turns into identity under the condition, if signal processing realizes one-to-one mapping of the signal:

$$x \underset{f^{-1}}{\overset{f}{\rightleftarrows}} y; \quad I[x] = I[y], \qquad (1.3.6a)$$

where $f \in G$, G is a signal mappings group.

The identity (1.3.6) will be formulated and proved in the form of the corresponding theorems in Chapters 2, 3, and 4. Note that some signal mappings not related to one-to-one transformations provide identity (1.3.6). We consider some of these issues in Chapter 4.

W.G. Tuller [10] formulated the idea of comparing information quantities in the input and output of a processing unit in the form of (1.3.6).

The methodological approach applied to construction of information theory by Shannon is based on elementary techniques and methods of probability theory. It ignores the basis of probability theory built on Boolean algebra with a measure. Thus, we should expect to use Boolean algebra with a measure to construct the foundations of signal processing theory and information theory. This approach could be a unified factor allowing us to provide the unity of theoretical foundations of aforementioned directions of mathematical science, imparting to them force, universality, and commonality.

Furthermore, the physical character of signal processing theory and its applied orientation may stimulate the development of information theory, probability theory, mathematical statistics, and other mathematical directions and application of their research methods to all branches of natural science.

The principal concept of the work is constructing the unified foundations of signal processing theory and information theory on the basis of the notion of signal space built upon generalized Boolean algebra with a measure. The last induces metric in this space, and simultaneously it is a measure of information quantity contained in the signals. That consideration provides the unity of a theoretical foundation of interrelated directions of mathematical science: probability theory, signal processing theory, and information theory, based upon the unified methodological basis: generalized Boolean algebra with a measure.

2

Information Carrier Space Built upon Generalized Boolean Algebra with a Measure

In Chapter 1, the requirements for signal space, i.e., the space of material carriers of information and the requirements for a measure of information quantity were formulated. In this chapter, we show that the space built upon Boolean algebra with a measure meets all the properties of signal space and a measure of information quantity in full.

The difficulties of attempting to define some general notion of scientific use are well known. So, the fundamental notion of a "set" has no direct definition, but this fact does not interfere with the study of mathematics; it is enough to know the main theses on set theory.

In this chapter, our attention will be concentrated on a fundamental, but more vague notion of "information". Even a very superficial analysis produces considerable difficulty in defining information. Undoubtedly, the science needs such a definition. Signal processing theory and information theory require distinctly formulated axioms and theorems describing, on the one hand, the properties of a space of material carriers of information (signal space), and on the other hand, the properties of information itself, its measure, and the peculiarities of processing of its carriers (signals).

Boolean algebra with a measure and metric space built upon Boolean algebra and induced by this measure are well investigated [173–176, 202–213]. Main results on generalized Boolean algebras and rings are contained in several works [214–218]. Study of interrelations between lattices and geometries begins from [219], [220], [221]. The papers [204] and [222] are the first to describe geometric properties of Boolean ring and Boolean algebra respectively.

Analysis of the development of signal processing theory and information theory suggests their independent and isolated existence along with their weak interactions. Often, one can gain the impression, that information carriers (signals) and principles of their processing exist per se, while transmitted information and the approaches intended to describe a wide range of informational processes exist apart from information carriers. This is shown by the fact that the founders of signal theory, signal processing theory, and information theory considered a signal space irrespective of information carried by the signals [122], [127], [143], and on the other hand, a measure of information quantity regardless of its material carriers (signals) [50], [51], [85]. This contradiction will inevitably question the necessity of unifying mathematical foundations of both signal processing theory and information theory. It is shown in this chapter that this theoretical difficulty can be overcome by utilizing generalized Boolean algebra with a measure to describe informational characteristics and properties of stochastic signals.

Generalized Boolean algebra with a measure is a wonderful model to unify mathematical foundations of information theory and signal processing theory, inasmuch as a measure determines information quantity transferred by the signals forming the signal space with specific properties, and also induces metric in this space.

2.1 Information Carrier Space

As for the notation system, one should note the following. Upper case Latin letters are used to denote the sets of elements and their Boolean analogues. The symbols $\bigcup$ and $\bigcap$ are used to denote operations of theoretical-set union and intersection, respectively. To denote the notion of Boolean sum, the symbols $+$ and Σ are used. The symbol $\bigcup$ is used to emphasize a structure (for instance, a continuous or discrete one), which is inherent to a system of elements. To denote the notion of Boolean product, the symbols $\cdot$ and Π are used. The symbol Δ is used to denote a sum operation of Boolean ring and also symmetric difference of Boolean algebra. Null element of Boolean algebra is denoted by $\mathbf{O}$. An *indexed set* $\{A_t\}_{t \in T}$ is a mapping that associates each element $t \in T$ with an element A_t.

In this section, the main material concerning lattice theory and Boolean algebras is expounded. More detailed information is contained in the known algebraic literature (see [175], [176], [221], [223]).

Universal algebra $\mathbf{A} = (X, T_\mathbf{A})$ with a signature $T_\mathbf{A} = \{T_n | n \geq 0\}$ (a system of operations) is a set X (algebra carrier), where for any $n \geq 0$ to every $t \in T_n$, a n-ary algebraic operation is associated on $\mathbf{A} = (X, T_\mathbf{A})$ [221], [223].

Algebraic structure $\mathbf{L} = (X, T_\mathbf{L})$, $T_\mathbf{L} = (+, \cdot)$ is a *lattice* with binary operations of addition $a + b = \sup_\mathbf{L}\{a, b\}$ and multiplication $a \cdot b = \inf_\mathbf{L}\{a, b\}; a, b \in X$, if it satisfies the following identities:

$$
\begin{array}{rclrcll}
(a + b) + c & = & a + (b + c), & (a \cdot b) \cdot c & = & a \cdot (b \cdot c) & \text{(associativity)} \\
a + b & = & b + a, & a \cdot b & = & b \cdot a & \text{(commutativity)} \\
a + a & = & a, & a \cdot a & = & a & \text{(idempotency)} \\
(a + b) \cdot a & = & a, & (a \cdot b) + a & = & a & \text{(absorption)}
\end{array}
$$

Lattice $\mathbf{L}$ is called *distributive* if it satisfies the identities called distributive laws:

$$a(b + c) = ab + ac, \qquad a + (bc) = (a + b)(a + c).$$

Over a lattice a *null* $\mathbf{O}$ (*unit* $\mathbf{I}$) element (also called *zero* and *unity*, respectively) can be defined:

$$a + \mathbf{O} = a, \; a \cdot \mathbf{O} = \mathbf{O}; \quad a + \mathbf{I} = \mathbf{I}, \; a \cdot \mathbf{I} = a.$$

Lattice $\mathbf{L}$ is called *lattice with relative complements*, if for any element a from any

2.1 Information Carrier Space

interval $[b, c]$ it can be found that element d, $d \in [b, c]$, and that $a + d = c$ and $a \cdot d = b$. Element d is called *relative complement* of element a in the interval $[b, c]$.

Lattice **L** with zero **O** and unity **I** is called *complemented lattice*, if every element has a relative complement in the interval $[\mathbf{O}, \mathbf{I}]$. Relative complements in the interval $[\mathbf{O}, \mathbf{I}]$ are simply called *complements*.

In distributive lattice **L** with zero **O** and unity **I**, every element a possessing the complement a' has also a relative complement $d = (a' + b)c$ in any interval $[b, c]$, $a \in [b, c]$.

Distributive lattice **L** with null element **O** and relative complements is called *generalized Boolean lattice* **BL** in which a relative complement of the element a in the interval $[\mathbf{O}, a + b]$ is called the *difference* of the elements b and a and is denoted by $b - a$. **BL** with unit element **I** is called *Boolean lattice* **BL**$'$. One can also say that Boolean lattice **BL**$'$ is a distributive lattice with complements, i.e., a distributive lattice **L** with zero **O** and unity **I**.

Generalized Boolean lattice **BL**, considered as a universal algebra $\mathbf{BL} = (X, T_{\mathbf{BL}})$ with the signature $T_{\mathbf{BL}} = (+, \cdot, -, \mathbf{O})$ of the type $(2, 2, 2, 0)$, is called *generalized Boolean algebra* $\mathbf{B} = (X, T_{\mathbf{B}})$, $T_{\mathbf{B}} \equiv T_{\mathbf{BL}}$. Let $a \Delta b = (a + b) - ab$ in **BL**; then we obtain *generalized Boolean ring* $\mathbf{BR} = (X, T_{\mathbf{BR}})$ with the signature $T_{\mathbf{BR}} = (\Delta, \cdot, \mathbf{O})$ of the type $(2, 2, 0)$. On the contrary, any generalized Boolean ring $\mathbf{BR} = (X, T_{\mathbf{BR}})$ with the signature $T_{\mathbf{BR}} = (\Delta, \cdot, \mathbf{O})$ of the type $(2, 2, 0)$ can be turned into generalized Boolean lattice $\mathbf{BL} = (X, T_{\mathbf{BL}})$ with the signature $T_{\mathbf{BL}} = (+, \cdot, -, \mathbf{O})$ of the type $(2, 2, 2, 0)$, assuming $a + b = a \Delta b \Delta ab$.

Generalized Boolean algebra **B** can be defined by the following system of identities:

$$
\begin{aligned}
(a + b) + c &= a + (b + c), & (a \cdot b) \cdot c &= a \cdot (b \cdot c) & \text{(associativity)} \\
a + b &= b + a, & a \cdot b &= b \cdot a & \text{(commutativity)} \\
a + a &= a, & a \cdot a &= a & \text{(idempotency)} \\
(a + b) \cdot a &= a, & (a \cdot b) + a &= a & \text{(absorption)} \\
a \cdot (b + c) &= ab + ac, & a + bc &= (a + b)(a + c) & \text{(distributivity)} \\
a \cdot (b - a) &= \mathbf{O}, & a + (b - a) &= a + b.
\end{aligned}
$$

Boolean lattice **BL**$'$, considered as universal algebra $\mathbf{BL}' = (X, T_{\mathbf{BL}'})$ with the signature $T_{\mathbf{BL}'} = (+, \cdot, ', \mathbf{O}, \mathbf{I})$ of the type $(2, 2, 1, 0, 0)$, is called *Boolean algebra* $\mathbf{B}' = (X, T_{\mathbf{B}'})$, $T_{\mathbf{B}'} \equiv T_{\mathbf{BL}'}$.

Boolean algebra $\mathbf{B}'$ can be defined by the following system of identities:

$$
\begin{aligned}
(a + b) + c &= a + (b + c), & (a \cdot b) \cdot c &= a \cdot (b \cdot c) & \text{(associativity)} \\
a + b &= b + a, & a \cdot b &= b \cdot a & \text{(commutativity)} \\
a + a &= a, & a \cdot a &= a & \text{(idempotency)} \\
(a + b)a &= a, & (a \cdot b) + a &= a & \text{(absorption)} \\
a(b + c) &= ab + ac, & a + bc &= (a + b)(a + c) & \text{(distributivity)} \\
a + \mathbf{O} &= a, & a \cdot \mathbf{I} &= a; \\
a \cdot a' &= \mathbf{O}, & a + a' &= \mathbf{I}.
\end{aligned}
$$

There are supplementary operations in Boolean algebra derived from the main ones. The most important ones are difference $a - b = ab'$ and symmetric difference $a\Delta b = (a - b) + (b - a)$.

Let $a\Delta b = ab' + a'b$ in Boolean algebra $\mathbf{B}'$; we obtain *Boolean ring* $\mathbf{BR}'$ with the signature $(\Delta, \cdot, \mathbf{O}, \mathbf{I})$ of the type $(2, 2, 0, 0)$. Conversely, any $\mathbf{BR}'$ with the signature $(\Delta, \cdot, \mathbf{O}, \mathbf{I})$ of the type $(2, 2, 0, 0)$ can be transformed into Boolean algebra $\mathbf{B}'$ with the signature $(+, \cdot,', \mathbf{O}, \mathbf{I})$ of the type $(2, 2, 1, 0, 0)$, assuming $a + b = a\Delta b\Delta ab$, $a' = a\Delta \mathbf{I}$. Thus, *Stone's duality* between Boolean algebras and Boolean rings is established.

Finite additive measure on generalized Boolean algebra $\mathbf{B}$ is a finite real function m on $\mathbf{B}$ satisfying the following conditions:

1. $m(a) \geq 0$ for any $a \in \mathbf{B}$
2. $m(a + b) = m(a) + m(b)$, if $ab = \mathbf{O}$

The conditions imply $m(a+b) \leq m(a) + m(b)$, $m(a) \leq m(b)$ on $a \leq b$ and $m(\mathbf{O}) = 0$.

Measure on generalized Boolean algebra $\mathbf{B}$ is called σ-*additive measure* (or σ-measure), if $m[\sum_{i=1}^{\infty} x_i] = \sum_{i=1}^{\infty} m(x_i)$ for any countable orthogonal system $\{x_i : x_i \cdot x_j = \mathbf{O}, i \neq j; i, j \in \{1, 2, \ldots\}\}$.

Further, to denote subalgebras of generalized Boolean algebra with a measure defined on the various sets, for instance on X and Y, $X \neq Y$, we write $\mathbf{B}(X)$ and $\mathbf{B}(Y)$, respectively.

We define the notion of information carrier space using generalized Boolean algebra $\mathbf{B}(\Omega)$ with a measure m.

Definition 2.1.1. *Information carrier space* Ω *is a set of the elements* $\{A, B, \ldots\}$: $A, B, \ldots \subset \Omega$ *called information carriers, that possesses the following properties:*

1. The space Ω is generalized Boolean algebra $\mathbf{B}$ with a measure m and the signature $(+, \cdot, -, \mathbf{O})$ of the type $(2, 2, 2, 0)$.

2. Any set A: $A \subset \Omega$ as an information carrier in Ω can be represented by a totality of the elements $\{A_\alpha\}$: $A = \bigcup_\alpha A_\alpha$; each of them is characterized by *normalized measure*: $m(A_\alpha) = 1$. Thus, any set A: $A \subset \Omega$ forms the system of the elements (of sets) with normalized measure.

3. Any set A of the space Ω is linearly ordered set by an index, i.e., any two of its elements A_α and A_β are comparable by the index: $\alpha < \beta \Rightarrow A_\alpha < A_\beta$. Systems of the elements $A, B, \ldots$ not possessing this property will not be considered below.

4. System of the elements $A = \{A_\alpha\}$ possesses the property of continuity, if for any arbitrary pair of the elements A_α and A_β of the system A, there exists such an element A_γ of the system A that $\alpha < \gamma < \beta$. This is a *system of elements with continuous structure*. Systems not possessing this property are *systems of elements with discrete structure*.

5. The following operations are defined in the space Ω over the elements A_α, $A_\alpha \in A$, $A = \sum_\alpha A_\alpha$; B_β, $B_\beta \in B$, $B = \sum_\alpha B_\beta$; $A, B \subset \Omega$:

(a) addition — $A + B = \sum_\alpha A_\alpha + \sum_\beta B_\beta$
(b) multiplication — $A_{\alpha 1} \cdot A_{\alpha 2}, B_{\beta 1} \cdot B_{\beta 2}, A_\alpha \cdot B_\beta, A \cdot B = (\sum_\alpha A_\alpha) \cdot (\sum_\beta B_\beta)$
(c) difference — $A_{\alpha 1} - A_{\alpha 2}, B_{\beta 1} - B_{\beta 2}, A_\alpha - B_\beta, A - B = A - (A \cdot B)$
(d) symmetric difference — $A_{\alpha 1} \Delta A_{\alpha 2}, B_{\beta 1} \Delta B_{\beta 2}, A_\alpha \Delta B_\beta, A \Delta B = (A - B) + (B - A)$
(e) null element $\mathbf{O}$: $A \Delta A = \mathbf{O}$; $A - A = \mathbf{O}$

6. A measure m on generalized Boolean algebra $\mathbf{B}(\Omega)$ introduces a *metric* defining a distance $\rho(A, B)$ between the sets A, B by the relationship:

$$\rho(A, B) = m(A \Delta B), \tag{2.1.1}$$

where $A \Delta B$ is a symmetric difference of sets A, B.

Analogously, the distance $\rho(A_{\alpha 1}, A_{\alpha 2})$ between the single elements $A_{\alpha 1}$ and $A_{\alpha 2}$ of the set A is defined:

$$\rho(A_{\alpha 1}, A_{\alpha 2}) = m(A_{\alpha 1} \Delta A_{\alpha 2}), \tag{2.1.2}$$

Thus, information carrier space (Ω, ρ) is a *metric space*.

7. Isomorphism Iso from the space Ω onto the space Ω' — $Iso: \Omega \stackrel{Iso}{\to} \Omega'$ preserving a measure m: $m(A) = m(A')$, $m(B) = m(B')$, $A, B \subset \Omega$, $A', B' \subset \Omega'$ also preserves a metric induced by this measure:

$$\rho(A, B) = m(A \Delta B) = m(A' \Delta B') = \rho(A', B')$$

Thus, the spaces (Ω, ρ) and (Ω', ρ) are isometric.

2.2 Geometrical Properties of Metric Space Built upon Generalized Boolean Algebra

In this section, we consider the questions concerning geometrical properties of metric space Ω based on generalized Boolean algebra $\mathbf{B}(\Omega)$ with a measure m.

2.2.1 Main Relationships between Elements of Metric Space Built upon Generalized Boolean Algebra with a Measure

Let $\mathbf{B}(\Omega)$ be a generalized Boolean algebra $\mathbf{B} = (\Omega, T_\mathbf{B})$ with a measure m defined on a set Ω. A measure m on $\mathbf{B}(\Omega)$ introduces a metric in Ω, determining a distance $\rho(A, B)$ between the elements $A, B \in \Omega$ by the formula [175, Section VI.1,(I)]:

$$\rho(A, B) = m(A \Delta B), \tag{2.2.1}$$

where $A \Delta B$ is the symmetric difference between the elements A, B of the space Ω.

Thus, introducing the metric, generalized Boolean algebra $\mathbf{B} = (\Omega, T_{\mathbf{B}})$ turns into the *metric space* Ω, whose topological properties are considered in [175]. Main relationships between the elements A, B of metric space Ω can be obtained easily from the relations of $\mathbf{B}(\Omega)$ with a measure m and null element $\mathbf{O}$.

The main group of identities:

$$m(A + B) = m(A) + m(B) - m(AB);$$

$$m(A + B) = m(A\Delta B) + m(AB);$$

$$m(A\Delta B) = m(A) + m(B) - 2m(AB).$$

The main group of inequalities:

$$m(A) + m(B) \geq m(A + B) \geq m(A\Delta B);$$

$$m(A) + m(B) \geq m(A + B) \geq \max[m(A), m(B)];$$

$$\max[m(A), m(B)] \geq \sqrt{m(A) \cdot m(B)} \geq \min[m(A), m(B)] \geq m(AB).$$

Relationships of orthogonality:
Two elements $X, Y \in \Omega$ of generalized Boolean algebra $\mathbf{B}(\Omega)$, $X, Y \in \Omega$ are called *orthogonal* $X \perp Y$, if the identity holds [175, Section VI.1,(I)]:

$$XY = \mathbf{O}, \quad (m(XY) = 0).$$

Then, for an arbitrary pair of the elements A, B, we have:

$$AB \perp A' \perp B' \perp AB \perp A\Delta B,$$

where $A' = A\Delta(AB)$, $B' = B\Delta(AB)$.

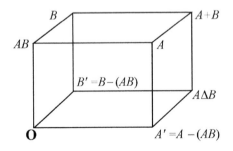 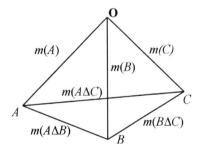

FIGURE 2.2.1 Hexahedron built upon a set of all subsets of element $A + B$

FIGURE 2.2.2 Tetrahedron built upon elements $\mathbf{O}, A, B, C$

Consider hexahedron $\mathbf{O}$, AB, A', B', A, B, $A\Delta B$, $A + B$ built upon a set of all the subsets of the element $A + B$ of generalized Boolean algebra $\mathbf{B}(\Omega)$ represented by $A + B = (AB)\Delta A'\Delta B'$ (see Fig. 2.2.1). One can try to perceive a geometry of the space Ω by the method of immersion of metric space Ω, formed by $\mathbf{B}(\Omega)$ with a measure m into Euclidean space. Consider the points of hexahedron $\mathbf{O}$, AB, A', B',

2.2 Geometrical Properties of Metric Space Built upon Generalized Boolean Algebra

$A, B, A\Delta B, A+B$ as the points of Euclidean space $\mathbf{R}^n$ of dimension $n = 3$, so that null element of the space Ω coincides with null element (zero) of Euclidean space $\mathbf{R}^3$, and axes x_1, x_2, x_3 pass through the points A', B', AB, respectively. Coordinates of hexahedron points $\mathbf{O}, AB, A', B', A, B, A\Delta B, A+B$ are, respectively, equal to:

$$\mathbf{O}\,(0,0,0), \quad AB\,(0,0,m(AB)), \quad A'\,(m(A'),0,0), \quad B'\,(0,m(B'),0);$$

$$A\,(m(A'),0,m(AB)), \quad B\,(0,m(B'),m(AB));$$

$$A\Delta B\,(m(A'),m(B'),0), \quad A+B\,(m(A'),m(B'),m(AB)).$$

The points $\mathbf{O}, AB, A', B', A, B, A\Delta B, A+B$ belong to a sphere $Sp(O_{AB}, R_{AB})$ in $\mathbf{R}^3$, so that the coordinates of the center O_{AB} and radius R_{AB} of a sphere are, respectively, equal to:

$$O_{AB}\,(m(A')/2, m(B')/2, m(AB)/2), \tag{2.2.2}$$

$$R_{AB} = \sqrt{(m(A')^2 + m(B')^2 + m(AB)^2)}/2. \tag{2.2.3}$$

Metric (2.2.1) between the points A, B in the space Ω is given by the relationship:

$$\rho(A,B) = |x_1^A - x_1^B| + |x_2^A - x_2^B| + |x_3^A - x_3^B|. \tag{2.2.4}$$

Metric $d(A,B)$ between the points A, B in Euclidean space $\mathbf{R}^3$ is defined by the known equation:

$$d(A,B) = [|x_1^A - x_1^B|^2 + |x_2^A - x_2^B|^2 + |x_3^A - x_3^B|^2]^{1/2}. \tag{2.2.5}$$

It is obvious that $\rho(A,B) \geq d(A,B)$. The sides of triangle $\Delta \mathbf{O}AB$ in the space Ω, according to Equation (2.2.4), are equal to:

$$\rho(\mathbf{O},A) = m(A') + m(AB); \quad \rho(\mathbf{O},B) = m(B') + m(AB);$$

$$\rho(A,B) = m(A') + m(B').$$

The sides of triangle $\Delta \mathbf{O}AB$ in the space $\mathbf{R}^3$, according to Equation (2.2.5), are equal to:

$$d(\mathbf{O},A) = [m^2(A') + m^2(AB)]^{1/2}; \quad d(\mathbf{O},B) = [m^2(B') + m^2(AB)]^{1/2};$$

$$d(A,B) = [m^2(A') + m^2(B')]^{1/2}.$$

For the sides of triangle $\Delta \mathbf{O}AB$, the relationships of sine rule (sine theorem) and cosine rule (cosine theorem) of a triangle in Euclidean space $\mathbf{R}^3$ hold.

For hexahedron $\mathbf{O}, AB, A', B', A, B, A\Delta B, A+B$, the following relationships hold.

The equalities between the opposite sides of hexahedron:

$$m(A') = m(A'\Delta \mathbf{O}) = m[A\Delta(AB)] = m[(A+B)\Delta B] = m[(A\Delta B)\Delta B'];$$

$$m(B') = m(B'\Delta \mathbf{O}) = m[B\Delta(AB)] = m[(A+B)\Delta A] = m[(A\Delta B)\Delta A'].$$

The equalities between diagonals of the opposite faces of hexahedron:

$$m(A\Delta B) = m[(A\Delta B)\Delta \mathbf{O}] = m(A'\Delta B') = m[(A+B)\Delta(AB)];$$

$$m(A) = m(A\Delta \mathbf{O}) = m[A'\Delta(AB)] = m[(A+B)\Delta B'] = m[(A\Delta B)\Delta B];$$

$$m(B) = m(B\Delta \mathbf{O}) = m[B'\Delta(AB)] = m[(A+B)\Delta A'] = m[(A\Delta B)\Delta A].$$

The equalities between large diagonals of hexahedron:

$$m(A+B) = m[(A+B)\Delta \mathbf{O}] = m[(A\Delta B)\Delta AB] = m(A'\Delta B) = m(A\Delta B').$$

For an arbitrary triplet of the elements A, B, C and null elements $\mathbf{O}$ of the space Ω, **the tetrahedron metric relationships** hold (see Fig. 2.2.2):

$$m(A\Delta B) = m(A\Delta \mathbf{O}) + m(\mathbf{O}\Delta B) - 2m[(A\Delta \mathbf{O})(\mathbf{O}\Delta B)];$$

$$m(B\Delta C) = m(B\Delta \mathbf{O}) + m(\mathbf{O}\Delta C) - 2m[(B\Delta \mathbf{O})(\mathbf{O}\Delta C)];$$

$$m(C\Delta A) = m(C\Delta \mathbf{O}) + m(\mathbf{O}\Delta A) - 2m[(C\Delta \mathbf{O})(\mathbf{O}\Delta A)];$$

$$m(A\Delta B) = m(A\Delta C) + m(C\Delta B) - 2m[(A\Delta C)(C\Delta B)];$$

$$m(B\Delta C) = m(B\Delta A) + m(A\Delta C) - 2m[(B\Delta A)(A\Delta C)];$$

$$m(C\Delta A) = m(C\Delta B) + m(B\Delta A) - 2m[(C\Delta B)(B\Delta A)].$$

2.2.2 Notion of Line in Metric Space Built upon Generalized Boolean Algebra with a Measure

The notion of a line (we shall mean a straight line here and below) in metric space Ω with metric (2.2.1) differs from the corresponding notion in Euclidean space and is introduced by the following definition [222]:

Definition 2.2.1. *Line l in the metric space Ω with metric (2.2.1) is a set containing at least three elements A, B, X: $A, B, X \in l \subset \Omega$, if the metric identity holds:*

$$\rho(A, B) = \rho(A, X) + \rho(X, B). \tag{2.2.6}$$

We say that the element X is situated between the elements A and B ($A \prec X \prec B$ or $A \succ X \succ B$). The condition (2.2.6) is equivalent to the identity:

$$m(X) = m(AX) + m(XB) - m(AB). \tag{2.2.7}$$

Equation (2.2.7) has two solutions with respect to X: 1) $X = A+B$, 2) $X = A \cdot B$. Then, obviously, the following lemmas hold.

Lemma 2.2.1. *For an arbitrary pair of the elements A, B of metric space Ω, the elements $A, X = A+B, B$ lie on the same line, so that the element $X = A+B$ is situated between the elements A and B:*

$$m(A\Delta X) + m(X\Delta B) = m(A\Delta B).$$

2.2 Geometrical Properties of Metric Space Built upon Generalized Boolean Algebra

Lemma 2.2.2. *For an arbitrary pair of the elements A, B of metric space Ω, the elements $A, Y = A \cdot B, B$ lie on the same line, so that the element $Y = A \cdot B$ is situated between the elements A and B:*

$$m(A \Delta Y) + m(Y \Delta B) = m(A \Delta B).$$

For the vertices of hexahedron built upon the points $\mathbf{O}$, AB, A', B', A, B, $A\Delta B$, $A + B$, the following relationships of belonging to the same line l_{ij} in Ω hold:

1. For the points $AB, B, A + B, A$:

$$m[(AB)\Delta(A+B)] = m[(AB)\Delta B] + m[B\Delta(A+B)] \Rightarrow AB, B, A+B \in l_{11};$$

$$m(B\Delta A) = m[B\Delta(A+B)] + m[(A+B)\Delta A] \Rightarrow B, A+B, A \in l_{12};$$

$$m[(A+B)\Delta(AB)] = m[(A+B)\Delta A] + m[A\Delta(AB)] \Rightarrow A+B, A, AB \in l_{13};$$

$$m[A\Delta B] = m[A\Delta(AB)] + m[(AB)\Delta B] \Rightarrow A, AB, B \in l_{14}.$$

2. For the points $\mathbf{O}, B', A\Delta B, A'$:

$$m[\mathbf{O}\Delta(A\Delta B)] = m[\mathbf{O}\Delta B'] + m[B'\Delta(A\Delta B)] \Rightarrow \mathbf{O}, B', A\Delta B \in l_{21};$$

$$m(B'\Delta A') = m[B'\Delta(A\Delta B)] + m[(A\Delta B)\Delta A'] \Rightarrow B', A\Delta B, A' \in l_{22};$$

$$m[(A\Delta B)\Delta \mathbf{O}] = m[(A\Delta B)\Delta A'] + m[A'\Delta \mathbf{O}] \Rightarrow A\Delta B, A', \mathbf{O} \in l_{23};$$

$$m[A'\Delta B'] = m[A'\Delta \mathbf{O}] + m[\mathbf{O}\Delta B'] \Rightarrow A', \mathbf{O}, B' \in l_{24}.$$

3. For the points $\mathbf{O}, AB, A, A'$:

$$m[\mathbf{O}\Delta A] = m[\mathbf{O}\Delta(AB)] + m[(AB)\Delta A] \Rightarrow \mathbf{O}, AB, A \in l_{31};$$

$$m[(AB)\Delta A'] = m[(AB)\Delta A] + m[A\Delta A'] \Rightarrow AB, A, A' \in l_{32};$$

$$m[A\Delta \mathbf{O}] = m[A\Delta A'] + m[A'\Delta \mathbf{O}] \Rightarrow A, A', \mathbf{O} \in l_{33};$$

$$m[A'\Delta(AB)] = m[A'\Delta \mathbf{O}] + m[\mathbf{O}\Delta(AB)] \Rightarrow A', \mathbf{O}, AB \in l_{34}.$$

4. For the points $\mathbf{O}, AB, B, B'$:

$$m[\mathbf{O}\Delta B] = m[\mathbf{O}\Delta(AB)] + m[(AB)\Delta B] \Rightarrow \mathbf{O}, AB, B \in l_{41};$$

$$m[(AB)\Delta B'] = m[(AB)\Delta B] + m[B\Delta B'] \Rightarrow AB, B, B' \in l_{42};$$

$$m[B\Delta \mathbf{O}] = m[B\Delta B'] + m[B'\Delta \mathbf{O}] \Rightarrow B, B', \mathbf{O} \in l_{43};$$

$$m[B'\Delta(AB)] = m[B'\Delta \mathbf{O}] + m[\mathbf{O}\Delta(AB)] \Rightarrow B', \mathbf{O}, AB \in l_{44}.$$

5. For the points $A, A', A\Delta B, A + B$:

$$m[A\Delta(A\Delta B)] = m[A\Delta A'] + m[A'\Delta(A\Delta B)] \Rightarrow A, A', A\Delta B \in l_{51};$$

$$m[A'\Delta(A + B)] = m[A'\Delta(A\Delta B)] + m[(A\Delta B)\Delta(A + B)] \Rightarrow$$
$$\Rightarrow A', A\Delta B, A + B \in l_{52};$$

$$m[(A\Delta B)\Delta A] = m[(A\Delta B)\Delta(A + B)] + m[(A + B)\Delta A] \Rightarrow$$
$$\Rightarrow A\Delta B, A + B, A \in l_{53};$$

$$m[(A + B)\Delta A'] = m[(A + B)\Delta A] + m[A\Delta A'] \Rightarrow A + B, A, A' \in l_{54}.$$

6. For the points $B, B', A\Delta B, A + B$:

$$m[B\Delta(A\Delta B)] = m[B\Delta B'] + m[B'\Delta(A\Delta B)] \Rightarrow B, B', A\Delta B \in l_{61};$$

$$m[B'\Delta(A + B)] = m[B'\Delta(A\Delta B)] + m[(A\Delta B)\Delta(A + B)] \Rightarrow$$
$$\Rightarrow B', A\Delta B, A + B \in l_{62};$$

$$m[(A\Delta B)\Delta B] = m[(A\Delta B)\Delta(A + B)] + m[(A + B)\Delta B] \Rightarrow$$
$$\Rightarrow A\Delta B, A + B, B \in l_{63};$$

$$m[(A + B)\Delta B'] = m[(A + B)\Delta B] + m[B\Delta B'] \Rightarrow A + B, B, B' \in l_{64}.$$

7. For the points $\mathbf{O}, AB, A + B, A\Delta B$:

$$m[\mathbf{O}\Delta(A+B)] = m[\mathbf{O}\Delta(AB)] + m[(AB)\Delta(A+B)] \Rightarrow \mathbf{O}, AB, A+B \in l_{71};$$

$$m[(AB)\Delta(A\Delta B)] = m[(AB)\Delta(A + B)] + m[(A + B)\Delta(A\Delta B)] \Rightarrow$$
$$\Rightarrow AB, A + B, A\Delta B \in l_{72};$$

$$m[(A + B)\Delta \mathbf{O}] = m[(A + B)\Delta(A\Delta B)] + m[(A\Delta B)\Delta \mathbf{O}] \Rightarrow$$
$$\Rightarrow A + B, A\Delta B, \mathbf{O} \in l_{73};$$

$$m[(A\Delta B)\Delta(AB)] = m[(A\Delta B)\Delta \mathbf{O}] + m[\mathbf{O}\Delta(AB)] \Rightarrow A\Delta B, \mathbf{O}, AB \in l_{74}.$$

8. For the points A', A, B', B:

$$m(A'\Delta B) = m(A'\Delta A) + m(A\Delta B) \Rightarrow A', A, B \in l_{81};$$

$$m(A\Delta B') = m(A\Delta B) + m(B\Delta B') \Rightarrow A, B, B' \in l_{82};$$

$$m(B\Delta A') = m(B\Delta B') + m(B'\Delta A') \Rightarrow B, B', A' \in l_{83};$$

$$m(B'\Delta A) = m(B'\Delta A') + m(A'\Delta A) \Rightarrow B', A', A \in l_{84}.$$

2.2 Geometrical Properties of Metric Space Built upon Generalized Boolean Algebra

9. For the points $\mathbf{O}, A, A+B, B'$:

$$m[\mathbf{O}\Delta(A+B)] = m(\mathbf{O}\Delta A) + m[A\Delta(A+B)] \Rightarrow \mathbf{O}, A, A+B \in l_{91};$$

$$m(A\Delta B') = m[A\Delta(A+B)] + m[(A+B)\Delta B'] \Rightarrow A, A+B, B' \in l_{92};$$

$$m[(A+B)\Delta \mathbf{O}] = m[(A+B)\Delta B'] + m(B'\Delta \mathbf{O}) \Rightarrow A+B, B', \mathbf{O} \in l_{93};$$

$$m(B'\Delta A) = m(B'\Delta \mathbf{O}) + m(\mathbf{O}\Delta A) \Rightarrow B', \mathbf{O}, A \in l_{94}.$$

10. For the points $AB, B', A\Delta B, A'$:

$$m[(AB)\Delta(A\Delta B)] = m[(AB)\Delta B] + m[B\Delta(A\Delta B)] \Rightarrow AB, B, A\Delta B \in l_{101};$$

$$m(B\Delta A') = m[B\Delta(A\Delta B)] + m[(A\Delta B)\Delta A'] \Rightarrow B, A\Delta B, A' \in l_{102};$$

$$m[(A\Delta B)\Delta(AB)] = m[(A\Delta B)\Delta A'] + m[A'\Delta(AB)] \Rightarrow A\Delta B, AB, A' \in l_{103};$$

$$m(A'\Delta B) = m[A'\Delta(AB)] + m[(AB)\Delta B] \Rightarrow A', AB, B \in l_{104}.$$

11. For the points $\mathbf{O}, B, A+B, A'$:

$$m[\mathbf{O}\Delta(A+B)] = m(\mathbf{O}\Delta B) + m[B\Delta(A+B)] \Rightarrow \mathbf{O}, B, A+B \in l_{111};$$

$$m(B\Delta A') = m[B\Delta(A+B)] + m[(A+B)\Delta A'] \Rightarrow B, A+B, A' \in l_{112};$$

$$m[(A+B)\Delta \mathbf{O}] = m[(A+B)\Delta A'] + m[A'\Delta \mathbf{O}] \Rightarrow A+B, A', \mathbf{O} \in l_{113};$$

$$m(A'\Delta B) = m(A'\Delta \mathbf{O}) + m(\mathbf{O}\Delta B') \Rightarrow A', \mathbf{O}, B' \in l_{114}.$$

12. For the points $AB, B', A\Delta B, A$:

$$m[(AB)\Delta(A\Delta B)] = m[(AB)\Delta B'] + m[B'\Delta(A\Delta B)] \Rightarrow AB, B', A\Delta B \in l_{121};$$

$$m(B'\Delta A) = m[B'\Delta(A\Delta B)] + m[(A\Delta B)\Delta A] \Rightarrow B', A\Delta B, A \in l_{122};$$

$$m[(A\Delta B)\Delta(AB)] = m[(A\Delta B)\Delta A] + m[A\Delta(AB)] \Rightarrow A\Delta B, A, AB \in l_{123};$$

$$m(A\Delta B') = m[A\Delta(AB)] + m[(AB)\Delta B'] \Rightarrow A', AB, B' \in l_{124}.$$

If a combination axiom of Hilbert's axiom system [224, Section 2] claiming that there exists the only line passing through two given points is used with respect to the group of identities, then we can obtain the following paradoxical conclusion. The relationships (1 through 12) imply that the following four points belong to the same line:

1. $AB, B, A+B, A \in l_1$
2. $\mathbf{O}, B', A\Delta B, A' \in l_2$
3. $\mathbf{O}, AB, A, A' \in l_3$
4. $\mathbf{O}, AB, B, B' \in l_4$
5. $A, A', A\Delta B, A+B \in l_5$

6. $B, B', A\Delta B, A+B \in l_6$
7. $\mathbf{O}, AB, A+B, A\Delta B \in l_7$
8. $A', A, B', B \in l_8$
9. $\mathbf{O}, A, A+B, B' \in l_9$
10. $AB, B', A\Delta B, A' \in l_{10}$
11. $\mathbf{O}, B, A+B, A' \in l_{11}$
12. $AB, B', A\Delta B, A \in l_{12}$

Therefore, line l_1 is identical to the lines $l_3 \ldots l_8, l_9, l_{11}$:

$l_1 \equiv l_3$ (since the points $AB, A \in l_1$ and $AB, A \in l_3$);
$l_1 \equiv l_4$ (since the points $AB, B \in l_1$ and $AB, B \in l_4$);
$l_1 \equiv l_5$ (since the points $A, A+B \in l_1$ and $A, A+B \in l_5$);
$l_1 \equiv l_6$ (since the points $B, A+B \in l_1$ and $B, A+B \in l_6$);
$l_1 \equiv l_7$ (since the points $AB, A+B \in l_1$ and $AB, A+B \in l_7$);
$l_1 \equiv l_8$ (since the points $A, B \in l_1$ and $A, B \in l_8$);
$l_1 \equiv l_9$ (since the points $A, A+B \in l_1$ and $A, A+B \in l_9$);
$l_1 \equiv l_{11}$ (since the points $B, A+B \in l_1$ and $B, A+B \in l_{11}$);

and the line l_2 is identical to the lines $l_3 \ldots l_8, l_{10}, l_{12}$:

$l_2 \equiv l_3$ (since the points $\mathbf{O}, A' \in l_2$ and $\mathbf{O}, A' \in l_3$);
$l_2 \equiv l_4$ (since the points $\mathbf{O}, B' \in l_2$ and $\mathbf{O}, B' \in l_4$);
$l_2 \equiv l_5$ (since the points $A', A\Delta B \in l_2$ and $A', A\Delta B \in l_5$);
$l_2 \equiv l_6$ (since the points $B', A\Delta B \in l_2$ and $B', A\Delta B \in l_6$);
$l_2 \equiv l_7$ (since the points $\mathbf{O}, A\Delta B \in l_2$ and $\mathbf{O}, A\Delta B \in l_7$);
$l_2 \equiv l_8$ (since the points $A', B' \in l_2$ and $A', B' \in l_8$);
$l_2 \equiv l_{10}$ (since the points $A', A\Delta B \in l_2$ and $A', A\Delta B \in l_{10}$);
$l_2 \equiv l_{12}$ (since the points $B', A\Delta B \in l_2$ and $B', A\Delta B \in l_{12}$).

This implies that all the vertices of hexahedron $\mathbf{O}$, AB, A', B', A, B, $A\Delta B$, $A+B$ belong to the same line l. However, that cannot be, inasmuch as the following triplets of the points A', B', AB; $A', B', A+B$; $A, B, A\Delta B$; $\mathbf{O}, A, B$; $AB, A', A+B$; $AB, B', A+B$ do not lie on the same line.

It is easy to see that axiomatics of metric space Ω has to differ from Hilbert's axiom system.

The following relationships of belonging of four points to the same circle c_i in $\mathbf{R}^3$ with metric (2.2.5) are inherent to the vertices of hexahedron $\mathbf{O}$, AB, A', B', A, B, $A\Delta B$, $A+B$:

2.2 Geometrical Properties of Metric Space Built upon Generalized Boolean Algebra

1. For the points $AB, B, A+B, A$:
$$d(A+B, AB)d(A, B) = d(AB, B)d(A+B, A) + d(AB, A)d(A+B, B) \Rightarrow$$
$$\Rightarrow AB, B, A+B, A \in c_1.$$

2. For the points $\mathbf{O}, B', A\Delta B, A'$:
$$d(A\Delta B, \mathbf{O})d(A', B') = d(B', \mathbf{O})d(A\Delta B, A') + d(A', \mathbf{O})d(A\Delta B, B') \Rightarrow$$
$$\Rightarrow \mathbf{O}, B', A\Delta B, A' \in c_2.$$

3. For the points $\mathbf{O}, AB, A, A'$:
$$d(A, \mathbf{O})d(A', AB) = d(AB, \mathbf{O})d(A, A') + d(A, AB)d(A', \mathbf{O}) \Rightarrow$$
$$\Rightarrow \mathbf{O}, AB, A, A' \in c_3.$$

4. For the points $\mathbf{O}, AB, B, B'$:
$$d(B, \mathbf{O})d(B', AB) = d(AB, \mathbf{O})d(B, B') + d(B, AB)d(B', \mathbf{O}) \Rightarrow$$
$$\Rightarrow \mathbf{O}, AB, B, B' \in c_4.$$

5. For the points $A, A', A\Delta B, A+B$:
$$d(A+B, A')d(A\Delta B, A) = d(A\Delta B, A')d(A+B, A) +$$
$$+ d(A+B, A\Delta B)d(A, A') \Rightarrow A, A', A\Delta B, A+B \in c_5.$$

6. For the points $B, B', A\Delta B, A+B$:
$$d(A+B, B')d(A\Delta B, B) = d(A\Delta B, B')d(A+B, B) +$$
$$+ d(A+B, A\Delta B)d(B, B') \Rightarrow B, B', A\Delta B, A+B \in c_6.$$

7. For the points $\mathbf{O}, AB, A+B, A\Delta B$:
$$d(A+B, \mathbf{O})d(A\Delta B, AB) = d(A+B, A\Delta B)d(AB, \mathbf{O}) +$$
$$+ d(A+B, AB)d(A\Delta B, \mathbf{O}) \Rightarrow \mathbf{O}, AB, A+B, A\Delta B \in c_7.$$

8. For the points A', A, B', B:
$$d(B, A')d(A, B') = d(B, B')d(A, A') + d(A, B)d(A', B') \Rightarrow$$
$$\Rightarrow A', A, B', B \in c_8.$$

9. For the points $\mathbf{O}, A, A+B, B'$:
$$d(A+B, \mathbf{O})d(A, B') = d(A, \mathbf{O})d(A+B, B') + d(B', \mathbf{O})d(A+B, A) \Rightarrow$$
$$\Rightarrow \mathbf{O}, A, A+B, B' \in c_9.$$

10. For the points $AB, B, A\Delta B, A'$:

$$d(A\Delta B, AB)d(A', B) = d(AB, B)d(A\Delta B, A') + d(AB, A')d(A\Delta B, B) \Rightarrow$$
$$\Rightarrow AB, B, A\Delta B, A' \in c_{10}.$$

11. For the points $\mathbf{O}, B, A+B, A'$:

$$d(A+B, \mathbf{O})d(A', B) = d(B, \mathbf{O})d(A+B, A') + d(A', \mathbf{O})d(A+B, B) \Rightarrow$$
$$\Rightarrow \mathbf{O}, B, A+B, A' \in c_{11}.$$

12. For the points $AB, B', A\Delta B, A$:

$$d(A\Delta B, AB)d(A, B') = d(AB, B')d(A\Delta B, A) + d(AB, A)d(A\Delta B, B') \Rightarrow$$
$$\Rightarrow AB, B', A\Delta B, A \in c_{12}.$$

Thus, the aforementioned quartets of the vertices of hexahedron $\mathbf{O}, AB, A', B', A, B, A\Delta B, A+B$ belong to the corresponding circles $c_1 \ldots c_{12}$ in Euclidean space $\mathbf{R}^3$, which will be identified with the lines in the space Ω, and all eight vertices of the hexahedron belong to the same sphere $Sp(O_{AB}, R_{AB})$ in $\mathbf{R}^3$.

There is a relation of order on generalized Boolean algebra $\mathbf{B}(\Omega)$ and on any lattice, so for the elements $A, B \in \Omega$ the relation $A \leq B$ means that $A = AB$: $A \leq B \Leftrightarrow A = AB$. The relation of a strict order $A < B$ between the elements $A, B \in \Omega$ means equivalence: $A < B \Leftrightarrow (A \leq B)$ & $(A \neq B)$.

Definition 2.2.2. If Q_1 and Q_2 are the elements of generalized Boolean algebra $\mathbf{B}(\Omega)$, then the set $[Q_1, Q_2] = \{X_\alpha : X_\alpha \in \Omega, Q_1 \leq X_a \leq Q_2\}$ is called a *closed interval* (or simply an interval).

Lemma 2.2.3. *The elements Q_1, X_α, Q_2 of the interval $[Q_1, Q_2]$, $X_\alpha \neq Q_{1,2}$ lie on the same line l in metric space Ω:*

$$Q_1, X_\alpha, Q_2 \in [Q_1, Q_2] \Rightarrow Q_1, X_\alpha, Q_2 \in l \Leftrightarrow Q_1 \prec X_\alpha \prec Q_2.$$

Proof. For the elements Q_1, X_α, Q_2 of the interval $[Q_1, Q_2]$, defined on generalized Boolean algebra $\mathbf{B}(\Omega)$ with a measure m, the following relations hold:

$$Q_1 < X_\alpha \Rightarrow Q_1 = Q_1 \cdot X_\alpha, \quad X_\alpha < Q_2 \Rightarrow X_\alpha = X_\alpha \cdot Q_2,$$

so, the metric identities hold:

$$m(Q_1 \Delta X_\alpha) = m(X_\alpha) - m(Q_1), \quad m(X_\alpha \Delta Q_2) = m(Q_2) - m(X_\alpha).$$

According to relationships, the metric identity, being the condition of belonging of three points to the same line, holds:

$$\rho(Q_1, Q_2) = \rho(Q_1, X_\alpha) + \rho(X_\alpha, Q_2) = m(Q_2) - m(Q_1),$$

where $\rho(Q_1, Q_2) = m(Q_1 \Delta Q_2)$ is metric between the elements Q_1, Q_2 in the space Ω. $\square$

2.2 Geometrical Properties of Metric Space Built upon Generalized Boolean Algebra

Definition 2.2.3. The mapping φ associating every element $\alpha \in \Phi$ with the element $A_\alpha \in \Omega$ is called the *indexed set* $\{A_\alpha\}_{\alpha \in \Phi}$ on generalized Boolean algebra $\mathbf{B}(\Omega)$ with a measure m: $\varphi : \alpha \to A_\alpha$.

Definition 2.2.4. If the indexed set $A = \{A_\alpha\}_{\alpha \in \Phi}$ is a subset of generalized Boolean algebra $\mathbf{B}(\Omega)$, whose elements are connected by a linear order, i.e., for an arbitrary pair of the elements $A_\alpha, A_\beta \in A$, $\alpha \leq \beta$, the inequality $A_\alpha \leq A_\beta$ holds, then the set A is a *linearly ordered indexed set* or indexed chain.

Theorem 2.2.1. *For the elements $\{A_j\}$, $j = 0, 1, \ldots, n$ of linearly ordered indexed set $A = \{A_j\}$, $A \subset \Omega$, defined on generalized Boolean algebra $\mathbf{B}(\Omega)$ with a measure m, the metric identity holds:*

$$m(A_0 \Delta A_n) = \sum_{j=1}^{n} m(A_{j-1} \Delta A_j). \quad (2.2.8)$$

Proof. Since for the elements of linearly ordered set $A = \{A_j\}$ defined on generalized Boolean algebra $\mathbf{B}(\Omega)$ with a measure m, an implication $A_{j-1} \leq A_j \Leftrightarrow A_{j-1} = A_{j-1} \cdot A_j$ and the metric identity hold:

$$m(A_{j-1} \Delta A_j) = m(A_j) - m(A_{j-1}). \quad (2.2.9)$$

Inserting the right part of the identity (2.2.9) into the sum of the right part of the identity (2.2.8), we get the value of the series' sum:

$$\sum_{j=1}^{n} m(A_{j-1} \Delta A_j) = \sum_{j=1}^{n} [m(A_j) - m(A_{j-1})] = m(A_0 \Delta A_n).$$

$\square$

Based on Theorem 2.2.1, the condition of three points belonging to the same line in the space Ω (2.2.6) can be extended to the case of arbitrary large numbers of points. We consider that the elements $\{A_j\}$ of the indexed set $A = \{A_j\}$, $j = 0, 1, \ldots, n$ lie on the same line in metric space Ω, if the metric identity holds (compare with [222]):

$$\rho(A_0 \Delta A_n) = \sum_{j=1}^{n} \rho(A_{j-1} \Delta A_j). \quad (2.2.10)$$

where $\rho(A_\alpha, A_\beta) = m(A_\alpha \Delta A_\beta)$ is metric in space Ω.
Then Theorem 2.2.1 has the following corollary.

Corollary 2.2.1. *The elements $\{A_j\}$ of linearly ordered indexed set $A = \{A_j\}$, $j = 0, 1, \ldots, n$, $A \subset \Omega$ belong to the same line l:*

$$A = \{A_j : \forall A_j \in l : A_0 \prec A_1 \prec \ldots \prec A_j \prec \ldots \prec A_n\}.$$

Definition 2.2.5. If all the elements of the line l in metric space Ω form a partially ordered set, then the line is called the *line with partially ordered elements*.

Definition 2.2.6. If all the elements of the line l in metric space Ω form a linearly ordered set, then the line is called the *line with linearly ordered elements*.

Thus, in metric space Ω, we shall differentiate lines with partially and linearly ordered elements.

Mutual situation of two distinct lines in the space Ω is characterized by the following feature in contrast with the lines in Euclidean space.

Example 2.2.1. Consider two linearly ordered indexed sets $A = \{A_j\}$, $j = 0, 1, \ldots, J$, $B = \{B_i\}$, $i = 0, 1, \ldots, I$ on generalized Boolean algebra $\mathbf{B}(\Omega)$: $A, B \subset \Omega$, whose subsets $A^* \subset A$, $B^* \subset B$ are identical to each other in the interval $[Q_0, Q_K]$:

$$\begin{cases} A^* = \{A_j : A_j = Q_k, A_j \in A, j = k + N_J, k = 0, 1, \ldots, K\}; \\ B^* = \{B_i : B_i = Q_k, B_i \in B, i = k + N_I, k = 0, 1, \ldots, K\}, \end{cases} \quad (2.2.11)$$

where $K, N_J, N_I \in \mathbf{N}$ are constants, so that $K + N_J < J$ and $K + N_I < I$. Condition (2.2.11) implies that the intersection of linearly ordered sets $A = \{A_j\}$, $B = \{B_i\}$ is the linearly ordered set $Q = \{Q_k\}$:

$$A \cap B = Q = \{Q_k\}, k = 0, 1, \ldots, K.$$

Then, according to the Corollary 2.2.1 of Theorem 2.2.1, the lines l_A and l_B, on which the elements $\{A_j\}$ of the set A and the elements $\{B_i\}$ of the set B are situated, respectively, intersect each other on the set Q:

$$l_A \cap l_B = Q = \{Q_k\}, k = 0, 1, \ldots, K. \quad \triangledown$$

This example implies that two distinct lines l_A and l_B in the space Ω could have an arbitrarily large number $K + 1$ of common points $\{Q_k\}$ belonging to the same interval $[Q_0, Q_K]$: $\{Q_k\} \subset [Q_0, Q_K]$.

Example 2.2.2. Consider two linearly ordered sets $A' = \{A'_j\}$, $j = 0, 1, \ldots, J$ and $A'' = \{A''_i\}$, $i = 0, 1, \ldots, I$ defined in the interval $[\mathbf{O}, A]$ of generalized Boolean algebra $\mathbf{B}(\Omega)$, which are such that $A' \subset A$, $A'' \subset A$ and the limits of the sequences of the sets $\{A'_j\}$ and $\{A''_i\}$ coincide with their largest and least elements, which are, respectively, equal to:

$$\begin{cases} A' = \{A'_j : A'_j \in [\mathbf{O}, A], \lim_{j \to 0} A'_j = \mathbf{O} \;\&\; \lim_{j \to \infty} A'_j = A\}; \\ A'' = \{A''_i : A''_i \in [\mathbf{O}, A], \lim_{i \to 0} A''_i = \mathbf{O} \;\&\; \lim_{i \to \infty} A''_i = A\}. \end{cases} \quad (2.2.12)$$

The condition (2.2.12) implies that the intersection of the sets $A' = \{A'_j\}$ and $A'' = \{A''_i\}$ contains two elements of generalized Boolean algebra $\mathbf{B}(\Omega)$, $\mathbf{O}$ and A: $A' \cap A'' = \{\mathbf{O}, A\}$. Then, according to Corollary 2.2.1 of Theorem 2.2.1, the lines l' and l'', on which the elements $\{A'_j\}$ of the set A' and the elements $\{A''_i\}$ of the set A'' are situated, intersect at two points of the space Ω: $l' \cap l'' = \{\mathbf{O}, A\}$. $\triangledown$

2.2 Geometrical Properties of Metric Space Built upon Generalized Boolean Algebra

This example implies that two distinct lines l' and l'' in space Ω, passing through the points of linearly ordered sets $A' = \{A'_j\}$, $A'' = \{A''_i\}$, respectively: $A' \subset l'$ and $A'' \subset l''$, being the subsets of the interval $[\mathbf{O}, A]$: $A' \subset [\mathbf{O}, A]$ and $A'' \subset [\mathbf{O}, A]$, intersect at two extreme points of this interval $\mathbf{O}, A$.

Definition 2.2.7. Linearly ordered indexed set $A = \{A_\alpha\}_{\alpha \in \Phi}$ is called an *everywhere dense set*, if for all $A_\alpha, A_\beta \in A$: $A_\alpha < A_\beta$ an element A_γ can be found, such that: $A_\alpha < A_\gamma < A_\beta$.

Definition 2.2.8. The element $Q \in \Omega$ is called the *limit of everywhere dense linearly ordered indexed set* $A = \{A_\alpha\}_{\alpha \in \Phi}$ defined on generalized Boolean algebra $\mathbf{B}(\Omega)$ with a measure m: $A \subset \Omega$, if any neighborhood $U_\varepsilon(Q)$ of the point Q can be associated with such an index $\alpha_U \in \Delta$, that for all $\alpha > \alpha_U$ the condition $A_\alpha \in U_\varepsilon(Q)$ holds:

$$Q = \lim_{\alpha \to \alpha_Q} A_\alpha \Leftrightarrow \forall \varepsilon > 0 : \exists \alpha_U : \forall \alpha > \alpha_U : m(A_\alpha \Delta Q) < \varepsilon.$$

Example 2.2.3. Consider two linearly ordered everywhere dense indexed sets $A = \{A_\alpha\}_{\alpha \in \Phi_1}$ and $B = \{B_\beta\}_{\beta \in \Phi_2}$ defined in the intervals $[A_0, Q_A]$ and $[B_0, Q_B]$ of generalized Boolean algebra $\mathbf{B}(\Omega)$, respectively: $A \subset [A_0, Q_A]$, $B \subset [B_0, Q_B]$, so that $[A_0, Q_A] \cap [B_0, Q_B] = [Q_1, Q_2]$, $A_0 < Q_1$ & $B_0 < Q_1$, $Q_2 < Q_A$ & $Q_2 < Q_B$, and the limits of the sequences A and B coincide with the extreme points of the interval $[Q_1, Q_2]$, respectively:

$$\begin{cases} \lim_{\alpha \to \alpha_1} A_\alpha = \lim_{\beta \to \beta_1} B_\beta = Q_1; \\ \lim_{\alpha \to \alpha_2} A_\alpha = \lim_{\beta \to \beta_2} B_\beta = Q_2. \end{cases} \quad (2.2.13)$$

Identities (2.2.13) imply that the intersection of linearly ordered sets $A = \{A_\alpha\}_{\alpha \in \Phi_1}$ and $B = \{B_\beta\}_{\beta \in \Phi_2}$ contains two elements of generalized Boolean algebra $\mathbf{B}(\Omega)$: Q_1 and Q_2: $A \cap B = \{Q_1, Q_2\}$.

Then, according to Corollary 2.2.1 of Theorem 2.2.1, the lines l_A and l_B, on which the elements $A = \{A_\alpha\}_{\alpha \in \Phi_1}$, $B = \{B_\beta\}_{\beta \in \Phi_2}$ of the sets A, B are situated, respectively, intersect at two points of the space Ω: $l_A \cap l_B = \{Q_1, Q_2\}$. ▽

This example implies that two distinct lines l_A and l_B, passing through the points of linearly ordered sets $A = \{A_\alpha\}_{\alpha \in \Phi_1}$, $B = \{B_\beta\}_{\beta \in \Phi_2}$, respectively: $\{A_\alpha\} \subset l_A$, $\{B_\beta\} \subset l_B$, can intersect at two or more points of the space Ω.

2.2.3 Notions of Sheet and Plane in Metric Space Built upon Generalized Boolean Algebra with a Measure

To characterize geometrical relations between a pair and a triplet of elements in metric space Ω not connected by the relation of an order, we introduce extra-axiomatic definitions of the notions of sheet and plane, respectively.

Definition 2.2.9. In metric space Ω, *sheet* L_{AB} is defined as a geometrical image of subalgebra $\mathbf{B}(A+B)$ of generalized Boolean algebra $\mathbf{B}(\Omega)$, $A, B \in \Omega$, represented by three-element partition:

$$A + B = (AB) + (A - B) + (B - A),$$

where the elements $AB, A - B, B - A$ are pairwise orthogonal and differ from null element:

$$AB \perp (A - B) \perp (B - A) \perp AB \Leftrightarrow$$
$$\Leftrightarrow (AB) \cdot (A - B) = \mathbf{O}, \quad (A - B) \cdot (B - A) = \mathbf{O}, \quad (B - A) \cdot (AB) = \mathbf{O};$$
$$AB \neq \mathbf{O}, \quad A - B \neq \mathbf{O}, \quad B - A \neq \mathbf{O}.$$

The coordinates of the points $\mathbf{O}$, AB, $A - B$, $B - A$, A, B, $A\Delta B$, $A + B$ of the sheet L_{AB} in $\mathbf{R}^3$ are equal to:

$$\mathbf{O}\,(0,0,0), \quad AB\,(0, 0, m(AB)), \quad A - B\,(m(A - B), 0, 0);$$
$$B - A\,(0, m(B - A), 0), \quad A\,(m(A - B), 0, m(AB)), \quad B\,(0, m(B - A), m(AB));$$
$$A\Delta B\,(m(A - B), m(B - A), 0), \quad A + B\,(m(A - B), m(B - A), m(AB)).$$

The points $\mathbf{O}$, AB, $A - B$, $B - A$, A, B, $A\Delta B$, $A + B$ of the sheet L_{AB} belong to the sphere $Sp(O_{AB}, R_{AB})$ in $\mathbf{R}^3$, so that the center O_{AB} and radius R_{AB} of the sphere are determined by the Equations (2.2.2) and (2.2.3), respectively.

Definition 2.2.10. In metric space Ω, *sheet* L_{AB} is a set of the points which contains two elements $A, B \in \Omega$ that are not connected by relation of an order ($A \neq AB$ & $B \neq AB$), lines passing through two given points; every line that passes through an arbitrary pair of the points on the lines passing through two given points.

By Definitions 2.2.9 and 2.2.10, the algebraic and geometric essences of a sheet are established.

Definition 2.2.11. Two points A and B are called *generator points of the sheet* L_{AB} in metric space Ω, if they are not connected by relation of an order on generalized Boolean algebra $\mathbf{B}(\Omega)$: $A \neq AB$ & $B \neq AB$.

Definitions 2.2.9 and 2.2.10 imply that the sheet L_{AB} given by generator points A, B contains the points $\mathbf{O}$, AB, $A - B$, $B - A$, A, B, $A\Delta B$, $A + B$, and also the points $\{X_\alpha\}$ of a linearly ordered indexed set X lying on the line passing through the extreme points $\mathbf{O}, X$ of the interval $[\mathbf{O}, X]$, where $X = Y * Z$, $Y = A * B$, $Z = A * B$, and the asterisk $*$ is an arbitrary signature operation of generalized Boolean algebra $\mathbf{B}(\Omega)$.

To denote one-to-one correspondence between an algebraic notion W and its geometric image $Geom(W)$, we shall write:

$$W \leftrightarrow Geom(W). \tag{2.2.14}$$

2.2 Geometrical Properties of Metric Space Built upon Generalized Boolean Algebra

For instance, the correspondence between the sheet L_{AB} and subalgebra $\mathbf{B}(A+B)$ of generalized Boolean algebra $\mathbf{B}(\Omega)$ will be denoted as $L_{AB} \leftrightarrow A+B$.

Applying the correspondence (2.2.14), one can define the relationships between the objects of metric space Ω with some given properties. For instance, the intersection of two sheets L_{AB} and L_{CD}, determined by the generator points A, B and C, D, respectively, is the sheet with generator points $(A+B)C$ and $(A+B)D$ or the sheet with generator points $(C+D)A$ and $(C+D)B$:

$$L_{AB} \cap L_{CD} \leftrightarrow (A+B)(C+D) =$$

$$= \begin{cases} (A+B)C + (A+B)D \leftrightarrow L_{(A+B)C,(A+B)D}; \\ (C+D)A + (C+D)B \leftrightarrow L_{(C+D)A,(C+D)B}. \end{cases}$$

Using the last relationship, if the sheet L_{AB} is a subset of the sheet L_{CD} we obtain:

$$L_{AB} \subset L_{CD} \leftrightarrow A+B \subset C+D \Rightarrow (A+B)(C+D) = A+B \leftrightarrow L_{AB},$$

then the intersection is the sheet L_{AB}: $L_{AB} \cap L_{CD} = L_{AB}$.

If the points A and C of the sheets L_{AB} and L_{CD} are identically equal: $A \equiv C$, then intersection of these sheets is the sheet $L_{A,BD}$ with generator points A and BD:

$$L_{AB} \cap L_{CD} \leftrightarrow (A+B)(C+D) = (A+B)(A+D) =$$

$$= A + AD + AB + BD = A + BD \leftrightarrow L_{A,BD},$$

i.e.: $A \equiv C \Rightarrow L_{AB} \cap L_{CD} = L_{A,BD}$.

But if the points A and C of the sheets L_{AB} and L_{CD} are identically equal: $A \equiv C$, and the elements B and D are orthogonal $BD = \mathbf{O}$, then intersection of these sheets is an element A:

$$L_{AB} \cap L_{CD} \leftrightarrow (A+B)(C+D) = (A+B)(A+D) =$$

$$= A + AD + AB + BD = A + BD = A,$$

i.e.: $(A \equiv C) \,\&\, (BD = \mathbf{O}) \Rightarrow L_{AB} \cap L_{CD} = A$.

Definition 2.2.12. In metric space Ω, *plane* α_{ABC} passing through three points A, B, C, which are not pairwise connected by relation of an order:

$$A \neq AB \,\&\, B \neq AB \,\&\, B \neq BC \,\&\, C \neq CB \,\&\, A \neq AC \,\&\, C \neq AC,$$

is a geometric image of subalgebra $\mathbf{B}(A+B+C)$ of generalized Boolean algebra $\mathbf{B}(\Omega)$.

Definition 2.2.13. In metric space Ω, *plane* α_{ABC} is a set of the points, which contains: the arbitrary three points A, B, C not lying on the same line; the lines passing through a pair of three points A, B, C; any line that passes through an arbitrary pair of the points on the lines passing through any pair of three given points A, B, C.

By Definitions 2.2.12 and 2.2.13, the algebraic and geometric essences of a plane are established, respectively.

Definition 2.2.14. Three points A, B, C are called *generator points of the plane* α_{ABC} in metric space Ω, if they are not pairwise connected by relation of an order:

$$A \neq AB \ \& \ B \neq AB \ \& \ B \neq BC \ \& \ C \neq CB \ \& \ A \neq AC \ \& \ C \neq AC.$$

Using the relation (2.2.14), we obtain that the intersection of two planes α_{ABC} and α_{DEF}, where the points A, B, C and D, E, F are not connected with each other by relation of order, is the plane with generator points $A(D+E+F)$, $B(D+E+F)$, $C(D+E+F)$, or the plane with generator points $D(A+B+C)$, $E(A+B+C)$, $F(A+B+C)$:

$$\alpha_{ABC} \cap \alpha_{DEF} \leftrightarrow (A+B+C)(D+E+F) =$$

$$= \begin{cases} A(D+E+F) + B(D+E+F) + C(D+E+F); \\ D(A+B+C) + E(A+B+C) + F(A+B+C). \end{cases} \leftrightarrow$$

$$\leftrightarrow \begin{cases} \alpha_{A(D+E+F), B(D+E+F), C(D+E+F)}; \\ \alpha_{D(A+B+C), E(A+B+C), F(A+B+C)}. \end{cases}$$

Using the last relation, if the plane α_{ABC} is a subset of the plane α_{DEF} we obtain:

$$\alpha_{ABC} \subset \alpha_{DEF} \leftrightarrow A+B+C \subset D+E+F \Rightarrow (A+B+C)(D+E+F) =$$

$$= A+B+C,$$

then the intersection of these planes is the plane α_{ABC}:

$$\alpha_{ABC} \subset \alpha_{DEF} \Rightarrow \alpha_{ABC} \cap \alpha_{DEF} = \alpha_{ABC}.$$

If the points A, B and D, E of the planes $\alpha_{ABC}, \alpha_{DEF}$ are pairwise identical $A \equiv D$, $B \equiv E$, then the intersection of these planes is the plane $\alpha_{A,B,CF}$ with generator points A, B, CF:

$$A \equiv D, B \equiv E \Rightarrow \alpha_{ABC} \cap \alpha_{DEF} \leftrightarrow (A+B+C)(D+E+F) =$$

$$= (A+B+C)(A+B+F) = A+B+CF \leftrightarrow \alpha_{A,B,CF},$$

i.e., $A \equiv D, B \equiv E \Rightarrow \alpha_{ABC} \cap \alpha_{DEF} = \alpha_{A,B,CF}$. If the points A, B and D, E of the planes $\alpha_{ABC}, \alpha_{DEF}$ are pairwise identical $A \equiv D$, $B \equiv E$, and the elements C and F are orthogonal $CF = \mathbf{O}$, then the intersection of these planes is the sheet L_{AB} with generator points A, B:

$$(A \equiv D, B \equiv E) \ \& \ (CF = \mathbf{O}) \Rightarrow \alpha_{ABC} \cap \alpha_{DEF} \leftrightarrow (A+B+C)(D+E+F) =$$

$$= (A+B+C)(A+B+F) = A+B+CF = A+B,$$

i.e., $(A \equiv D, B \equiv E) \ \& \ (CF = \mathbf{O}) \Rightarrow \alpha_{ABC} \cap \alpha_{DEF} = L_{AB}$.

2.2 Geometrical Properties of Metric Space Built upon Generalized Boolean Algebra

But if the points A and D of the planes α_{ABC}, α_{DEF} are identical $A \equiv D$, and the elements B, C and $E+F$ are pairwise orthogonal $B(E+F) = \mathbf{O}$, $C(E+F) = \mathbf{O}$, then the intersection of these planes is the element A:

$$(A \equiv D) \ \& \ (B(E+F) = \mathbf{O}) \ \& \ (C(E+F) = \mathbf{O}) \Rightarrow \alpha_{ABC} \cap \alpha_{DEF} \leftrightarrow$$

$$\leftrightarrow (A+B+C)(D+E+F) = A(A+E+F) + B(A+E+F) + C(A+E+F) =$$

$$= A + BA + CA + B(E+F) + C(E+F) = A,$$

i.e., $(A \equiv D) \ \& \ (B(E+F) = \mathbf{O}) \ \& \ (C(E+F) = \mathbf{O}) \Rightarrow \alpha_{ABC} \cap \alpha_{DEF} = A$.

Example 2.2.4. Let the elements A and D of the planes α_{ABC}, α_{DEF} be identical $A \equiv D$, and the elements B, C and $E+F$ are pairwise orthogonal $B(E+F) = \mathbf{O}$, $C(E+F) = \mathbf{O}$. Then the intersection of these planes is the element A:

$$(A \equiv D) \ \& \ (B(E+F) = \mathbf{O}) \ \& \ (C(E+F) = \mathbf{O}) \Rightarrow \alpha_{ABC} \cap \alpha_{DEF} = A. \ \triangledown$$

Example 2.2.5. Consider the sheets L_{AB}, L_{AE} with the generator points A, B and A, E belonging to the planes α_{ABC}, α_{AEF}, respectively. Let the lines l_{AB}, l_{AE} passing through the generator points A, B and A, E of these sheets, respectively, intersect each other A, so that:

$$\begin{cases} l_{AB} \cap l_{AE} = A; \\ A, B \in l_{AB} \subset L_{AB} \subset \alpha_{ABC}; \\ A, E \in l_{AE} \subset L_{AE} \subset \alpha_{AEF}. \end{cases} \triangledown$$

This example implies that two planes α_{ABC}, α_{AEF}, two sheets L_{AB}, L_{AE}, and two lines l_{AB}, l_{AE} belonging to these planes α_{ABC}, α_{AEF}, respectively, can intersect each other in a single point in metric space Ω.

Example 2.2.6. Consider three planes $\{\alpha_i\}$, $i = 1, 2, 3$ in metric space Ω, each of them with the generator points $\{A_i, B_i, C_i\}$, $i = 1, 2, 3$ connected by relation of order: $A_1 < A_2 < A_3$, $B_1 < B_2 < B_3$, $C_1 < C_2 < C_3$. Then, according to Corollary 2.2.1 of Theorem 2.2.1, the points $\{A_i\}, \{B_i\}, \{C_i\}$, $i = 1, 2, 3$ lie on the lines l_A, l_B, l_C, respectively: $A_i \in l_A$, $B_i \in l_B$, $C_i \in l_C$. It should be noted that $A_1 < A_2$, hence, the points A_1, A_2 belong to the plane α_2: $A_1, A_2 \in \alpha_2$, however, $A_1 < A_2 < A_3$, hence, the point A_3 does not belong to the plane α_2. This example implies that in metric space Ω among three points A_1, A_2, A_3 belonging to the same line l_A, there are the points A_1, A_2 that belong to the plane α_2. However, the point A_3 does not belong to this plane. It is elucidated by the fact, that in metric space Ω, the axiom of "absolute geometry" that "if two points of a line belong to a plane, then all the points of this line belong to this plane" [224, Section 2] does not hold. $\triangledown$

Example 2.2.7. Consider, as well as in the previous example, three planes $\{\alpha_i\}$, $i = 1, 2, 3$ in metric space Ω, each of them with generator points $\{A_i, B_i, C_i\}$, $i = 1, 2, 3$ connected with each other by relation of order: $A_1 < A_2 < A_3$, $B_1 < B_2 < B_3$, $C_1 < C_2 < C_3$. Every pair of the generator points $\{A_i, B_i\}$, $i = 1, 2, 3$ determines the sheets $\{L_{ABi}\}$ belonging to the corresponding planes $\{\alpha_i\}$. The points $\{A_i\}, \{B_i\}, \{C_i\}$, $i = 1, 2, 3$ lie on the lines l_A, l_B, l_C, respectively: $A_i \in l_A$, $B_i \in l_B$, $C_i \in l_C$. Let also the lines l_A, l_B intersect in the point AB: $l_A \cap l_B = AB$, so that $AB = A_i B_i$, $i = 1, 2, 3$. It should be noted that $AB < A_1 < A_2$, $AB < B_1 < B_2$, hence, the lines l_{A2}, l_{B2} pass through the points AB, A_1, A_2; AB, B_1, B_2, respectively, $AB, A_1, A_2 \in l_{A2}$; $AB, B_1, B_2 \in l_{B2}$ belong to the plane α_2, so that $l_{A2} \cap l_{B2} = AB$. Therefore two crossing lines l_{A2}, l_{B2} belong to the same plane α_2; however, two crossing lines l_A, l_B do not belong to the same plane α_2. In metric space Ω the corollary from axioms of connection of "absolute geometry" that "two crossing lines belong to the same plane" [224, Section 2] does not hold. Thus, in metric space Ω, two crossing lines can belong or not belong to the same plane. ▽

If the generator points $\{A_i, B_i, C_i\}$ of the planes $\{\alpha_i\}$ are connected by relation of order: $A_{i-1} < A_i < A_{i+1}$, $B_{i-1} < B_i < B_{i+1}$, $C_{i-1} < C_i < C_{i+1}$ and, therefore, lie on the same line respectively: $A_{i-1}, A_i, A_{i+1} \in l_A$; $B_{i-1}, B_i, B_{i+1} \in l_B$; $C_{i-1}, C_i, C_{i+1} \in l_C$, then we shall consider that the planes $\alpha_{i-1}, \alpha_i, \alpha_{i+1}$ belong to one another: $\alpha_{i-1} \subset \alpha_i \subset \alpha_{i+1}$.

If the generator points $\{A_i, B_i\}$ of the sheets $\{L_i\}$ are connected by relation of order: $A_{i-1} < A_i < A_{i+1}$, $B_{i-1} < B_i < B_{i+1}$ and, therefore, lie on the same line respectively: $A_{i-1}, A_i, A_{i+1} \in l_A$; $B_{i-1}, B_i, B_{i+1} \in l_B$, then we shall consider that the sheets L_{i-1}, L_i, L_{i+1} belong to one another: $L_{i-1} \subset L_i \subset L_{i+1}$.

If the generator points $\{A_i, B_i, C_i\}$ of the planes $\{\alpha_i\}$ are connected by relation of order: $A_{i-1} < A_i < A_{i+1}$, $B_{i-1} < B_i < B_{i+1}$, $C_{i-1} < C_i < C_{i+1}$ and, therefore, lie on the same line, respectively: $A_{i-1}, A_i, A_{i+1} \in l_A$; $B_{i-1}, B_i, B_{i+1} \in l_B$; $C_{i-1}, C_i, C_{i+1} \in l_C$, then we shall consider that the lines $l_{Ai-1}, l_{Ai}, l_{Ai+1}$:

$$l_{Ai} = l_A \cap \alpha_i = \{A_\alpha : (\forall A_\alpha \in A_i) \ \& \ (\forall A_\alpha \in l_A)\},$$

belong to one another, respectively: $l_{Ai-1} \subset l_{Ai} \subset l_{Ai+1}$.

2.2.4 Axiomatic System of Metric Space Built upon Generalized Boolean Algebra with a Measure

To formulate the axiomatic system of metric space Ω built upon generalized Boolean algebra $\mathbf{B}(\Omega)$ with a measure m, we shall use Hilbert's idea that every space is a set of the objects of three types—points, lines, and planes, so that there are relations of mutual containment (connection or incidence); betweenness relations (i.e., relations of an order); relations of congruence, continuity, and parallelism, and descriptions of these relations are given by axioms [224, Section 1]. Consider the axiomatic system of metric space Ω starting from the properties of its main objects — points, lines, sheets, and planes.

2.2 Geometrical Properties of Metric Space Built upon Generalized Boolean Algebra

Axioms of connection (containment)

A_1^1. There exists at least one line passing through two given points.

A_2^1. There exist at least three points that belong to every line.

A_3^1. There exist at least three points that do not belong to the same line.

A_4^1. Two crossing lines have at least one point in common.

A_5^1. There exists only one sheet passing through null element and two points that do not lie with null element on the same line.

A_6^1. A null element and at least two points belong to each sheet.

A_7^1. There exist at least three points that do not belong to the same sheet and differ from the null element.

A_8^1. Two different crossing sheets have at least one point in common, which differs from null element.

A_9^1. There exists only one plane passing through null element and three points that do not lie on the same sheet.

A_{10}^1. A null element and at least three points belong to every plane.

A_{11}^1. There exist at least four points that do not belong to the same plane and differ from the null element.

A_{12}^1. Two different crossing planes have at least one point in common that differs from the null element.

Corollaries from axioms of connection

C_1^1. There exists only one plane passing through a sheet and a point that does not belong to it (from A_9^1 and A_5^1).

C_2^1. There exists only one plane passing through two distinct sheets crossing in a point which differs from the null element (from A_9^1 and A_5^1).

Axioms of order

A_1^2. Three points A, B, X of metric space Ω belong to some line l, and X is situated between the points A and B, if the metric identity holds:

$$\rho(A, B) = \rho(A, X) + \rho(X, B) \Rightarrow A \prec X \prec B \text{ (or } A \succ X \succ B),$$

where $\rho(A, B) = m(A \Delta B)$ is a metric between the elements A, B in the space Ω.

A_2^2. Among any three points on a line, there exists no more than one point lying between the other two.

A_3^2. Two elements A, B of metric space Ω, which are not connected by relation of order $((A \neq AB) \;\&\; (B \neq AB))$ lie on the same line l_1 along with the elements AB and $A + B$: $A, B, AB, A + B \in l_1$, and also lie on the same line l_2 along with the elements $A - B$ and $B - A$: $A, B, B - A, A - B \in l_2$, so that the following relations of geometric order hold:

$$\forall A, B : (A \neq AB) \;\&\; (B \neq AB) \Rightarrow A, B, AB, A + B \in l_1 :$$

$$AB \prec A \prec A + B \prec B \prec AB \quad \text{or} \quad AB \succ A \succ A + B \succ B \succ AB;$$

$$\forall A, B : (A \neq AB) \;\&\; (B \neq AB) \Rightarrow A, B, B - A, A - B \in l_2 :$$

$$A \prec B \prec B - A \prec A - B \prec A \quad \text{or} \quad A \succ B \succ B - A \succ A - B \succ A.$$

A_4^2. The elements $\{A_j\}$ of linearly ordered indexed set $A = \{A_j\}$, $j = 0, 1, \ldots, n$, $A_j < A_{j+1}$ in metric space Ω lie on the same line l, so that the metric identity holds:

$$\rho(A_0 \Delta A_n) = \sum_{j=1}^{n} \rho(A_{j-1} \Delta A_j) \Leftrightarrow$$

$$\Leftrightarrow A_0 \prec A_1 \prec \ldots \prec A_j \prec \ldots \prec A_n \Leftrightarrow \{A_j\} \subset l.$$

Definition 2.2.15. In the space Ω, *triangle* ΔABC is a set of the elements belonging to the union of intersections of each of three sheets L_{AB}, L_{BC}, L_{CA} of the plane α_{ABC} with the collections of the lines $\{l_{AB}^i\}$, $\{l_{BC}^j\}$, $\{l_{CA}^k\}$ passing through the generator points A, B; B, C; C, A of these sheets respectively:

$$\Delta ABC = [L_{AB} \cap (\bigcup_i l_{AB}^i)] \cup [L_{BC} \cap (\bigcup_j l_{BC}^j)] \cup [L_{CA} \cap (\bigcup_k l_{CA}^k)].$$

Definition 2.2.16. *Angle* $\angle B_{\Delta ABC}$ of the triangle ΔABC in the space Ω is a set of the elements belonging to the union of intersections of each of two sheets L_{AB} L_{BC} of the plane α_{ABC} with the collections of the lines $\{l_{AB}^i\}$, $\{l_{BC}^j\}$ passing through the generator points A, B; B, C of these sheets respectively:

$$\angle B_{\Delta ABC} = [L_{AB} \cap (\bigcup_i l_{AB}^i)] \cup [L_{BC} \cap (\bigcup_j l_{BC}^j)].$$

Axiom of congruence

A_1^3. If, in a group G of mappings of the space Ω into itself, there exists the mapping $g \in G$ preserving a measure m, at the same time the points $A, B, C \in \Omega$, determining the plane α_{ABC}, and pairwise determining the sheets L_{AB}, L_{BC}, L_{CA}, are mapped into the points $A', B', C' \in \Omega$; the triangle ΔABC is mapped into the triangle $\Delta A'B'C'$; the sheets

2.2 Geometrical Properties of Metric Space Built upon Generalized Boolean Algebra

L_{AB}, L_{BC}, L_{CA} are mapped into the sheets $L_{A'B'}$, $L_{B'C'}$, $L_{C'A'}$, and the plane α_{ABC} is mapped into the plane $\alpha_{A'B'C'}$ respectively:

$$A \xrightarrow{g} A',\ B \xrightarrow{g} B',\ C \xrightarrow{g} C';\ \Delta ABC \xrightarrow{g} \Delta A'B'C';$$

$$L_{AB} \xrightarrow{g} L_{A'B'},\ L_{BC} \xrightarrow{g} L_{B'C'},\ L_{CA} \xrightarrow{g} L_{C'A'};\ \alpha_{ABC} \xrightarrow{g} \alpha_{A'B'C'},$$

then corresponding sheets, and also planes, triangles, and angles formed by the sheets are congruent:

$$L_{AB} \cong L_{A'B'},\ L_{BC} \cong L_{B'C'},\ L_{CA} \cong L_{C'A'};$$

$$\alpha_{ABC} \cong \alpha_{A'B'C'};\ \Delta ABC \cong \Delta A'B'C';$$

$$\angle A_{\Delta ABC} \cong \angle A'_{\Delta A'B'C'},\ \angle B_{\Delta ABC} \cong \angle B'_{\Delta A'B'C'},\ \angle C_{\Delta ABC} \cong \angle C'_{\Delta A'B'C'}.$$

Corollary from axiom of congruence

C_1^3. If triangles ΔABC and $\Delta A'B'C'$ are congruent in the space Ω, then there exists a mapping $f \in G$ preserving a measure m, which maps the triangle ΔABC into the triangle $\Delta A'B'C'$:

$$\Delta ABC \xrightarrow{f} \Delta A'B'C'.$$

Axiom of continuity

A_1^4. (Cantor's axiom). If, within the interval $X^* = [\mathbf{O}, X]$, $X^* \subset \Omega$:

$$X^* = \{x : \mathbf{O} \leq x \leq X\},$$

there is an infinite system of intervals $\{[a_i, b_i]\}$, where each following interval is contained within the previous one:

$$(\mathbf{O} < a_i < a_{i+1} < b_{i+1} < b_i < X),$$

at the same time, on a set X^*, there is no interval that lies within all the intervals of the system, then there exists the only element that belongs to all these intervals.

Axioms of parallels
Axioms of parallels for a sheet

A_1^5(a) In a given sheet L_{AB}, through a point $C \in L_{AB}$ that does not belong to a given line l_{AB} passing through the points A, B, there can be drawn at least one line l_C not crossing a given one l_{AB}, under the condition that the point C is the element of linearly ordered indexed set $\{C_\gamma\}_{\gamma \in \Gamma}$, $C \in \{C_\gamma\}_{\gamma \in \Gamma}$.

$A_2^5(a)$ In a given sheet L_{AB}, through a point $C \in L_{AB}$ that does not belong to a given line l_{AB} passing through the points A, B, there can be drawn only one line l_C not crossing a given l_{AB} under the condition that the point C is not the element of linearly ordered indexed set $\{C_\gamma\}_{\gamma \in \Gamma}$, $C \notin \{C_\gamma\}_{\gamma \in \Gamma}$.

$A_3^5(a)$ In a given sheet L_X, through a point $C \in L_X$ that does not belong to a given line l_{AB} passing through the points A, B, one cannot draw a line not crossing a given l_{AB} under the condition that the points A, B belong to an interval $[\mathbf{O}, X]: A, B \in [\mathbf{O}, X]$, so that:

$$\mathbf{O} < A < B < X \Rightarrow \mathbf{O}, A, B, X \in l_{AB},$$

and the point C is not the element of linearly ordered indexed set $\{C_\gamma\}_{\gamma \in \Gamma}$, $C \notin \{C_\gamma\}_{\gamma \in \Gamma}$ while C takes the values from the set $\{B \Delta A, X \Delta B, X \Delta A, X \Delta B \Delta A\}$.

Axioms of parallels for a plane

$A_1^5(b)$ In a given plane α_{ABC}, through a point C that does not belong to a given line l_{AB} passing through the points A, B, there can be drawn at least one line l_C not crossing a given l_{AB} under the condition that the point C and the line l_{AB} belong to the distinct sheets of the plane α_{ABC}:

$$(l_{AB} \in L_{AB}) \ \& \ [(C \in L_{BC}) \text{or} (C \in L_{CA})];$$

$$L_{AB} \neq L_{BC} \neq L_{CA} \neq L_{AB}.$$

$A_2^5(b)$ In a given plane α_{ABC}, through a point C that does not belong to a given line l_{AB} passing through the points A, B, there can be drawn only one line l_C not crossing a given l_{AB}, under the condition that the point C is not the element of linearly ordered indexed set $\{C_\gamma\}_{\gamma \in \Gamma}$, $C \notin \{C_\gamma\}_{\gamma \in \Gamma}$, and along with the line l_{AB}, it belongs to the same sheet L_{AB}.

$A_3^5(b)$ In a given plane α_{ABC}, through a point C that does not belong to a given line l_{AB} passing through the points A, B, one cannot draw a line not crossing a given l_{AB} under the condition that the points A, B belong to an interval $[\mathbf{O}, X]: A, B \in [\mathbf{O}, X]$, so that:

$$\mathbf{O} < A < B < X \Rightarrow \mathbf{O}, A, B, X \in l_{AB},$$

and the point C is not the element of linearly ordered indexed set $\{C_\gamma\}_{\gamma \in \Gamma}$, $C \notin \{C_\gamma\}_{\gamma \in \Gamma}$ while C takes the values from the set $\{B \Delta A, X \Delta B, X \Delta A, X \Delta B \Delta A\}$.

A given axiomatic system of the space Ω, built upon generalized Boolean algebra $\mathbf{B}(\Omega)$ with a measure m implies that the axioms of connection and the axioms of parallels are characterized with essentially the lesser constraints than the axioms of analogous groups of Euclidean space. Thus, it may be assumed that geometry of generalized Boolean algebra $\mathbf{B}(\Omega)$ with a measure m contains in itself some other geometries. In particular, axioms of parallels both for a sheet and for a plane contain the axioms of parallels of hyperbolic, elliptic, and parabolic (Euclidean) geometries.

2.2.5 Metric and Trigonometrical Relationships in Space Built upon Generalized Boolean Algebra with a Measure

Consider metric and trigonometrical relationships in space built upon generalized Boolean algebra $\mathbf{B}(\Omega)$ with a measure m. The distinction between the metric (2.2.1) of the space Ω built upon generalized Boolean algebra $\mathbf{B}(\Omega)$ with a measure m and the metric of Euclidean space predetermines the presence in the space Ω of geometric properties that differ from the properties of Euclidean space. The axiom A_1^1 from axioms of connection of geometry of generalized Boolean algebra $\mathbf{B}(\Omega)$ with a measure m, asserting that "there exists at least one line passing through two given points", however, does not mean that the distance between points A and B is an arbitrary quantity. Metric $\rho(A, B)$ for a pair of arbitrary points A and B of the space Ω is a strictly determined quantity and does not depend on a line, passing through the points A and B, along which this metric is evaluated. This statement poses a question: what should be considered as the sides and the angles of a triangle built upon three arbitrary points A, B, C of the space Ω, if through two of triangle's vertices there can be drawn at least one line? The answer to this question is contained in the Definitions 2.2.15 and 2.2.16.

Main metric relationships of a triangle $\triangle ABC$ of the space Ω will be obtained on the base of the following argument. For an arbitrary triplet of the elements A, B, C in the space Ω, we denote:

$$a = A - (B + C), \ b = B - (A + C), \ c = C - (A + B), \ d = ABC; \quad (2.2.15)$$

$$a' = BC - ABC, \ b' = AC - ABC, \ c' = AB - ABC. \quad (2.2.16)$$

Then:
$$m(A\Delta B) = m(a) + m(b) + m(a') + m(b'); \quad (2.2.17)$$
$$m(B\Delta C) = m(b) + m(c) + m(b') + m(c'); \quad (2.2.18)$$
$$m(C\Delta A) = m(c) + m(a) + m(c') + m(a'); \quad (2.2.19)$$
$$m[(A\Delta C)(C\Delta B)] = m(c) + m(c'); \quad (2.2.20)$$
$$m[(B\Delta A)(A\Delta C)] = m(a) + m(a'); \quad (2.2.21)$$
$$m[(C\Delta B)(B\Delta A)] = m(b) + m(b'). \quad (2.2.22)$$

Theorem 2.2.2. *Cosine theorem of the space* Ω. *For a triangle* $\triangle ABC$ *in the space* Ω, *the following relationships hold:*

$$m(A\Delta B) = m(A\Delta C) + m(C\Delta B) - 2m[(A\Delta C)(C\Delta B)]; \quad (2.2.23)$$

$$m(B\Delta C) = m(B\Delta A) + m(A\Delta C) - 2m[(B\Delta A)(A\Delta C)]; \quad (2.2.24)$$

$$m(C\Delta A) = m(C\Delta B) + m(B\Delta A) - 2m[(C\Delta B)(B\Delta A)]. \quad (2.2.25)$$

Proof of the theorem follows directly from Equations (2.2.17) through (2.2.22).

It should be noted that the identity holds:

$$m[(A\Delta C) + (C\Delta B)] = m[(B\Delta A) + (A\Delta C)] = m[(C\Delta B) + (B\Delta A)] =$$

$$= m(A + B + C) - m(ABC), \quad (2.2.26)$$

where $m(A+B+C) - m(ABC) = p(A, B, C)$ is *half-perimeter* of triangle $\triangle ABC$:

$$p(A, B, C) = [m(A\Delta B) + (B\Delta C) + m(C\Delta A)]/2 =$$

$$= m(a) + m(b) + m(c) + m(a') + m(b') + m(c').$$

To evaluate the proximity measure of the sides AC and CB of triangle $\triangle ABC$, an *angular measure* φ_{AB} may be introduced that possesses the following properties:

$$\cos^2 \varphi_{AB} = \frac{m[(A\Delta C)(C\Delta B)]}{m[(A\Delta C) + (C\Delta B)]} = \quad (2.2.27)$$

$$= \frac{m(c) + m(c')}{m(a) + m(b) + m(c) + m(a') + m(b') + m(c')};$$

$$\sin^2 \varphi_{AB} = \frac{m[A\Delta B]}{m[(A\Delta C) + (C\Delta B)]} = \quad (2.2.28)$$

$$= \frac{m(a) + m(b) + m(a') + m(b')}{m(a) + m(b) + m(c) + m(a') + m(b') + m(c')};$$

$$\cos^2 \varphi_{AB} + \sin^2 \varphi_{AB} = 1.$$

Theorem 2.2.3. *Sine theorem of the space* Ω. *For a triangle* $\triangle ABC$ *in the space* Ω, *the following relationship holds:*

$$\frac{m[A\Delta B]}{\sin^2 \varphi_{AB}} = \frac{m[B\Delta C]}{\sin^2 \varphi_{BC}} = \frac{m[C\Delta A]}{\sin^2 \varphi_{CA}} = \quad (2.2.29)$$

$$= m(A + B + C) - m(ABC) = p(A, B, C).$$

Proof of theorem follows directly from the Equations (2.2.26) and (2.2.28).

Theorem 2.2.4. *Theorem on* cos-*invariant of the space* Ω. *For a triangle* $\triangle ABC$ *in the space* Ω, *the identity that determines* cos-*invariant of angular measures of this triangle holds:*

$$\cos^2 \varphi_{AB} + \cos^2 \varphi_{BC} + \cos^2 \varphi_{CA} = 1. \quad (2.2.30)$$

2.2 Geometrical Properties of Metric Space Built upon Generalized Boolean Algebra

Proof of theorem follows from the Equations (2.2.20) through (2.2.22) and (2.2.26), and also from the Equation (2.2.27).

There is two-sided inequality that follows from the equality (2.2.30) and determines constraint relationships between the angles of a triangle in the space Ω:

$$3\arccos(1/\sqrt{3}) \leq \varphi_{AB} + \varphi_{BC} + \varphi_{CA} \leq 3\pi + 3\arccos(1/\sqrt{3}). \quad (2.2.31)$$

According to (2.2.31), one can emphasize four of the most essential intervals of domain of definition of a triangle angle sum in the space Ω that determine the belonging of a triangle to a space with hyperbolic, Euclidean, elliptic, and other geometries, respectively:

$$3\arccos(1/\sqrt{3}) \leq \varphi_{AB} + \varphi_{BC} + \varphi_{CA} < \pi \quad \text{(hyperbolic space)};$$

$$\varphi_{AB} + \varphi_{BC} + \varphi_{CA} = \pi \quad \text{(Euclidean space)};$$

$$\pi < \varphi_{AB} + \varphi_{BC} + \varphi_{CA} < 3\pi \quad \text{(elliptic space)};$$

$$3\pi \leq \varphi_{AB} + \varphi_{BC} + \varphi_{CA} \leq 3\pi + 3\arccos(1/\sqrt{3}).$$

Thus, commonality of the properties of the space Ω built upon generalized Boolean algebra $\mathbf{B}(\Omega)$ with a measure m extends on three well studied geometries—hyperbolic, Euclidean, and elliptic.

Theorem 2.2.5. *Theorem on* sin*-invariant of the space* Ω. *For a triangle* $\triangle ABC$ *in the space* Ω, *the identity that determines* sin*-invariant of angular measures of this triangle holds:*

$$\sin^2 \varphi_{AB} + \sin^2 \varphi_{BC} + \sin^2 \varphi_{CA} = 2. \quad (2.2.32)$$

Proof of theorem follows from the Equations (2.2.17) through (2.2.19) and (2.2.26), and also from the Equation (2.2.28).

Theorem 2.2.6. *Function* (2.2.28) *is a metric.*

Proof. Write an obvious inequality:

$$2\sin^2 \varphi_{CA} \leq 2. \quad (2.2.33)$$

According to the identity (2.2.32), we substitute 2 in the right part of the inequality (2.2.33) for the sum of squared sines of triangle's angles:

$$\sin^2 \varphi_{AB} + \sin^2 \varphi_{BC} + \sin^2 \varphi_{CA} \geq 2\sin^2 \varphi_{CA}.$$

The next inequality follows from the obtained one:

$$\sin^2 \varphi_{AB} + \sin^2 \varphi_{BC} \geq \sin^2 \varphi_{CA}. \quad (2.2.34)$$

Equation (2.2.28) implies that $\sin^2 \varphi_{AB} = 0$ only if $m(A \Delta B) = 0$, or if $A \equiv B$:

$$A \equiv B \Rightarrow \sin^2 \varphi_{AB} = 0. \quad (2.2.35)$$

Taking into account the symmetry of the function $\sin^2 \varphi_{AB} = \sin^2 \varphi_{BA}$ and also the properties (2.2.34) and (2.2.35), it is easy to conclude that the function (2.2.28) satisfies all the axioms of the metric. □

Nevertheless, it is too problematic to use the function (2.2.28) as a metric in the space Ω because of the presence of the third element C. It is possible to overcome the the problem introducing the following metric in the space Ω:

$$\mu(A, B) = \frac{m(A\Delta B)}{m(A+B)}. \qquad (2.2.36)$$

Formally, this result could be obtained from (2.2.28) if let $C = \mathbf{O}$, where $\mathbf{O}$ is a null element of the space Ω. Thus, while analyzing the metric relationships in a plane in the space Ω, instead of a triangle $\triangle ABC$, one should consider a tetrahedron $\mathbf{O}ABC$. Metric (2.2.36) was introduced in [225].

2.2.6 Properties of Metric Space with Normalized Metric Built upon Generalized Boolean Algebra with a Measure

It is convenient, with a help of a metric (2.2.26), to measure the distances between the elements of some collection $\{\mathbf{O}, A_1, A_2, \ldots, A_n\}$, for instance, n-dimensional *simplex* $Sx(A)$: $A = \sum_{j=0}^{n} A_j$, $A_0 \equiv \mathbf{O}$, when it is not necessary to know absolute values of the distances $\rho(A_i, A_j) = m(A_i \Delta A_j)$ between the pairs of elements A_i, A_j.

Consider a set A that is a sum of the elements $\{A_j\}$: $A = \sum_{j=0}^{n} A_j$. The elements $\{A_j\}$ of a set A in metric space Ω are situated in vertices of n-dimensional simplex $Sx(A)$. Further, separate three elements A_i, A_j and A_k from a set A: $A_i, A_j, A_k \in A$ and we introduce the following designations:

$$A_i + A_j = A_{ij};\ A_j + A_k = A_{jk};\ A_k + A_i = A_{ki};\ A_i + A_j + A_k = A_{ijk}.$$

Then, according to Lemma 2.2.1, the elements A_{ij}, A_{jk}, A_{ki} are situated on the corresponding sides of a triangle $\triangle A_i A_j A_k$ in metric space Ω, as shown in Fig. 2.2.3, so that:

$$A_i, A_{ij}, A_j \in l_{ij};\ A_j, A_{jk}, A_k \in l_{jk};\ A_k, A_{ki}, A_i \in l_{ki},$$

where l_{ij}; l_{jk}; l_{ki} are the lines of metric space Ω.

In the plane of triangle $\triangle A_i A_j A_k$, the point $A_{ijk} = A_i + A_j + A_k$ is situated upon the intersection of lines l_{ijk}; l_{jki}; l_{kij}, so that the following relations of belonging hold:

$$A_{ij}, A_{ijk}, A_k \in l_{ijk};\ A_{jk}, A_{ijk}, A_i \in l_{jki};\ A_{ki}, A_{ijk}, A_j \in l_{kij}.$$

Definition 2.2.17. The point $A_{ijk} = A_i + A_j + A_k$ of a two-dimensional face (triangle) $\triangle A_i A_j A_k$ of n-dimensional simplex $Sx(A)$: $A = \sum_{j=0}^{n} A_j$, $A_0 \equiv \mathbf{O}$ is called *barycenter* of two-dimensional simplex (triangle) $\triangle A_i A_j A_k$:

$$A_{ijk} = bc[Sx(A_i + A_j + A_k)].$$

2.2 Geometrical Properties of Metric Space Built upon Generalized Boolean Algebra

Extending the introduced notion upon three-dimensional simplex (tetrahedron) $Sx(A_i + A_j + A_k + A_m)$ formed by a collection of elements $\{A_i, A_j, A_k, A_m\}$ of a set A in metric space Ω, we obtain that a set $A_{ijkm} = A_i + A_j + A_k + A_m$ is a barycenter of simplex $Sx(A_i + A_j + A_k + A_m)$:

$$A_{ijkm} = bc[Sx(A_i + A_j + A_k + A_m)],$$

so that in metric space Ω, the point A_{ijkm} lies on the intersection of four lines passing through barycenters of four two-dimensional faces of tetrahedron $Sx(A_i + A_j + A_k + A_m)$ and also through the opposite vertices:

$$A_{jkm}, A_{ijkm}, A_i \in l_i; \ A_{ikm}, A_{ijkm}, A_j \in l_j; \ A_{ijm}, A_{ijkm}, A_k \in l_k; A_{ijk}, A_{ijkm}, A_m \in l_m$$

where

$$A_{jkm} = bc[Sx(A_j + A_k + A_m)]; \ A_{ikm} = bc[Sx(A_i + A_k + A_m)];$$
$$A_{ijm} = bc[Sx(A_i + A_j + A_m)]; \ A_{ijk} = bc[Sx(A_i + A_j + A_k)];$$
$$A_{ijkm} = l_i \bigcap l_j \bigcap l_k \bigcap l_m.$$

Similarly, a set $A = \sum_{j=0}^{n} A_j$, $A_0 \equiv \mathbf{O}$ is *barycenter* of n-dimensional simplex $Sx(A)$ in metric space Ω:

$$A = bc\left[Sx\left(A = \sum_{j=0}^{n} A_j\right)\right].$$

Lemma 2.2.4. *In the space Ω, all the elements $\{A_j\}$ of n-dimensional simplex $Sx(A)$: $A = \sum_{j=0}^{n} A_j$, $A_0 \equiv \mathbf{O}$ belong to some n-dimensional sphere $Sp(O_A, R_A)$ with center O_A and radius R_A: $\forall A_j \in Sx(A): A_j \in Sp(O_A, R_A)$.*

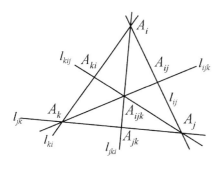

FIGURE 2.2.3 Simplex $Sx(A_i + A_j + A_k)$ in metric space Ω

In the space Ω, n-dimensional sphere $Sp(O_A, R_A)$ can be drawn around an arbitrary n-dimensional simplex $Sx(A)$. It would be a good thing here to draw an analogy with an Euclidean sphere in $\mathbf{R}^3$, where a distance between a pair of points A, B is determinated by an Euclidean metric (2.2.5), and by angular measure of an arc of a large circle passing through a given pair of the points A, B and the center of the sphere. Distances between the elements $\{A_j\}$ of n-dimensional simplex $Sx(A)$, lying on n-dimensional sphere $Sp(O_A, R_A)$ in the space Ω, are uniquely determined by metric $\mu(A_i, A_j)$ (36), which is the function of angular measure φ_{ij} between the elements A_i and A_j:

$$\mu(A_i, A_j) = \sin^2 \varphi_{ij} = \frac{m(A_i \Delta A_j)}{m(A_i + A_j)}. \quad (2.2.37)$$

Consider the main properties of metric space w with metric $\mu(A, B)$ (2.2.36), which is a subset of metric space Ω with metric $\rho(A, B) = m(A\Delta B)$. At the same time, we require that three elements A, X, B, belonging to the same line l in the space Ω: $A, X, B \in l$, $l \in \Omega$, so that the element X lies between the elements A and B, would also belong to the same line l' in the space w: $A, X, B \in l'$, $l' \in w$. Then two metric identities jointly hold:

$$\begin{cases} \rho(A, B) = \rho(A, X) + \rho(X, B); \\ \mu(A, B) = \mu(A, X) + \mu(X, B). \end{cases} \quad (2.2.38)$$

According to the definitions of metrics $\rho(A, B)$ and $\mu(A, B)$ of spaces Ω and w, respectively, the system (2.2.38) can be written as follows:

$$\begin{cases} m(A\Delta B) = m(A\Delta X) + m(X\Delta B); \\ \dfrac{m(A\Delta B)}{m(A + B)} = \dfrac{m(A\Delta X)}{m(A + X)} + \dfrac{m(X\Delta B)}{m(X + B)}. \end{cases} \quad (2.2.39)$$

The first equation of the system (2.2.39), as noted above (see the identities (2.2.6) and (2.2.7)), has two solutions with respect to X: (1) $X = A+B$ and (2) $X = A \cdot B$. Substituting these values of X into the second equation of (2.2.39), we notice that it turns into an identity on $X = A + B$, but it does not hold on $X = A \cdot B$. This implies that the following equivalent requirement to the space w: w should be closed under the operation of addition of generalized Boolean algebra $\mathbf{B}(\Omega)$ with a measure m.

We now give a general definition of the space w.

Definition 2.2.18. *Metric space w with metric $\rho_w(A, B) = m(A\Delta B)/m(A + B)$ is the subset of the elements $\{A, B, \ldots\}$ of metric space Ω with metric $\rho(A, B) = m(A\Delta B)$ built upon generalized Boolean algebra $\mathbf{B}(\Omega)$ with a measure m, which: (1) includes null element $\mathbf{O}$ of generalized Boolean algebra $\mathbf{B}(\Omega)$, $\mathbf{O} \in w$; (2) is closed under operation of addition of generalized Boolean algebra $\mathbf{B}(\Omega)$, $A, B \in w \Rightarrow A + B \in w$; (3) along with every its element A, contains all the elements contained in A: $A \in w$, $X \in A \Rightarrow X \in w$.*

Summarizing this definition briefly, metric space w with metric $\rho_w(A, B) = m(A\Delta B)/m(A+B)$ is the set of principal ideals $\{J_{A+B} = A+B, J_{B+C} = B+C, \ldots\}$ of generalized Boolean algebra $\mathbf{B}(\Omega)$ with a measure m formed by the pairs of nonnull elements A, B; B, C; $\ldots$ of metric space Ω with metric $\rho(A, B) = m(A\Delta B)$.

Generally, mapping of one space into another (in this case, the space Ω into the space w: $\Omega \to w$) according to Klein's terminology [197], can be interpreted by two distinct methods: "active" and "passive". While using the passive method of interpretation, mapping $\Omega \to w$ is represented as a change of coordinate system, i.e., for a fixed element, for instance $A \in \Omega$, which has coordinates $x = [x_1, x_2, \ldots, x_n]$, new coordinates $x' = [x'_1, x'_2, \ldots, x'_n]$ are assigned in space w. From this approach, the Definition 2.2.18 of the space w is given. Conversely, with active understanding, the coordinate system is fixed, and the space is transformed. Each element A of the space Ω with coordinates $x = [x_1, x_2, \ldots, x_n]$ is associated with the element a

2.2 Geometrical Properties of Metric Space Built upon Generalized Boolean Algebra

of the space ω with coordinates $x' = [x'_1, x'_2, \ldots, x'_n]$; thus, some transformation of the elements of the space is established; it is that method of interpretation we shall deal with henceforth while considering the properties of the space ω.

We now give one more definition of the space ω.

Definition 2.2.19. *Metric space ω with metric $\rho_\omega(A, B)$ is the set of the elements $\{a, b, \ldots\}$ in which each pair a, b is the result of mapping f of the corresponding pair of elements A, B from metric space Ω:*

$$f : A \to a, \ B \to b; \ A, B \in \Omega; \ a, b \in \omega,$$

and the distance $\rho_\omega(A, B)$ between the elements a and b in the space ω is equal to $m(A\Delta B)/m(A+B)$, and three elements A, X, B belonging to the same line l in the space Ω: $A, X, B \in l,\ l \in \Omega$, so that the element X lies between the elements A and B, mapped into three elements a, x, b respectively:

$$f : A \to a, \ X \to x, \ B \to b;$$

thus, they belong to the same line l' in the space ω: $a, x, b \in l',\ l' \in \omega$, and the element x lies between the elements a and b.

We now summarize the properties of the space (ω, ρ_ω).

Group of geometric properties

1. Space (ω, ρ_ω) is a metric space with metric $\rho_\omega(A, B) = m(A\Delta B)/m(A+B)$, where $f_{AB}\colon A \to a,\ B \to b;\ A, B \in \Omega;\ a, b \in \omega$, and indices A, B of mapping f_{AB} point to the fact that it is applied to a pair of elements A, B only.

2. Under mapping $f_{AB}\colon A \to a,\ B \to b$ of a pair of elements A, B of the space Ω with coordinates $x^A = [x_1^A, \ldots, x_n^A]$, $x^B = [x_1^B, \ldots, x_n^B]$ into the elements a, b of the space ω with coordinates $x^a = [x_1^a, \ldots, x_n^a]$, $x^b = [x_1^b, \ldots, x_n^b]$, respectively, coordinates x^A, x^B of the elements A, B and coordinates x^a, x^b of the elements a, b are interconnected by the relationships:

$$x^a = k \cdot x^A, \quad x^b = k \cdot x^B, \quad k = 1/m(A+B). \tag{2.2.40}$$

   (a) Under pairwise mapping of three elements A, B, C of the space Ω into the elements a, b, c of the space ω, respectively, the connection between coordinates x^A, x^B, x^C of the elements A, B, C and coordinates x^a, x^b, x^c of the elements a, b, c is determined by the relationships:

$$f_{AB} : A \to a',\ B \to b',\ x^{a'} = k_{ab} x^A,\ x^{b'} = k_{ab} x^B;$$

$$f_{BC} : B \to b'',\ C \to c'',\ x^{b''} = k_{bc} x^B,\ x^{c''} = k_{bc} x^C;$$

$$f_{CA} : C \to c''',\ A \to a''',\ x^{c'''} = k_{ca} x^C,\ x^{a'''} = k_{ca} x A,$$

   where $k_{ab} = 1/m(A+B)$; $k_{bc} = 1/m(B+C)$; $k_{ca} = 1/m(C+A)$.

(b) Under mapping f_{AB} of a pair of elements A, B of the space Ω into the elements a, b of the space ω: f_{AB}: $A \to a$, $B \to b$, every element $x \in \omega$: f_{AB}: $X \to x$, $X \in A + B$ is characterized by *normalized measure* $\mu(x)$:

$$\mu(x) = m(X)/m(A+B). \qquad (2.2.41)$$

3. Under mapping f: $A \to a$, $X \to x$, $B \to b$ of three elements $A, X, B \in \Omega$ belonging to the same line l in the space Ω, where the element X lies between the elements A and B, the elements a, x, b also belong to some line l' in the space ω: $a, x, b \in l'$, $l' \in \omega$, and the element x lies between the elements a and b.

4. Under mapping f: $Sx(A) \to Sx(a)$, $Sx(A) \in \Omega$, $Sx(a) \in \omega$ of the simplex $Sx(A)$: $A = \sum_{j=0}^{n} A_j$, $A_0 \equiv \mathbf{O}$ into the simplex $Sx(a)$: $a = \sum_{j=0}^{n} a_j$, $a_0 \equiv \mathbf{O}$, the barycenters of p-dimensional faces of the simplex $Sx(A)$, $p \leq n$ are mapped into the barycenters of the corresponding faces of the simplex $Sx(a)$:

$$A_{ij} = bc[Sx(A_i + A_j)] \to a_{ij} = bc[Sx(a_i + a_j)];$$
$$A_{ijk} = bc[Sx(A_i + A_j + A_k)] \to a_{ijk} = bc[Sx(a_i + a_j + a_k)];$$
$$A_{ijkm} = bc[Sx(A_i + A_j + A_k + A_m)] \to a_{ijkm} = bc[Sx(a_i + a_j + a_k + a_m)];$$
$$\dots\dots\dots\dots\dots\dots\dots\dots\dots\dots\dots\dots\dots\dots\dots\dots\dots\dots\dots;$$
$$A = bc[Sx(A)] \to a = bc[Sx(a)].$$

5. Diameter $\mathrm{diam}(\omega)$ of the space ω is equal to 1:

$$\mathrm{diam}(\omega) = \sup_{a,\,b \in \omega} \mu(a,b) = \mu(a, \mathbf{O}) = 1.$$

Group of algebraic properties

1. Metric space (ω, ρ_ω) is a set of principal ideals $\{J_{A+B} = A+B,\ J_{B+C} = B+C, \dots\}$ of generalized Boolean algebra $\mathbf{B}(\Omega)$ with a measure m, which are formed by the corresponding pairs of non-null elements A, B; B, C; ... of metric space Ω with metric $\rho(A,B) = m(A \Delta B)$.

   (a) The principal ideal J_{A+B} of generalized Boolean algebra $\mathbf{B}(\Omega)$ with a measure m, generated by a pair of non-null elements A and B, is an independent generalized Boolean algebra $\mathbf{B}(A+B)$ with a measure m.

   (b) The mapping f_{AB}: $A \to a$, $B \to b$ of a pair of elements A, B of the space Ω into a pair of elements a, b of the space ω determines an isomorphism between generalized Boolean algebras $\mathbf{B}(A+B)$ with a measure m and $\mathbf{B}(a+b)$ with a measure μ, so that measures m and μ are connected by the relationship (2.2.41).

2. Metric space ω contains the null element $\mathbf{O}$ of generalized Boolean algebra $\mathbf{B}(\Omega)$ with a measure m.

3. Metric space ω is closed under the operation of addition of generalized Boolean algebra $\mathbf{B}(\Omega)$ with a measure m. Taking into account closure, associativity, and commutativity of operation of addition, one can conclude that the space ω is an additive Abelian semigroup $\omega = (\mathbf{SG}, +)$.

4. Metric space ω is closed under operation of multiplication of generalized Boolean algebra $\mathbf{B}(\Omega)$ with a measure m. Taking into account closure, associativity, and commutativity of operation of multiplication, one can conclude that the space ω is a multiplicative Abelian semigroup $\omega = (\mathbf{SG}, \cdot)$.

5. Stone's duality between generalized Boolean algebra $\mathbf{B}(\Omega)$ with the signature $(+, \cdot, -, \mathbf{O})$ of the type $(2, 2, 2, 0)$ and generalized Boolean ring $\mathbf{BR}(\Omega)$ with the signature $(\Delta, \cdot, \mathbf{O})$ of the type $(2, 2, 0)$ induces the following correspondence. If ω is an ideal of generalized Boolean algebra $\mathbf{B}(\Omega)$, then ω is an ideal of generalized Boolean ring $\mathbf{BR}(\Omega)$.

6. Metric space ω is closed under an operation of addition Δ of generalized Boolean ring $\mathbf{BR}(\Omega)$ with a measure m. Taking into account closure, associativity, and commutativity of addition operation Δ, the presence of null element $\mathbf{O} \in \omega$: $a \Delta \mathbf{O} = \mathbf{O} \Delta a = a$, and the presence of inverse element $a^{-1} = a$: $a \Delta a^{-1} = \mathbf{O}$ for each element $a \in \omega$, one can conclude that the space ω is an additive Abelian group: $\omega = (\mathbf{G}, \Delta)$.

2.3 Informational Properties of Information Carrier Space

The content of this section is based on the definition of information carrier space built upon generalized Boolean algebra with a measure and introduced in section 2.1. The narration is arranged in such a manner that at first we consider main relationships between information carriers $A, B \subset \Omega$ (sets of elements) of information carrier space Ω (the description of the space Ω at a macrolevel), and then we consider main relationships within a separate information carrier (a set of the elements) A of the space Ω (the description of the space Ω at a microlevel). Any set of the elements $A = \{A_\alpha\}$, $A \subset \Omega$ is considered simultaneously as a subalgebra $\mathbf{B}(A)$ of generalized Boolean algebra $\mathbf{B}(\Omega)$ with signature $(+, \cdot, -, \mathbf{O})$ of the type $(2, 2, 2, 0)$. At a final part of the section, we introduce theorems characterizing main informational relationships on the most interesting (from the view of signal processing theory) mappings of the elements of information carrier space. Informational properties of information carrier space Ω are introduced by the following axiom.

Axiom 2.3.1. *Axiom of a measure of a binary operation. Measure $m(a)$ of the element a: $a = b \circ c$; $a, b, c \in \Omega$, considered as a result of a binary operation $\circ$ of* generalized Boolean algebra $\mathbf{B}(\Omega)$ with a measure m, determines the quantity of information $I_a = m(a)$ corresponding to the result of the given operation.

Based on the axiom, a measure m of the element a of carrier space Ω determines a quantitative aspect of information, while a binary operation $\circ$ of generalized Boolean algebra determines a qualitative aspect of this information. Within the formulated axiomatic statement 2.3.1, depending on a type of relations between the sets $A = \{A_\alpha\}$, $B = \{B_\beta\}$ and/or their elements, we shall distinguish the following types of information quantity introduced by the following definitions.

Definition 2.3.1. *Quantity of overall information* $I_{A_{\alpha 1} + A_{\alpha 2}}$, contained in an arbitrary pair of elements $A_{\alpha 1}, A_{\alpha 2}$: $A_{\alpha 1}, A_{\alpha 2} \in A$ of the set A considered as subalgebra of generalized Boolean algebra $\mathbf{B}(\Omega)$, is an information quantity equal to the measure of the sum of these elements:

$$I_{A_{\alpha 1} + A_{\alpha 2}} = m(A_{\alpha 1} + A_{\alpha 2}).$$

Definition 2.3.2. *Quantity of mutual information* $I_{A_{\alpha 1} \cdot A_{\alpha 2}}$, contained in an arbitrary pair of elements $A_{\alpha 1}, A_{\alpha 2}$: $A_{\alpha 1}, A_{\alpha 2} \in A$ of the set A considered as subalgebra of generalized Boolean algebra $\mathbf{B}(\Omega)$, is an information quantity equal to the measure of the product of these elements:

$$I_{A_{\alpha 1} \cdot A_{\alpha 2}} = m(A_{\alpha 1} \cdot A_{\alpha 2}).$$

Definition 2.3.3. *Quantity of absolute information* I_{A_α}, contained in arbitrary element A_α: $A_\alpha \in A$ of the set A considered as subalgebra of generalized Boolean algebra $\mathbf{B}(\Omega)$, is an information quantity equal to 1, according to the property 2 of information carrier space Ω (see Definition 2.1.1):

$$I_{A_\alpha} = m(A_\alpha \cdot A_\alpha) = m(A_\alpha + A_\alpha) = m(A_\alpha) = 1.$$

Remark 2.3.1. Quantity of absolute information I_{A_α} may be considered the quantity of overall information $I_{A_\alpha + A_\alpha} = m(A_\alpha + A_\alpha)$ or the quantity of mutual information $I_{A_\alpha \cdot A_\alpha} = m(A_\alpha \cdot A_\alpha)$ contained in the element A_α with respect to itself.

Definition 2.3.4. *Quantity of particular relative information*, which is contained in the element $A_{\alpha 1}$ with respect to the element $A_{\alpha 2}$ (or vice versa, is contained in the element $A_{\alpha 2}$ with respect to the element $A_{\alpha 1}$), $A_{\alpha 1}, A_{\alpha 2} \in A$, is an information quantity equal to a measure of a difference of these elements:

$$I_{A_{\alpha 1} - A_{\alpha 2}} = m(A_{\alpha 1} - A_{\alpha 2}); \quad I_{A_{\alpha 2} - A_{\alpha 1}} = m(A_{\alpha 2} - A_{\alpha 1}).$$

Definition 2.3.5. *Quantity of relative information* $I_{A_{\alpha 1} \Delta A_{\alpha 2}}$, which is contained by the element $A_{\alpha 1}$ with respect to the element $A_{\alpha 2}$ and vice versa, $A_{\alpha 1}, A_{\alpha 2} \in A$, is an information quantity equal to a measure of a symmetric difference of these elements:

$$I_{A_{\alpha 1} \Delta A_{\alpha 2}} = m(A_{\alpha 1} \Delta A_{\alpha 2}) = I_{A_{\alpha 1} - A_{\alpha 2}} + I_{A_{\alpha 2} - A_{\alpha 1}}.$$

2.3 Informational Properties of Information Carrier Space

Remark 2.3.2. Quantity of relative information $I_{A_{\alpha 1} \Delta A_{\alpha 2}}$ is identically equal to the introduced metric (2.1.1) between the elements $A_{\alpha 1}$ and $A_{\alpha 2}$ of the set A considered a subalgebra of generalized Boolean algebra $\mathbf{B}(\Omega)$:

$$I_{A_{\alpha 1} \Delta A_{\alpha 2}} = m(A_{\alpha 1} \Delta A_{\alpha 2}) = \rho(A_{\alpha 1} \Delta A_{\alpha 2}).$$

Definition 2.3.6. *Quantity of absolute information* I_A (I_B) is an information quantity contained in a set A (B) considered a subalgebra of generalized Boolean algebra $\mathbf{B}(\Omega)$ with a collection of the elements $\{A_\alpha\}$, ($\{B_\beta\}$), in the consequence of structural diversity of set A (B):

$$I_A = m(A) = m(\bigcup_\alpha A_\alpha) \equiv m(\sum_\alpha A_\alpha);$$

$$I_B = m(B) = m(\bigcup_\beta B_\beta) \equiv m(\sum_\beta B_\beta).$$

Definition 2.3.7. *Quantity of overall information* I_{A+B} contained in an arbitrary pair of sets A and B is an information quantity equal to a measure of sum of these two sets A and B; each is considered a subalgebra of generalized Boolean algebra $\mathbf{B}(\Omega)$ (see Fig. 2.3.1a):

$$I_{A+B} = m(A+B).$$

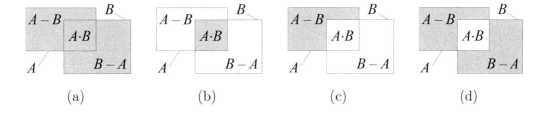

FIGURE 2.3.1 Information quantities between two sets A and B: (a) quantity of overall information; (b) quantity of mutual information; (c) quantity of particular relative information; (d) quantity of relative information

Definition 2.3.8. *Quantity of mutual information* $I_{A \cdot B}$ contained in an arbitrary pair of sets A and B is an information quantity equal to a measure of product of sets A and B; each is considered a subalgebra of generalized Boolean algebra $\mathbf{B}(\Omega)$ (see Fig. 2.3.1b):

$$I_{A \cdot B} = m(A \cdot B).$$

Remark 2.3.3. Quantity of absolute information I_A may be considered a quantity of overall information $I_{A+A} = m(A + A)$ or a quantity of mutual information $I_{A \cdot A} = m(A \cdot A)$ contained in a set A with respect to itself.

Definition 2.3.9. *Quantity of particular relative information* contained in the set A with respect to the set B, i.e. I_{A-B} (or vice versa, in the set B with respect to the set A, i.e. I_{B-A}), is an information quantity equal to a measure of the difference of the sets A and B, respectively (see Fig. 2.3.1c):

$$I_{A-B} = m(A - B), \quad I_{B-A} = m(B - A).$$

Definition 2.3.10. *Quantity of relative information* $I_{A\Delta B}$ contained in the set A with respect to the set B and vice versa is an information quantity equal to a measure of the symmetric difference of the sets A and B (see Fig. 2.3.1d):

$$I_{A\Delta B} = m(A \Delta B) = I_{A-B} + I_{B-A}.$$

Remark 2.3.4. Quantity of relative information $I_{A\Delta B}$ is identically equal to the introduced metric (2.1.1) between the elements A and B of the space Ω:

$$I_{A\Delta B} = m(A\Delta B) = \rho(A\Delta B). \tag{2.3.1}$$

Metric space Ω allows us to give the following geometric interpretation of the main introduced notions.

In the information carrier space Ω (see Definition 2.1.1), all the elements of the set $A = \bigcup_\alpha A_\alpha$ are situated on a surface of some n-dimensional sphere $Sp(\mathbf{O}, R)$, whose center is a null element $\mathbf{O}$ of generalized Boolean algebra $\mathbf{B}(\Omega)$ and radius R is equal to 1 (see Fig. 2.3.2):

$$R = m(A_\alpha \Delta \mathbf{O}) = m(A_\alpha) = 1.$$

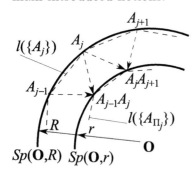

FIGURE 2.3.2 Elements of set $A = \bigcup_\alpha A_\alpha$ situated on surface of n-dimensional sphere $Sp(\mathbf{O}, R)$

A distance from null element $\mathbf{O}$ of the space Ω to some arbitrary set A is equal to a measure of a set $m(A)$ or is equal to the quantity of absolute information I_A, which is contained in a given set:

$$\rho(A\Delta \mathbf{O}) = m(A\Delta \mathbf{O}) = m(A) = I_A.$$

It is obvious that quantity of absolute information I_A of a set A is equal to quantity of relative information $I_{A\Delta \mathbf{O}}$ contained in a set A with respect to null element $\mathbf{O}$:

$$I_A = m(A) = m(A\Delta \mathbf{O}) = I_{A\Delta \mathbf{O}}.$$

Measure $m(A)$ of a set A of the elements or a quantity of absolute information I_A contained in a given set may be interpreted as a length of a set in metric space Ω. Quantity of relative information $I_{A\Delta B}$ contained in a set A with respect to a set B, and vice versa, as noted in Remark 2.3.2, is identically equal to a distance between sets A and B in the space Ω (2.3.1).

Main relationships between the sets A and B in metric space Ω and geometric properties of Ω are considered in section 2.2. From these relationships, one can easily

2.3 Informational Properties of Information Carrier Space

obtain the corresponding interrelations between various types of information. Here we bring the main ones only.

Group of the identities:

$$I_{A+B} = I_A + I_B - I_{AB}; \quad I_{A+B} = I_{A\Delta B} + I_{AB}; \quad I_{A\Delta B} = I_A + I_B - 2I_{AB};$$

Group of the inequalities:

$$I_A + I_B \geq I_{A+B} \geq I_{A\Delta B}; \quad I_A + I_B \geq I_{A+B} \geq \max[I_A, I_B];$$

$$\max[I_A, I_B] \geq \sqrt{I_A \cdot I_B} \geq \min[I_A, I_B] \geq I_{A\cdot B}.$$

Informational relationships for an arbitrary triplet of sets A, B, C and null element $\mathbf{O}$ of the space Ω, that are equivalent to metric relationships of tetrahedron (see Fig. 2.2.2), look like:

$$I_{A\Delta B} = I_A + I_B - 2I_{AB}; \quad I_{B\Delta C} = I_B + I_C - 2I_{BC}; \quad I_{C\Delta A} = I_C + I_A - 2I_{CA};$$

$$I_{A\Delta B} = I_{A\Delta C} + I_{C\Delta B} - 2I_{(A\Delta C)(C\Delta B)}.$$

Now one should consider the unities of information quantity measurement. Mathematical apparatus for constructing the stated foundations of both signal processing theory and information theory is Boolean algebra with a measure. As a unit of information quantity, we take a quantity of absolute information I_{A_α}, which is contained in a single element A_α of a set $A = \bigcup_\alpha A_\alpha$ with normalized measure ($m(A_\alpha) = 1$), and according to Definition 2.3.3, it is equal to 1:

$$I_{A_\alpha} = m(A_\alpha)|_{A_\alpha \in A} = 1.$$

Definition 2.3.11. Quantity of absolute information I_{A_α} contained in a single element A_α: $A_\alpha \in A \subset \Omega$ of a set A, $A = \bigcup_\alpha A_\alpha$ is called abit (*abit* is abbreviation for *absolute unit*): $I_{A_\alpha} = m(A_\alpha) = 1$ abit.

Now we finish the consideration of the main relationships between information carriers (sets of elements) of the space Ω and consider the main relationships within the framework of a single information carrier (a set of the elements) of the space Ω.

As noted above, all the elements $\{A_\alpha\}$ of an arbitrary set $A = \bigcup_\alpha A_\alpha$ in information carrier space are situated on a surface of some n-dimensional sphere $Sp(\mathbf{O}, R)$, whose center is a null element $\mathbf{O}$ of generalized Boolean algebra $\mathbf{B}(\Omega)$ and radius R is equal to 1: $R = 1$ (see Fig. 2.3.2).

Evidently, all the relationships between the sets of the space Ω hold for arbitrary elements $\{A_\alpha\}$ of a single set $A = \bigcup_\alpha A_\alpha$.

For all the elements $\{A_\alpha\}$ with normalized measure $m(A_\alpha) = 1$ of an arbitrary set $A = \bigcup_\alpha A_\alpha$, one may introduce a metric between the elements A_α and A_β, that is equivalent to the metric (2.1.2):

$$d(A_\alpha, A_\beta) = \frac{1}{2}m(A_\alpha \Delta A_\beta) = \frac{1}{2}[m(A_\alpha) + m(A_\beta) - 2m(A_\alpha \cdot A_\beta)] = 1 - m(A_\alpha \cdot A_\beta);$$

$$d(A_\alpha, A_\beta) = \frac{1}{2}\rho(A_\alpha, A_\beta). \tag{2.3.2}$$

The set of the elements with a discrete structure $\{A_j\}$ in metric space Ω is represented by the vertices $\{A_j\}$ of polygonal line $l(\{A_j\})$ that lie upon a sphere $Sp(\mathbf{O}, R)$, and at the same time, are the vertices of an n-dimensional simplex $Sx(A)$ inscribed into the sphere $Sp(\mathbf{O}, R)$. The length (perimeter) $P(\{A_j\})$ of a closed polygonal line $l(\{A_j\})$ is determined by the expression:

$$P(\{A_j\}) = \sum_{j=0}^{n} d(A_j, A_{j+1}) = \frac{1}{2}\sum_{j=0}^{n} \rho(A_j, A_{j+1}),$$

where $d(A_j, A_k)$, $\rho(A_j, A_k)$ are metrics determined by the relationships (2.3.2) and (2.1.2), respectively.

Here, the values of indices are denoted modulo $n+1$, i.e., $A_{n+1} \equiv A_0$.

The set of the elements with a continuous structure $\{A_\alpha\}$ in metric space Ω is a continuous closed line $l(\{A_\alpha\})$ situated upon a sphere $Sp(\mathbf{O}, R)$, which at the same time is a fragment of n-dimensional simplex $Sx(A)$ inscribed into a sphere $Sp(\mathbf{O}, R)$, and in series connects the vertices $\{A_j\}$ of a given simplex.

On the base of these notions, one can distinguish two forms of structural diversity of a set $A = \bigcup_\alpha A_\alpha$ and correspondingly, two measures of collections of the elements of a set A, i.e., *overall quantity of information* and *relative quantity of information* introduced by the following definitions.

Definition 2.3.12. *Overall quantity of information* $I(A)$, which is contained in a set A considered a collection of the elements $\{A_\alpha\}$, is an information quantity equal to the measure of their sum:

$$I(A) = m(A) = m(\bigcup_\alpha A_\alpha) = m(\sum_\alpha A_\alpha). \tag{2.3.3}$$

Evidently, this measure of structural diversity is identical to the quantity of absolute information I_A contained in a set A:

$$I(A) = I_A.$$

Definition 2.3.13. *Relative quantity of information* $I_\Delta(A)$, which is contained in a set A in the consequence of distinctions between the elements of a collection $\{A_\alpha\}$, is an information quantity equal to the measure of their symmetric difference:

$$I_\Delta(A) = m(\underset{\alpha}{\Delta} A_\alpha), \tag{2.3.4}$$

where $\underset{\alpha}{\Delta} A_\alpha = \ldots \Delta A_\alpha \Delta A_\beta \ldots$ is a symmetric difference between the elements of collection $\{A_\alpha\}$ of a set A.

The presence of two measures of structural diversity of a collection $\{A_\alpha\}$ of the set elements is stipulated by Stone's duality between generalized Boolean algebra $\mathbf{B}(\Omega)$ with signature $(+, \cdot, -, \mathbf{O})$ of the type $(2, 2, 2, 0)$ and a generalized Boolean ring $\mathbf{BR}(\Omega)$ with signature $(\Delta, \cdot, \mathbf{O})$ of the type $(2,2,0)$ that are isomorphic between each other [176], [215].

Information, contained in a set A of elements, is revealed at the same time in similarity and distinction between the single elements of collection $\{A_\alpha\}$. If a set A consists of the identical elements $\ldots = A_\alpha = A_\beta = \ldots$ of a collection $\{A_\alpha\}$, then the overall quantity of information $I(A)$ contained in a set A is equal to 1: $I(A) = 1$, and relative quantity of information $I_\Delta(A)$ contained in a set A is equal to zero: $I_\Delta(A) = 0$. This means that overall quantity of information contained in a set A consisting of a collection of identical elements $\{A_\alpha\}$ is equal to a measure of an element: $I(A) = m(A_\alpha) = 1$, whereas an information quantity that can be extracted from such a set is determined by relative quantity of information $I_\Delta(A)$ and equal to zero.

Consider the main relationships that characterize the introduced measures of collections of the elements $m(\sum_j A_j)$ and $m(\underset{j}{\Delta} A_j)$ for a set A with discrete structure $\{A_j\}$:

$$m(\sum_{j=0}^n A_j) = \sum_{j=0}^n m(A_j) - \sum_{0 \leq j < k \leq n} m(A_j A_k) + \sum_{0 \leq j < k < l \leq n} m(A_j A_k A_l) - \ldots \quad (2.3.5)$$

$$\ldots + (-1)^n m(\prod_{j=0}^n A_j);$$

$$m(\underset{j=0}{\overset{n}{\Delta}} A_j) = m(\sum_{j=0}^n A_j) - m(\sum_{0 \leq j < k \leq n}^n A_j A_k) + m(\sum_{0 \leq j < k < l \leq n}^n A_j A_k A_l) - \ldots \quad (2.3.6)$$

$$\ldots + (-1)^n m(\prod_{j=0}^n A_j).$$

The identities (2.3.5) and (2.3.6) imply double inequality:

$$m(\underset{j}{\Delta} A_j) \leq m(\sum_j A_j) \leq \sum_j m(A_j), \quad (2.3.7)$$

which implies, in its turn, double informational inequality:

$$I_\Delta(A) \leq I_A \leq \sum_j m(A_j). \quad (2.3.8)$$

If a set A consists of disjoint elements $(A_j \cdot A_k = \mathbf{O})$, then the inequality (2.3.8) transforms to the identity:

$$I_\Delta(A) = I_A = \sum_j m(A_j). \quad (2.3.9)$$

It can be shown that for a set A with discrete structure $\{A_j\}$, $j = 0, 1, \ldots, n$, the following relationship holds:

$$m(\sum_j A_j) = P(\{A_j\}) + m(\prod_j A_j), \quad (2.3.10)$$

where $P(\{A_j\}) = \frac{1}{2} \sum_{j=0}^{n} m(A_j \Delta A_{j+1}) = \sum_{j=0}^{n} d(A_j, A_{j+1})$ is perimeter of a closed polygonal line $l(\{A_j\})$ that in series connects the ordered elements $\{A_j\}$ of a set A in metric space of information carriers Ω. Here the values of indices are denoted modulo $n+1$, i.e., $A_{n+1} \equiv A_0$.

The relationships (2.3.6) and (2.3.10) imply the equality:

$$m(\underset{j}{\Delta} A_j) = P(\{A_j\}) - P(\{A_j \cdot A_{j+1}\}) + P(\{A_j \cdot A_{j+1} \cdot A_{j+2}\}) - \ldots \quad (2.3.11)$$

$$+ (-1)^k P(\{\prod_{i=j}^{j+k} A_i\}) + \ldots + m(\prod_j A_j) \cdot \mathrm{mod}_2(n-1),$$

where $\mathrm{mod}_2(n-1) = \begin{cases} 1, & n = 2k, \ k \in \mathbf{N}; \\ 0, & n = 2k+1, \end{cases}$

$P(\{A_j\}) = \sum_{j=0}^{n} d(A_j, A_{j+1})$ is perimeter of a closed polygonal line $l(\{A_j\})$ that in series connects the ordered elements $\{A_j\}$ of a set A in metric space of information carriers Ω (see Fig. 2.3.2);

$P(\{A_j \cdot A_{j+1}\}) = \sum_{j=0}^{n} d[(A_{j-1} \cdot A_j), (A_j \cdot A_{j+1})]$ is perimeter of a closed polygonal line $l(\{A_{\Pi j}\})$ that in series connects the ordered elements $\{A_{\Pi j}\}$: $A_{\Pi j} = A_j \cdot A_{j+1}$ of a set $A_\Pi = \bigcup_j A_{\Pi j}$ in metric space Ω (see Fig. 2.3.2);

$P(\{\prod_{i=j}^{j+k} A_i\}) = \sum_{j=0}^{n} d[\prod_{i=j}^{j+k} A_i, \prod_{i=j+1}^{j+k+1} A_i]$ is perimeter of a closed polygonal line $l(\{A_{\Pi j, j+k}\})$ that in series connects the ordered elements $\{A_{\Pi j,k}\}$: $A_{\Pi j,k} = \prod_{i=j}^{j+k} A_i$ of a set $A_{\Pi k} = \bigcup_j A_{\Pi j,k}$ in metric space Ω.

The relationship (2.3.10) implies that overall quantity of information $I(A)$ is an information quantity contained in a set A in the consequence of a presence of metric distinctions between the elements of a collection $\{A_j\}$. This overall quantity of information is determined by perimeter $P(\{A_j\})$ of a closed polygonal line $l(\{A_j\})$ in metric space Ω and also by quantity of mutual information defined by a measure of the product of the elements $m(\prod_j A_j)$.

The relationship (2.3.11) implies that relative quantity of information $I_\Delta(A)$ is an information quantity contained in a set A owing to the presence of metric distinctions between the elements of a collection $\{A_j\}$. But unlike overall quantity of information $I(A)$, relative quantity of information $I_\Delta(A)$ is determined by perimeter $P(\{A_j\})$ of a closed polygonal line $l(\{A_j\})$ in metric space Ω, taking into account the influence of metric distinctions between the products $\prod_{i=j}^{j+k} A_i$ of the elements of a collection $\{A_j\}$.

Both measures of structural diversity of a set of the elements $A = \bigcup_j A_j$

2.3 Informational Properties of Information Carrier Space

($m(\sum_j A_j)$ and $m(\underset{j}{\Delta} A_j)$) are the functions of perimeter $P(\{A_j\})$ of the ordered structure of a set A in the form of a closed polygonal line $l(\{A_j\})$ that in series connects the elements of a collection $\{A_j\}$ in metric space Ω with metric $d(A_j, A_k) = \frac{1}{2}m(A_j \Delta A_k)$. The first of these measures, i.e., overall quantity of information $I(A)$, has a sense of entire quantity of information contained in a set A considered a collection of elements with an ordered structure. Relative quantity of information $I_\Delta(A)$ has a sense of information quantity that may be extracted from a set by a proper processing.

Under discretization, a set A with continuous structure $\{A_\alpha\}$ is represented by a finite (countable) set $\{A_j\}$. In its turn, a set of the elements A' with discrete structure $\{A_j\}$ may be associated with the first one:

$$A = \bigcup_\alpha A_\alpha \to A' = \bigcup_j A_j,$$

so that each element A_j of a set A' with discrete structure is, at the same time, the element of a set A with continuous structure: $A_j \in A$.

Thus, a discretization $D : A \to A'$ of a set of the elements A with continuous structure $\{A_\alpha\}$ may be considered a mapping of a set $A = \bigcup_\alpha A_\alpha$ into a set $A' = \bigcup_j A_j$ with discrete structure $\{A_j\}$, so that each element of a set A' with discrete structure is, at the same time, the element of a set A with continuous structure, and the distinct elements $A_\alpha, A_\beta \in A, A_\alpha \neq A_\beta$ of a set A are mapped into the distinct elements $A_j, A_k \in A', A_j \neq A_k$ of a set A':

$$D : A \to A'; \tag{2.3.12}$$
$$A_\alpha, A_\beta \in A, A_\alpha \neq A_\beta, \ A_\alpha \to A_j, A_\beta \to A_k, \ A_j, A_k \in A', A_j \neq A_k. \tag{2.3.12a}$$

The mapping D possessing the property (2.3.12a) is an injective one. Homomorphism (2.3.12) maps generalized Boolean algebra $\mathbf{B}(A)$ with the signature $(+, \cdot, -, \mathbf{O})$ of the type $(2, 2, 2, 0)$ into subalgebra $\mathbf{B}(A')$ preserving the operations:

$A_\alpha + A_\beta = A_j + A_k$ (addition)

$A_\alpha \cdot A_\beta = A_j \cdot A_k$ (multiplication)

$A_\alpha - A_\beta = A_j - A_k$ (difference / obtaining a relative complement)

$\mathbf{O}_A \equiv \mathbf{O}_{A'} \equiv \mathbf{O}$ (identity of null element)

In common case, discretization of a set of the elements A with continuous structure $\{A_\alpha\}$ is not an isomorphic mapping that preserves both measures of structural diversity $I(A)$ (2.3.3) and $I_\Delta(A)$ (2.3.4):

$$I(A) \neq I(A'); \quad I_\Delta(A) \neq I_\Delta(A'), \tag{2.3.13}$$

where $I(A)$ is overall quantity of information contained in a set of the elements A with continuous structure $\{A_\alpha\}$; $I(A')$ is overall quantity of information contained in a set of the elements A' with discrete structure $\{A_j\}$; $I_\Delta(A)$ is relative quantity of information contained in a set of the elements A with continuous structure $\{A_\alpha\}$; $I_\Delta(A')$ is relative quantity of information contained in a set of the elements A' with discrete structure $\{A_j\}$.

Consider the main informational relationships characterizing the representation of a set of the elements A with continuous structure $\{A_\alpha\}$ by a finite (countable) set $\{A_j\}$.

Under discretization $D: A \to A'$ (2.3.12), the following informational inequalities hold:

$$I(A) \geq I(A'), \qquad (2.3.14a)$$

$$I(A') \geq I_\Delta(A'). \qquad (2.3.14b)$$

It should be noted, that the first inequality (2.3.14a) is stipulated by Axiom 1.3.1. The second inequality is stipulated by the relationship (2.3.6) between overall quantity of information and relative quantity of information contained in a result of discretization $\{A_j\}$ of a set of the elements A with continuous structure $\{A_\alpha\}$.

Overall quantity of information $I(A)$, which is contained in a set of the elements A with continuous structure $\{A_\alpha\}$, is equal to the sum of overall quantity of information $I(A')$ contained in a set A' as a result of discretization of a set A and a quantity of information I'_L, arising from curvature of a structure of a set A and is lost under discretization:

$$I(A) = I(A') + I'_L. \qquad (2.3.15)$$

Information quantity I'_L is called *information losses of the first genus*. For a set of the elements A with continuous structure $\{A_\alpha\}$ in the space Ω, we can introduce a measure of a structure *curvature* denoted as $c(A)$ and characterizing a deflection of structure locus $\{A_\alpha\}$ of a set A from a line:

$$c(A) = \frac{I(A) - I(A')}{I(A)}. \qquad (2.3.16)$$

Obviously, curvature $c(A)$ of a set of the elements A can be varied within $0 \leq c(A) \leq 1$. In the case of an arbitrary pair of adjacent elements $A_j, A_{j+1} \in A'$ of a discrete set A' and the element $A_\alpha \in A$, $A_\alpha \in [A_j A_{j+1}, A_j + A_{j+1}]$, the metric identity holds:

$$d[A_j, A_{j+1}] = d[A_j, A_\alpha] + d[A_\alpha, A_{j+1}],$$

where $d[A_\alpha, A_\beta] = \frac{1}{2}m(A_\alpha \Delta A_\beta)$ is metric in the space Ω, curvature $c(A)$ of a set of the elements A with continuous structure is equal to zero.

Overall quantity of information $I(A')$ contained in a set A' with discrete structure $\{A_j\}$, is equal to the sum of relative quantity of information $I_\Delta(A')$ of this set and quantity of redundant information I''_L contained in a set A' owing to nonempty pairwise intersections (due to mutual information) between its elements:

$$I(A') = I_\Delta(A') + I''_L. \qquad (2.3.17)$$

Information quantity I''_L is called *information losses of the second genus*. For the result of discretization, i.e., for a set A' of the elements with discrete structure $\{A_j\}$

2.3 Informational Properties of Information Carrier Space

of the space Ω, one can introduce a measure of informational redundancy $r(A')$ that characterizes informational interrelations between the elements of discrete structure $\{A_j\}$ of a set A' (or simply, mutual information between them):

$$r(A') = \frac{I(A') - I_\Delta(A')}{I(A')}. \qquad (2.3.18)$$

In the case of an arbitrary pair of adjacent elements, a measure of their product is equal to zero and the measure of informational redundancy $r(A')$ of a structure $\{A_j\}$ is also equal to zero.

Substituting equality (2.3.17) into (2.3.15), we obtain the following relation:

$$I(A) = I_\Delta(A') + I'_L + I''_L. \qquad (2.3.19)$$

The sense of the expression (2.3.19) can be elucidated in the following way: from a set of the elements A with continuous structure, one cannot extract more information quantity than a value of relative quantity of information $I_\Delta(A')$ contained in A owing to diversity between the elements of its structure. Information losses I'_L and I''_L take place, on the one hand, owing to curvature $c(A)$ of a structure $\{A_\alpha\}$ of a set A in metric space Ω, and on the other hand, as a consequence of some informational redundancy $r(A')$ of a discrete set A' (mutual information between its elements $\{A_j\}$).

Discretization $D: A \to A'$, according to (2.3.19), must provide a maximum of the ratio of relative quantity of information $I_\Delta(A')$, contained in a set A' with discrete structure, to overall quantity of information $I(A)$ contained in a set A with continuous structure:

$$\frac{I_\Delta(A')}{I(A)} \to \max, \qquad (2.3.20)$$

or is equivalent to provide a minimum sum of information losses of the first I'_L and the second I''_L genus:

$$I'_L + I''_L \to \min.$$

In this case, the following theorem can be formulated.

Theorem 2.3.1. *Theorem on equivalent (from the standpoint of preserving overall quantity of information) representation of a set of the elements A with continuous structure $\{A_\alpha\}$ by a set A' with discrete structure $\{A_j\}$. In the information carrier space Ω, a set of the elements $A = \bigcup_\alpha A_\alpha$, $A \subset \Omega$ with continuous structure $\{A_\alpha\}$ and a finite measure $m(A) < \infty$ can be represented equivalently by a finite set $A' = \bigcup_j A_j$ without any information losses under the condition, if a set A of the elements with continuous structure $\{A_\alpha\}$ can be represented by a partition of its orthogonal elements $\{A_j\}$: $A_j \cdot A_k = \mathbf{O}$, so that between relative quantity of information $I_\Delta(A')$ contained in a set A' and overall quantity of information $I(A)$ contained in a set A, the equality holds:*

$$I(A) = I_\Delta(A'). \qquad (2.3.21)$$

This identity provides the possibility of extracting the same information quantity contained in a set A, from a finite set A' without any information losses. Such a representation of a set A with continuous structure by a finite set A' with discrete structure, when the relation (2.3.21) holds, is equivalent from the standpoint of preserving overall quantity of information.

In this case, between both measures of structural diversity of the sets of the elements with continuous $\{A_\alpha\}$ and discrete $\{A_j\}$ structures respectively, i.e., between measures of sets A and A', the identity holds:

$$m(\sum_\alpha A_\alpha) \equiv m(\sum_j A_j),$$

and also the equality between aforementioned measures and measure of symmetric difference of a set $m(\underset{j}{\Delta} A_j)$ holds:

$$m(\sum_\alpha A_\alpha) = m(\sum_j A_j) = m(\underset{j}{\Delta} A_j),$$

and from an informational standpoint, is equivalent to the identity:

$$I(A) = I(A') = I_\Delta(A') = I_\Delta(A).$$

Fulfillment of these identities is provided by orthogonality of the elements $\{A_j\}$, so a measure of a set A' with discrete structure $\{A_j\}$ is equal to a measure of symmetric difference of the elements of a given set A:

$$m(\sum_j A_j) = m(\underset{j}{\Delta} A_j),$$

i.e., overall quantity of information $I(A')$ of a set A' is equal to relative quantity of information $I_\Delta(A')$ of this set: $I(A') \equiv I_\Delta(A')$. This means that under discrete representation of a set of the elements A with continuous structure $\{A_\alpha\}$ possessing the mentioned property, information losses of the first I''_L and second I'''_L genus are equal to zero: $I'_L = I''_L = 0$.

The structural diversity measure of a set of the elements A with continuous structure $\{A_\alpha\}$, based on a measure of symmetric difference between the elements (2.3.4), allows us to formulate the following variant of a theorem on equivalent representation.

Theorem 2.3.2. *Theorem on equivalent (from the standpoint of preserving relative quantity of information) representation of a set of the elements A with continuous structure $\{A_\alpha\}$ by a set A' with discrete structure $\{A_j\}$. In the information carrier space Ω, a set of the elements $A = \bigcup_\alpha A_\alpha$, $A \subset \Omega$ with continuous structure $\{A_\alpha\}$ and a finite measure $m(A) < \infty$ can be represented equivalently by a finite set $A' = \bigcup_j A_j$, if between relative quantity of information $I_\Delta(A)$, contained in a set A of the elements with continuous structure $\{A_\alpha\}$, and relative quantity of information*

2.3 Informational Properties of Information Carrier Space

$I_\Delta(A')$, contained in a set of the elements A' with discrete structure, the equality holds:

$$I_\Delta(A) = I_\Delta(A'). \tag{2.3.22}$$

This identity provides the possibility of extracting the same relative quantity of information contained in a set of the elements $A = \bigcup_\alpha A_\alpha$ with continuous structure $\{A_\alpha\}$, from a finite set $A' = \bigcup_j A_j$ with discrete structure $\{A_j\}$. Such a representation of a set A with continuous structure by a finite set A' with discrete structure, while the relation (2.3.22) holds, is equivalent from the standpoint of preserving relative quantity of information.

Theorem 2.3.3. *Theorem on isomorphic mapping preserving a measure.* If in metric space Ω, a set of the elements $A = \bigcup_\alpha A_\alpha$ with continuous (discrete) structure $\{A_\alpha\}$ and a finite measure $m(A) < \infty$ is isomorphic to a set of the elements $A' = \bigcup_\alpha A'_\alpha$ with continuous (discrete) structure $\{A'_\alpha\}$, $A, A' \subset \Omega$:

$$A_\alpha \underset{g^{-1}}{\overset{g}{\rightleftarrows}} A'_\alpha;\ A \underset{g^{-1}}{\overset{g}{\rightleftarrows}} A';\ g \in G, \tag{2.3.23}$$

where G is an automorphism group of generalized Boolean algebra $\mathbf{B}(\Omega)$ with a measure m, then the mapping g is a measure preserving isomorphism [175]:

$$A_{\alpha 1} + A_{\alpha 2} = A'_{\alpha 1} + A'_{\alpha 2}; \tag{2.3.24a}$$

$$A_{\alpha 1} \cdot A_{\alpha 2} = A'_{\alpha 1} \cdot A'_{\alpha 2}; \tag{2.3.24b}$$

$$A_{\alpha 1} - A_{\alpha 2} = A'_{\alpha 1} - A'_{\alpha 2}; \tag{2.3.24c}$$

$$A_{\alpha 1} \Delta A_{\alpha 2} = A'_{\alpha 1} \Delta A'_{\alpha 2}. \tag{2.3.24d}$$

Corollary 2.3.1. *Isomorphic mapping g (2.3.23) of a set A with continuous (discrete) structure $A = \bigcup_\alpha A_\alpha$ into a set A' with continuous (discrete) structure $A' = \bigcup_\alpha A'_\alpha$ preserves the measures of all binary operations between an arbitrary pair of the elements $A_{\alpha 1}, A_{\alpha 2} \in A$ of a set A considered as subalgebra $\mathbf{B}(A)$ of generalized Boolean algebra $\mathbf{B}(\Omega)$ with signature $(+, \cdot, -, \mathbf{O})$ of the type $(2, 2, 2, 0)$:*

$$m(A_{\alpha 1} + A_{\alpha 2}) = m(A'_{\alpha 1} + A'_{\alpha 2}); \tag{2.3.25a}$$

$$m(A_{\alpha 1} \cdot A_{\alpha 2}) = m(A'_{\alpha 1} \cdot A'_{\alpha 2}); \tag{2.3.25b}$$

$$m(A_{\alpha 1} - A_{\alpha 2}) = m(A'_{\alpha 1} - A'_{\alpha 2}); \tag{2.3.25c}$$

$$m(A_{\alpha 1} \Delta A_{\alpha 2}) = m(A'_{\alpha 1} \Delta A'_{\alpha 2}). \tag{2.3.25d}$$

The group of the identities (2.3.25a through d) implies the following corollary.

Corollary 2.3.2. *Isomorphic mapping g (2.3.23) preserves all the sorts of information quantities between an arbitrary pair of the elements $A_{\alpha 1}, A_{\alpha 2} \in A$ of a set A considered as subalgebra $\mathbf{B}(A)$ of generalized Boolean algebra $\mathbf{B}(\Omega)$ with signature $(+, \cdot, -, \mathbf{O})$ of the type $(2, 2, 2, 0)$:*

$$I_{A_{\alpha 1} + A_{\alpha 2}} = I_{A'_{\alpha 1} + A'_{\alpha 2}}; \qquad (2.3.26a)$$

$$I_{A_{\alpha 1} \cdot A_{\alpha 2}} = I_{A'_{\alpha 1} \cdot A'_{\alpha 2}}; \qquad (2.3.26b)$$

$$I_{A_{\alpha 1} - A_{\alpha 2}} = I_{A'_{\alpha 1} - A'_{\alpha 2}}; \qquad (2.3.26c)$$

$$I_{A_{\alpha 1} \Delta A_{\alpha 2}} = I_{A'_{\alpha 1} \Delta A'_{\alpha 2}}. \qquad (2.3.26d)$$

Corollary 2.3.3. *Isomorphic mapping g (2.3.23) preserves overall quantity of information $I(A)$ contained in a set A considered as a collection of elements $\{A_\alpha\}$:*

$$I(A) = m(A) = m(\sum_\alpha A_\alpha) = m(\sum_\alpha A'_\alpha) = m(A') = I(A').$$

Corollary 2.3.4. *Isomorphic mapping g (2.3.23) preserves relative quantity of information $I_\Delta(A)$ contained in a set A owing to distinctions between the elements $\{A_\alpha\}$:*

$$I_\Delta(A) = m(\underset{\alpha}{\Delta} A_\alpha) = m(\underset{\alpha}{\Delta} A'_\alpha) = I_\Delta(A').$$

The results of Theorem 2.3.3 can be generalized upon isomorphic mapping of a pair of sets that are information carriers.

Theorem 2.3.4. *Theorem on isomorphic mapping. Under isomorphic mapping g (2.3.23) of a pair of sets A, B with continuous (discrete) structure $A = \bigcup_\alpha A_\alpha$, $B = \bigcup_\beta B_\beta$ into the sets A', B' with continuous (discrete) structure $A' = \bigcup_\alpha A'_\alpha$, $B' = \bigcup_\beta B'_\beta$ in information carrier space Ω, respectively, $A, A' \subset \Omega$; $B, B' \subset \Omega$:*

$$A \underset{g^{-1}}{\overset{g}{\rightleftarrows}} A'; \; B \underset{g^{-1}}{\overset{g}{\rightleftarrows}} B'; \; g \in G, \qquad (2.3.27)$$

the measures of all binary operations between a pair of sets $A, B \subset A \cup B$, whose union $A \cup B$ is considered a subalgebra $\mathbf{B}(A + B)$ of a generalized Boolean algebra $\mathbf{B}(\Omega)$ with signature $(+, \cdot, -, \mathbf{O})$ of the type $(2, 2, 2, 0)$, are preserved:

$$m(A + B) = m(A' + B'); \qquad (2.3.28a)$$

$$m(A \cdot B) = m(A' \cdot B'); \qquad (2.3.28b)$$

$$m(A - B) = m(A' - B'); \qquad (2.3.28c)$$

$$m(A \Delta B) = m(A' \Delta B'). \qquad (2.3.28d)$$

The group of identities (2.3.28a) through (2.3.28d) implies the following corollary.

2.3 Informational Properties of Information Carrier Space

Corollary 2.3.5. *Isomorphic mapping g (2.3.27) preserves all the sorts of information quantity between a pair of sets $A, B \subset A \cup B$, whose union $A \cup B$ is considered a subalgebra $\mathbf{B}(A+B)$ of generalized Boolean algebra $\mathbf{B}(\Omega)$ with signature $(+, \cdot, -, \mathbf{O})$ of the type $(2, 2, 2, 0)$:*

$$I_{A+B} = I_{A'+B'}; \qquad (2.3.29a)$$

$$I_{A \cdot B} = I_{A' \cdot B'}; \qquad (2.3.29b)$$

$$I_{A-B} = I_{A'-B'}; \qquad (2.3.29c)$$

$$I_{A \triangle B} = I_{A' \triangle B'}. \qquad (2.3.29d)$$

By the relationships (2.3.26a) through (2.3.26d) and (2.3.29a) through (2.3.29d), Corollaries 2.3.2 and 2.3.5 establish invariance properties of the quantities of overall, mutual, particular relative, and relative information, respectively. Corollaries 2.3.3 and 2.3.4 define invariance properties of overall and relative quantities of information, respectively. Besides, Corollary 2.3.3 sets up invariance property of the quantity of absolute information.

3

Informational Characteristics and Properties of Stochastic Processes

The physical nature of real signals participating in the processes of transmitting, receiving, and processing information can vary widely. However, despite the diversity of signals, there is a class of mathematical models embracing main properties of real signals of various types. Such a class is represented, in a general case, by multivariate stochastic functions, but within this work the models of the signals are bounded by scalar stochastic processes only.

All fundamental inferences of information theory and signal processing theory are based on probabilistic-statistical descriptions of the signals, their interactions and transformations. This approach, being closely interrelated with Boolean algebra, remains a determinative one and is used in this chapter and subsequent ones.

The essential role of the notion of space in recent natural science research was underlined in Section 1.1. The notion of space performs an important function in probability theory. Correct problem statements concerning a measure of closeness between random variables must reveal what should be understood by a measure of closeness of random variables: a closeness of their realizations upon a probabilistic space or a closeness of distribution functions of these random variables, and so on. All the results concerning a distance evaluation upon the sets of random variables, as well as the methods of their use for applied problems of various types, form the subject of theory of probabilistic metric spaces [226].

A lot of publications are devoted to investigation of metric characteristics of random variables and stochastic processes in metric spaces [226], [227], [228], [229], [230].

It was also noted in Section 1.1 that nonlinearity as a general property of various sorts of matter, provides the diversity of material forms of the outer world. The processes of signal receiving and signal processing in electronic systems of various functionalities are not an exception in this sense. Use of invariance principles is considered one of the basic methods of description of nonlinear phenomena of various types.

There are such invariants of groups of mappings of stochastic processes described in this work that characterize probabilistic-statistical interrelations between their instantaneous values (samples), and are based upon metric relations between them.

3.1 Normalized Function of Statistical Interrelationship of Stochastic Processes

Stochastic process $\xi(t)$ is represented by a set of random variables under the fixed values of argument t, so the same probabilistic characteristics are used for description are usually used for random variables: probability density functions, cumulative distribution functions, characteristic functions, moment and cumulant functions [114], [115], [113].

To describe characteristics of a pair of stochastic processes $\xi(t)$ and $\eta(t')$ we shall use different designations of time parameter t and t', assuming that these stochastic processes exist in a different reference systems connected by location b and scale a ($a \neq 0$) transformations:

$$\begin{cases} t' = a \cdot t + b; \\ t = (t' - b)/a. \end{cases} \tag{3.1.1}$$

Normalized correlation function $r_\xi(t_1, t_2)$ and normalized cross-correlation function $r_{\xi\eta}(t_1, t'_2)$ characterize only linear statistical interrelationships (dependences) of random variables, i.e., the samples $\xi(t_1)$, $\xi(t_2)$ and $\xi(t_1)$, $\eta(t'_2)$ of stochastic processes $\xi(t)$ and $\eta(t')$ at the instants t_1, t_2, t'_2, respectively. *Correlation ratio* (*normalized variance function*) is considered to be more complete characteristic of statistical interrelation between two samples of stochastic process is defined by the following expression [231], [232]:

$$\theta_\xi^2(t_1, t_2) = \frac{1}{D_\xi(t_1)} \mathbf{M}_{\xi(t_2)}\{[\mathbf{M}\{\xi(t_1)/\xi(t_2)\} - \mathbf{M}\{\xi(t_1)\}]^2\},$$

where $D_\xi(t)$ is a variance of stochastic process $\xi(t)$; $\mathbf{M}(*)$ denotes a symbol of expectation; $\mathbf{M}\{\xi(t)\}$ is an expectation of stochastic process $\xi(t)$; $\mathbf{M}\{\xi(t_1)/\xi(t_2)\}$ is a conditional expectation of stochastic process $\xi(t)$;

$$\mathbf{M}_{\xi(t_2)}\{[\mathbf{M}\{\xi(t_1)/\xi(t_2)\} - \mathbf{M}\{\xi(t_1)\}]^2\} =$$

$$= \int_{-\infty}^{\infty} [\mathbf{M}\{\xi(t_1)/\xi(t_2)\} - \mathbf{M}\{\xi(t_1)\}]^2 p(x_2, t_2) dx_2,$$

where $p(x, t)$ is a probability density function (PDF).

It should be noted that correlation ratio does not possess the symmetry property: $\theta_\xi(t_1, t_2) \neq \theta_\xi(t_2, t_1)$, and it is not invariant with respect to one-to-one mappings of stochastic processes inasmuch as it is not preserved while realizing the last ones.

To define a complete measure of statistical interrelationship between two samples $\xi(t_j)$ and $\xi(t_k)$ of stochastic process $\xi(t)$, and also to define an invariant of bijective mappings of stochastic process, it is logical to introduce the function

3.1 Normalized Function of Statistical Interrelationship of Stochastic Processes

$\psi_\xi(t_j, t_k)$, which is characterized by metric $d(p, p_0)$ between joint PDF of the samples $p_2(x_j, x_k; t_j, t_k)$ and the product of their univariate PDFs $p_1(x_j, t_j)$, $p_1(x_k, t_k)$:

$$\psi_\xi(t_j, t_k) = d(p, p_0), \qquad (3.1.2)$$

where metric $d(p, p_0)$ is defined by the following expression:

$$d(p, p_0) = \frac{1}{2} \int_{-\infty}^{\infty} \int_{-\infty}^{\infty} |p_2(x_j, x_k; t_j, t_k) - p_1(x_j, t_j)p_1(x_k, t_k)| dx_j dx_k. \qquad (3.1.2a)$$

Index 0 attached to p_0 shows the peculiarity that the product of univariant PDFs corresponds to bivariate PDF of statistically independent samples.

Definition 3.1.1. The function $\psi_\xi(t_j, t_k)$, which is defined by the expression (3.1.2) and characterizes a measure of statistical interrelationship between two samples $\xi(t_j)$, $\xi(t_k)$ of stochastic process $\xi(t)$, is called the *normalized function of statistical interrelationship* (NFSI).

The NFSI $\psi_\xi(t_j, t_k)$ of Gaussian stochastic process, as it follows from (3.1.2), is entirely defined by normalized correlation function $r_\xi(t_j, t_k)$:

$$\psi_\xi(t_j, t_k) = g[r_\xi(t_j, t_k)], \qquad (3.1.3)$$

where $g[x]$ is some deterministic function, which is rather exactly (with an error not larger than 1% in the interval $x \in [0, 2/3]$ and not larger than 10% in the interval $x \in [2/3, 1]$) approximated by the following expression:

$$g[x] \approx 1 - \sqrt{1 - \frac{2}{\pi} \arcsin[||x||]}. \qquad (3.1.4)$$

In a similar manner, we define a measure of statistical interrelationship between two samples $\xi(t_j)$ and $\eta(t'_k)$ of stochastic processes $\xi(t)$ and $\eta(t')$, introducing the function:

$$\psi_{\xi\eta}(t_j, t'_k) = d_{\xi\eta}(p, p_0), \qquad (3.1.5)$$

$$d_{\xi\eta}(p, p_0) = \frac{1}{2} \int_{-\infty}^{\infty} \int_{-\infty}^{\infty} |p_2(x_j, y_k; t_j, t'_k) - p_1(x_j, t_j)p_1(y_k, t'_k)| dx_j dy_k, \qquad (3.1.5a)$$

where $d_{\xi\eta}(p, p_0)$ is metric between joint PDF $p_2(x_j, y_k; t_j, t_k)$ of the samples $\xi(t_j)$ and $\eta(t'_k)$ and the product of their univariate PDFs $p_1(x_j, t_j)$ and $p_1(y_k, t'_k)$.

Definition 3.1.2. The function $\psi_{\xi\eta}(t_j, t'_k)$, which is defined by (3.1.5) and characterizes a measure of statistical interrelationship between two samples $\xi(t_j)$, $\eta(t'_k)$ of stochastic processes $\xi(t)$ and $\eta(t')$, respectively, is called *mutual normalized function of statistical interrelationship* (mutual NFSI).

Mutual NFSI $\psi_{\xi\eta}(t_j, t'_k)$ of a pair of Gaussian stochastic processes $\xi(t)$ and $\eta(t')$, from Equation (3.1.5), is defined by normalized cross-correlation function $r_{\xi\eta}(t_j, t'_k)$:

$$\psi_{\xi\eta}(t_j, t'_k) = g[r_{\xi\eta}(t_j, t'_k)], \tag{3.1.6}$$

where the function $g[x]$ is approximated by Equation (3.1.4).

Main properties of NFSI and mutual NFSI are determined by informational properties of stochastic processes and will be considered in the next section.

We shall call the function $d_{\xi\eta}(t_j, t'_k)$ connected with metric $d_{\xi\eta}(p, p_0)$ by the relationship:

$$d_{\xi\eta}(t_j, t'_k) = 1 - d_{\xi\eta}(p, p_0) \tag{3.1.7}$$

the *metric between two samples* $\xi(t_j)$ and $\eta(t'_k)$ of stochastic processes $\xi(t)$ and $\eta(t')$ respectively.

Then, according to the formulas (3.1.5) and (3.1.7), the functions $\psi_{\xi\eta}(t_j, t'_k)$ and $d_{\xi\eta}(t_j, t'_k)$ are connected by the identity:

$$\psi_{\xi\eta}(t_j, t'_k) + d_{\xi\eta}(t_j, t'_k) = 1. \tag{3.1.8}$$

Components $\psi_{\xi\eta}(t_j, t'_k)$ and $d_{\xi\eta}(t_j, t'_k)$ appearing in formula (3.1.8) are closeness measure and distance measure between two samples $\xi(t_j)$ and $\eta(t'_k)$ of stochastic processes $\xi(t)$ and $\eta(t')$, respectively. Using these functions, one can introduce similar measures for a pair of stochastic processes $\xi(t)$ and $\eta(t')$ by determining the quantities $\psi_{\xi\eta}$ and $d_{\xi\eta}$ in the following way:

$$\psi_{\xi\eta} = \sup_{t_j, t'_k \in]-\infty, \infty[} \psi_{\xi\eta}(t_j, t'_k); \tag{3.1.9a}$$

$$d_{\xi\eta} = \inf_{t_j, t'_k \in]-\infty, \infty[} d_{\xi\eta}(t_j, t'_k). \tag{3.1.9b}$$

We have closeness measure and distance measure for a pair of stochastic processes $\xi(t)$ and $\eta(t')$. The quantity $\psi_{\xi\eta}$ determined by the formula (3.1.9a) we shall call *coefficient of statistical interrelation*, and the quantity $d_{\xi\eta}$ determined by the formula (3.1.9b) we shall call *metric between stochastic processes* $\xi(t)$ and $\eta(t')$. Coefficient of statistical interrelation and metric between stochastic processes $\xi(t)$ and $\eta(t')$ are connected by a relationship similar to (3.1.8):

$$\psi_{\xi\eta} + d_{\xi\eta} = 1. \tag{3.1.10}$$

Based on the known and introduced characteristics of stochastic processes, one can classify them taking into account constraints imposed upon their probabilistic characteristics. Stochastic process is considered weakly stationary (stationary in a narrow sense), if its NFSI is determined exclusively by the time difference $\tau = t_2 - t_1$ between a pair of its samples.

Fourier transform $\mathcal{F}[*]$ of NFSI $\psi_\xi(\tau)$ of stationary stochastic process $\xi(t)$ allows determining a new characteristic.

3.1 Normalized Function of Statistical Interrelationship of Stochastic Processes

Definition 3.1.3. Fourier transform $\mathcal{F}$ of NFSI $\psi_\xi(\tau)$ of stationary stochastic process $\xi(t)$ is called *hyperspectral density* $\sigma_\xi(\omega)$:

$$\sigma_\xi(\omega) = \mathcal{F}[\psi_\xi(\tau)] = \frac{1}{2\pi} \int_{-\infty}^{\infty} \psi_\xi(\tau) \exp[-i\omega\tau] d\tau. \quad (3.1.11)$$

According to (3.1.11),

$$\sigma_\xi(0) = \frac{1}{2\pi} \int_{-\infty}^{\infty} \psi_\xi(\tau) d\tau. \quad (3.1.12)$$

Inverse Fourier transform $\mathcal{F}^{-1}$ allows restoring NFSI $\psi_\xi(\tau)$ of stochastic process $\xi(t)$ on the base of its hyperspectral density $\sigma_\xi(\omega)$:

$$\psi_\xi(\tau) = \mathcal{F}^{-1}[\sigma_\xi(\omega)] = \int_{-\infty}^{\infty} \sigma_\xi(\omega) \exp[i\tau\omega] d\omega. \quad (3.1.13)$$

Enumerate the main properties of hyperspectral density (HSD) $\sigma_\xi(\omega)$ of stationary stochastic process $\xi(t)$:

1. $\sigma_\xi(\omega)$ is a continuous and real function.
2. $\sigma_\xi(\omega)$ is a nonnegative function: $\sigma_\xi(\omega) \geq 0$.
3. $\sigma_\xi(\omega)$ is an even function: $\sigma_\xi(\omega) = \sigma_\xi(-\omega)$.
4. $\sigma_\xi(\omega)$ is a normalized function: $\int_{-\infty}^{\infty} \sigma_\xi(\omega) d\omega = 1$.

For NFSI $\psi_\xi(\tau)$ of stationary stochastic process $\xi(t)$, we introduce a quantity $\Delta\tau$ characterizing effective width of its NFSI:

$$\Delta\tau: \quad \psi_\xi(0)\Delta\tau = \int_0^{\infty} \psi_\xi(\tau) d\tau;$$

thus, this relationship implies that *effective width of NFSI* $\Delta\tau$ is equal to:

$$\Delta\tau = \int_0^{\infty} \psi_\xi(\tau) d\tau. \quad (3.1.14)$$

For HSD $\sigma_\xi(\omega)$ of stationary stochastic process $\xi(t)$, we introduce a quantity $\Delta\omega$ characterizing effective width of its HSD:

$$\Delta\omega: \quad \sigma_\xi(0)\Delta\omega = \int_0^{\infty} \sigma_\xi(\omega) d\omega;$$

thus, this relationship implies that *effective width of HSD* $\Delta\omega$ is equal to:

$$\Delta\omega = \frac{1}{\sigma_\xi(0)} \int_0^\infty \sigma_\xi(\omega)d\omega = \frac{1}{2\sigma_\xi(0)}. \quad (3.1.15)$$

The product of effective width of NFSI $\Delta\tau$ (3.1.14) and effective width of HSD $\Delta\omega$ (3.1.15) of stationary stochastic process $\xi(t)$, taking into account the relationship (3.1.12), is equal to:

$$\Delta\tau \cdot \Delta\omega = \frac{1}{2\sigma_\xi(0)} \int_0^\infty \psi_\xi(\tau)d\tau = \pi/2. \quad (3.1.16)$$

The expression (3.1.16) characterizes the known uncertainty relation for the functions connected by Fourier transform; the larger is an effective width of NFSI $\Delta\tau$ the smaller is an effective width of HSD $\Delta\omega$, and vice versa.

Let f be bijective mapping of stochastic process $\xi(t)$, determined in the interval $T_\xi = [t_0, t_*]$, into stochastic process $\eta(t')$, determined in the interval $T_\eta = [t'_0, t'_*]$, from the group of mappings G, $f \in G$, assuming the inverse mapping f^{-1} exists:

$$f: \xi(t) \to \eta(t'); \quad f^{-1}: \eta(t') \to \xi(t); \quad f^{-1} \cdot f = 1, \quad (3.1.17)$$

where 1 is the unity of the group G.

Then the following theorem holds.

Theorem 3.1.1. *Theorem on invariance of NFSI $\psi_\xi(t_j, t_k)$ under bijective mapping of stochastic process. Bijective mapping (3.1.17), which maps stochastic process $\xi(t)$ into a process $\eta(t')$ preserves its NFSI:*

$$\psi_\xi(t_j, t_k) = \psi_\eta(t'_j, t'_k),$$

where $\psi_\xi(t_j, t_k)$, $\psi_\eta(t'_j, t'_k)$ are NFSIs of stochastic processes $\xi(t)$ and $\eta(t')$ respectively.

Proof. We again describe the probabilistic characteristics of a pair of stochastic processes $\xi(t)$ and $\eta(t')$ by using different designations of time parameter t and t', assuming that these stochastic processes exist in different reference systems.

According to the Definition 3.1.1, NFSI is the function:

$$\psi_\xi(t_j, t_k) = d(p, p_0), \quad (3.1.18)$$

$$d(p, p_0) = \frac{1}{2} \int_{-\infty}^\infty |p_\xi(x_j, x_k; t_j, t_k) - p_\xi(x_j, t_j)p_\xi(x_k, t_k)| dx_j dx_k, \quad (3.1.19)$$

where $d(p, p_0)$ is metric between joint PDF $p_\xi(x_j, x_k; t_j, t_k)$ of the samples $\xi(t_j)$, $\eta(t'_k)$ and the product of their univariate PDFs $p_\xi(x_j, t_j)$, $p_\xi(x_k, t_k)$.

We know that under bijective mappings of random variables:

$$f: \xi(t_{j,k}) \to \eta(t'_{j,k}),$$

the invariance property of probability differential holds:

$$p_\xi(x_j, x_k; t_j, t_k)dx_j dx_k = p_\eta(y_j, y_k; t'_j, t'_k)dy_j dy_k;$$

$$p_\xi(x_j, t_j)dx_j = p_\eta(y_j, t'_j)dy_j; \quad p_\xi(x_k, t_k)dx_k = p_\eta(y_k, t'_k)dy_k.$$

These relations imply, that under bijective mappings of stochastic processes (3.1.17), the metric (3.1.19) is preserved, and therefore, NFSI (3.1.18) is preserved too:

$$\psi_\xi(t_j, t_k) = \psi_\eta(t'_j, t'_k). \tag{3.1.20}$$

□

Corollary 3.1.1. *Under the mapping (3.1.17) preserving the domain of definition of stochastic process $T_\xi = [t_0, t_*] = T_\eta$ the identity holds:*

$$\psi_\xi(t_j, t_k) = \psi_\eta(t_j, t_k). \tag{3.1.21}$$

Corollary 3.1.2. *Under the bijective mapping (3.1.17) of stationary stochastic process $\xi(t)$, when the condition (3.1.21) holds, besides the fact that NFSI is preserved: $\psi_\xi(\tau) = \psi_\eta(\tau)$; HSD is also preserved $\sigma_\xi(\omega)$:*

$$\sigma_\xi(\omega) = \mathcal{F}[\psi_\xi(\tau)] = \mathcal{F}[\psi_\eta(\tau)] = \sigma_\eta(\omega), \tag{3.1.22}$$

where $\mathcal{F}[*]$ denotes Fourier transform.

Thus, under bijective mapping of the stationary stochastic process $\xi(t)$ into the stochastic process $\eta(t)$, their NFSI and HSD are preserved.

Example 3.1.1. If $\xi(t)$ is Gaussian stochastic process, then the notions of NFSI $\psi_\xi(t_j, t_k)$ and normalized correlation function $r_\xi(t_j, t_k)$ of a pair of samples $\xi(t_j)$ and $\xi(t_k)$ coincide with accuracy to some deterministic function $g[x]$ (see (3.1.3)):

$$\psi_\xi(t_j, t_k) = g[r_\xi(t_j, t_k)],$$

where a function $g[x]$ is approximated by (3.1.4). ▽

Example 3.1.2. The extension of power spectral density $S_\xi(\omega)$ under nonlinear mapping of stationary stochastic process $\xi(t)$ is well known in statistical radiophysics (radioengineering):

$$f: \xi(t) \to \eta(t); \quad f^{-1}: \eta(t) \to \xi(t).$$

This means that under nonlinear mapping f of stochastic process $\xi(t)$, the power spectral density $S_\eta(\omega)$ of the process $\eta(t)$ is distributed within a wider frequency band than power spectral density $S_\xi(\omega)$ of the process $\xi(t)$ is distributed.

Similarly, if one more nonlinear mapping of stochastic process $\eta(t)$: $h: \eta(t) \to$

$\eta^*(t)$ is realized, then power spectral density $S_\eta^*(\omega)$ of the process $\eta^*(t)$ is distributed within a wider frequency band than power spectral density $S_\eta(\omega)$ of the process $\eta(t)$ is distributed. However, if the stochastic process $\eta(t)$ is transformed with the inverse function h (with respect to the initial mapping f: $h = f^{-1}$), then we obtain the initial stochastic process $\xi(t)$: $\eta^*(t) = \xi(t)$.

It is obvious that power spectral density of the processes $\eta^*(t)$ and $\xi(t)$ should be identically equal: $S_\eta^*(\omega) = S_\xi(\omega)$, but that contradicts an initial statement concerning an extension of power spectral density under nonlinear mapping of a stochastic process.

The obtained paradox can be easily elucidated with the use of HSD $\sigma_\xi(\omega)$ of stochastic process $\xi(t)$ and Corollary 3.1.2 of Theorem 3.1.1 on its invariance under one-to-one transformation of stochastic process. In the mapping $f : \xi(t) \to \eta(t)$, we have an identity between HSDs of the initial $\xi(t)$ and the resultant $\eta(t)$ processes: $\sigma_\xi(\omega) = \sigma_\eta(\omega)$, and under the mapping $h : \eta(t) \to \eta^*(t)$, the similar identity between HSDs of the processes $\eta(t)$ and $\eta^*(t)$ holds: $\sigma_\eta(\omega) = \sigma_\eta^*(\omega)$. Thus, HSD $\sigma_\eta^*(\omega)$ of stochastic process $\eta^*(t)$ is identically equal to HSD $\sigma_\xi(\omega)$ of initial stochastic process $\xi(t)$: $\sigma_\eta^*(\omega) = \sigma_\xi(\omega)$, and if the secondary mapping is inverse with respect to the initial one: $h = f^{-1}$, then no paradoxical conclusions appear. $\triangledown$

3.2 Normalized Measure of Statistical Interrelationship of Stochastic Processes

In Section 3.1, the normalized function of statistical interrelationship (NFSI) and mutual NFSI were introduced by the Definitions 3.1.1 and 3.1.2 respectively. NFSI and mutual NFSI characterize a measure of statistical interrelationship between two instantaneous values (samples) of stochastic signals (processes) at distinct moments of time. This measure is based on metric relationships between the samples of stochastic signals (processes) and is defined by a metric between joint probability density function (PDF) and the product of univariate PDFs of the samples. This could impact obtaining accurate values of NFSI and mutual NFSI of the samples, even if their distributions are known. Nevertheless, often for practical applications, it is necessary to know an amount of dependence between instantaneous values of two distinct stochastic signals (processes) interacting at the same instant. Information concerning the distributions of the processed signals is absent. In order to establish such a characteristic of a pair of interacting samples of two stochastic signals (processes) it is useful to take into account the considerations listed below.

Any pair of stochastic signals (processes) $\xi(t), \eta(t)$ may be considered a partially ordered set $\mathbf{\Gamma}$, where for two instantaneous values (samples) $\xi_{t_1} = \xi(t_1), \eta_{t_2} = \eta(t_2)$ of the processes $\xi(t), \eta(t) \in \mathbf{\Gamma}$ at each instant $t \in T$, the relation of order $\xi_t \leq \eta_t$ (or $\xi_t \geq \eta_t$) is defined.

The partially ordered set $\mathbf{\Gamma}$ is a lattice with operations of least upper bound and greatest lower bound (of join and meet), respectively: $\xi_t \vee \eta_t = \sup_{\mathbf{\Gamma}}\{\xi_t, \eta_t\}$,

3.2 Normalized Measure of Statistical Interrelationship of Stochastic Processes

$\xi_t \wedge \eta_t = \inf_{\Gamma}\{\xi_t, \eta_t\}$, and if $\xi_t \leq \eta_t$, then $\xi_t \wedge \eta_t = \xi_t$ and $\xi_t \vee \eta_t = \eta_t$ [221], [223]:

$$\xi_t \leq \eta_t \Leftrightarrow \begin{cases} \xi_t \wedge \eta_t = \xi_t; \\ \xi_t \vee \eta_t = \eta_t. \end{cases}$$

Upon partially ordered set Γ, we can naturally define an operation of addition $\xi_t + \eta_t$ between two samples ξ_t, η_t of the processes $\xi(t), \eta(t) \in \Gamma$ at each instant $t \in T$. Then the partially ordered set Γ becomes a *lattice-ordered group* $\Gamma(+, \vee, \wedge)$ (or *L*-group).

In any *L*-group, the following statements hold [221], [233]:

1. $\Gamma(+)$ is an additive group.
2. $\Gamma(\vee, \wedge)$ is a lattice.
3. For arbitrary elements ξ_t, η_t, a_t, b_t from $\Gamma(+, \vee, \wedge)$, the following identities hold:

$$a_t + (\xi_t \wedge \eta_t) + b_t = (a_t + \xi_t + b_t) \wedge (a_t + \eta_t + b_t);$$
$$a_t + (\xi_t \vee \eta_t) + b_t = (a_t + \xi_t + b_t) \vee (a_t + \eta_t + b_t).$$

There exists a neutral element (zero) 0 in the additive group $\Gamma(+)$ of *L*–group $\Gamma(+, \vee, \wedge)$: $0 \in \Gamma(+)$, such that for $\forall x \in \Gamma(+)$, the inverse element $(-x)$ exists, and the identity holds: $x + (-x) = 0$.

Most cases of signal processing problems deal with stochastic signals (processes) $\xi(t), \eta(t)$ with symmetric (even) univariate PDFs $p_\xi(x), p_\eta(y)$ of the following kind: $p_\xi(x) = p_\xi(-x); p_\eta(y) = p_\eta(-y)$. The characteristics of statistical interrelation of a pair of instantaneous values (samples) ξ_t, η_t of stochastic signals (processes) $\xi(t), \eta(t) \in \Gamma$ introduced in this section and main results formulated in the form of theorems are predominantly oriented on the class of the signals with those exact properties, i.e., with even univariate PDFs $p_\xi(x), p_\eta(y)$; and these signals (processes) interact in partially ordered sets with the properties of lattice-ordered group $\Gamma(+, \vee, \wedge)$ (*L*-group).

Theorem 3.2.1. *For the samples ξ_t, η_t of stochastic processes $\xi(t), \eta(t)$ interacting in L-group Γ, $\xi_t, \eta_t \in \Gamma$, the functions $\mu(\xi_t, \eta_t)$, $\mu'(\xi_t, \eta_t)$ equal to:*

$$\mu(\xi_t, \eta_t) = 2(\mathbf{P}[\xi_t \vee \eta_t > 0] - \mathbf{P}[\xi_t \wedge \eta_t > 0]); \tag{3.2.1}$$
$$\mu'(\xi_t, \eta_t) = 2(\mathbf{P}[\xi_t \vee \eta_t < 0] - \mathbf{P}[\xi_t \wedge \eta_t < 0]), \tag{3.2.1a}$$

are metrics.

Proof. Consider the probabilities $\mathbf{P}[\xi_t \wedge \eta_t > 0]$, $\mathbf{P}[\xi_t \vee \eta_t > 0]$, $\mathbf{P}[\xi_t \wedge \eta_t < 0]$, $\mathbf{P}[\xi_t \vee \eta_t < 0]$, which, according to the formulas [115, (3.2.80)] and [115, (3.2.85)], are equal to:

$$\mathbf{P}[\xi_t \wedge \eta_t > 0] = 1 - (F_\xi(0) + F_\eta(0) - F_{\xi\eta}(0,0)); \tag{3.2.2a}$$
$$\mathbf{P}[\xi_t \vee \eta_t > 0] = 1 - F_{\xi\eta}(0,0); \tag{3.2.2b}$$
$$\mathbf{P}[\xi_t \wedge \eta_t < 0] = F_\xi(0) + F_\eta(0) - F_{\xi\eta}(0,0); \tag{3.2.2c}$$

$$\mathbf{P}[\xi_t \vee \eta_t < 0] = F_{\xi\eta}(0,0), \tag{3.2.2d}$$

where $F_{\xi\eta}(x,y)$ is joint cumulative distribution function (CDF) of the samples ξ_t, η_t; $F_\xi(x)$, $F_\eta(y)$ are univariate CDFs of the samples ξ_t, η_t.

Then the function $\mathbf{P}[\xi_t > 0]$, $\mathbf{P}[\eta_t > 0]$ defined on the lattice $\mathbf{\Gamma}$ is called valuation if the identity holds [221, § X.1 (V1)]:

$$\mathbf{P}[\xi_t > 0] + \mathbf{P}[\eta_t > 0] = \mathbf{P}[\xi_t \vee \eta_t > 0] + \mathbf{P}[\xi_t \wedge \eta_t > 0]. \tag{3.2.3}$$

Valuation $\mathbf{P}[\xi_t > 0]$ is isotonic, inasmuch as an implication holds [221, § X.1 (V2)]:

$$\xi_t \geq \xi'_t \Rightarrow \mathbf{P}[\xi_t > 0] \geq \mathbf{P}[\xi'_t > 0]. \tag{3.2.4}$$

Joint fulfillment of the equations (3) and (4), according to theorem 1 [221, § X.1], implies the quantity $\mu(\xi_t, \eta_t)$ that is equal to:

$$\mu(\xi_t, \eta_t) = 2(\mathbf{P}[\xi_t \vee \eta_t > 0] - \mathbf{P}[\xi_t \wedge \eta_t > 0]) =$$

$$= 2(F_\xi(0) + F_\eta(0)) - 4F_{\xi\eta}(0,0), \tag{3.2.5}$$

is metric.

Substituting Equations (3.2.2c) and (3.2.2d) into the formula (3.2.1a), we obtain the expression for the function $\mu'(\xi_t, \eta_t)$:

$$\mu'(\xi_t, \eta_t) = 2(F_\xi(0) + F_\eta(0)) - 4F_{\xi\eta}(0,0), \tag{3.2.6}$$

i.e., the functions $\mu(\xi_t, \eta_t)$, $\mu'(\xi_t, \eta_t)$ defined by Equations (3.2.1) and (3.2.1a) are identically equal to:

$$\mu(\xi_t, \eta_t) = \mu'(\xi_t, \eta_t),$$

and the function $\mu'(\xi_t, \eta_t)$ is also metric. □

Definition 3.2.1. *The quantity $\mu(\xi_t, \eta_t)$ defined by Equation (3.2.1) is called a metric between two samples ξ_t, η_t of stochastic processes $\xi(t)$, $\eta(t)$ interacting in L-group $\mathbf{\Gamma}$.*

Thus, partially ordered set $\mathbf{\Gamma}$ with operations $\chi(t) = \xi(t) \vee \eta(t)$, $\tilde{\chi}(t) = \xi(t) \wedge \eta(t)$, $t \in T$ is metric space $(\mathbf{\Gamma}, \mu)$ with respect to metric μ (3.2.1).

Then any pair of stochastic processes $\xi(t), \eta(t) \in \mathbf{\Gamma}$ with even univariate PDFs can be associated with the following normalized measure between the samples ξ_t, η_t.

Definition 3.2.2. *Normalized measure of statistical interrelationship* (NMSI) *between the samples ξ_t, η_t of stochastic processes $\xi(t)$, $\eta(t)$ with even univariate PDFs is the quantity $\nu(\xi_t, \eta_t)$ equal to:*

$$\nu(\xi_t, \eta_t) = 1 - \mu(\xi_t, \eta_t), \tag{3.2.7}$$

where $\mu(\xi_t, \eta_t)$ is a metric between two samples ξ_t, η_t of stochastic processes $\xi(t)$, $\eta(t)$, which is determined by Equation (3.2.1).

The last relationship and formula (3.2.5) together imply that NMSI $\nu(\xi_t, \eta_t)$ is determined by joint CDF $F_{\xi\eta}(x,y)$ and univariate CDFs $F_\xi(x)$, $F_\eta(y)$ of the samples ξ_t, η_t:

$$\nu(\xi_t, \eta_t) = 1 + 4F_{\xi\eta}(0,0) - 2(F_\xi(0) + F_\eta(0)). \tag{3.2.8}$$

Theorem 3.2.2. *For a pair of stochastic processes $\xi(t)$, $\eta(t)$ with even univariate PDFs in L–group Γ: $\xi(t), \eta(t) \in \Gamma$, $t \in T$, the metric $\mu(\xi_t, \eta_t)$ between the samples ξ_t, η_t is an invariant of a group H of continuous mappings $\{h_{\alpha,\beta}\}$, $h_{\alpha,\beta} \in H$; $\alpha, \beta \in A$ of stochastic processes, which preserve zero (neutral/null element) 0 of the group $\Gamma(+)$: $h_{\alpha,\beta}(0) = 0$:*

$$\mu(\xi_t, \eta_t) = \mu(\xi'_t, \eta'_t); \tag{3.2.9}$$

$$h_\alpha : \xi(t) \to \xi'(t), \quad h_\beta : \eta(t) \to \eta'(t); \tag{3.2.9a}$$

$$h_\alpha^{-1} : \xi'(t) \to \xi(t), \quad h_\beta^{-1} : \eta'(t) \to \eta(t), \tag{3.2.9b}$$

where ξ'_t, η'_t are the samples of stochastic processes $\xi'(t), \eta'(t)$ in L–group Γ': $h_{\alpha,\beta}$: $\Gamma \to \Gamma'$.

Proof. Under bijective mappings $\{h_{\alpha,\beta}\}$, $h_{\alpha,\beta} \in H$ (9a,b), the invariance property of probability differential holds, which implies the identity between joint CDF $F_{\xi\eta}(x,y)$, $F_{\xi'\eta'}(x',y')$ and univariate CDFs $F_\xi(x)$, $F'_\xi(x')$; $F_\eta(y)$, $F'_\eta(y')$ of the pairs of the samples ξ_t, η_t; ξ'_t, η'_t respectively:

$$F_{\xi'\eta'}(x',y') = F_{\xi\eta}(x,y); \tag{3.2.10}$$

$$F'_\xi(x') = F_\xi(x); \tag{3.2.10a}$$

$$F'_\eta(y') = F_\eta(y). \tag{3.2.10b}$$

Thus, taking into account (3.2.5), Equations (3.2.10), (3.2.10a), and (3.2.10b) imply the identity (3.2.9). □

Corollary 3.2.1. *For a pair of stochastic processes $\xi(t)$, $\eta(t)$ with even univariate PDFs in L–group Γ: $\xi(t), \eta(t) \in \Gamma$, $t \in T$, NMSI $\nu(\xi_t, \eta_t)$ of the samples ξ_t, η_t is an invariant of a group H of continuous mappings $\{h_{\alpha,\beta}\}$, $h_{\alpha,\beta} \in H$; $\alpha, \beta \in A$ of stochastic processes, which preserve zero 0 of group $\Gamma(+)$: $h_{\alpha,\beta}(0) = 0$:*

$$\nu(\xi_t, \eta_t) = \nu(\xi'_t, \eta'_t); \tag{3.2.11}$$

$$h_\alpha : \xi(t) \to \xi'(t), \quad h_\beta : \eta(t) \to \eta'(t); \tag{3.2.11a}$$

$$h_\alpha^{-1} : \xi'(t) \to \xi(t), \quad h_\beta^{-1} : \eta'(t) \to \eta(t), \tag{3.2.11b}$$

where ξ'_t and η'_t are the samples of stochastic processes $\xi'(t), \eta'(t)$ in L–group Γ': $h_{\alpha,\beta}$: $\Gamma \to \Gamma'$.

Theorem 3.2.3. *For a pair of Gaussian centered stochastic processes $\xi(t)$ and $\eta(t)$ with correlation coefficient $\rho_{\xi\eta}$ between the samples ξ_t, η_t, NMSI $\nu(\xi_t, \eta_t)$ is equal to:*

$$\nu(\xi_t, \eta_t) = \frac{2}{\pi} \arcsin[\rho_{\xi\eta}]. \tag{3.2.12}$$

Proof. Find an expression for joint CDF $F_{\xi\eta}(x,y)$ of the processes $\xi(t)$ and $\eta(t)$ at the point $x=0$, $y=0$, which, according to Equation (12) of Appendix II of the work [113], is determined by double integral $K_{00}(\alpha)$:

$$F_{\xi\eta}(0,0) = \left(\frac{\sqrt{1-\rho_{\xi\eta}^2}}{\pi}\right) K_{00}(\alpha),$$

where $\alpha = \pi - \arccos(\rho_{\xi\eta})$, $K_{00}(\alpha) = \alpha/(2\sin\alpha)$ (see Equation (14) of the Appendix II in [113]), $\sin\alpha = \sqrt{1-\rho_{\xi\eta}^2}$. Then, after necessary transformations we obtain the resultant expression for $F_{\xi\eta}(0,0)$:

$$F_{\xi\eta}(0,0) = \left(1 + \frac{2}{\pi}\arcsin[\rho_{\xi\eta}]\right)/4.$$

Substituting the last expression into (3.2.8), we obtain the required identity (3.2.12). □

Joint fulfillment of the relationships (3.1.18) of Theorem 3.1.1, (3.2.12) of Theorem 3.2.3, and Equation (3.1.6) implies that mutual NFSI $\psi_{\xi\eta}(t,t)$ and NMSI $\nu(\xi_t,\eta_t)$ between two samples ξ_t and η_t of stochastic processes $\xi(t)$ and $\eta(t)$, which are introduced by Definitions 3.1.2 and 3.2.2, respectively, are connected by approximate equality:

$$\psi_{\xi\eta}(t,t) \approx 1 - \sqrt{1 - \nu(\xi_t,\eta_t)}.$$

Theorem 3.2.4. *For Gaussian centered stochastic processes $\xi(t)$ and $\eta(t)$, which additively interact in L–group Γ: $\chi(t) = \xi(t) + \eta(t)$, $t \in T$, the following relationship between NMSIs $\nu(\xi_t,\chi_t)$, $\nu(\eta_t,\chi_t)$, and $\nu(\xi_t,\eta_t)$ of the corresponding pairs of their samples ξ_t, χ_t; η_t, χ_t; ξ_t, η_t holds:*

$$\nu(\xi_t,\chi_t) + \nu(\eta_t,\chi_t) - \nu(\xi_t,\eta_t) = 1. \tag{3.2.13}$$

Proof. Let $q^2 = D_\eta/D_\xi$ be a ratio of the variances D_η, D_ξ, and $\rho_{\xi\eta}$ is a correlation coefficient between the samples ξ_t, η_t of Gaussian processes $\xi(t)$, $\eta(t)$. Then correlation coefficients $\rho_{\xi\chi}$, $\rho_{\eta\chi}$ of the pairs of samples ξ_t, χ_t and η_t, χ_t are correspondingly equal to:

$$\rho_{\xi\chi} = (1 + \rho_{\xi\eta}q)/\sqrt{1 + 2\rho_{\xi\eta}q + q^2}, \tag{3.2.14a}$$

$$\rho_{\eta\chi} = (q + \rho_{\xi\eta})/\sqrt{1 + 2\rho_{\xi\eta}q + q^2}. \tag{3.2.14b}$$

As for Gaussian stochastic signals, NMSI is defined by Equation (3.2.12). So the identity holds:

$$\nu(\xi_t,\chi_t) + \nu(\eta_t,\chi_t) - \nu(\xi_t,\eta_t) =$$

$$= \frac{2}{\pi}\arcsin[\rho_{\xi\chi}] + \frac{2}{\pi}\arcsin[\rho_{\eta\chi}] - \frac{2}{\pi}\arcsin[\rho_{\xi\eta}]. \tag{3.2.15}$$

3.2 Normalized Measure of Statistical Interrelationship of Stochastic Processes 87

Using the relationships [234, (I.3.5)], it is easy to obtain a sum of the first two items of the right side of Equation (3.2.15):

$$\frac{2}{\pi}\arcsin[\rho_{\xi\chi}] + \frac{2}{\pi}\arcsin[\rho_{\eta\chi}] = \frac{2}{\pi}(\pi - \arcsin[c]) =$$

$$= \frac{2}{\pi}(\pi - \arcsin[\sqrt{1 - \rho_{\xi\eta}^2}]),$$

where $c = \rho_{\xi\chi}\sqrt{1 - \rho_{\eta\chi}^2} + \rho_{\eta\chi}\sqrt{1 - \rho_{\xi\chi}^2}$.

Substituting the sum of arcsines into the right side of (3.2.15), we calculate a sum of NMSIs of the pairs of samples $\xi_t, \chi_t; \eta_t, \chi_t; \xi_t, \eta_t$:

$$\nu(\xi_t, \chi_t) + \nu(\eta_t, \chi_t) - \nu(\xi_t, \eta_t) =$$

$$= \frac{2}{\pi}(\pi - \arcsin[\sqrt{1 - \rho_{\xi\eta}^2}]) - \frac{2}{\pi}\arcsin[\rho_{\xi\eta}] = 1.$$

□

Theorem 3.2.4 has the following corollary.

Corollary 3.2.2. *For Gaussian centered stochastic processes $\xi(t), \eta(t)$, which additively interact in partially ordered set Γ: $\chi(t) = \xi(t) + \eta(t)$, $t \in T$, and their corresponding pairs of the samples $\xi_t, \eta_t; \xi_t, \chi_t; \chi_t, \eta_t$, the metric identity holds:*

$$\mu(\xi_t, \eta_t) = \mu(\xi_t, \chi_t) + \mu(\chi_t, \eta_t). \tag{3.2.16}$$

Joint fulfillment of the identities (3.2.7) and (3.2.13) implies this metric identity.

Thus, Theorem 3.2.4 establishes invariance relationship (3.2.13) for NMSI of the pairs of the samples $\xi_t, \chi_t; \eta_t, \chi_t; \xi_t, \eta_t$ of additively interacting Gaussian stochastic processes $\xi(t), \eta(t)$. This identity does not depend on their energetic characteristics despite the fact that NMSIs $\nu(\xi_t, \chi_t)$ and $\nu(\eta_t, \chi_t)$ are the functions of the variances D_ξ and D_η of the samples ξ_t and η_t of Gaussian processes $\xi(t)$ and $\eta(t)$.

The results (3.2.13) and (3.2.16) of Theorem 3.2.4 could be generalized upon additively interacting processes $\xi(t)$ and $\eta(t)$, not demanding a Gaussian property for their distributions. This generalization is provided by the following theorem.

Theorem 3.2.5. *For nonGaussian stochastic processes $\xi(t)$ and $\eta(t)$ with even univariate PDFs, which additively interact with each other in L-group Γ: $\chi(t) = \xi(t) + \eta(t)$, $t \in T$, and also for their corresponding pairs of the samples $\xi_t, \eta_t; \xi_t, \chi_t; \chi_t, \eta_t$, the metric identity holds:*

$$\mu(\xi_t, \eta_t) = \mu(\xi_t, \chi_t) + \mu(\chi_t, \eta_t). \tag{3.2.17}$$

Proof. According to (3.2.5), the metric is equal to:

$$\mu(\xi_t, \eta_t) = 2(F_\xi(0) + F_\eta(0)) - 4F_{\xi\eta}(0, 0).$$

Metrics $\mu(\xi_t, \chi_t)$ and $\mu(\eta_t, \chi_t)$, according to the identity (3.2.5), are determined by the following relationships:

$$\mu(\xi_t, \chi_t) = 2(F_\xi(0) + F_\chi(0)) - 4F_{\xi\chi}(0,0);$$

$$\mu(\eta_t, \chi_t) = 2(F_\eta(0) + F_\chi(0)) - 4F_{\eta\chi}(0,0).$$

Note that under the additive interaction $\chi(t) = \xi(t) + \eta(t)$, the following relationship between CDFs of the samples ξ_t, η_t, χ_t holds:

$$F_{\xi\chi}(0,0) + F_{\eta\chi}(0,0) = F_{\xi\eta}(0,0) + F_\chi(0).$$

Taking into account the last identity, the sum of metrics $\mu(\xi_t, \chi_t)$, $\mu(\eta_t, \chi_t)$ is equal to:

$$\mu(\xi_t, \chi_t) + \mu(\eta_t, \chi_t) = 2(F_\xi(0) + F_\eta(0)) - 4F_{\xi\eta}(0,0) = \mu(\xi_t, \eta_t).$$

$\square$

Corollary 3.2.3. *For nonGaussian stochastic processes $\xi(t)$ and $\eta(t)$ with even univariate PDFs, which additively interact with each other in L–group Γ: $\chi(t) = \xi(t) + \eta(t)$, $t \in T$, the following relationship between NMSIs $\nu(\xi_t, \eta_t)$, $\nu(\xi_t, \chi_t)$, $\nu(\eta_t, \chi_t)$ of the corresponding pairs of the samples ξ_t, η_t; ξ_t, χ_t; η_t, χ_t holds:*

$$\nu(\xi_t, \chi_t) + \nu(\eta_t, \chi_t) - \nu(\xi_t, \eta_t) = 1. \tag{3.2.18}$$

The metric identity (3.2.17) implies a conclusion that the samples ξ_t, χ_t, η_t of stochastic processes $\xi(t)$, $\chi(t)$, $\eta(t)$, respectively, lie on the same line in metric space (Γ, μ).

For stochastic processes $\xi(t)$ and $\eta(t)$ with even univariate PDFs $p_\xi(x)$ and $p_\eta(y)$: $p_\xi(x) = p_\xi(-x)$; $p_\eta(y) = p_\eta(-y)$, which interact in L–group Γ with operations of join $\chi(t) = \xi(t) \vee \eta(t)$ and meet $\tilde\chi(t) = \xi(t) \wedge \eta(t)$, $t \in T$, there are some peculiarities between metrics and NMSIs of the corresponding pairs of their samples established by the following theorems.

Theorem 3.2.6. *For the samples ξ_t and η_t of stochastic processes $\xi(t)$ and $\eta(t)$ with even univariate PDFs, which interact in L–group Γ with operations of join and meet, respectively: $\chi(t) = \xi(t) \vee \eta(t)$, $\tilde\chi(t) = \xi(t) \wedge \eta(t)$, $t \in T$, the functions $\mu(\xi_t, \chi_t)$ and $\mu(\xi_t, \tilde\chi_t)$ that are equal according to (3.2.1):*

$$\mu(\xi_t, \chi_t) = 2(\mathbf{P}[\xi_t \vee \chi_t > 0] - \mathbf{P}[\xi_t \wedge \chi_t > 0]); \tag{3.2.19a}$$

$$\mu(\xi_t, \tilde\chi_t) = 2(\mathbf{P}[\xi_t \vee \tilde\chi_t > 0] - \mathbf{P}[\xi_t \wedge \tilde\chi_t > 0]), \tag{3.2.19b}$$

are metrics between the corresponding samples ξ_t, χ_t and ξ_t, $\tilde\chi_t$.

Proof of theorem is realized by testing of the valuation identity (3.2.3) and isotonic condition for the functions (3.2.19a) and (3.2.19b).

3.2 Normalized Measure of Statistical Interrelationship of Stochastic Processes

Theorem 3.2.7. *For stochastic processes $\xi(t)$ and $\eta(t)$ with even univariate PDFs, which interact in L–group Γ with operations of join and meet, respectively: $\chi(t) = \xi(t) \vee \eta(t)$, $\tilde{\chi}(t) = \xi(t) \wedge \eta(t)$, $t \in T$, between metrics $\mu(\xi_t, \chi_t)$, $\mu(\xi_t, \tilde{\chi}_t)$ of the corresponding pairs of the samples ξ_t, χ_t and ξ_t, $\tilde{\chi}_t$, the following relationships hold:*

$$\mu(\xi_t, \chi_t) = 2[F_\xi(0) - F_{\xi\eta}(0,0)]; \tag{3.2.20a}$$

$$\mu(\xi_t, \tilde{\chi}_t) = 2[F_\eta(0) - F_{\xi\eta}(0,0)]. \tag{3.2.20b}$$

Proof. According to the lattice absorption property, the following identities hold:

$$\xi_t \wedge \chi_t = \xi_t \wedge (\xi_t \vee \eta_t) = \xi_t; \tag{3.2.21a}$$

$$\xi_t \vee \tilde{\chi}_t = \xi_t \vee (\xi_t \wedge \eta_t) = \xi_t. \tag{3.2.21b}$$

According to the lattice idempotency property, the following identities hold:

$$\xi_t \vee \chi_t = \xi_t \vee (\xi_t \vee \eta_t) = \xi_t \vee \eta_t; \tag{3.2.22a}$$

$$\xi_t \wedge \tilde{\chi}_t = \xi_t \wedge (\xi_t \wedge \eta_t) = \xi_t \wedge \eta_t. \tag{3.2.22b}$$

According to Definition 3.2.1, metric $\mu(\xi_t, \chi_t)$ is equal to:

$$\mu(\xi_t, \chi_t) = 2(\mathbf{P}[\xi_t \vee \chi_t > 0] - \mathbf{P}[\xi_t \wedge \chi_t > 0]). \tag{3.2.23}$$

According to Equations (3.2.22a) and (3.2.2b), the equality holds:

$$\mathbf{P}[\xi_t \vee \chi_t > 0] = \mathbf{P}[\xi_t \vee \eta_t > 0] = 1 - F_{\xi\eta}(0,0), \tag{3.2.24}$$

and according to (3.2.21a), the equality holds:

$$\mathbf{P}[\xi_t \wedge \chi_t > 0] = \mathbf{P}[\xi_t > 0] = 1 - F_\xi(0), \tag{3.2.25}$$

where $F_{\xi\eta}(x,y)$ is the joint CDF of the samples ξ_t and η_t; $F_\xi(x)$ is the univariate CDF of the sample ξ_t.

Substituting the values of probabilities (3.2.24), (3.2.25) into (3.2.23), we obtain:

$$\mu(\xi_t, \chi_t) = 2[F_\xi(0) - F_{\xi\eta}(0,0)]. \tag{3.2.26}$$

Similarly, according to Definition 3.2.1, metric $\mu(\xi_t, \tilde{\chi}_t)$ is equal to:

$$\mu(\xi_t, \tilde{\chi}_t) = 2(\mathbf{P}[\xi_t \vee \tilde{\chi}_t > 0] - \mathbf{P}[\xi_t \wedge \tilde{\chi}_t > 0]). \tag{3.2.27}$$

According to (3.2.21b), the equality holds:

$$\mathbf{P}[\xi_t \vee \tilde{\chi}_t > 0] = \mathbf{P}[\xi_t > 0] = 1 - F_\xi(0), \tag{3.2.28}$$

and according to the relationships (3.2.22b) and (3.2.2a), the equality holds:

$$\mathbf{P}[\xi_t \wedge \tilde{\chi}_t > 0] = \mathbf{P}[\xi_t \wedge \eta_t > 0] = 1 - [F_\xi(0) + F_\eta(0) - F_{\xi\eta}(0,0)], \tag{3.2.29}$$

where $F_{\xi\eta}(x,y)$ is joint CDF of the samples ξ_t and η_t; $F_\xi(x)$ and $F_\eta(y)$ are univariate CDFs of the samples ξ_t and η_t.

Substituting the values of probabilities (3.2.28) and (3.2.29) into (3.2.27), we obtain:

$$\mu(\xi_t, \tilde{\chi}_t) = 2[F_\eta(0) - F_{\xi\eta}(0,0)]. \tag{3.2.30}$$

□

Theorem 3.2.7 implies several corollaries.

Corollary 3.2.4. *For stochastic processes $\xi(t)$ and $\eta(t)$, which interact in L–group Γ with operations: $\chi(t) = \xi(t) \vee \eta(t)$, $\tilde{\chi}(t) = \xi(t) \wedge \eta(t)$, $t \in T$, and for their NMSIs $\nu(\xi_t, \chi_t)$, $\nu(\xi_t, \tilde{\chi}_t)$ of the corresponding pairs of their samples ξ_t, χ_t and ξ_t, $\tilde{\chi}_t$, the following relationships hold:*

$$\nu(\xi_t, \chi_t) = 1 + 2[F_{\xi\eta}(0,0) - F_\xi(0)]; \qquad (3.2.31a)$$

$$\nu(\xi_t, \tilde{\chi}_t) = 1 + 2[F_{\xi\eta}(0,0) - F_\eta(0)]. \qquad (3.2.31b)$$

Proof of the corollary follows directly from (3.2.7).

Corollary 3.2.5. *For stochastic processes $\xi(t)$, $\eta(t)$, which interact in L–group Γ with operations: $\chi(t) = \xi(t) \vee \eta(t)$, $\tilde{\chi}(t) = \xi(t) \wedge \eta(t)$, $t \in T$, and also for metrics $\mu(\xi_t, \eta_t)$, $\mu(\xi_t, \chi_t)$, $\mu(\eta_t, \chi_t)$, $\mu(\xi_t, \tilde{\chi}_t)$, $\mu(\eta_t, \tilde{\chi}_t)$ between the corresponding pairs of their samples ξ_t, η_t; ξ_t, χ_t; η_t, χ_t; $\xi_t, \tilde{\chi}_t$; $\eta_t, \tilde{\chi}_t$, the metric identities hold:*

$$\mu(\xi_t, \eta_t) = \mu(\xi_t, \chi_t) + \mu(\chi_t, \eta_t); \qquad (3.2.32a)$$

$$\mu(\xi_t, \eta_t) = \mu(\xi_t, \tilde{\chi}_t) + \mu(\tilde{\chi}_t, \eta_t). \qquad (3.2.32b)$$

Proof. Joint fulfillment of the relationships (3.2.26) and (3.2.5):

$$\mu(\xi_t, \chi_t) = 2[F_\xi(0) - F_{\xi\eta}(0,0)];$$

$$\mu(\eta_t, \chi_t) = 2[F_\eta(0) - F_{\xi\eta}(0,0)];$$

$$\mu(\xi_t, \eta_t) = 2(F_\xi(0) + F_\eta(0)) - 4F_{\xi\eta}(0,0),$$

implies the identity ((3.2.32)a).

Similarly, joint fulfillment of the relationships (3.2.30) and (3.2.5):

$$\mu(\xi_t, \tilde{\chi}_t) = 2[F_\eta(0) - F_{\xi\eta}(0,0)];$$

$$\mu(\eta_t, \tilde{\chi}_t) = 2[F_\xi(0) - F_{\xi\eta}(0,0)];$$

$$\mu(\xi_t, \eta_t) = 2(F_\xi(0) + F_\eta(0)) - 4F_{\xi\eta}(0,0),$$

implies the identity (3.2.32b). $\square$

The identities (3.2.32a) and (3.2.32b) imply that the samples ξ_t, χ_t, η_t (ξ_t, $\tilde{\chi}_t$, η_t) of stochastic processes $\xi(t)$, $\eta(t)$, $\chi(t)$, $\tilde{\chi}(t)$, respectively, lie on the same line in metric space Γ.

Corollary 3.2.6. *For stochastic processes $\xi(t)$ and $\eta(t)$, which interact in L–group Γ with operations: $\chi(t) = \xi(t) \vee \eta(t)$, $\tilde{\chi}(t) = \xi(t) \wedge \eta(t)$, $t \in T$, and also for their NMSIs $\nu(\xi_t, \eta_t)$, $\nu(\xi_t, \chi_t)$, $\nu(\eta_t, \chi_t)$, $\nu(\xi_t, \tilde{\chi}_t)$, $\nu(\eta_t, \tilde{\chi}_t)$ of the corresponding pairs of the samples ξ_t, η_t; ξ_t, χ_t; η_t, χ_t; $\xi_t, \tilde{\chi}_t$; $\eta_t, \tilde{\chi}_t$, the following relationships hold:*

$$\nu(\xi_t, \chi_t) + \nu(\eta_t, \chi_t) - \nu(\xi_t, \eta_t) = 1; \qquad (3.2.33a)$$

$$\nu(\xi_t, \tilde{\chi}_t) + \nu(\eta_t, \tilde{\chi}_t) - \nu(\xi_t, \eta_t) = 1. \qquad (3.2.33b)$$

3.2 Normalized Measure of Statistical Interrelationship of Stochastic Processes

Proof of the corollary follows directly from the relationships (3.2.32a), (3.2.32b), and (3.2.7).

Corollary 3.2.7. *For stochastic processes $\xi(t)$ and $\eta(t)$, which interact in L–group $\mathbf{\Gamma}$ with operations: $\chi(t) = \xi(t) \vee \eta(t)$, $\tilde{\chi}(t) = \xi(t) \wedge \eta(t)$, $t \in T$, the relationships between metrics $\mu(\xi_t, \chi_t)$, $\mu(\xi_t, \tilde{\chi}_t)$ (3.2.20a), (3.2.20b) and between NMSIs $\nu(\xi_t, \chi_t)$, $\nu(\xi_t, \tilde{\chi}_t)$ (3.2.31a), (3.2.31b) of the corresponding pairs of their samples ξ_t, χ_t, and ξ_t, $\tilde{\chi}_t$ are invariants of a group H of continuous mappings $\{h_{\alpha,\beta}\}$, $h_{\alpha,\beta} \in H$; $\alpha, \beta \in A$ of stochastic processes (3.2.9a), (3.2.9b):*

$$\begin{cases} \mu(\xi_t, \chi_t) = \mu(\xi'_t, \chi'_t); \\ \mu(\xi_t, \tilde{\chi}_t) = \mu(\xi'_t, \tilde{\chi}'_t); \end{cases} \quad (3.2.34)$$

$$\begin{cases} \nu(\xi_t, \chi_t) = \nu(\xi'_t, \chi'_t); \\ \nu(\xi_t, \tilde{\chi}_t) = \nu(\xi'_t, \tilde{\chi}'_t), \end{cases} \quad (3.2.35)$$

where ξ'_t and η'_t are the samples of stochastic processes $\xi'(t)$ and $\eta'(t)$, which interact with each other in L–group $\mathbf{\Gamma}'$, $h_{\alpha,\beta}: \mathbf{\Gamma} \to \mathbf{\Gamma}'$ with operations: $\chi'(t) = \xi'(t) \vee \eta'(t)$, $\tilde{\chi}'(t) = \xi'(t) \wedge \eta'(t)$, $t \in T$; χ'_t, $\tilde{\chi}'_t$ are the samples of the results $\chi'(t)$ and $\tilde{\chi}'(t)$ of interactions of stochastic processes $\xi'(t)$ and $\eta'(t)$.

Proof of the corollary follows directly from the identities (3.2.20) and (3.2.31), and also from CDF invariance relations (3.2.10).

Corollary 3.2.8. *For statistically independent stochastic processes $\xi(t)$ and $\eta(t)$ with even univariate PDFs, which interact in L–group $\mathbf{\Gamma}$ with operations: $\chi(t) = \xi(t) \vee \eta(t)$, $\tilde{\chi}(t) = \xi(t) \wedge \eta(t)$, $t \in T$, between metrics $\mu(\xi_t, \chi_t)$, $\mu(\xi_t, \tilde{\chi}_t)$ (3.2.20), and also between NMSIs $\nu(\xi_t, \chi_t)$, $\nu(\xi_t, \tilde{\chi}_t)$ (3.2.31) of the corresponding pairs of their samples ξ_t, χ_t and ξ_t, $\tilde{\chi}_t$, the following equations hold:*

$$\begin{cases} \mu(\xi_t, \chi_t) = 1 - 2F_{\xi\eta}(0,0); \\ \mu(\xi_t, \tilde{\chi}_t) = 1 - 2F_{\xi\eta}(0,0), \end{cases} \quad (3.2.36)$$

$$\begin{cases} \nu(\xi_t, \chi_t) = 2F_{\xi\eta}(0,0); \\ \nu(\xi_t, \tilde{\chi}_t) = 2F_{\xi\eta}(0,0). \end{cases} \quad (3.2.37)$$

Proof of the corollary is realized by a direct substitution of CDF's values into the relationships (3.2.20) and (3.2.31): $F_\xi(0) = 0.5$, $F_\eta(0) = 0.5$.

Corollary 3.2.9. *For statistically independent stochastic processes $\xi(t)$ and $\eta(t)$ with even univariate PDFs, which interact in L–group $\mathbf{\Gamma}$ with operations: $\chi(t) = \xi(t) \vee \eta(t)$, $\tilde{\chi}(t) = \xi(t) \wedge \eta(t)$, $t \in T$, between metrics $\mu(\xi_t, \chi_t)$, $\mu(\xi_t, \tilde{\chi}_t)$ (3.2.20) and also between NMSI $\nu(\xi_t, \chi_t)$, $\nu(\xi_t, \tilde{\chi}_t)$ (3.2.31) of the corresponding pairs of their samples ξ_t, χ_t and ξ_t, $\tilde{\chi}_t$, the following relationships hold:*

$$\begin{cases} \mu(\xi_t, \chi_t) = 0.5; \\ \mu(\xi_t, \tilde{\chi}_t) = 0.5; \end{cases} \quad (3.2.38a)$$

$$\begin{cases} \nu(\xi_t, \chi_t) = 0.5; \\ \nu(\xi_t, \tilde{\chi}_t) = 0.5. \end{cases} \quad (3.2.38b)$$

Proof of the corollary is realized by a direct substitution of CDF values into the relationships (3.2.20) and (3.2.31): $F_\xi(0) = 0.5$, $F_\eta(0) = 0.5$, $F_{\xi\eta}(0,0) = F_\xi(0)F_\eta(0) = 0.25$.

Corollary 3.2.10. *For statistically independent stochastic processes $\xi(t)$ and $\eta(t)$ with even univariate PDFs, which interact in L-group Γ with operations of join and meet: $\chi(t) = \xi(t) \vee \eta(t)$, $\tilde\chi(t) = \xi(t) \wedge \eta(t)$, $t \in T$, metric $\mu(\chi_t, \tilde\chi_t)$ and NMSI $\nu(\chi_t, \tilde\chi_t)$ between the samples χ_t and $\tilde\chi_t$ are, respectively, equal to:*

$$\mu(\chi_t, \tilde\chi_t) = \mu(\xi_t, \eta_t) = 1; \qquad (3.2.39a)$$

$$\nu(\chi_t, \tilde\chi_t) = \nu(\xi_t, \eta_t) = 0. \qquad (3.2.39b)$$

Proof. Consider the probabilities $\mathbf{P}[\chi_t \vee \tilde\chi_t > 0]$, $\mathbf{P}[\chi_t \wedge \tilde\chi_t > 0]$, which for statistically independent stochastic processes $\xi(t)$ and $\eta(t)$ are, respectively, equal to:

$$\mathbf{P}[\chi_t \vee \tilde\chi_t > 0] = \mathbf{P}[(\xi_t \vee \eta_t) \vee (\xi_t \wedge \eta_t) > 0] = \mathbf{P}[\xi_t \vee \eta_t > 0];$$

$$\mathbf{P}[\chi_t \wedge \tilde\chi_t > 0] = \mathbf{P}[(\xi_t \vee \eta_t) \wedge (\xi_t \wedge \eta_t) > 0] = \mathbf{P}[\xi_t \wedge \eta_t > 0].$$

Substituting the last two relationships into definition of metric (3.2.1), we obtain the initial identity (3.2.39a). The last identity and the coupling Equation (3.2.7) between metric and NMSI of a pair of samples implies the initial equality (3.2.39b). □

Thus, Theorem 3.2.7 establishes invariance relations for metrics and NMSIs of the pairs of the samples ξ_t, χ_t; η_t, χ_t and $\xi_t, \tilde\chi_t$; $\eta_t, \tilde\chi_t$ of stochastic signals (processes) $\xi(t)$, $\eta(t)$ that interact in L-group Γ with operations: $\chi(t) = \xi(t) \vee \eta(t)$, $\tilde\chi(t) = \xi(t) \wedge \eta(t)$, $t \in T$, so that these identities do not depend on the energetic relationships between interacting processes.

The obtained results allow us to generalize the geometric properties of metric signal space (Γ, μ) with metric μ (1).

Stochastic signals (processes) $\xi(t)$, $\eta(t)$, $t \in T$ interact with each other in metric signal space (Γ, μ) with the properties of L-group $\Gamma(+, \vee, \wedge)$ and metric μ (1), so that the results of their interaction $\chi_+(t)$, $\chi_\vee(t)$, $\chi_\wedge(t)$ are described by binary operations of addition $+$, join $\vee$ and meet $\wedge$, respectively:

$$\chi_+(t) = \xi(t) + \eta(t); \qquad (3.2.40a)$$

$$\chi_\vee(t) = \xi(t) \vee \eta(t); \qquad (3.2.40b)$$

$$\chi_\wedge(t) = \xi(t) \wedge \eta(t). \qquad (3.2.40c)$$

Similarly, the samples ξ_t, η_t of stochastic signals (processes) $\xi(t)$, $\eta(t)$, $t \in T$ interact in metric signal space (Γ, μ) with L-group properties and metric μ (3.2.1), so that the results of their interaction χ_t^+, $\chi_t^\vee$, $\chi_t^\wedge$ are described by the same binary operations, respectively:

$$\chi_t^+ = \xi_t + \eta_t; \qquad (3.2.41a)$$

$$\chi_t^\vee = \xi_t \vee \eta_t; \qquad (3.2.41b)$$

3.2 Normalized Measure of Statistical Interrelationship of Stochastic Processes

$$\chi_t^\wedge = \xi_t \wedge \eta_t. \tag{3.2.41c}$$

Consider a number of theorems, which can elucidate the main geometric properties of metric signal space $(\mathbf{\Gamma}, \mu)$ with the properties of L–group $\mathbf{\Gamma}(+, \vee, \wedge)$ and metric μ (1).

Theorem 3.2.8. *In metric signal space $(\mathbf{\Gamma}, \mu)$ with metric μ (3.2.1), a pair of samples ξ_t and η_t of stochastic signals (processes) $\xi(t)$, $\eta(t)$, $t \in T$ and the result of their interaction χ_t^+, defined by operation of addition (3.2.41a), lie on the same line l_+: $\xi_t, \chi_t^+, \eta_t \in l_+$, so that the result of interaction χ_t^+ lies between the samples ξ_t and η_t, and the metric identity holds:*

$$\mu(\xi_t, \eta_t) = \mu(\xi_t, \chi_t^+) + \mu(\chi_t^+, \eta_t). \tag{3.2.42}$$

Theorem 3.2.8 is an overformulation of Theorem 5 through the notion of line and ternary relation of "betweenness".

Theorem 3.2.9. *In metric signal space $(\mathbf{\Gamma}, \mu)$ with metric μ (3.2.1), a pair of samples ξ_t and η_t of stochastic signals (processes) $\xi(t)$, $\eta(t)$, $t \in T$ and the result of their interaction $\chi_t^\vee$, defined by operation of join (3.2.41b), lie on the same line $l_\vee$: $\xi_t, \chi_t^\vee, \eta_t \in l_\vee$, so that the result of interaction $\chi_t^\vee$ lies between the samples ξ_t and η_t, and the metric identity holds:*

$$\mu(\xi_t, \eta_t) = \mu(\xi_t, \chi_t^\vee) + \mu(\chi_t^\vee, \eta_t). \tag{3.2.43}$$

Theorem 3.2.10. *In metric signal space $(\mathbf{\Gamma}, \mu)$ with metric μ (3.2.1), a pair of samples ξ_t and η_t of stochastic signals (processes) $\xi(t)$ and $\eta(t)$, $t \in T$ and the result of their interaction $\chi_t^\wedge$, defined by operation of meet (3.2.41c), lie on the same line $l_\wedge$: $\xi_t, \chi_t^\wedge, \eta_t \in l_\wedge$, so that the result of interaction $\chi_t^\wedge$ lies between the samples ξ_t and η_t, and the metric identity holds:*

$$\mu(\xi_t, \eta_t) = \mu(\xi_t, \chi_t^\wedge) + \mu(\chi_t^\wedge, \eta_t). \tag{3.2.44}$$

Theorems 3.2.9 and 3.2.10 are an overformulation of Corollary 3.2.5 of Theorem 3.2.7 through the notions of line and ternary relation of "betweenness". These two theorems could be united by the following one.

Theorem 3.2.11. *In metric signal space $(\mathbf{\Gamma}, \mu)$ with metric μ (3.2.1), a pair of samples ξ_t and η_t of stochastic signals (processes) $\xi(t)$, $\eta(t)$, $t \in T$ and the results of their interaction $\chi_t^\vee/\chi_t^\wedge$, defined by operations of join (3.2.41b) and meet (3.2.41c), lie on the same line $l_{\vee,\wedge}$: $\xi_t, \chi_t^\vee, \chi_t^\wedge, \eta_t \in l_{\vee,\wedge}$: $l_{\vee,\wedge} = l_\vee = l_\wedge$, so that the result of interaction $\chi_t^\vee/\chi_t^\wedge$ lies between the samples ξ_t, η_t, and each sample of a pair ξ_t and η_t lies between the samples $\chi_t^\vee$ and $\chi_t^\wedge$, and the metric identities between the samples $\xi_t, \chi_t^\vee, \chi_t^\wedge, \eta_t$ hold:*

$$\mu(\xi_t, \chi_t^\wedge) = \mu(\chi_t^\vee, \eta_t); \tag{3.2.45a}$$

$$\mu(\xi_t, \chi_t^\vee) = \mu(\chi_t^\wedge, \eta_t); \tag{3.2.45b}$$

$$\mu(\xi_t, \eta_t) = \mu(\chi_t^\vee, \chi_t^\wedge); \tag{3.2.45c}$$

$$\mu(\chi_t^\vee, \chi_t^\wedge) = \mu(\chi_t^\vee, \xi_t) + \mu(\xi_t, \chi_t^\wedge); \tag{3.2.45d}$$

$$\mu(\chi_t^\vee, \chi_t^\wedge) = \mu(\chi_t^\vee, \eta_t) + \mu(\eta_t, \chi_t^\wedge). \tag{3.2.45e}$$

Proof. Theorem 3.2.7 implies the following metric identities between the samples $\xi_t, \chi_t^\vee, \chi_t^\wedge, \eta_t$:

$$\begin{cases} \mu(\xi_t, \chi_t^\vee) = 2[F_\xi(0) - F_{\xi\eta}(0,0)]; \\ \mu(\xi_t, \chi_t^\wedge) = 2[F_\eta(0) - F_{\xi\eta}(0,0)]; \end{cases}$$

$$\begin{cases} \mu(\eta_t, \chi_t^\vee) = 2[F_\eta(0) - F_{\xi\eta}(0,0)]; \\ \mu(\eta_t, \chi_t^\wedge) = 2[F_\xi(0) - F_{\xi\eta}(0,0)], \end{cases}$$

and these systems imply the identities (3.2.45a) and (3.2.45b). Besides, the following triplets of the samples lie on the same lines, respectively: $\xi_t, \chi_t^\vee, \eta_t$; $\eta_t, \chi_t^\wedge, \xi_t$; $\chi_t^\wedge, \xi_t, \chi_t^\vee$; $\chi_t^\vee, \eta_t, \chi_t^\wedge$. Hence, all four samples $\xi_t, \chi_t^\vee, \chi_t^\wedge, \eta_t$ belong to the same line.

Summing pairwise the values of metrics in the first and in the second equality systems respectively, we obtain the value of metric $\mu(\chi_t^\vee, \chi_t^\wedge)$ between the samples $\chi_t^\vee, \chi_t^\wedge$:

$$\mu(\chi_t^\vee, \chi_t^\wedge) = \mu(\chi_t^\vee, \xi_t) + \mu(\xi_t, \chi_t^\wedge) = \mu(\chi_t^\vee, \eta_t) + \mu(\eta_t, \chi_t^\wedge) =$$
$$= 2[F_\xi(0) + F_\eta(0)] - 4F_{\xi\eta}(0,0).$$

The last relationship is identically equal to metric $\mu(\xi_t, \eta_t)$ between the samples ξ_t and η_t, which is defined by the expression (3.2.5), so the initial equality (3.2.45c) holds. Summing pairwise the values of metrics from the first and the second equalities of both equality systems respectively, we obtain metric identities (3.2.45d) and (3.2.45e). □

Theorem 3.2.12. *In metric signal space* (Γ, μ) *with metric* μ *(3.2.1), the results of interaction* $\chi_t^+, \chi_t^\vee, \chi_t^\wedge$ *of the samples* ξ_t *and* η_t *of stochastic signals (processes)* $\xi(t)$, $\eta(t)$, $t \in T$, *defined by operations in (3.2.41), lie on the same line* l: $\chi_t^\vee, \chi_t^+, \chi_t^\wedge \in l$, *so that the element* χ_t^+ *lies between the samples* $\chi_t^\vee, \chi_t^\wedge$, *and the metric identity holds:*

$$\mu(\chi_t^\vee, \chi_t^\wedge) = \mu(\chi_t^\vee, \chi_t^+) + \mu(\chi_t^+, \chi_t^\wedge). \tag{3.2.46}$$

Proof. The samples of triplet $\chi_t^+, \chi_t^\vee, \chi_t^\wedge$ are connected with each other by operation of addition of group $\Gamma(+)$: $\chi_t^+ = \chi_t^\vee + \chi_t^\wedge$, so Theorem 3.2.5 implies the identity (3.2.46). □

The following visual interpretation could be given for Theorems 3.2.8 through 3.2.12. The images of lines of metric signal space (Γ, μ) with metric μ (3.2.1) in $\mathbf{R}^3$ are circles, so the main content of Theorems 3.2.8 through 3.2.12 is represented in Fig. 3.2.1a through c, respectively, by denoting the triplets of points lying on the corresponding lines (circles in $\mathbf{R}^3$).

For neutral element (zero) 0 of group $\Gamma(+)$, and also for a pair of samples ξ_t and η_t of stochastic signals (processes) $\xi(t)$, $\eta(t)$, $t \in T$ and the result of their interaction χ_t^+ defined by binary operation (3.2.41a) of addition of the group, the following metric identities hold:

$$\mu(0, \xi_t) = 1; \tag{3.2.47a}$$

$$\mu(0, \eta_t) = 1; \tag{3.2.47b}$$

$$\mu(0, \chi_t^+) = 1. \tag{3.2.47c}$$

3.2 Normalized Measure of Statistical Interrelationship of Stochastic Processes

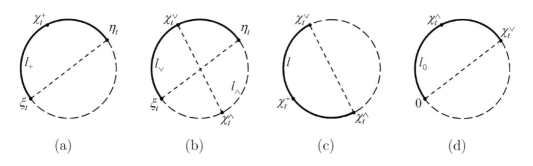

FIGURE 3.2.1 Metric relationships between samples of signals in metric signal space (Γ, μ) that correspond to the relationships (a) (3.2.42); (b) (3.2.43), (3.2.44), (3.2.45); (c) (3.2.46); and (d) (3.2.48)

Theorem 3.2.13. *In metric signal space (Γ, μ) with metric μ (3.2.1), the results of interaction $\chi_t^\vee$, $\chi_t^\wedge$ of a pair of samples ξ_t and η_t of stochastic signals (processes) $\xi(t), \eta(t), t \in T$, defined by the operations (3.2.41b) and (3.2.41c), and also neutral element 0 of the group $\Gamma(+)$ lie on the same line l_0: $\chi_t^\vee, \chi_t^\wedge, 0 \in l_0$, so that the element $\chi_t^\wedge$ lies between the elements $\chi_t^\vee$, 0; and the metric identity holds:*

$$\mu(0, \chi_t^\vee) = \mu(0, \chi_t^\wedge) + \mu(\chi_t^\wedge, \chi_t^\vee). \qquad (3.2.48)$$

Proof. According to the equality (3.2.1), the values of metrics $\mu(0, \chi_t^\vee)$, $\mu(0, \chi_t^\wedge)$, $\mu(\chi_t^\wedge, \chi_t^\vee)$ are determined by the following expressions:

$$\mu(0, \chi_t^\vee) = 2(\mathbf{P}[(\xi_t \vee \eta_t) \vee 0 > 0] - \mathbf{P}[(\xi_t \vee \eta_t) \wedge 0 > 0]) = 2\mathbf{P}[\xi_t \vee \eta_t > 0];$$

$$\mu(0, \chi_t^\wedge) = 2(\mathbf{P}[(\xi_t \wedge \eta_t) \vee 0 > 0] - \mathbf{P}[(\xi_t \wedge \eta_t) \wedge 0 > 0]) = 2\mathbf{P}[\xi_t \wedge \eta_t > 0];$$

$$\mu(\chi_t^\wedge, \chi_t^\vee) = 2(\mathbf{P}[(\xi_t \vee \eta_t) \vee (\xi_t \wedge \eta_t) > 0] - \mathbf{P}[(\xi_t \vee \eta_t) \wedge (\xi_t \wedge \eta_t) > 0]) =$$
$$= 2(\mathbf{P}[\xi_t \vee \eta_t > 0] - \mathbf{P}[\xi_t \wedge \eta_t > 0]),$$

and their joint fulfillment directly implies the initial identity (3.2.48). $\square$

The content of Theorem 3.2.13 is visually represented in Fig. 3.2.1(d).

Let the samples ξ_t, η_t of the stochastic signals (processes) $\xi(t), \eta(t), t \in T$ interact with each other in metric signal space (Γ, μ) with metric μ (3.2.1), so that the result of their interaction χ_t is described by some signature operation of L–group $\Gamma(+, \vee, \wedge)$:

$$\chi_t = \xi_t \oplus \eta_t. \qquad (3.2.49)$$

Then the following lemma holds.

Lemma 3.2.1. *In metric signal space (Γ, μ) with metric μ (3.2.1), under the mapping h_α from the group H of continuous mappings $h_\alpha \in H = \{h_\alpha\}$ preserving neutral element 0 of the group $\Gamma(+)$: $h_\alpha(0) = 0$, for the result of interaction χ_t (3.2.49) of a pair of samples ξ_t and η_t of stochastic signals (processes) $\xi(t), \eta(t),$*

$t \in T$, which is defined by some operations (3.2.41), metric between the elements χ_t and $h_\alpha(\chi_t)$ is equal to zero:

$$\mu(\chi_t, h_\alpha(\chi_t)) = 0; \qquad (3.2.50)$$
$$h_\alpha : \chi(t) \to \chi'(t) = h_\alpha(\chi(t)); \qquad (3.2.50a)$$
$$h_\alpha^{-1} : \chi'(t) \to \chi(t). \qquad (3.2.50b)$$

Proof. According to Theorem 3.2.2, metric μ (3.2.1) is invariant of the group H of mappings $\{h_{\alpha,\beta}\}$, $h_{\alpha,\beta} \in H$; $\alpha, \beta \in A$ of stochastic signals (processes) (3.2.9a) and (3.2.9b):

$$\mu(\xi_t, \eta_t) = \mu(\xi'_t, \eta'_t);$$
$$h_\alpha : \xi(t) \to \xi'(t), \quad h_\beta : \eta(t) \to \eta'(t);$$
$$h_\alpha^{-1} : \xi'(t) \to \xi(t), \quad h_\beta^{-1} : \eta'(t) \to \eta(t).$$

Thus, if the relations hold:

$$h_1 : \chi(t) \to \chi(t), \quad h_\alpha : \chi(t) \to \chi'(t) = h_\alpha(\chi(t)),$$

where h_1 is an identity mapping (the unity of the group H): $h_\alpha^{-1} \cdot h_\alpha = h_1$, then the identity holds:

$$\mu(\chi_t, \chi_t) = \mu(\chi_t, h_\alpha(\chi_t)) = 0.$$

$\square$

The lemma is necessary to prove the following theorem.

Theorem 3.2.14. *In metric signal space (Γ, μ) with metric μ (3.2.1), under an arbitrary mapping f of the result of interaction χ_t (3.2.49) of a pair of samples ξ_t and η_t of stochastic signals (processes) $\xi(t)$, $\eta(t)$, $t \in T$, which is defined by some of operations (3.2.41), the metric inequalities hold:*

$$\mu(\xi_t, \chi_t) \leq \mu(\xi_t, f(\chi_t)); \qquad (3.2.51a)$$
$$\mu(\eta_t, \chi_t) \leq \mu(\eta_t, f(\chi_t)), \qquad (3.2.51b)$$

so that the inequalities (3.2.51) turn into equalities, if and only if the mapping f belongs to a group $H = \{h_\alpha\}$ of continuous mappings $f \in H$ preserving neutral element 0 of the additive group $\Gamma(+)$: $h_\alpha(0) = 0$, $h_\alpha \in H$.

Proof. At first we prove the second part of theorem. If over the result of interaction χ_t (3.2.49) of the samples ξ_t and η_t, the mapping f is realized such that it belongs to a group H of odd mappings $f \in H$, then according to Lemma 3.2.1, the identity $\mu(\chi_t, f(\chi_t)) = 0$ implies the identity $\chi_t \equiv f(\chi_t)$ between the result of interaction χ_t (3.2.49) of the samples ξ_t, η_t of stochastic signals (processes) $\xi(t)$, $\eta(t)$, $t \in T$ and the result of its mapping $f(\chi_t)$. Then, according to metric identities (3.2.42) through (3.2.44) of Theorems 3.2.8 through 3.2.10, respectively, the samples ξ_t, $\chi_t = f(\chi_t)$, η_t lie on the same line $l_\oplus$ (see Fig. 3.2.2), and the identities hold:

$$\mu(\xi_t, \eta_t) = \mu(\xi_t, \chi_t) + \mu(\chi_t, \eta_t);$$

3.2 Normalized Measure of Statistical Interrelationship of Stochastic Processes

$$\mu(\xi_t, \eta_t) = \mu(\xi_t, f(\chi_t)) + \mu(f(\chi_t), \eta_t);$$
$$\mu(\xi_t, \chi_t) = \mu(\xi_t, f(\chi_t));$$
$$\mu(\eta_t, \chi_t) = \mu(\eta_t, f(\chi_t)).$$

The second part of theorem has been proved.

If over the result of interaction χ_t (3.2.49) of the samples ξ_t and η_t, the mapping f is realized such that it does not belong to a group H of continuous mappings $f \notin H$ preserving neutral element 0 of the group $\mathbf{\Gamma}(+)$: $h_\alpha(0) = 0$, $h_\alpha \in H$, then $f(\chi_t) \neq \chi_t$, and correspondingly, $f(\chi_t)$ does not belong to the line $l_\oplus$, on which the samples ξ_t, η_t, χ_t lie (see Fig. 3.2.2).

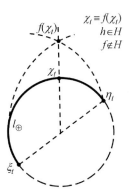

FIGURE 3.2.2 Metric relationships between samples of signals in metric space $(\mathbf{\Gamma}, \mu)$ elucidating Theorem 3.2.14

Then the strict inequalities hold:

$$\mu(\xi_t, \eta_t) < \mu(\xi_t, f(\chi_t)) + \mu(f(\chi_t), \eta_t);$$
$$\mu(\xi_t, \chi_t) < \mu(\xi_t, f(\chi_t));$$
$$\mu(\eta_t, \chi_t) < \mu(\eta_t, f(\chi_t)).$$

□

Theorem 3.2.14 has a consequence that is important for both signal processing theory and information theory. We formulate it in the form of the following theorem.

Theorem 3.2.15. *In metric signal space $(\mathbf{\Gamma}, \mu)$ with metric μ (3.2.1), under an arbitrary mapping f of the result of interaction χ_t (3.2.49) of the samples ξ_t and η_t of stochastic signals (processes) $\xi(t)$, $\eta(t)$, $t \in T$, which is defined by some of operations (3.2.41), the inequalities between NMSIs hold:*

$$\nu(\xi_t, f(\chi_t)) \leq \nu(\xi_t, \chi_t); \qquad (3.2.52a)$$
$$\nu(\eta_t, f(\chi_t)) \leq \nu(\eta_t, \chi_t), \qquad (3.2.52b)$$

so that inequalities (3.2.52) turn into identities, if and only if the mapping f belongs to a group $H = \{h_\alpha\}$ of continuous mappings $f \in H$ preserving neutral element 0 of the additive group $\mathbf{\Gamma}(+)$: $h_\alpha(0) = 0$, $h_\alpha \in H$.

Theorem 3.2.15 is a substantial analogue of the known theorem on data processing inequality formulated for mutual information $I(X,Y)$ between the signals X, Y, Z, which form a Markov chain [70]:

$$X \to Y \to Z: \quad I(X,Y) \geq I(X,Z).$$

As well as a normalized measure of statistical interrelationship, mutual information $I(X,Y)$ is a measure of statistical dependence between the signals, which is preserved under their bijective mappings. The essential distinction between them lies in the fact that normalized measure of statistical interrelationship is completely defined by metric (3.2.1) through the relation (3.2.7), requiring the evenness of univariate PDFs of stochastic signals (processes) instead of Markov chain presence.

3.3 Probabilistic Measure of Statistical Interrelationship of Stochastic Processes

Let ξ_t and η_t be the instantaneous values (samples) of stochastic signals (processes) $\xi(t)$, $\eta(t)$, $t \in T$ with joint cumulative distribution function (CDF) $F_{\xi\eta}(x,y)$, joint probability density function (PDF) $p_{\xi\eta}(x,y)$ and also with even univariate PDFs $p_\xi(x)$ and $p_\eta(y)$ of the kind: $p_\xi(x) = p_\xi(-x)$; $p_\eta(y) = p_\eta(-y)$.

Then each pair of stochastic processes $\xi(t), \eta(t) \in \Gamma$ can be associated with the following normalized measure between the samples ξ_t and η_t based on probabilistic relationships between them.

Definition 3.3.1. *Probabilistic measure of statistical interrelationship* (PMSI) between the samples ξ_t, η_t of stochastic processes $\xi(t), \eta(t) \in \Gamma$ is the quantity $\nu_\mathbf{P}(\xi_t, \eta_t)$ equal to:

$$\nu_\mathbf{P}(\xi_t, \eta_t) = 3 - 4\mathbf{P}[\xi_t \vee \eta_t > 0], \tag{3.3.1}$$

where $\mathbf{P}[\xi_t \vee \eta_t > 0]$ is the probability that the random variable $\xi_t \vee \eta_t$, which is equal to join of the samples ξ_t and η_t, takes a value greater than zero.

Theorem 3.3.1. *PMSI $\nu_\mathbf{P}(\xi_t, \eta_t)$ is defined by the joint CDF $F_{\xi\eta}(x,y)$ of the samples ξ_t, η_t:*

$$\nu_\mathbf{P}(\xi_t, \eta_t) = 4F_{\xi\eta}(0,0) - 1. \tag{3.3.2}$$

Proof. According to the formula [115, (3.2.85)], random variable $\xi_t \vee \eta_t$ is characterized by CDF $F_{\xi\vee\eta}(z)$ equal to $F_{\xi\vee\eta}(z) = F_{\xi\eta}(z,z)$. So, the probability $\mathbf{P}[\xi_t \vee \eta_t > 0]$ is equal to $\mathbf{P}[\xi_t \vee \eta_t > 0] = 1 - F_{\xi\vee\eta}(0) = 1 - F_{\xi\eta}(0,0)$. $\square$

As follows from the equalities (3.2.8) and (3.3.2), for stochastic signals (processes) $\xi(t), \eta(t) \in \Gamma$ with joint CDF $F_{\xi\eta}(x,y)$ and even univariate PDFs $p_\xi(x)$, $p_\eta(y)$ of a sort: $p_\xi(x) = p_\xi(-x)$; $p_\eta(y) = p_\eta(-y)$, the notions of NMSI $\nu(\xi_t, \eta_t)$ and PMSI $\nu_\mathbf{P}(\xi_t, \eta_t)$ coincide:

$$\nu(\xi_t, \eta_t) = \nu_\mathbf{P}(\xi_t, \eta_t).$$

3.3 Probabilistic Measure of Statistical Interrelationship of Stochastic Processes

Thus the theorems listed below except special cases, repeat the corresponding theorems of the previous section where the properties of NMSI $\nu(\xi_t, \eta_t)$ are considered and remain valid for PMSI $\nu_{\mathbf{P}}(\xi_t, \eta_t)$.

Theorem 3.3.2. *PMSI $\nu_{\mathbf{P}}(\xi_t, \eta_t)$ can be defined by the following way:*

$$\nu_{\mathbf{P}}(\xi_t, \eta_t) = 3 - 4\mathbf{P}[\xi_t \wedge \eta_t < 0] = 4F_{\xi\eta}(0,0) - 1, \quad (3.3.3)$$

where $\mathbf{P}[\xi_t \wedge \eta_t < 0]$ is the probability that random variable $\xi_t \wedge \eta_t$, which is equal to meet of the samples ξ_t and η_t, takes a value less than zero.

Proof. According to the formula [115, (3.2.82)], random variable $\xi_t \wedge \eta_t$ is characterized by CDF $F_{\xi \wedge \eta}(z)$ equal to $F_{\xi \wedge \eta}(z) = F_\xi(z) + F_\eta(z) - F_{\xi\eta}(z, z)$, where $F_\xi(u)$ and $F_\eta(v)$ are univariate CDFs of the samples ξ_t and η_t, respectively. So, the probability $\mathbf{P}[\xi_t \wedge \eta_t < 0]$ is equal to:

$$\mathbf{P}[\xi_t \wedge \eta_t < 0] = F_\xi(0) + F_\eta(0) - F_{\xi\eta}(0,0) = 1 - F_{\xi\eta}(0,0).$$

$\square$

Theorem 3.3.3. *For a pair of stochastic processes $\xi(t)$, $\eta(t)$ with even univariate PDFs in L–group $\boldsymbol{\Gamma}$: $\xi(t), \eta(t) \in \boldsymbol{\Gamma}$, $t \in T$, PMSI $\nu_{\mathbf{P}}(\xi_t, \eta_t)$ of a pair of samples ξ_t, η_t is invariant of a group H of continuous mappings $\{h_{\alpha,\beta}\}$, $h_{\alpha,\beta} \in H$; $\alpha, \beta \in A$ of stochastic processes, which preserve neutral element 0 of the group $\boldsymbol{\Gamma}(+)$: $h_{\alpha,\beta}(0) = 0$:*

$$\nu_{\mathbf{P}}(\xi_t, \eta_t) = \nu_{\mathbf{P}}(\xi'_t, \eta'_t); \quad (3.3.4)$$

$$h_\alpha : \xi(t) \to \xi'(t), \quad h_\beta : \eta(t) \to \eta'(t); \quad (3.3.4a)$$

$$h_\alpha^{-1} : \xi'(t) \to \xi(t), \quad h_\beta^{-1} : \eta'(t) \to \eta(t), \quad (3.3.4b)$$

where ξ'_t and η'_t are the samples of stochastic processes $\xi'(t)$ and $\eta'(t)$ in L–group $\boldsymbol{\Gamma}'$: $h_{\alpha,\beta}$: $\boldsymbol{\Gamma} \to \boldsymbol{\Gamma}'$.

The proof of theorem is the same as the proof of Theorem 3.2.2.

Theorem 3.3.4. *For Gaussian centered stochastic processes $\xi(t)$ and $\eta(t)$ with correlation coefficient $\rho_{\xi\eta}$ between the samples ξ_t and η_t, their PMSI $\nu_{\mathbf{P}}(\xi_t, \eta_t)$ is equal to:*

$$\nu_{\mathbf{P}}(\xi_t, \eta_t) = \frac{2}{\pi} \arcsin[\rho_{\xi\eta}]. \quad (3.3.5)$$

The proof of theorem is the same as well as the proof of Theorem 3.2.3.

Joint fulfillment of the relationship (3.3.5) of Theorem 3.3.4 and Equation (3.1.6) implies, that mutual NFSI $\psi_{\xi\eta}(t, t)$ and PMSI $\nu_{\mathbf{P}}(\xi_t, \eta_t)$ between two samples ξ_t and η_t of stochastic processes $\xi(t)$ and $\eta(t)$ introduced by Definitions 3.1.2 and 3.3.1, respectively, are rather exactly connected by the following approximation:

$$\psi_{\xi\eta}(t, t) \approx 1 - \sqrt{1 - \nu_{\mathbf{P}}(\xi_t, \eta_t)}.$$

Theorem 3.3.5. *For Gaussian centered stochastic processes $\xi(t)$ and $\eta(t)$, which additively interact in L–group Γ: $\chi(t) = \xi(t)+\eta(t)$, $t \in T$, between PMSIs $\nu_{\mathbf{P}}(\xi_t, \chi_t)$, $\nu_{\mathbf{P}}(\eta_t, \chi_t)$, $\nu_{\mathbf{P}}(\xi_t, \eta_t)$ of the corresponding pairs of their samples ξ_t, χ_t; η_t, χ_t; ξ_t, η_t, the following relationship holds:*

$$\nu_{\mathbf{P}}(\xi_t, \chi_t) + \nu_{\mathbf{P}}(\eta_t, \chi_t) - \nu_{\mathbf{P}}(\xi_t, \eta_t) = 1. \tag{3.3.6}$$

The proof of theorem is the same as the proof of Theorem 3.2.4.

Thus, Theorem 3.3.5 establishes an invariance relationship for PMSIs of the pairs of the samples ξ_t, χ_t; η_t, χ_t; ξ_t, η_t of additively interacting Gaussian stochastic processes $\xi(t)$, $\eta(t)$, and besides, this identity does not depend on their energetic relationships, despite the fact that PMSIs $\nu_{\mathbf{P}}(\xi_t, \chi_t)$, $\nu_{\mathbf{P}}(\eta_t, \chi_t)$ are the functions of variances D_ξ, D_η of the samples ξ_t, η_t of Gaussian signals $\xi(t)$, $\eta(t)$.

The result (3.3.6) of Theorem 3.3.5 could be generalized upon additively interacting processes $\xi(t)$, $\eta(t)$ with rather arbitrary probabilistic-statistical properties, whose diversity is bounded by cardinality of the group of mappings $H = \{h_{\alpha,\beta}\}$, $\alpha, \beta \in A$, and besides, every mapping $h_{\alpha,\beta} \in H$ possesses the property of the mappings (3.3.4a,b). The required generalization is provided by the following theorem.

Theorem 3.3.6. *For nonGaussian stochastic processes $\xi(t)$ and $\eta(t)$, which additively interact with each other in L–group Γ: $\chi(t) = \xi(t) + \eta(t)$, $t \in T$, between PMSIs $\nu_{\mathbf{P}}(\xi_t, \eta_t)$, $\nu_{\mathbf{P}}(\xi_t, \chi_t)$, $\nu_{\mathbf{P}}(\eta_t, \chi_t)$ of the corresponding pairs of the samples ξ_t, η_t; ξ_t, χ_t; χ_t, η_t, that are invariants of the group H of odd mappings $H = \{h_{\alpha,\beta}\}$, $h_{\alpha,\beta} \in H$, $\alpha, \beta \in A$ of stochastic processes (3.3.4), the relationship holds:*

$$\nu_{\mathbf{P}}(\xi_t, \chi_t) + \nu_{\mathbf{P}}(\eta_t, \chi_t) - \nu_{\mathbf{P}}(\xi_t, \eta_t) = 1. \tag{3.3.7}$$

The proof of the theorem is the same as the proof of Theorem 3.2.5.

For partially ordered set Γ with lattice properties, where the processes $\xi(t)$ and $\eta(t)$ interact with the results of interaction $\chi(t)$, $\tilde{\chi}(t)$: $\chi(t) = \xi(t) \vee \eta(t)$, $\tilde{\chi}(t) = \xi(t) \wedge \eta(t)$, $t \in T$ in the form of lattice binary operations $\vee$, $\wedge$, the notion of normalized measure between the samples ξ_t and η_t of stochastic processes $\xi(t)$ and $\eta(t)$ needs refinement. It is stipulated by the fact that the sameness of the PMSI definition through the functions (3.3.1) and (3.3.3) does not hold in partially ordered set Γ with lattice properties because of violation of evenness property of PDFs of the samples χ_t and $\tilde{\chi}_t$.

Definition 3.3.2. *Probabilistic measures of statistical interrelationship* (PMSIs) *between the samples ξ_t, χ_t and ξ_t, $\tilde{\chi}_t$ of stochastic processes $\xi(t)$, $\chi(t)$, $\tilde{\chi}(t) \in \Gamma$: $\chi(t) = \xi(t) \vee \eta(t)$, $\tilde{\chi}(t) = \xi(t) \wedge \eta(t)$, $t \in T$ in partially ordered set Γ with lattice properties are the quantities $\nu_{\mathbf{P}}(\xi_t, \chi_t)$, $\nu_{\mathbf{P}}(\xi_t, \tilde{\chi}_t)$ that are respectively equal to:*

$$\nu_{\mathbf{P}}(\xi_t, \chi_t) = 3 - 4\mathbf{P}[\xi_t \wedge \chi_t < 0]; \tag{3.3.8a}$$

$$\nu_{\mathbf{P}}(\xi_t, \tilde{\chi}_t) = 3 - 4\mathbf{P}[\xi_t \vee \tilde{\chi}_t > 0], \tag{3.3.8b}$$

where $\mathbf{P}[\xi_t \wedge \chi_t < 0]$ and $\mathbf{P}[\xi_t \vee \tilde{\chi}_t > 0]$ are the probabilities that random variables $\xi_t \wedge \chi_t / \xi_t \vee \tilde{\chi}_t$, which are equal to meet and join of the samples ξ_t, χ_t and ξ_t, $\tilde{\chi}_t$, take the values less and more than zero, respectively.

Theorem 3.3.7. *For stochastic processes $\xi(t)$ and $\eta(t)$, which interact in partially ordered set Γ with lattice properties: $\chi(t) = \xi(t) \vee \eta(t)$, $\tilde{\chi}(t) = \xi(t) \wedge \eta(t)$, $t \in T$, and for their PMSIs $\nu_{\mathbf{P}}(\xi_t, \chi_t)$, $\nu_{\mathbf{P}}(\eta_t, \chi_t)$; $\nu_{\mathbf{P}}(\xi_t, \tilde{\chi}_t)$, $\nu_{\mathbf{P}}(\eta_t, \tilde{\chi}_t)$ of the corresponding pairs of their samples ξ_t, χ_t; η_t, χ_t and $\xi_t, \tilde{\chi}_t$; $\eta_t, \tilde{\chi}_t$, the following relationships hold:*

$$\nu_{\mathbf{P}}(\xi_t, \chi_t) = 1, \quad \nu_{\mathbf{P}}(\eta_t, \chi_t) = 1; \tag{3.3.9a}$$

$$\nu_{\mathbf{P}}(\xi_t, \tilde{\chi}_t) = 1, \quad \nu_{\mathbf{P}}(\eta_t, \tilde{\chi}_t) = 1. \tag{3.3.9b}$$

Proof. According to absorption axiom of lattice, the identities hold:

$$\xi_t \wedge \chi_t = \xi_t \wedge (\xi_t \vee \eta_t) = \xi_t; \tag{3.3.10a}$$

$$\eta_t \wedge \chi_t = \eta_t \wedge (\xi_t \vee \eta_t) = \eta_t; \tag{3.3.10b}$$

$$\xi_t \vee \tilde{\chi}_t = \xi_t \vee (\xi_t \wedge \eta_t) = \xi_t; \tag{3.3.10c}$$

$$\eta_t \vee \tilde{\chi}_t = \eta_t \vee (\xi_t \wedge \eta_t) = \eta_t. \tag{3.3.10d}$$

Substituting the results of (3.3.10) into the formula (3.3.8) and taking into account the evenness of univariate PDFs $p_\xi(x)$ and $p_\eta(y)$ of the samples ξ_t, η_t, the probabilities in (3.3.8) are determined by univariate CDFs $F_\xi(x)$, $F_\eta(y)$ on $x = y = 0$:

$$\begin{cases} \mathbf{P}[\xi_t \wedge \chi_t < 0] = F_\xi(0) = 1/2; \\ \mathbf{P}[\eta_t \wedge \chi_t < 0] = F_\eta(0) = 1/2; \end{cases} \tag{3.3.11a}$$

$$\begin{cases} \mathbf{P}[\xi_t \vee \tilde{\chi}_t > 0] = F_\xi(0) = 1/2; \\ \mathbf{P}[\eta_t \vee \tilde{\chi}_t > 0] = F_\eta(0) = 1/2. \end{cases} \tag{3.3.11b}$$

Substituting the results (3.3.11) into the formulas (3.3.8) we obtain (3.3.9). □

Thus, Theorem 3.3.7 establishes invariance relationships for PMSIs (3.3.9) of the pairs of samples ξ_t, χ_t; η_t, χ_t and $\xi_t, \tilde{\chi}_t$; $\eta_t, \tilde{\chi}_t$ of stochastic processes $\xi(t)$ and $\eta(t)$, which interact in partially ordered set Γ with lattice properties: $\chi(t) = \xi(t) \vee \eta(t)$, $\tilde{\chi}(t) = \xi(t) \wedge \eta(t)$, $t \in T$, and these identities do not depend on probabilistic distributions of interacting signals and their energetic relationships.

3.4 Information Distribution Density of Stochastic Processes

Each sample $\xi(t_j)$ of stochastic process $\xi(t)$ with normalized function of statistical interrelationship (NFSI) $\psi_\xi(t_j, t_k)$, introduced in the previous section, can be associated with some function $i_\xi(t_j, t)$ connected with NFSI by the relationship:

$$\psi_\xi(t_j, t_k) = d(p, p_0) = 1 - d[i_\xi(t_j, t); i_\xi(t_k, t)]; \tag{3.4.1}$$

$$d[i_\xi(t_j, t); i_\xi(t_k, t)] = \frac{1}{2} \int_{-\infty}^{\infty} |i_\xi(t_j, t) - i_\xi(t_k, t)| dt, \tag{3.4.1a}$$

where $d(p, p_0)$ is metric determined by (3.1.2a); $d[i_\xi(t_j, t); i_\xi(t_k, t)]$ is the metric between the functions $i_\xi(t_j, t)$ and $i_\xi(t_k, t)$ of the samples $\xi(t_j)$ and $\xi(t_k)$, respectively.

Definition 3.4.1. The function $i_\xi(t_j, t)$ connected with NFSI $\psi_\xi(t_j, t_k)$ by the relationship (3.4.1) is called *information distribution density* (IDD) of stochastic process $\xi(t)$.

For stationary stochastic process $\xi(t)$ with NFSI $\psi_\xi(\tau)$, the expression (3.4.1) takes the form:

$$\psi_\xi(\tau) = 1 - d[i_\xi(t_j, t); i_\xi(t_j + \tau, t)]; \qquad (3.4.2)$$

$$d[i_\xi(t_j, t); i_\xi(t_j + \tau, t)] = \frac{1}{2} \int_{-\infty}^{\infty} |i_\xi(t_j, t) - i_\xi(t_j + \tau, t)| dt, \qquad (3.4.2a)$$

The IDD of stationary stochastic process $\xi(t)$ will be denoted as $i_\xi(\tau)$ if it is not necessary to fix the relation of IDD to a single sample $\xi(t_j)$.

Consider the main properties of the IDD of stochastic process $\xi(t)$ carrying a finite information quantity $I(T_\xi) < \infty$ on the domain of definition T_ξ.

1. IDD is a nonnegative and bounded function: $0 \le i_\xi(t_j, t) \le i_\xi(t_j, t_j)$.
2. IDD is a normalized function: $\int_{T_\xi} i_\xi(t_j, t) dt = 1$.
3. IDD is a symmetric function with respect to the line $t = t_j$:

$$i_\xi(t_j, t_j - \tau) = i_\xi(t_j, t_j + \tau).$$

For a stationary stochastic process with a normalized function of statistical interrelationship (NFSI) $\psi_\xi(\tau)$, one can obtain a coupling equation for IDD $i_\xi(\tau)$ that is dual with respect to the expression (3.4.2a) directly from its consequence:

$$1 - \int_{-\infty}^{\tau} i_\xi(x) dx = \frac{1}{2} \psi_\xi(2\tau), \tau > 0.$$

Taking into account the evenness of the functions $i_\xi(\tau)$, $\psi_\xi(\tau)$ and the property of an integral with variable upper limit of integration, it is easy to obtain the coupling equation between IDD $i_\xi(\tau)$ and NFSI $\psi_\xi(\tau)$ of stochastic process:

$$i_\xi(\tau) = \begin{cases} \frac{1}{2}\psi'_-(2\tau), & \tau < 0; \\ -\frac{1}{2}\psi'_+(2\tau), & \tau \ge 0, \end{cases} \qquad (3.4.3)$$

where $\psi'_-(\tau)$ and $\psi'_+(\tau)$ are the derivatives of NFSI $\psi_\xi(\tau)$ on the left and on the right, respectively.

For Gaussian stationary stochastic process, IDD $i_\xi(\tau)$ is completely defined by the normalized correlation function $r_\xi(\tau)$:

$$i_\xi(\tau) = \begin{cases} \frac{1}{2}g'_-[r_\xi(2\tau)], & \tau < 0; \\ -\frac{1}{2}g'_+[r_\xi(2\tau)], & \tau \ge 0, \end{cases} \qquad (3.4.4)$$

3.4 Information Distribution Density of Stochastic Processes

where $g'_-[r_\xi(\tau)]$ and $g'_+[r_\xi(\tau)]$ are derivatives of deterministic functions $g[x]$ on the left and on the right, respectively (see Equations (3.1.3) and (3.1.4)).

The IDD $i_\xi(\tau)$ of a stochastic process, as follows from (3.4.3), has clear physical sense that IDD $i_\xi(\tau)$ characterizes a rate of change of statistical interrelationships between the samples of a stochastic process.

The formula (3.1.3) implies that, for stochastic process $\xi(t)$ in the form of white Gaussian noise, NFSI $\psi_\xi(\tau)$ is equal to:

$$\psi_\xi(\tau) = \begin{cases} 1, & \tau = 0; \\ 0, & \tau \neq 0, \end{cases}$$

and formula (3.4.3) implies that IDD $i_\xi(t_j, t)$ of an arbitrary sample $\xi(t_j)$ has a form of delta-function $i_\xi(t_j, t) = \delta(t - t_j)$. This means that a single sample $\xi(t_j)$ of white Gaussian noise does not carry any information regarding the other samples of this stochastic process.

It is easy to determine the IDD of stochastic process with help of the coupling equation (3.4.3), if its NFSI is known.

Example 3.4.1. Let the NFSI $\psi(\tau)$ of a stochastic process be determined by the function:

$$\psi(\tau) = \begin{cases} 1 - \dfrac{|\tau|}{a}, & |\tau| \leq a; \\ 0, & |\tau| \geq a. \end{cases}$$

Then its IDD $i(\tau)$ is equal to:

$$i(\tau) = \frac{1}{a}\left[1\left(\tau + \frac{a}{2}\right) - 1\left(\tau - \frac{a}{2}\right)\right],$$

where $1(\tau)$ is the Heaviside step function. ▽

Example 3.4.2. Let the NFSI $\psi(\tau)$ of a stochastic process be determined by the function:

$$\psi(\tau) = \left(1 - \frac{|\tau|}{2a}\right)^2, \quad |\tau| \leq 2a.$$

Then its IDD $i(\tau)$ is equal to:

$$i(\tau) = \begin{cases} \dfrac{1}{a}\left[1 - \dfrac{|\tau|}{a}\right], & |\tau| \leq a; \\ 0, & |\tau| \geq a. \end{cases} \quad \triangledown$$

Example 3.4.3. Let the NFSI $\psi(\tau)$ of a stochastic process be determined by the function:

$$\psi(\tau) = \exp(-\lambda|\tau|).$$

Then its IDD $i(\tau)$ is equal to:

$$i(\tau) = \lambda \exp(-2\lambda|\tau|). \quad \triangledown$$

Considering the relation between NFSI and IDD, one should note that IDD is a primary characteristic of this pair (IDD and NFSI) for the stochastic processes that are capable of carrying information. The IDD determines NFSI, not vice versa. This means that the expression (3.4.3) has a sense when known beforehand that the IDD of stochastic process exists, or the stochastic process can carry information. Naturally, there are some classes of stochastic processes for which the use of (3.4.3) is impossible or is possible but with a proviso. Based on the simplicity of consideration, we consider Gaussian stochastic processes as the examples of stochastic processes of various kinds.

Analytical stochastic processes form a wide class of stochastic processes, for which the use of (3.4.3) makes no sense [115], [4].

Stochastic process $\xi(t)$ is called *analytical* in the interval $[t_0, t_0 + T]$, if most realizations of the process assume an analytical continuation in this interval. Analyticity of a process $\xi(t)$ in a neighborhood of the point t_0 implies a possibility of representation of this process realization by Taylor's series with random coefficients:

$$\xi(t) = \sum_{k=0}^{\infty} \xi^{(k)}(t_0) \frac{(t-t_0)^k}{k!}, \qquad (3.4.5)$$

where $\xi^{(k)}(t)$ is a k-th order derivative of stochastic process $\xi(t)$ in mean square sense.

The following theorem determines necessary and sufficient conditions of analyticity of the Gaussian stochastic process [115].

Theorem 3.4.1. *Let the normalized correlation function $r_\xi(t_1, t_2)$ of Gaussian stochastic process $\xi(t)$ be analytical function of two variables in a neighborhood of the point (t_0, t_0). Then stochastic process $\xi(t)$ is analytical in a neighborhood of this point.*

Below are examples of analytical stochastic processes.

Example 3.4.4. Consider a Gaussian stationary stochastic process with a constant power spectral density $S(\omega) = N_0 = \text{const}$ bounded by the band $[-\Delta\omega/2, \Delta\omega/2]$:

$$S(\omega) = \begin{cases} N_0, & \omega \in [-\Delta\omega/2, \Delta\omega/2]; \\ 0, & \omega \notin [-\Delta\omega/2, \Delta\omega/2]. \end{cases} \qquad (3.4.6)$$

According to this, its normalized correlation function $r(\tau)$ is determined by the function:

$$r(\tau) = \frac{\sin(\Delta\omega\tau/2)}{\Delta\omega\tau/2}. \quad \triangledown \qquad (3.4.7)$$

Example 3.4.5. Consider a Gaussian stationary stochastic process with the normalized correlation function:

$$r(\tau) = \exp(-\mu\tau^2). \qquad (3.4.8)$$

In this case, its power spectral density $S(\omega)$ is determined by the expression:

$$S(\omega) = \sqrt{\frac{\pi}{\mu}} \exp\left[-\frac{\omega^2}{4\mu}\right]. \quad \triangledown \qquad (3.4.9)$$

3.4 Information Distribution Density of Stochastic Processes

Example 3.4.6. Consider a Gaussian stationary stochastic process with the normalized correlation function:

$$r(\tau) = \left[1 + \left(\frac{\tau}{\alpha}\right)^2\right]^{-1}. \quad (3.4.10)$$

In this case, its power spectral density $S(\omega)$ is determined by the expression:

$$S(\omega) = A \exp(-\alpha|\omega|). \quad \triangledown \quad (3.4.11)$$

All the values of realization of analytical stochastic process may be restored on the base of an indefinitely small interval of its realization while using the methods of analytical continuation. So, normalized correlation function and power spectral density of ergodic analytical stationary process could be determined on an indefinitely small interval of its realization. The realization of an analytical process may be extrapolated arbitrarily into the future with a given accuracy on the base of a small interval of past realization. In this sense, the realizations of analytical processes are similar to usual deterministic functions. Thus, the analytical stochastic processes cannot carry information.

The stochastic processes, shown by Examples 3.4.4 through 3.4.6, may be obtained by passing white Gaussian noise through the forming filter with amplitude-frequency characteristic (frequency response) $K(\omega)$, whose squared module is equal to the power spectral density of the output stochastic process (3.4.6), (3.4.9), and (3.4.11):

$$|K(\omega)|^2 = S(\omega). \quad (3.4.12)$$

However, filters with such amplitude-frequency characteristics are not physically realizable ones because they do not satisfy the corresponding *Paley-Wiener condition* [235]:

$$\int_{-\infty}^{\infty} \frac{\left|\ln |K(\omega)|^2\right|}{1 + \omega^2} d\omega < \infty.$$

A wide class of stochastic processes indicates that the use of the formula (3.4.3) is possible with a proviso; it is formed by narrowband Gaussian stationary stochastic processes with oscillated normalized correlation function:

$$\rho(\tau) = r(\tau) \cos \omega_0 \tau, \quad (3.4.13)$$

where ω_0 is a central frequency of a power spectral density of a process; $r(\tau)$ is an envelope of normalized correlation function:

$$r(\tau) = \sqrt{\rho^2(\tau) + \rho_\perp^2(\tau)};$$

and $\rho_\perp(\tau) = \frac{1}{\pi} \int_{-\infty}^{\infty} \frac{\rho(x)}{\tau - x} dx$ is a function connected with an initial one by Hilbert transform.

A narrowband Gaussian stationary stochastic process $\xi(t)$ with normalized correlation function (3.4.13) can carry information if its envelope is not an analytical function.

The following theorem defines the necessary condition, according to which a stochastic process has the ability to carry information.

Theorem 3.4.2. *For stochastic process $\xi(t)$ to possess the ability to carry information, it is necessary that its NFSI $\psi_\xi(t_j, t_k)$ satisfy the following requirements:*

1. *NFSI $\psi_\xi(t_j, t_k)$ is not a differentiable function in the point (t_j, t_j), and its derivatives in this point on the left $\psi'_\xi(t_j, t_j - 0)$ and on the right $\psi'_\xi(t_j, t_j + 0)$ have unlike signs and the same modules:*

$$\psi'_\xi(t_j, t_j - 0) = -\psi'_\xi(t_j, t_j + 0);$$

2. *Modules of derivatives of NFSI in the point (t_j, t_j) on the left and on the right are not equal to zero:*

$$|\psi'_\xi(t_j, t_j - 0)| = |\psi'_\xi(t_j, t_j + 0)| \neq 0.$$

Proof. Let $\xi(t)$ be an arbitrary stochastic process possessing the ability to carry information. This means that for any its sample $\xi(t_j)$, there exists IDD $i_\xi(t_j, t)$ with the properties 1, 2, 3 listed above on page 102. We can find the derivatives of NFSI $\psi_\xi(t_j, t_k)$ in the point (t_j, t_j) on both the left and right.

The definition of the derivative implies:

$$\psi'_\xi(t_j, t_j - 0) = \lim_{\Delta t \to 0} \frac{\Delta \psi_\xi(t_j, t_j - 0)}{\Delta t} = \lim_{\Delta t \to 0} \frac{\psi_\xi(t_j, t_j) - \psi_\xi(t_j, t_j - 0)}{\Delta t}; \quad (3.4.14a)$$

$$\psi'_\xi(t_j, t_j + 0) = \lim_{\Delta t \to 0} \frac{\Delta \psi_\xi(t_j, t_j + 0)}{\Delta t} = \lim_{\Delta t \to 0} \frac{\psi_\xi(t_j, t_j + 0) - \psi_\xi(t_j, t_j)}{\Delta t}. \quad (3.4.14b)$$

Taking into account the relationship (3.4.1), the expressions (3.4.14) take the form:

$$\psi'_\xi(t_j, t_j - 0) = \lim_{\Delta t \to 0} \frac{d[i(t_j, t); i(t_j - \Delta t, t)]}{\Delta t}; \quad (3.4.15a)$$

$$\psi'_\xi(t_j, t_j + 0) = -\lim_{\Delta t \to 0} \frac{d[i(t_j, t); i(t_j + \Delta t, t)]}{\Delta t}, \quad (3.4.15b)$$

where

$$d[i_\xi(t_j, t); i_\xi(t_j \pm \Delta t, t)] = \frac{1}{2} \int_{-\infty}^{\infty} |i_\xi(t_j, t) - i_\xi(t_j \pm \Delta t, t)| dt, \quad (3.4.16)$$

and

$$d[i_\xi(t_j, t); i_\xi(t_j \pm \Delta t, t)] \neq 0. \quad (3.4.17)$$

According to the definition of metric and the property of IDD evenness, the identity holds:

$$|i_\xi(t_j, t) - i_\xi(t_j - \Delta t, t)| = |i_\xi(t_j, t) - i_\xi(t_j + \Delta t, t)|. \quad (3.4.18)$$

3.4 Information Distribution Density of Stochastic Processes

Joint fulfillment of the formulas (3.4.15), (3.4.16), and (3.4.18) implies that derivatives of NFSI on the left and on the right have unlike signs in the same module:

$$\psi'_\xi(t_j, t_j - 0) = -\psi'_\xi(t_j, t_j + 0),$$

and NFSI $\psi_\xi(t_j, t_k)$ of stochastic process in the point (t_j, t_j) is nondifferentiable.

Besides, the relationship (3.4.17) implies that modules of derivatives of NFSI on the left and on the right are not equal to zero:

$$|\psi'_\xi(t_j, t_j - 0)| = |\psi'_\xi(t_j, t_j + 0)| \neq 0.$$

□

The theorem can be described on a qualitative level in the following way. For stochastic process $\xi(t)$ to possess the ability to carry information, it is necessary for the NFSI $\psi_\xi(t_j, t_k)$ to be characterized by the narrowed peak in a neighborhood of the point (t_j, t_j). Moreover, the sharper the peak of NFSI $\psi_\xi(t_j, t_k)$ (i.e., the larger the module of derivative $|\psi'_\xi(t_j, t_j)|$), the larger the maximum $i_\xi(t_j, t_j)$ of IDD $i_\xi(t_j, t)$, and correspondingly, the larger the overall quantity of information $I(T)$ that can be carried by stochastic process within time interval $[t_0, t_0 + T]$. Thus, Theorem 3.4.2 states that not all stochastic processes possess the ability to carry information.

Example 3.4.7. A wide class of stochastic processes can be obtained by passing white Gaussian noise through the forming *Butterworth filters* with squared module of amplitude-frequency characteristic $K_n(\omega)$ [236]:

$$|K_n(\omega)|^2 = \left[1 + \left(\frac{\omega}{W}\right)^{2n}\right]^{-1},$$

where W is Butterworth filter bandwidth at 0.5 power level; n is an order of Butterworth filter.

Butterworth filters satisfy the Paley-Wiener condition of physical realizability. Power spectral density $S_n(\omega)$ of Gaussian Butterworth stochastic processes of n-th order is determined by an expression similar to the previous one:

$$S_n(\omega) = \left[1 + \left(\frac{\omega}{W}\right)^{2n}\right]^{-1},$$

where n is an order of Butterworth stochastic process.

The existence of all the derivatives $R^{(2k)}(t_1, t_2)$ of covariation function $R(t_1, t_2)$ up to $2N$-th order inclusively is a necessary and sufficient condition of N-times differentiability of Gaussian stochastic process. This condition is equivalent to the following one. Power spectral density $S_n(\omega)$ decay faster than ω^{-2N-1} is necessary and sufficient condition of N-times differentiability of Gaussian stochastic process.

The formulated condition implies that Butterworth stochastic process of n-th order is $n - 1$-times differentiable, i.e., Butterworth stochastic processes of $n > 1$-th order are differentiable, and according to Theorem 3.4.2, they do not possess the ability to carry information. On the large values of an order n, Butterworth stochastic process asymptotically tends to quasi-white Gaussian process with constant power spectral density bounded by a bandwidth W. Thus, Butterworth stochastic processes of large orders, with respect to their properties, are close to the properties of analytical stochastic processes.

Conversely, Butterworth stochastic process of an order $n=1$ with power spectral density $S_1(\omega)$:

$$S_1(\omega) = \left[1 + \left(\frac{\omega}{W}\right)^2\right]^{-1},$$

and correlation function $R(\tau)$:

$$R(\tau) = \frac{W}{2}\exp[-W|\tau|],$$

is nondifferentiable, and has the ability to carry information. ▽

It should be noted that informational properties of stochastic processes with discrete time domain and informational properties of continuous stochastic processes may be described on the base of IDD. The use of Dirac delta function $\delta(t)$ allows IDD of stochastic process with discrete time domain to be represented in the form:

$$i_\xi(t_j, t) = \sum_{k=-n}^{n} i_{j+k} \cdot \delta(t - t_{j+k}), \quad \sum_{k=-n}^{n} i_{j+k} = 1.$$

Thus, the aforementioned properties 1, 2, 3 listed on page 102 hold for IDD $i_\xi(t_j, t)$. While considering informational properties of stochastic processes, the essential distinctions between their domains are not drawn as shown below.

3.5 Informational Characteristics and Properties of Stochastic Processes

By the mapping $\varphi[i_\xi(t_j, t)]$, information distribution density (IDD) $i_\xi(t_j, t)$ of an arbitrary sample $\xi(t_j)$ of stochastic process $\xi(t)$ with domain of definition T_ξ can be associated with its *ordinate set* X_j [237]:

$$X_j = \varphi[i_\xi(t_j, t)] = \{(t, y) : t \in T_\xi,\ 0 \leq y \leq i_\xi(t_j, t)\}. \tag{3.5.1}$$

Note that for an arbitrary pair of samples $\xi(t_j)$ and $\xi(t_k)$ of stochastic process $\xi(t)$ with IDDs $i_\xi(t_j, t)$ and $i_\xi(t_k, t)$, respectively (see Fig. 3.5.1), the metric identity holds:

$$d_{jk} = \frac{1}{2}\int_{T_\xi} |i_\xi(t_j, t) - i_\xi(t_k, t)|dt = \frac{1}{2}m[X_j \Delta X_k], \tag{3.5.2}$$

3.5 Informational Characteristics and Properties of Stochastic Processes

where $d_{jk} = \frac{1}{2}\int_{T_\xi} |i_\xi(t_j,t) - i_\xi(t_k,t)|dt$ is the metric between IDD $i_\xi(t_j,t)$ of the sample $\xi(t_j)$ and IDD $i_\xi(t_k,t)$ of the sample $\xi(t_k)$; $X_j = \varphi[i_\xi(t_j,t)]$, $X_k = \varphi[i_\xi(t_k,t)]$; $m[X_j\Delta X_k]$ is a measure of symmetric difference of the sets X_j and X_k.

The mapping (3.5.1) transfers IDDs $i_\xi(t_j,t)$, $i_\xi(t_k,t)$ into corresponding equivalent sets X_j, X_k and defines isometric mapping of set (I,d) of IDDs $\{i_\xi(t_\alpha,t)\}$ with metric d_{jk} (3.5.2) into the metric space (X,d^*) of the sets $\{X_\alpha\}$ with metric $d_{jk}^* = \frac{1}{2}m[X_j\Delta X_k]$.

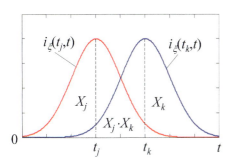

FIGURE 3.5.1 IDDs $i_\xi(t_j,t)$, $i_\xi(t_k,t)$ of samples $\xi(t_j)$, $\xi(t_k)$ of stochastic process $\xi(t)$

The normalization property 2 of IDD (see Section 3.4, page 102) provides the normalization of a measure $m(X_j)$ of an arbitrary sample $\xi(t_j)$ of stochastic process $\xi(t)$: $m(X_j) = \int_{T_\xi} i_\xi(t_j,t)dt = 1$.

The mapping (3.5.1) allows describing any stochastic process $\xi(t)$ possessing the ability to carry information as a collection of the samples $\{\xi(t_\alpha)\}$ and also as a collection of the sets $\{X_\alpha\}$: $X = \bigcup_\alpha X_\alpha$. Note that a measure of a single element X_α possesses the normalization property: $m(X_\alpha) = 1$.

Thus, any stochastic process (signal) may be considered a system of the elements $\{X_\alpha\}$ of metric space (X,d^*) with metric d_{jk}^* between a pair of elements X_j and X_k: $d_{jk}^* = \frac{1}{2}m[X_j\Delta X_k]$.

The mapping (3.5.1) allows considering an arbitrary stochastic process $\xi(t)$ as a collection of statistically dependent samples $\{\xi(t_\alpha)\}$ and also as subalgebra $\mathbf{B}(X)$ of generalized Boolean algebra $\mathbf{B}(\Omega)$ with a measure m and signature $(+,\cdot,-,\mathbf{O})$ of the type $(2,2,2,0)$ with the following operations over the elements $\{X_\alpha\} \subset X$, $X \in \Omega$:

1. addition $X_{\alpha 1} + X_{\alpha 2}$
2. multiplication $X_{\alpha 1} \cdot X_{\alpha 2}$
3. difference $X_{\alpha 1} - X_{\alpha 2}$
4. symmetric difference $X_{\alpha 1} \Delta X_{\alpha 2}$
5. null element $\mathbf{O}$: $X_\alpha \Delta X_\alpha = \mathbf{O}$; $X_\alpha - X_\alpha = \mathbf{O}$

Informational characteristics of stochastic process $\xi(t)$, considered a subalgebra $\mathbf{B}(X)$ of generalized Boolean algebra $\mathbf{B}(\Omega)$ with measure m, are introduced by the following axiom that is similar to Axiom 2.3.1.

Axiom 3.5.1. *Axiom of a measure of binary operation. Measure $m(X_\alpha)$ of the element X_α, $X_\alpha = X_\beta \circ X_\gamma$; $X_\alpha, X_\beta, X_\gamma \in X$, considered a result of binary operation "$\circ$" of subalgebra $\mathbf{B}(X)$ of generalized Boolean algebra $\mathbf{B}(\Omega)$ with a measure m, defines information quantity $I(X_\alpha) = m(X_\alpha)$ that corresponds to the result of this operation.*

Based on the axiom, the measure $m(X_\alpha)$ of the element X_α of the space (X, d^*) defines a quantitative aspect of information it contains whereas binary operation "∘" of generalized Boolean algebra $\mathbf{B}(\Omega)$ defines a qualitative aspect of information. Within the framework of the formulated Axiom 3.5.1 depending on the kinds of relations between the elements $\{\xi(t_\alpha)\}$ of stochastic process $\xi(t)$, we shall distinguish the following types of information quantities that according to Axiom 3.5.1, are defined by the corresponding binary operation of general Boolean algebra $\mathbf{B}(\Omega)$ (see Definitions 2.3.1 through 2.3.5).

Definition 3.5.1. *Quantity of overall information I_{jk}^+ contained in an arbitrary pair of samples $\xi(t_j)$ and $\xi(t_k)$ of stochastic process $\xi(t)$ is an information quantity equal to a measure of sum $m(X_j + X_k)$ of the elements X_j and X_k:*

$$I_{jk}^+ = m(X_j + X_k) = m(X_j) + m(X_k) - m(X_j \cdot X_k) = 2 - \psi_\xi(t_j, t_k),$$

where $X_j = \varphi[i_\xi(t_j, t)]$; $X_k = \varphi[i_\xi(t_k, t)]$; $m(X_j \cdot X_k)$ is a measure of product of the elements X_j and X_k; $\psi_\xi(t_j, t_k)$ is a normalized function of statistical interrelationship (NFSI) between the samples $\xi(t_j)$ and $\xi(t_k)$ of stochastic process $\xi(t)$.

Definition 3.5.2. *Quantity of mutual information I_{jk} contained simultaneously in both the sample $\xi(t_j)$ and the sample $\xi(t_k)$ of stochastic process $\xi(t)$ is an information quantity equal to a measure of product $m(X_j \cdot X_k)$ of the elements X_j and X_k:*

$$I_{jk} = m(X_j \cdot X_k) = \psi_\xi(t_j, t_k).$$

Definition 3.5.3. *Quantity of absolute information I_j contained in an arbitrary sample $\xi(t_j)$ of stochastic process $\xi(t)$ is a quantity of mutual information I_{jj} equal to 1:*

$$I_j = I_{jj} = \psi_\xi(t_j, t_j) = m(X_j \cdot X_j) = m(X_j) = 1.$$

Thus, a measure $m(X_j)$ of the element X_j defines an information quantity contained in the sample $\xi(t_j)$ concerning itself.

Definition 3.5.4. *Quantity of particular relative information I_{jk}^- contained in the sample $\xi(t_j)$ with respect to the sample $\xi(t_k)$ is an information quantity equal to a measure of difference $m(X_j - X_k)$ between the elements X_j and X_k:*

$$I_{jk}^- = m(X_j - X_k) = m(X_j) - m(X_j \cdot X_k) = 1 - \psi_\xi(t_j, t_k).$$

Definition 3.5.5. *Quantity of relative information I_{jk}^Δ contained in the sample $\xi(t_j)$ with respect to the sample $\xi(t_k)$ and, vice versa, contained in the sample $\xi(t_k)$ with respect to the sample $\xi(t_j)$, is an information quantity equal to a measure of symmetric difference $m(X_j \Delta X_k)$ between the elements X_j and X_k:*

$$I_{jk}^\Delta = m(X_j \Delta X_k) = m(X_j - X_k) + m(X_k - X_j) = I_{jk}^- + I_{kj}^- = 2(1 - \psi_\xi(t_j, t_k)).$$

We should consider a unit of information quantity (see Definition 2.3.11). Mathematical apparatus designed to construct the stated fundamentals of signal processing theory and information theory is generalized Boolean algebra with a normalized measure. So, as an information unit, we take the quantity of absolute information $I[\xi(t_\alpha)]$ contained in a single element $\xi(t_\alpha)$ of stochastic process $\xi(t)$ with a domain of definition T_ξ (in the form of a finite set or a continuum) and information distribution density (IDD) $i_\xi(t_\alpha, t)$, and according to the property of IDD normalization, the result is equal to:

$$I[\xi(t_\alpha)] = m(X_\alpha)|_{t_\alpha \in T_\xi} = \int_{T_\xi} i_\xi(t_\alpha; t)dt = 1 \text{ abit.} \qquad (3.5.3)$$

Definition 3.5.6. Quantity of absolute information contained in a single element (sample) $\xi(t_\alpha)$ of stochastic process $\xi(t)$, considered a subalgebra $\mathbf{B}(X)$ of generalized Boolean algebra $\mathbf{B}(\Omega)$ with a measure m, is called an <u>absolute unit</u>, or abit.

On the base of the mapping (3.5.1) and the approach that allows considering an arbitrary stochastic process $\xi(t)$ as a collection of statistically dependent samples $\{\xi(t_\alpha)\}$ and as a subalgebra $\mathbf{B}(X)$ of generalized Boolean algebra $\mathbf{B}(\Omega)$ with a measure m and the signature $(+, \cdot, -, \mathbf{O})$ of the type $(2, 2, 2, 0)$, one can determine two forms of structural diversity of the stochastic process $\xi(t)/X = \bigcup_\alpha X_\alpha$ and, correspondingly, two measures of collection of the elements $\{\xi(t_\alpha)\}$ of the signal $\xi(t)$: *overall quantity of information* and *relative quantity of information*, which are introduced by the following definitions.

Definition 3.5.7. *Overall quantity of information* $I[\xi(t)]$ contained in a stochastic process $\xi(t)$ considered a collection of the elements $\{X_\alpha\}$ is an information quantity equal to a measure of their sum:

$$I[\xi(t)] = m(\bigcup_\alpha X_\alpha) = m(\sum_\alpha X_\alpha). \qquad (3.5.4)$$

It is obvious that a given measure of structural diversity is identical to the quantity of absolute information I_X contained in a stochastic process $\xi(t)$: $I[\xi(t)] = I_X$.

Definition 3.5.8. *Relative quantity of information* $I_\Delta[\xi(t)]$ contained in a stochastic process $\xi(t)$ owing to distinctions between the elements of the collection $\{X_\alpha\}$ is an information quantity equal to a measure of symmetric difference of these elements:

$$I_\Delta[\xi(t)] = m(\underset{\alpha}{\Delta} X_\alpha), \qquad (3.5.5)$$

where $\underset{\alpha}{\Delta} X_\alpha = \ldots \Delta X_\alpha \Delta X_\beta \Delta \ldots$ is the symmetric difference of the elements of the collection $\{X_\alpha\}$ of stochastic process $\xi(t)/X$.

Overall quantity of information $I[\xi(t)]$ contained in stochastic process $\xi(t)$ with IDD $i_\xi(t_\alpha, t)$ and strictly distributed in its domain of definition $T_\xi = [t_0, t_0 + T]$, according to the Definition 3.5.7, is equal to:

$$I[\xi(t)] = m(\sum_\alpha X_\alpha)|_{t_\alpha \in T_\xi} = \int_{T_\xi} \sup_{t_\alpha \in T_\xi} [i_\xi(t_\alpha, t)]dt \quad \text{(abit).} \qquad (3.5.6)$$

Overall quantity of information $I[\xi(t)]$ contained in stationary stochastic process $\xi(t)$ with IDD $i_\xi(\tau)$ in its domain of definition $T_\xi = [t_0, t_0 + T]$ is determined by the identity:

$$I_\xi(T_\xi) = i_\xi(0) \cdot T \quad \text{(abit)}, \qquad (3.5.7)$$

where $i_\xi(0)$ is a maximum value of IDD $i_\xi(\tau)$ of stochastic process $\xi(t)$.

Information quantity $\Delta I(t_j, t \in [a,b])$ contained in a sample $\xi(t_j)$ concerning the instantaneous values of stochastic process $\xi(t)$ within time interval $[a,b] \subset T_\xi$ is equal to:

$$\Delta I(t_j, t \in [a,b]) = \int_a^b i_\xi(t_j, t) dt. \qquad (3.5.8)$$

Thus, IDD $i_\xi(t_j, t)$ of the sample $\xi(t_j)$ of stochastic process $\xi(t)$ may be defined as the limit of the ratio of information quantity $\Delta I(t_j, t \in [t_0, t_0 + \Delta t])$, contained in the sample $\xi(t_j)$ concerning the instantaneous values of stochastic process $\xi(t)$ within the interval $[t_0, t_0 + \Delta t]$, to the value of this interval Δt, while the last one tends to zero:

$$i_\xi(t_j, t) = \lim_{\Delta t \to 0} \frac{\Delta I(t_j, t \in [t_0, t_0 + \Delta t])}{\Delta t}.$$

The last relationship implies that IDD $i_\xi(t_\alpha, t)$ is a dimensional function, and its dimensionality is inversely proportional to the dimensionality of the time parameter.

Thus, one can draw the following conclusion: on the one hand, IDD $i_\xi(t_\alpha, t)$ of stochastic process $\xi(t)$ characterizes a distribution of information, contained in a single sample $\xi(t_\alpha)$ concerning all the instantaneous values of stochastic process $\xi(t)$. On the other hand, IDD $i_\xi(t_\alpha, t)$ characterizes a distribution of information along the samples $\{\xi(t_\alpha)\}$ of stochastic process $\xi(t)$ concerning the sample $\xi(t_\alpha)$.

IDD is inherent to any sample of stochastic process owing to a sample is an element of a collection of statistically interconnected other samples $\{\xi(t_\alpha)\}$ forming as the stochastic process $\xi(t)$ and cannot be associated with an arbitrary random variable as opposed to the sample $\xi(t_\alpha)$ of stochastic process $\xi(t)$. So, a single random variable considered outside the statistical collection of random variables does not carry any information.

We can generalize the properties of quantity of mutual information I_{jk} of a pair of samples $\xi(t_j)$ and $\xi(t_k)$ of stochastic process $\xi(t)$:

1. Quantity of mutual information I_{jk} is nonnegative and bounded:

$$0 \leq I_{jk} \leq I_{jj} = 1.$$

2. Quantity of mutual information I_{jk} is symmetric: $I_{jk} = I_{kj}$.

3. Quantity of mutual information I_{jk} is identical to normalized function of statistical interrelationship (NFSI) $\psi_\xi(t_j, t_k)$ between the samples $\xi(t_j)$ and $\xi(t_k)$ of stochastic process $\xi(t)$:

$$I_{jk} = \psi_\xi(t_j, t_k).$$

3.5 Informational Characteristics and Properties of Stochastic Processes

The last identity allows us to define NFSI $\psi_\xi(t_j, t_k)$, introduced earlier by the relationship (3.1.2), based on quantity of mutual information I_{jk} of a pair of samples $\xi(t_j)$ and $\xi(t_k)$ of stochastic process $\xi(t)$. It is obvious that the first three properties of NFSI are similar to the properties of quantity of mutual information on the list above, and a list of the general properties of NFSI $\psi_\xi(t_j, t_k)$ of stochastic processes possessing the ability to carry information looks like:

1. NFSI $\psi_\xi(t_j, t_k)$ is nonnegative and is bounded:
$$0 \le \psi_\xi(t_j, t_k) \le \psi_\xi(t_j, t_j) = 1.$$

2. NFSI $\psi_\xi(t_j, t_k)$ is symmetric: $\psi_\xi(t_j, t_k) = \psi_\xi(t_k, t_j)$.

3. NFSI $\psi_\xi(t_j, t_k)$ is not differentiable in the point (t_j, t_j), and its derivatives in this point on the left $\psi'_\xi(t_j, t_j - 0)$ and on the right $\psi'_\xi(t_j, t_j + 0)$ are different in sign and the same in module:
$$\psi'_\xi(t_j, t_j - 0) = -\psi'_\xi(t_j, t_j + 0).$$

4. Modules of derivatives of NFSI in the point (t_j, t_j) on the left and on the right are not equal to zero:
$$|\psi'_\xi(t_j, t_j - 0)| = |\psi'_\xi(t_j, t_j + 0)| \ne 0.$$

Normalized functions of statistical interrelation (NFSIs) $\psi_\xi(\tau)$ of stationary stochastic processes possess the following properties:

1. $0 \le \psi_\xi(\tau) \le \psi_\xi(0)$ (nonnegativity and boundedness).
2. $\psi_\xi(\tau) = \psi_\xi(-\tau)$ (evenness).
3. $\lim_{\tau \to \infty} \psi_\xi(\tau) = 0$.
4. NFSI $\psi_\xi(\tau)$ is not a differentiable function in the point $\tau = 0$, and its derivatives in this point on the left $\psi'_\xi(\tau - 0)$ and on the right $\psi'_\xi(\tau + 0)$ are different in sign and the same in module:
$$\psi'_\xi(\tau - 0) = -\psi'_\xi(\tau + 0);$$

5. Modules of derivatives of NFSI $\tau = 0$ on the left and on the right are not equal to zero:
$$|\psi'_\xi(\tau - 0)| = |\psi'_\xi(\tau + 0)| \ne 0.$$

6. Fourier transform $\mathcal{F}$ of NFSI $\psi_\xi(\tau)$ of stationary stochastic process $\xi(t)$ is a nonnegative function:
$$\mathcal{F}[\psi_\xi(\tau)] = \frac{1}{2\pi} \int_{-\infty}^{\infty} \psi_\xi(\tau) \exp[-i\omega\tau] d\tau \ge 0.$$

Taking into account the aforementioned considerations, the corollary list of the Theorem 3.1.1 may be continued.

Corollary 3.5.1. *Under bijective mapping (3.1.17), the measures of all sorts of information contained in a pair of samples $\xi(t_j)$ and $\xi(t_k)$ of stochastic process $\xi(t)$ are preserved:*
Quantity of mutual information I_{jk}:

$$I_{jk} = m(X_j \cdot X_k) = \psi_\xi(t_j, t_k) = \psi_\eta(t'_j, t'_k) = m(Y_j \cdot Y_k); \quad (3.5.9)$$

Quantity of absolute information I_j:

$$I_j = m(X_j) = \psi_\xi(t_j, t_j) = \psi_\eta(t'_j, t'_j) = m(Y_j) = 1; \quad (3.5.10)$$

Quantity of overall information I_{jk}^+:

$$I_{jk}^+ = m(X_j + X_k) = 2 - \psi_\xi(t_j, t_k) = 2 - \psi_\eta(t'_j, t'_k) = m(Y_j + Y_k); \quad (3.5.11)$$

Quantity of particular relative information I_{jk}^-:

$$I_{jk}^- = m(X_j - X_k) = 1 - \psi_\xi(t_j, t_k) = 1 - \psi_\eta(t'_j, t'_k) = m(Y_j - Y_k); \quad (3.5.12)$$

Quantity of relative information I_{jk}^Δ:

$$I_{jk}^\Delta = m(X_j \Delta X_k) = 2(1 - \psi_\xi(t_j, t_k)) = 2(1 - \psi_\eta(t'_j, t'_k)) = m(Y_j \Delta Y_k), \quad (3.5.13)$$

where $X_j = \varphi[i_\xi(t_j, t)]$; $X_k = \varphi[i_\xi(t_k, t)]$; $Y_j = \varphi[i_\eta(t'_j, t')]$; $Y_k = \varphi[i_\eta(t'_k, t')]$ are the ordinate sets of IDDs $i_\xi(t_\alpha, t)$, $i_\eta(t_\beta, t)$ of stochastic processes $\xi(t)$ and $\eta(t)$, respectively, defined by the formula (3.5.1).

Corollary 3.5.2. *Under bijective mapping of stochastic processes $\xi(t)$ (3.1.17), the differential of information quantity $i_\xi(t_j, t)dt$ is preserved:*

$$i_\xi(t_j, t)dt = i_\eta(t'_j, t')dt'. \quad (3.5.14)$$

Corollary 3.5.3. *Under bijective mapping (3.1.17), overall quantity of information $I[\xi(t)]$ contained in stochastic process $\xi(t)$ considered a collection of the elements $X = \bigcup_\alpha X_\alpha$ where every element X_α corresponds to IDD $i_\xi(t_\alpha, t)$ of the sample $\xi(t_\alpha)$ is preserved:*

$$I[\xi(t)] = m(X) = m[\bigcup_\alpha X_\alpha] = m[\bigcup_\beta Y_\beta] = m(Y) = I[\eta(t')], \quad (3.5.15)$$

where X is the result of mapping (3.5.1) of stochastic process $\xi(t)$ into the set of ordinate sets $\{X_\alpha\}$: $X = \bigcup_\alpha X_\alpha$.

3.5 Informational Characteristics and Properties of Stochastic Processes

Corollary 3.5.4. *Under bijective mapping (3.1.17), relative quantity of information* $I_\Delta[\xi(t)]$ *contained in stochastic process* $\xi(t)$ *considered a collection of elements* $X = \bigcup_\alpha X_\alpha$, *where every element* X_α *corresponds to IDD* $i_\xi(t_\alpha, t)$ *of the sample* $\xi(t_\alpha)$, *is preserved:*

$$I_\Delta[\xi(t)] = m(\underset{\alpha}{\Delta} X_\alpha) = m(\underset{\beta}{\Delta} Y_\beta) = I_\Delta[\eta(t')], \tag{3.5.16}$$

where X is the result of mapping (3.5.1) of stochastic process $\xi(t)$ into the set of ordinate sets $\{X_\alpha\}$: $X = \bigcup_\alpha X_\alpha$.

We now introduce the characterization of informational properties of two stochastic processes $\xi(t)$ and $\eta(t')$.

Each pair of samples $\xi(t_j)$, $\eta(t'_k)$ of stochastic processes $\xi(t)$ and $\eta(t')$ may be associated with two functions $i_{\xi\eta}(t'_k, t)$ and $i_{\eta\xi}(t_j, t')$ that we shall call *mutual information distribution density* (mutual IDD).

Mutual IDD $i_{\xi\eta}(t'_k, t)$ characterizes the distribution of information along the time parameter t (along the samples $\{\xi(t_\alpha)\}$ of stochastic process $\xi(t)$) concerning the sample $\eta(t'_k)$ of stochastic process $\eta(t')$ and is connected with mutual NFSI $\psi_{\xi\eta}(t_j, t'_k)$ (see Definition 3.1.2) by the relationship:

$$\psi_{\xi\eta}(t_j, t'_k) = 1 - d[i_{\xi\eta}(t'_k, t); i_\xi(t_j, t)]; \tag{3.5.17}$$

$$d[i_{\xi\eta}(t'_k, t); i_\xi(t_j, t)] = \frac{1}{2} \int_{-\infty}^{\infty} |i_{\xi\eta}(t'_k, t) - i_\xi(t_j, t)| dt, \tag{3.5.17a}$$

where $d[i_{\xi\eta}(t'_k, t); i_\xi(t_j, t)]$ is the metric between mutual IDD $i_{\xi\eta}(t'_k, t)$ of the sample $\eta(t'_k)$ and IDD $i_\xi(t_j, t)$ of the sample $\xi(t_j)$.

Definition 3.5.9. *The function* $i_{\xi\eta}(t'_k, t)$ *connected with mutual NFSI* $\psi_{\xi\eta}(t_j, t'_k)$ *and IDD* $i_\xi(t_j, t)$ *by the relationship (3.5.17) is called mutual information distribution density* (mutual IDD).

The function $d[i_{\xi\eta}(t'_k, t); i_\xi(t_j, t)]$ may be considered a metric between the samples $\xi(t_j)$, $\eta(t'_k)$ of stochastic processes $\xi(t)$ and $\eta(t')$, respectively. According to (3.5.17), the functions $i_{\xi\eta}(t'_k, t)$, $i_\xi(t_j, t)$ and $\psi_{\xi\eta}(t_j, t'_k)$, $d[i_{\xi\eta}(t'_k, t); i_\xi(t_j, t)]$ are connected by the identity:

$$\psi_{\xi\eta}(t_j, t'_k) + d[i_{\xi\eta}(t'_k, t); i_\xi(t_j, t)] = 1. \tag{3.5.18}$$

Defining the quantities $\psi_{\xi\eta}$ and $\rho_{\xi\eta}$ by:

$$\psi_{\xi\eta} = \sup_{t_j, t'_k \in]-\infty, \infty[} \psi_{\xi\eta}(t_j, t'_k); \tag{3.5.19a}$$

$$\rho_{\xi\eta} = \inf_{t_j, t'_k \in]-\infty, \infty[} d[i_{\xi\eta}(t'_k, t); i_\xi(t_j, t)], \tag{3.5.19b}$$

we obtain a closeness measure and a distance measure for stochastic processes $\xi(t)$ and $\eta(t')$, respectively.

The quantity $\psi_{\xi\eta}$ determined by the formulas (3.1.9a) and (3.5.19a) we shall call *coefficient of statistical interrelation*, and the quantity $\rho_{\xi\eta}$ determined by the formula (3.5.1b), we shall call *metric between stochastic processes* $\xi(t)$ and $\eta(t')$. As against the metric $d_{\xi\eta}$ defined by the formula (3.1.9b), the metric $\rho_{\xi\eta}$ (3.5.19b) is based on a distance between mutual IDD and IDD of the samples of stochastic processes $\xi(t)$ and $\eta(t')$.

Coefficient of statistical interrelation $\psi_{\xi\eta}$ of stochastic processes $\xi(t)$, $\eta(t')$ and the metric $\rho_{\xi\eta}$ between them are connected by a relationship similar to the identity (3.1.8):

$$\psi_{\xi\eta} + \rho_{\xi\eta} = 1. \tag{3.5.20}$$

Mutual IDD $i_{\xi\eta}(t'_k, t)$ along with IDD $i_\eta(t'_k, t')$ are characteristics of a single sample $\eta(t'_k)$ of the sample collection $\{\eta(t'_\beta)\}$ of stochastic process $\eta(t')$, but unlike IDD, the sample $\eta(t'_k)$ is considered within its relation with the sample collection $\{\xi(t_\alpha)\}$ of another stochastic process $\xi(t)$.

Mutual IDD $i_{\eta\xi}(t_j, t')$ characterizes the distribution of information along the time parameter t' (along the samples $\{\eta(t'_\beta)\}$ of stochastic process $\eta(t')$ concerning the sample $\xi(t_j)$ of stochastic process $\xi(t)$) and is connected with mutual NFSI $\psi_{\eta\xi}(t'_k, t_j)$ by the equation:

$$\psi_{\eta\xi}(t'_k, t_j) = 1 - d[i_{\eta\xi}(t_j, t'); i_\eta(t'_k, t')], \tag{3.5.21}$$

where $d[i_{\eta\xi}(t_j, t'); i_\eta(t'_k, t')] = \frac{1}{2} \int\limits_{-\infty}^{\infty} |i_{\eta\xi}(t_j, t') - i_\eta(t'_k, t')| dt'$ is the metric between mutual IDD $i_{\eta\xi}(t_j, t')$ of the sample $\xi(t_j)$ and IDD $i_\eta(t'_k, t')$ of the sample $\eta(t'_k)$.

Mutual IDD $i_{\eta\xi}(t_j, t')$, also as IDD $i_\xi(t_j, t)$, is a characteristic of a single sample $\xi(t_j)$ of the sample collection $\{\xi(t_\alpha)\}$ of stochastic process $\xi(t)$, but, in this case, unlike IDD, the sample $\xi(t_j)$ is considered within its relation with the sample collection $\{\eta(t'_\beta)\}$ of another stochastic process $\eta(t')$.

The values of the functions $\psi_{\xi\eta}(t_j, t'_k)$ and $\psi_{\eta\xi}(t'_k, t_j)$ determined by the formulas (3.5.17) and (3.5.21) characterize the quantity of mutual information I_{jk} ($I_{jk} = I_{kj}$) contained in both the sample $\xi(t_j)$ and the sample $\eta(t'_k)$ of stochastic processes $\xi(t)$ and $\eta(t')$:

$$\psi_{\xi\eta}(t_j, t'_k) = \psi_{\eta\xi}(t'_k, t_j) = I_{jk}. \tag{3.5.22}$$

The relations between IDDs $i_\xi(t_j, t)$, $i_\eta(t'_k, t')$ and mutual IDDs $\psi_{\xi\eta}(t_j, t'_k)$, $\psi_{\eta\xi}(t'_k, t_j)$ of the samples $\xi(t_j)$, $\eta(t'_k)$, respectively, are defined by the consequence from the identity (3.5.22):

$$\int\limits_{-\infty}^{\infty} |i_{\xi\eta}(t'_k, t) - i_\xi(t_j, t)| dt = \int\limits_{-\infty}^{\infty} |i_{\eta\xi}(t_j, t') - i_\eta(t'_k, t')| dt'. \tag{3.5.23}$$

Consider the main properties of mutual IDD $i_{\xi\eta}(t'_k, t)$ of a pair of stochastic processes $\xi(t)$ and $\eta(t')$.

1. Mutual IDD is nonnegative: $i_{\xi\eta}(t'_k, t) \geq 0$.

2. Mutual IDD is normalized: $\int\limits_{-\infty}^{\infty} i_{\xi\eta}(t'_k, t)dt = 1$.

3. In general case, mutual IDD is not even: $i_{\xi\eta}(t'_k, t) \neq i_{\xi\eta}(t'_k, -t)$.

4. Information quantity $\Delta I(t'_k, t \in [a, b])$ contained in the sample $\eta(t'_k)$ concerning the instantaneous values of stochastic process $\xi(t)$ within the time interval $[a, b]$ is equal to $\Delta I(t'_k, t \in [a, b]) = \int\limits_{a}^{b} i_{\xi\eta}(t'_k, t)dt$.

5. Metric relationships between IDDs $i_\xi(t_j, t)$, $i_\eta(t'_k, t')$ and mutual IDDs $i_{\eta\xi}(t_j, t')$, $i_{\xi\eta}(t'_k, t)$ of the samples $\xi(t_j)$, $\eta(t'_k)$ of stochastic processes $\xi(t)$ and $\eta(t')$, respectively, are determined by the formula (3.5.23).

6. For a pair of stochastic processes $\xi(t)$ and $\eta(t')$, each sample $\xi(t_j)$ of stochastic process $\xi(t)$ may be associated with the sample $\eta(t'_k)$ of stochastic process $\eta(t')$, which is such that metric $d[i_{\xi\eta}(t'_k, t); i_\xi(t_j, t)]$ between mutual IDD $i_{\xi\eta}(t'_k, t)$ of the sample $\eta(t'_k)$ and IDD $i_\xi(t_j, t)$ of the sample $\xi(t_j)$ is minimal:

$$d[i_{\xi\eta}(t'_k, t); i_\xi(t_j, t)] = \frac{1}{2} \int\limits_{-\infty}^{\infty} |i_{\xi\eta}(t'_k, t) - i_\xi(t_j, t)|dt \to \min. \quad (3.5.24)$$

7. For stationary coupled stochastic processes $\xi(t)$ and $\eta(t')$, the metric (3.5.24) between mutual IDD $i_{\xi\eta}(t'_k, t)$ of the sample $\eta(t'_k)$ and IDD $i_\xi(t_j, t)$ of the sample $\xi(t_j)$ depends only on the time difference $\tau = t'_k - t_j$ between the samples $\eta(t'_k)$ and $\xi(t_j)$ of stochastic processes $\eta(t')$ and $\xi(t)$:

$$d[i_{\xi\eta}(t'_k, t); i_\xi(t_j, t)] = \frac{1}{2} \int\limits_{-\infty}^{\infty} |i_{\xi\eta}(t_j + \tau, t) - i_\xi(t_j, t)|dt =$$
$$= d[i_{\xi\eta}(t_j + \tau, t); i_\xi(t_j, t)] = d[\tau].$$

(a) For stationary coupled stochastic processes $\xi(t)$ and $\eta(t')$, the mutual NFSI $\psi_{\xi\eta}(t_j, t'_k)$ depends only on the time difference $\tau = t'_k - t_j$ between the samples $\eta(t'_k)$ and $\xi(t_j)$ of stochastic processes $\eta(t')$ and $\xi(t)$ respectively:

$$\psi_{\xi\eta}(t_j, t'_k) = \psi_{\xi\eta}(\tau) = 1 - d[\tau].$$

8. Let $\xi(t)$ and $\eta(t')$ be functionally interconnected stochastic processes:

$$\eta(t'_k) = f[\xi(t)]; \quad \xi(t) = f^{-1}[\eta(t'_k)], \quad (3.5.25)$$

where $f[*]$ is a one-to-one function.

Then for each sample $\eta(t'_k)$ of stochastic process $\eta(t')$, there exists such a sample $\xi(t_j)$ of stochastic process $\xi(t)$, that mutual IDD $i_{\xi\eta}(t'_k, t)$ of the sample $\eta(t'_k)$ and IDD $i_\xi(t_j, t)$ of the sample $\xi(t_j)$ are identically equal:

$$i_{\xi\eta}(t'_k, t) \equiv i_\xi(t_j, t).$$

(a) Let $\xi(t)$ and $\eta(t')$ are functionally interconnected stochastic processes (3.5.25). Then for each sample $\eta(t'_k)$ of stochastic process $\eta(t')$, there exists such a sample $\xi(t_j)$ of stochastic process $\xi(t)$, that the metric (3.5.24) between mutual IDD $i_{\xi\eta}(t'_k,t)$ of the sample $\eta(t'_k)$ and IDD $i_\xi(t_j,t)$ of the sample $\xi(t_j)$ is equal to zero:

$$d[i_{\xi\eta}(t'_k,t); i_\xi(t_j,t)] = 0.$$

The eight properties of mutual IDD directly imply general properties of mutual NFSI.

1. Mutual NFSI $\psi_{\xi\eta}(t_j,t'_k)$ is nonnegative and bounded: $0 \leq \psi_{\xi\eta}(t_j,t'_k) \leq 1$.
2. Mutual NFSI $\psi_{\xi\eta}(t_j,t'_k)$ is symmetric:

$$\psi_{\xi\eta}(t_j,t'_k) = \psi_{\eta\xi}(t'_k,t_j).$$

3. The equality $\psi_{\xi\eta}(t_j,t'_k) = 1$ holds if and only if stochastic processes $\xi(t)$, $\eta(t')$ are connected by one-to-one correspondence (3.5.25).
4. The equality $\psi_{\xi\eta}(t_j,t'_k) = 0$ holds if and only if stochastic processes $\xi(t)$, $\eta(t')$ are statistically independent.

Let each IDD $i_\xi(t_j,t)$, $i_\eta(t'_k,t')$ and also each mutual IDD $i_{\eta\xi}(t_j,t')$, $i_{\xi\eta}(t'_k,t)$ of an arbitrary pair of samples $\xi(t_j)$, $\eta(t'_k)$ of statistically dependent (in general case) stochastic processes $\xi(t)$, $\eta(t')$ be associated with their *ordinate sets* X_j, Y_k, X_j^Y, Y_k^X (see Fig. 3.5.2) by the mapping φ (3.5.1):

$$X_j = \varphi[i_\xi(t_j,t)] = \{(t,y) : t \in T_\xi;\ 0 \leq y \leq i_\xi(t_j,t)\}; \quad (3.5.26a)$$

$$Y_k = \varphi[i_\eta(t'_k,t')] = \{(t',z) : t' \in T_\eta;\ 0 \leq z \leq i_\eta(t'_k,t')\}; \quad (3.5.26b)$$

$$X_j^Y = \varphi[i_{\eta\xi}(t_j,t')] = \{(t',z) : t \in T_\xi,\ t' \in T_\eta;\ 0 \leq z \leq i_{\eta\xi}(t_j,t')\}; \quad (3.5.26c)$$

$$Y_k^X = \varphi[i_{\xi\eta}(t'_k,t)] = \{(t,y) : t \in T_\xi,\ t' \in T_\eta;\ 0 \leq y \leq i_{\xi\eta}(t'_k,t)\}, \quad (3.5.26d)$$

where $T_\xi = [t_0, t_0 + T_x]$, $T_\eta = [t'_0, t'_0 + T_y]$ are the domains of definitions of stochastic processes $\xi(t)$, $\eta(t')$, respectively.

For an arbitrary pair of the samples $\xi(t_j)$, $\eta(t'_k)$ of statistically dependent stochastic processes $\xi(t)$, $\eta(t')$ with IDDs $i_\xi(t_j,t)$, $i_\eta(t'_k,t')$ and mutual IDDs $i_{\eta\xi}(t_j,t')$, $i_{\xi\eta}(t'_k,t)$, the metric identities hold:

$$d_{jk}^{\xi\eta} = d[i_\xi(t_j,t); i_{\xi\eta}(t'_k,t)] =$$

$$= \frac{1}{2}\int_{T_\xi} |i_\xi(t_j,t) - i_{\xi\eta}(t'_k,t)|dt = \frac{1}{2}m(X_j \Delta Y_k^X); \quad (3.5.27a)$$

$$d_{kj}^{\eta\xi} = d[i_\eta(t'_k,t'); i_{\eta\xi}(t_j,t')] =$$

$$= \frac{1}{2}\int_{T_\eta} |i_\eta(t'_k,t') - i_{\eta\xi}(t_j,t')|dt' = \frac{1}{2}m(X_j^Y \Delta Y_k), \quad (3.5.27b)$$

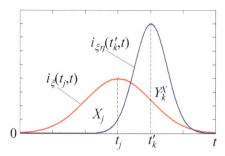

FIGURE 3.5.2 IDD $i_\xi(t_j, t)$ and mutual IDD $i_{\xi\eta}(t_k^0, t)$ of samples $\xi(t_j)$, $\eta(t_k')$ of stochastic processes $\xi(t)$, $\eta(t')$

where $d_{jk}^{\xi\eta}/d_{kj}^{\eta\xi}$ are the metrics between mutual IDD of the sample $\eta(t_k')/\xi(t_j)$ and IDD of the sample $\xi(t_j)/\eta(t_k')$, respectively.

According to the equality (3.5.23), the metric identity holds:
$$\frac{1}{2}m(X_j \Delta Y_k^X) = \frac{1}{2}m(X_j^Y \Delta Y_k).$$

Besides, for an arbitrary pair of the samples $\xi(t_j)/\eta(t_k')$ and $\xi(t_l)/\eta(t_m')$ of stochastic process $\xi(t)/\eta(t')$, the metric identities hold (see relationship (3.5.2)):

$$d_{jl}^\xi = d[i_\xi(t_j, t); i_\xi(t_l, t)] =$$
$$= \frac{1}{2} \int_{T_\xi} |i_\xi(t_j, t) - i_\xi(t_l, t)| dt = \frac{1}{2} m(X_j \Delta X_l); \quad (3.5.28a)$$

$$d_{km}^\eta = d[i_\eta(t_k', t'); i_\eta(t_m', t')] =$$
$$= \frac{1}{2} \int_{T_\eta} |i_\eta(t_k', t') - i_\eta(t_m', t')| dt' = \frac{1}{2} m(Y_k \Delta Y_m). \quad (3.5.28b)$$

The mappings (3.5.26) transform IDDs $i_\xi(t_j, t)$, $i_\eta(t_k', t')$ and mutual IDDs $i_{\eta\xi}(t_j, t')$, $i_{\xi\eta}(t_k', t)$ into the corresponding equivalent sets X_j, Y_k, X_j^Y, Y_k^X, defining isometric mapping of the function space $(I, d_{jk}^{\xi\eta}, d_{jl}^\xi, d_{km}^\eta)$ of stochastic processes $\xi(t)$, $\eta(t')$ with metrics $d_{jk}^{\xi\eta}$ (3.5.27) and d_{jl}^ξ, d_{km}^η (3.5.28) into the set space $(\Omega, d_{jk}^{\xi\eta}, d_{jl}^\xi, d_{km}^\eta)$ with metrics $\frac{1}{2}m(X_j \Delta Y_k^X)$, $\frac{1}{2}m(X_j \Delta X_l)$, $\frac{1}{2}m(Y_k \Delta Y_m)$.

The normalization property of IDD and mutual IDD of a pair of samples $\xi(t_j)$, $\eta(t_k')$ of stochastic processes $\xi(t)$, $\eta(t')$ provides the normalization of the measures $m(X_j)$, $m(Y_k)$, $m(X_j^Y)$, $m(Y_k^X)$:

$$m(X_j) = \int_{T_\xi} i_\xi(t_j, t) dt = 1, \qquad m(X_j^Y) = \int_{T_\eta} i_{\eta\xi}(t_j, t') dt' = 1;$$

$$m(Y_k) = \int_{T_\eta} i_\eta(t_k', t') dt' = 1, \qquad m(Y_k^X) = \int_{T_\xi} i_{\xi\eta}(t_k', t) dt = 1.$$

The mappings (3.5.26) allow us to consider stochastic processes (signals) $\xi(t)$, $\eta(t')$ with IDDs $i_\xi(t_j, t)$, $i_\eta(t'_k, t')$ and mutual IDDs $i_{\eta\xi}(t_j, t')$, $i_{\xi\eta}(t'_k, t)$ as the collections of statistically dependent samples $\{\xi(t_\alpha)\}$, $\{\eta(t'_\beta)\}$, and also as the collections of sets $X = \{X_\alpha\}$, $Y = \{Y_\beta\}$:

$$\varphi : \xi(t) \to X = \bigcup_\alpha X_\alpha; \quad \eta(t') \to Y = \bigcup_\beta Y_\beta.$$

Besides, the sets $X \subset \Omega$, $Y \subset \Omega$ form the subalgebra $\mathbf{B}(X + Y)$ of generalized Boolean algebra $\mathbf{B}(\Omega)$ with a measure m and signature $(+, \cdot, -, \mathbf{O})$ of the type $(2, 2, 2, 0)$, with operations of addition, multiplication, subtraction (relative complement operation), and null element $\mathbf{O}$ [223], [175], [176].

On the base of mappings (3.5.26) and considering an arbitrary pair of stochastic processes $\xi(t)$, $\eta(t')$ a collection of statistically dependent samples $\{\xi(t_\alpha)\}$, $\{\eta(t'_\beta)\}$ and also a subalgebra $\mathbf{B}(X+Y)$ of generalized Boolean algebra $\mathbf{B}(\Omega)$ with a measure m and signature $(+, \cdot, -, \mathbf{O})$ of the type $(2, 2, 2, 0)$, one may define a measure of a collection of the elements of this pair of signals $\xi(t)$, $\eta(t')$ as a *quantity of mutual information* introduced by the following definition.

Definition 3.5.10. *Quantity of mutual information* $I[\xi(t), \eta(t')] = I_{XY}$ *contained in an arbitrary pair of signals* $\xi(t)$ *and* $\eta(t')$ *(sets* X *and* Y*, respectively) is an information quantity contained in a set that is the product of the elements* X *and* Y *of generalized Boolean algebra* $\mathbf{B}(\Omega)$ *with a measure* m:

$$I[\xi(t), \eta(t')] = I_{XY} = m(X \cdot Y) = m[\sum_\alpha (X_\alpha^Y \cdot \sum_\beta Y_\beta)] = m[\sum_\beta (Y_\beta^X \cdot \sum_\alpha X_\alpha)],$$

where X and Y are the results of mappings (3.5.1) of stochastic processes $\xi(t)$ and $\eta(t')$ into the sets of the ordinate sets $\{X_\alpha\}$, $\{Y_\beta\}$: $X = \bigcup_\alpha X_\alpha$, $Y = \bigcup_\beta Y_\beta$; X_α, Y_β^X are the results of mappings (3.5.26) of IDD $i_\xi(t_\alpha, t)$ and mutual IDD $i_{\xi\eta}(t'_\beta, t)$ of the samples $\xi(t_\alpha)$ and $\eta(t'_\beta)$ of stochastic processes $\xi(t)$ and $\eta(t')$.

The properties 1 through 4 of mutual NFSI (see page 118) of stochastic processes $\xi(t)$ and $\eta(t')$ may be supplemented with the following theorem.

Theorem 3.5.1. *Let* f *be a bijective mapping of stochastic processes* $\xi(t)$ *and* $\eta(t')$*, defined in the intervals* $T_\xi = [t_{0\xi}, t_\xi]$*,* $T_\eta = [t_{0\eta}, t_\eta]$*, into stochastic processes* $\alpha(t'')$ *and* $\beta(t''')$ *defined in the intervals* $T_\alpha = [t_{0\alpha}, t_\alpha]$*,* $T_\beta = [t_{0\beta}, t_\beta]$*, respectively:*

$$\alpha(t'') = f[\xi(t)], \quad \xi(t) = f^{-1}[\alpha(t'')]; \quad (3.5.29a)$$

$$\beta(t''') = f[\eta(t')], \quad \eta(t') = f^{-1}[\beta(t''')]. \quad (3.5.29b)$$

Then the function f *that maps the stochastic processes* $\xi(t)$ *and* $\eta(t')$ *into stochastic processes* $\alpha(t'')$ *and* $\beta(t''')$*, respectively, preserves mutual NFSI:*

$$\psi_{\xi\eta}(t_j, t'_k) = \psi_{\alpha\beta}(t''_j, t'''_k),$$

where $\psi_{\xi\eta}(t_j, t'_k)$, $\psi_{\alpha\beta}(t''_j, t'''_k)$ are mutual NFSIs of the pairs of stochastic processes $\xi(t)$, $\eta(t')$ and $\alpha(t'')$, $\beta(t''')$, respectively.

3.5 Informational Characteristics and Properties of Stochastic Processes

The proof of the Theorem 3.5.1 is similar to the proof of the Theorem 3.1.1.

Corollary 3.5.5. *Under bijective mappings (3.5.29), the measures of all sorts of information contained in a pair of the samples $\xi(t_j)$, $\eta(t'_k)$ of stochastic processes $\xi(t)$ and $\eta(t')$ are preserved:*
Quantity of mutual information I_{jk}:

$$I_{jk} = m(X_j \cdot Y_k^X) = \psi_{\xi\eta}(t_j, t'_k) = \psi_{\alpha\beta}(t''_j, t'''_k) = m(A_j \cdot B_k^A); \quad (3.5.30)$$

Quantity of overall information I_{jk}^+:

$$I_{jk}^+ = m(X_j + Y_k^X) = 2 - \psi_{\xi\eta}(t_j, t'_k) = 2 - \psi_{\alpha\beta}(t''_j, t'''_k) = m(A_j + B_k^A); \quad (3.5.31)$$

Quantity of particular relative information I_{jk}^-:

$$I_{jk}^- = m(X_j - Y_k^X) = 1 - \psi_{\xi\eta}(t_j, t'_k) = 1 - \psi_{\alpha\beta}(t''_j, t'''_k) = m(A_j - B_k^A); \quad (3.5.32)$$

Quantity of relative information I_{jk}^Δ:

$$I_{jk}^\Delta = m(X_j \Delta Y_k^X) = 2(1-\psi_{\xi\eta}(t_j, t'_k)) = 2(1-\psi_{\alpha\beta}(t''_j, t'''_k)) = m(A_j \Delta B_k^A), \quad (3.5.33)$$

where $X_j = \varphi[i_\xi(t_j, t)]$; $Y_k^X = \varphi[i_{\xi\eta}(t'_k, t)]$; $A_j = \varphi[i_\alpha(t''_j, t'')]$; $B_k^A = \varphi[i_{\alpha\beta}(t'''_k, t''')]$ are ordinate sets of IDDs $i_\xi(t_j, t)$, $i_{\xi\eta}(t'_k, t')$ of stochastic processes $\xi(t)$ and $\eta(t')$ and also IDDs $i_\alpha(t''_j, t'')$, $i_{\alpha\beta}(t'''_k, t''')$ of stochastic processes $\alpha(t'')$ and $\beta(t''')$ respectively, defined by the relationships (3.5.26).

Corollary 3.5.6. *Under bijective mappings (3.5.29), the quantity of mutual information $I[\xi(t), \eta(t')]$ contained in a pair of stochastic processes $\xi(t)$ and $\eta(t')$, is preserved:*

$$I[\xi(t), \eta(t')] = m(X, Y) = m[\sum_\beta (Y_\beta^X \cdot \sum_\alpha X_\alpha)] =$$
$$= m[\sum_\delta (B_\delta^A \cdot \sum_\gamma A_\gamma)] = m(A, B) = I[\alpha(t''), \beta(t''')], \quad (3.5.34)$$

where X_α, Y_β^X are the results of mappings (3.5.26) of IDD $i_\xi(t_\alpha, t)$ and mutual IDD $i_{\xi\eta}(t'_\beta, t)$ of the samples $\xi(t_\alpha)$ and $\eta(t'_\beta)$ of stochastic processes $\xi(t)$ and $\eta(t')$; X, Y are the results of mappings (3.5.1) of stochastic processes $\xi(t)$ and $\eta(t')$ into the sets of ordinate sets $\{X_\alpha\}$, $\{Y_\beta\}$: $X = \bigcup_\alpha X_\alpha$, $Y = \bigcup_\beta Y_\beta$; A_γ, B_δ^A are the results of mappings (3.5.26) of IDD $i_\alpha(t''_\gamma, t'')$ and mutual IDD $i_{\alpha\beta}(t'''_\delta, t'')$ of the samples $\alpha(t''_\gamma)$ and $\beta(t'''_\delta)$ of stochastic processes $\alpha(t'')$ and $\beta(t''')$; A, B are the results of mappings (3.5.1) of stochastic processes $\alpha(t'')$ and $\beta(t''')$ into the sets of ordinate sets $\{A_\gamma\}$, $\{B_\delta\}$: $A = \bigcup_\gamma A_\gamma$, $B = \bigcup_\delta B_\delta$.

Summarizing the aforementioned considerations, the mapping φ defined by the relationships (3.5.26) realizes morphism of *physical signal space* $\boldsymbol{\Gamma} = \{\xi(t), \eta(t'), \ldots\}$ into *informational signal space* $\boldsymbol{\Omega} = \{X, Y, \ldots\}$:

$$\varphi: \xi(t) \to X = \bigcup_\alpha X_\alpha; \quad \eta(t') \to Y = \bigcup_\beta Y_\beta; \quad \boldsymbol{\Gamma} \to \boldsymbol{\Omega}, \qquad (3.5.35)$$

whose properties are the subject of Chapter 4.

Invariants of bijective mappings of stochastic processes in the form of NFSI, mutual NFSI, IDD and mutual IDD, introduced on the base of Definitions 3.1.1, 3.1.2, 3.4.1, and 3.5.9, respectively, with the help of the mappings (3.5.26), allow us to use generalized Boolean algebra with a measure to describe physical and informational signal interactions in physical and informational signal spaces, respectively. Measure m on generalized Boolean algebra $\mathbf{B}(\Omega)$, on the one hand, is a measure of information quantity transferred by the signals forming the informational signal space $\boldsymbol{\Omega}$, and on the other hand, induces a metric in this space.

Invariants of a group of mappings of stochastic processes in the form of IDD and mutual IDD will be used to investigate characteristics and properties of signal space built upon generalized Boolean algebra with a measure, and also to establish the main quantitative regularities characterizing the signal functional transformations with special properties used in signal processing.

On the basis of invariants of a group of mappings of stochastic processes in the form of IDD, mutual IDD, NFSI and PMSI, the informational and metric relationships between the signals, which interact in signal spaces with the distinct algebraic properties, will be established.

NFSI and IDD will be used to evaluate the quantities of information transferred by discrete and continuous signals and determine noiseless communication channel capacity.

NMSI will be used to investigate the potential quality indices of signal processing in signal spaces with distinct algebraic properties.

A pair of invariants, such as NFSI and HSD connected by Fourier transform, may be used for analysis of nonlinear transformations of stochastic processes.

Invariants of group of mappings of stochastic processes in the form of mutual NFSI, NMSI and PMSI may be used while synthesizing algorithms and units of multichannel signal processing under prior uncertainty conditions in the presence of interference (noise).

4
Signal Spaces with Lattice Properties

The principal ideas and results of Chapters 2 and 3 taking into account the requirements to both a signal space concept and a measure of information quantity formulated in Section 1.3, allow using the apparatus of generalized Boolean algebra to construct the models of spaces where informational and real physical signal interactions occur. On the other hand, these ideas and results permit us to establish the main regularities and relationships of signal interactions in such spaces.

However, one should clearly distinguish these two models of signal spaces and their main geometrical and informational properties. Within the framework of the signal space model, it is necessary to obtain results that define quantitative informational relationships between the signals based on their homomorphic and isomorphic mappings. It is interesting to investigate the quantitative informational relationships between the signals and the results of their interactions in a signal space with properties of additive commutative group (in linear space) and also in signal spaces with other algebraic properties.

4.1 Physical and Informational Signal Spaces

The notions of information distribution density (IDD) and mutual IDD, introduced in Sections 3.4 and 3.5, provide an entire informational description of stochastic process (signal). Such a description is complete in the sense that it characterizes a stochastic process (a signal) within its interrelation with other stochastic processes (signals) from *physical signal space* $\Gamma = \{\xi(t), \eta(t), \ldots\}$. The main informational relationships between an arbitrary pair of signals $\xi(t)$ and $\eta(t)$ from the space Γ are determined by IDD and mutual IDD of their samples $\{\xi(t_\alpha)\}$ and $\{\eta(t_\beta)\}$, respectively.

Definition 4.1.1. The set Γ of stochastic signals $\{\xi(t), \eta(t), \ldots\}$ interacting with each other forms *the physical signal space* that is a commutative semigroup $\mathbf{SG} = (\Gamma, \oplus)$ with respect to a binary operation $\oplus$ with the following properties:

$$\xi(t) \oplus \eta(t) \in \Gamma \quad \text{(closure)};$$

$$\xi(t) \oplus \eta(t) = \eta(t) \oplus \xi(t) \quad \text{(commutativity)};$$

$$\xi(t) \oplus (\eta(t) \oplus \zeta(t)) = (\xi(t) \oplus \eta(t)) \oplus \zeta(t) \quad \text{(associativity)}.$$

Definition 4.1.2. Binary operation $\oplus$: $\chi(t) = \xi(t) \oplus \eta(t)$ between the elements $\xi(t)$ and $\eta(t)$ of semigroup $\mathbf{SG} = (\mathbf{\Gamma}, \oplus)$ is called *physical interaction* of the signals $\xi(t)$ and $\eta(t)$ in physical signal space $\mathbf{\Gamma}$.

The mappings (3.5.26) allow us to consider the stochastic processes (signals) $\xi(t)$ and $\eta(t')$ with IDDs $i_\xi(t_j, t)$ and $i_\eta(t'_k, t')$ and mutual IDDs $i_{\eta\xi}(t_j, t')$ and $i_{\xi\eta}(t'_k, t)$ as the collections of statistically dependent samples $\{\xi(t_\alpha)\}$, $\{\eta(t'_\beta)\}$ and also as the collections of sets $X = \{X_\alpha\}$, $Y = \{Y_\beta\}$:

$$\varphi: \xi(t) \to X = \bigcup_\alpha X_\alpha; \quad \eta(t') \to Y = \bigcup_\beta Y_\beta. \tag{4.1.1}$$

The sets $X \subset \Omega$, $Y \subset \Omega$ form the subalgebra $\mathbf{B}(X+Y)$ of generalized Boolean algebra $\mathbf{B}(\Omega)$ with a measure m and signature $(+, \cdot, -, \mathbf{O})$ of the type $(2,2,2,0)$ with the operations of addition, multiplication, subtraction (relative complement operation), and null element $\mathbf{O}$ [175], [176], [223].

The mapping φ (4.1.1) realizes a morphism of *physical signal space* $\mathbf{\Gamma} = \{\xi(t), \eta(t'), \ldots\}$ into *informational signal space* $\mathbf{\Omega} = \{X, Y, \ldots\}$:

$$\varphi: \mathbf{\Gamma} \to \mathbf{\Omega}. \tag{4.1.2}$$

The mapping φ (4.1.1) describes the stochastic processes (signals) from general algebraic positions on the basis of the notion of information carrier space, whose content is elucidated in Chapter 2.

In order to fix the interrelation (4.1.2) between physical and informational signal spaces while denoting the signals $\xi(t)$ and $\eta(t)$ and their images X and Y, connected by the relationship (4.1.1), the notations $\xi(t)/X$ and $\eta(t)/Y$ will be used, respectively.

Generally, in physical signal space $\mathbf{\Gamma} = \{\xi(t), \eta(t), \ldots\}$, the instantaneous values $\{\xi(t_\alpha)\}$ and $\{\eta(t_\beta)\}$ of the signals $\xi(t)$ and $\eta(t)$ interact in the same coordinate dimension according to the Definition 4.1.2: $\xi(t_\alpha) \oplus \eta(t_\alpha)$. Since the interrelations between informational characteristics of the signals are considered in physical and informational signal spaces, to provide the commonality of approach, we suppose that the interaction of instantaneous values $\{\xi(t_\alpha)\}$, $\{\eta(t'_\beta)\}$ of the signals $\xi(t), \eta(t')$ in physical signal space, in a general case, may be realized in the distinct reference systems $\xi(t_\alpha) \oplus \eta(t'_\alpha)$.

We define the notion of *informational signal space* $\mathbf{\Omega}$, using the analogy with the notion of information carrier space (see Section 2.1), on the basis of apparatus of Boolean algebra $\mathbf{B}(\Omega)$ with a measure m.

Definition 4.1.3. *Informational signal space* $\mathbf{\Omega}$ is a set of the elements $\{X, Y, \ldots\}$: $\{X, Y, \ldots\} \subset \mathbf{\Omega}$: $\{\varphi: \xi(t) \to X, \eta(t') \to Y, \ldots\}$ characterized by the following general properties.

1. The space $\mathbf{\Omega}$ is generalized Boolean algebra $\mathbf{B}(\Omega)$ with a measure m and signature $(+, \cdot, -, \mathbf{O})$ of the type $(2,2,2,0)$.

4.1 Physical and Informational Signal Spaces

2. In the space $\mathbf{\Gamma}/\mathbf{\Omega}$, the signal $\xi(t)/X$ with IDD $i_\xi(t_\alpha, t)$ and NFSI $\psi_\xi(t_\alpha, t_\beta)$ forms a collection of the elements $\{\xi(t_\alpha)\}/\{X_\alpha\}$ with normalized measure:

$$m(X_\alpha) = m(X_\alpha \cdot X_\alpha) = \psi_\xi(t_\alpha, t_\alpha) = 1.$$

3. In the space $\mathbf{\Gamma}/\mathbf{\Omega}$, the signal $\xi(t)/X$ with IDD $i_\xi(t_\alpha, t)$ possesses the property of continuity if, for an arbitrary pair of the elements $\xi(t_\alpha), \xi(t_\beta)/X_\alpha, X_\beta$, there exists such an element $\xi(t_\gamma)/X_\gamma$, that: $t_\alpha < t_\gamma < t_\beta$ ($\alpha < \gamma < \beta$). We call this a signal with continuous structure. The signals that do not possess this property are called the signals with discrete structure.

4. In the space $\mathbf{\Omega}$, the following operations are defined over the elements $\{X_\alpha\} \subset X$, $\{Y_\beta\} \subset Y$, $X \in \mathbf{\Omega}$, $Y \in \mathbf{\Omega}$ that characterize *informational interrelations* between them:

   (a) Addition — $X + Y = \sum_\alpha X_\alpha + \sum_\beta Y_\beta$;
   (b) Multiplication — $X_{\alpha 1} \cdot X_{\alpha 2}$, $Y_{\beta 1} \cdot Y_{\beta 2}$, $X_\alpha \cdot Y_\beta$, $X \cdot Y = (\sum_\alpha X_\alpha) \cdot (\sum_\beta Y_\beta)$;
   (c) Difference — $X_{\alpha 1} - X_{\alpha 2}$, $Y_{\beta 1} - Y_{\beta 2}$, $X_\alpha - Y_\beta$, $X - Y = X - (X \cdot Y)$;
   (d) Symmetric difference — $X_{\alpha 1} \Delta X_{\alpha 2}$, $Y_{\beta 1} \Delta Y_{\beta 2}$, $X_\alpha \Delta Y_\beta$, $X \Delta Y = (X - Y) + (Y - X)$;
   (e) Null element $\mathbf{O}$: $X \Delta X = \mathbf{O}$; $X - X = \mathbf{O}$.

5. A measure m introduces a metric upon generalized Boolean algebra $\mathbf{B}(\mathbf{\Omega})$ defining the distance $\rho(X, Y)$ between the signals $\xi(t)/X$ and $\eta(t)/Y$ (between the sets $X, Y \in \mathbf{\Omega}$) by the relationship:

$$\rho(X, Y) = m(X \Delta Y) = m(X) + m(Y) - 2m(XY), \qquad (4.1.3)$$

where $X \Delta Y$ is a symmetric difference of the sets X and Y.

Similarly, we define the distance $\rho(X_\alpha, X_\beta)$ between the single elements X_α and X_β of the signal $\xi(t)/X$: $\rho(X_\alpha, X_\beta) = m(X_\alpha \Delta X_\beta)$ and the distance $\rho(X_\alpha, Y_\beta)$ between the single elements X_α and Y_β of the signals $\xi(t)/X$ and $\eta(t')/Y$, respectively:

$$\rho(X_\alpha, Y_\beta) = m(X_\alpha \Delta Y_\beta^X) = m(X_\alpha^Y \Delta Y_\beta).$$

Thus, the signal space $\mathbf{\Omega}$ is *metric space*.

6. Isomorphism Iso of signal space $\mathbf{\Omega}$ into the signal space $\mathbf{\Omega}'$ — Iso: $\mathbf{\Omega} \stackrel{Iso}{\rightarrow} \mathbf{\Omega}'$, preserving a measure m: $m(X) = m(X')$, $m(Y) = m(Y')$, $X, Y \subset \mathbf{\Omega}$, $X', Y' \subset \mathbf{\Omega}'$, also preserves metric induced by this measure:

$$\rho(X, Y) = m(X \Delta Y) = m(X' \Delta Y') = \rho(X', Y').$$

Thus, the spaces $\mathbf{\Omega}$ and $\mathbf{\Omega}'$ are isometric.

Informational properties of signal space Ω are introduced by Axiom 4.1.1 that is analogous to Axiom 3.5.1 concerning the signal $\xi(t)/X$ considered a subalgebra $\mathbf{B}(X)$ of generalized Boolean algebra $\mathbf{B}(\Omega)$ with a measure m.

Axiom 4.1.1. *Axiom of a measure of binary operation. Measure $m(Z)$ of the element $Z\colon Z = X \circ Y;\ X, Y, Z \in \Omega$ considered as the result of binary operation $\circ$ of generalized Boolean algebra $\mathbf{B}(\Omega)$ with a measure m defines the information quantity $I_Z = m(Z)$ that corresponds to the result of this operation.*

The axiom implies that a measure m of the element Z of the space Ω defines a quantitative aspect of information contained in this element, while a binary operation "$\circ$" of generalized Boolean algebra $\mathbf{B}(\Omega)$ defines a qualitative aspect of this information.

Within the framework of Axiom 4.1.1, depending on relations between the signals $\xi(t)/X = \bigcup_{\alpha} X_\alpha$ and $\eta(t')/Y = \bigcup_{\beta} Y_\beta$, we shall distinguish the following types of information quantities. The same information quantities are denoted in a twofold manner, for instance, as in Definitions 3.5.1 through 3.5.5, with respect to the signals $\xi(t)$ and $\eta(t')$ directly, and on the other hand, with respect to the sets $X = \bigcup_{\alpha} X_\alpha$, $Y = \bigcup_{\beta} Y_\beta$ associated with the signals by the mapping (4.1.1).

Definition 4.1.4. *Quantity of absolute information $I[\xi(t)] = I_X$ (or $I[\eta(t')] = I_Y$) is an information quantity contained in the signal $\xi(t)$ (or $\eta(t')$) considered as a collection of the elements $\{X_\alpha\}$ (or $\{Y_\beta\}$) in the consequence of its structural diversity:*

$$I[\xi(t)] = I_X = m(X) = m(\sum_{\alpha} X_\alpha);$$

$$I[\eta(t')] = I_Y = m(Y) = m(\sum_{\beta} Y_\beta),$$

where $m(\sum_{\alpha} X_\alpha)$ and $m(\sum_{\beta} Y_\beta)$ are measures of sums of the elements $\{X_\alpha\}$ and $\{Y_\beta\}$ of the signals $\xi(t)$ and $\eta(t')$, respectively.

Quantity of absolute information with respect to its content is identical to overall quantity of information (see Definition 3.5.7) and is calculated by the formula (3.5.6).

Definition 4.1.5. *Quantity of overall information $I^+[\xi(t), \eta(t')] = I_{X+Y}$ contained in an arbitrary pair of the signals $\xi(t)$ and $\eta(t')$ (in the sets X and Y, respectively) is an information quantity contained in the set that is the sum of the elements X and Y of generalized Boolean algebra $\mathbf{B}(\Omega)$:*

$$I^+[\xi(t), \eta(t')] = I_{X+Y} = m(X + Y) = m(X) + m(Y) - m(XY).$$

Definition 4.1.6. *Quantity of mutual information $I[\xi(t), \eta(t')] = I_{XY}$ contained in an arbitrary pair of signals $\xi(t)$ and $\eta(t')$ (in the sets X and Y, respectively) is* an information quantity contained in the set that is the product of the elements X and Y of generalized Boolean algebra $\mathbf{B}(\Omega)$:

$$I[\xi(t),\eta(t')] = I_{XY} = m(X \cdot Y) = m[\sum_\alpha (X_\alpha^Y \cdot \sum_\beta Y_\beta)] = m[\sum_\beta (Y_\beta^X \cdot \sum_\alpha X_\alpha)],$$

where X and Y are the results of the mapping (4.1.1) of stochastic processes (signals) $\xi(t)$ and $\eta(t')$ into the sets of ordinate sets $\{X_\alpha\}$ and $\{Y_\beta\}$: $X = \bigcup_\alpha X_\alpha$, $Y = \bigcup_\beta Y_\beta$; X_α and Y_β^X are the results of mappings (3.5.26) of IDD $i_\xi(t_\alpha,t)$ and mutual IDD $i_{\xi\eta}(t'_\beta,t)$ of the samples $\xi(t_\alpha)$ and $\eta(t'_\beta)$ of stochastic processes (signals) $\xi(t)$ and $\eta(t')$, respectively.

Remark 4.1.1. Quantity of absolute information I_X contained in the signal $\xi(t)$ may be interpreted as the quantity of overall information $I_{X+X} = m(X+X)$, or as the quantity of mutual information $I_{XX} = m(X \cdot X)$ contained in the signal $\xi(t)$ with respect to itself.

Definition 4.1.7. *Quantity of particular relative information* $I^-[\xi(t),\eta(t')] = I_{X-Y}$ *contained in the signal* $\xi(t)/X$ *with respect to the signal* $\eta(t')/Y$ *(or vice versa, contained in the signal* $\eta(t')/Y$ *with respect to the signal* $\xi(t)/X$, *i.e.,* $I^-[\eta(t'),\xi(t)] = I_{Y-X}$), *is an information quantity contained in the difference between the elements* X *and* Y *of generalized Boolean algebra* $\mathbf{B}(\Omega)$:

$$I^-[\xi(t),\eta(t')] = I_{X-Y} = m(X-Y) = m(X) - m(XY);$$
$$I^-[\eta(t'),\xi(t)] = I_{Y-X} = m(Y-X) = m(Y) - m(XY).$$

Definition 4.1.8. *Quantity of relative information* $I^\Delta[\xi(t),\eta(t')] = I_{X\Delta Y}$ *contained in the signal* $\xi(t)/X$ *with respect to the signal* $\eta(t')/Y$ *is an information quantity contained in the symmetric difference between the elements* X *and* Y *of generalized Boolean algebra* $\mathbf{B}(\Omega)$:

$$I^\Delta[\xi(t),\eta(t')] = I_{X\Delta Y} = m(X\Delta Y) = m(X-Y) + m(Y-X) = I_{X-Y} + I_{Y-X}.$$

Quantity of relative information $I_{X\Delta Y}$ is identically equal to an introduced metric:

$$I_{X\Delta Y} = \rho(X,Y).$$

Regarding the units of information quantity, as a unit of information quantity in signal space Ω, we take the quantity of absolute information $I[\xi(t_\alpha)]$ contained in a single element $\xi(t_\alpha)$ of the signal $\xi(t)$ with a domain of definition T_ξ (in the form of a discrete set or continuum) and information distribution density (IDD) $i_\xi(t_\alpha,t)$, and according to the relationship (3.5.3), it is equal to:

$$I[\xi(t_\alpha)] = m(X_\alpha)|_{t_\alpha \in T_\xi} = \int_{T_\xi} i_\xi(t_\alpha;t)dt = 1\mathrm{abit}. \qquad (4.1.4)$$

So, the unit of information quantity in signal space Ω is introduced by definition that is similar to the Definition 3.5.6.

Definition 4.1.9. In signal space Ω, *a unit of information quantity* is the quantity of absolute information contained in a single element (the sample) $\xi(t_\alpha)$ of stochastic process (signal) $\xi(t)$ considered a subalgebra $\mathbf{B}(X)$ of generalized Boolean algebra $\mathbf{B}(\Omega)$ with a measure m, and it is called *absolute unit* or abit (*absolute unit*).

The metric signal space Ω is an informational space and allows us to give the following geometrical interpretation of the main introduced notions.

In signal space Ω, all the elements of the signal $\xi(t)/X = \bigcup_\alpha X_\alpha$ are situated on the surface of some n-dimensional sphere $Sp(\mathbf{O}, R)$, whose center is null element $\mathbf{O}$ of the space Ω, and its radius R is equal to 1:

$$R = m(X_\alpha \Delta \mathbf{O}) = m(X_\alpha) = 1.$$

The distance from null element $\mathbf{O}$ of the space Ω to an arbitrary signal $\xi(t)/X$ is equal to a measure of this signal $m(X)$ or a quantity of absolute information I_X contained in this signal $\xi(t)$:

$$\rho(X \Delta \mathbf{O}) = m(X \Delta \mathbf{O}) = m(X) = I_X.$$

It is obvious that quantity of absolute information I_X of the signal $\xi(t)/X$ is equivalent to the quantity of relative information $I_{X\Delta\mathbf{O}}$ contained in the signal $\xi(t)/X$ with respect to null element $\mathbf{O}$:

$$I_X = m(X) = m(X \Delta \mathbf{O}) = I_{X\Delta\mathbf{O}}.$$

Measure $m(X)$ of the signal $\xi(t)/X$, or a quantity of absolute information I_X contained in this signal, may be interpreted as a length of the signal $\xi(t)$ in metric space Ω. Quantity of relative information $I_{X\Delta Y}$ contained in the signal $\xi(t)$ with respect to the signal $\eta(t')$, has a sense of a distance between the signals $\xi(t)$ and $\eta(t')$ (between the sets X and Y) in signal space Ω:

$$I_{X\Delta Y} = m(X\Delta Y) = \rho(X, Y).$$

In informational signal space Ω, informational properties of the signals are characterized by the following main relationships.

Group of the identities for an arbitrary pair of signals $\xi(t)/X$ and $\eta(t')/Y$:

$$I_{X+Y} = I_X + I_Y - I_{XY};$$

$$I_{X+Y} = I_{X\Delta Y} + I_{XY};$$

$$I_{X\Delta Y} = I_X + I_Y - 2I_{XY};$$

Group of the inequalities:

$$I_X + I_Y \geq I_{X+Y} \geq I_{X\Delta Y};$$

$$I_X + I_Y \geq I_{X+Y} \geq \max[I_X, I_Y];$$

$$\max[I_X, I_Y] \geq \sqrt{I_X I_Y} \geq \min[I_X, I_Y] \geq I_{XY}.$$

Informational relationships for an arbitrary triplet of the signals $\xi(t)/X$, $\eta(t')/Y$, $\zeta(t'')/Z$ and null elements $\mathbf{O}$ of signal space Ω, that are equivalent to metric relationships of tetrahedron (see the relationships described in Subsection 2.2.1 on page 32), hold:

$$I_{X\Delta Y} = I_X + I_Y - 2I_{XY};$$

$$I_{Y\Delta Z} = I_Y + I_Z - 2I_{YZ};$$

$$I_{Z\Delta X} = I_Z + I_X - 2I_{ZX};$$

$$I_{X\Delta Y} = I_{X\Delta Z} + I_{Z\Delta Y} - 2I_{(X\Delta Z)(Z\Delta Y)}.$$

To conclude discussion of the main relationships characterizing an informational interaction between the signals in the space Ω, we begin considering the main relationships within a single signal of the space Ω.

It is obvious that all the relationships between the elements of the signal space hold with respect to arbitrary elements (samples) $\{X_\alpha\}$ of the signal $\xi(t)/X = \bigcup_\alpha X_\alpha$. For all the elements $\{X_\alpha\}$ of the signal $\xi(t)/X$ with normalized measure $m(X_\alpha) = 1$, it is convenient to introduce a metric between the elements X_α and X_β that is equivalent to the metric (4.1.3):

$$d(X_\alpha, X_\beta) = \frac{1}{2} m(X_\alpha \Delta X_\beta) = \frac{1}{2}[m(X_\alpha) + m(X_\beta) - 2m(X_\alpha \cdot X_\beta)] = 1 - m(X_\alpha \cdot X_\beta);$$

$$d(X_\alpha, X_\beta) = \frac{1}{2}\rho(X_\alpha, X_\beta). \tag{4.1.5}$$

Signal $\xi(t)/X = \bigcup_j X_j$ with discrete structure $\{X_j\}$ (with discrete time domain) in metric signal space Ω is represented by the vertices $\{X_j\}$ of polygonal line $l(\{X_j\})$ that lie upon a sphere $Sp(\mathbf{O}, R)$ and at the same time are the vertices of n-dimensional simplex $Sx(X)$ inscribed into the sphere $Sp(\mathbf{O}, R)$. The length (perimeter) $P(\{X_j\})$ of the closed polygonal line $l(\{X_j\})$ is determined by the expression:

$$P(\{X_j\}) = \sum_{j=0}^{n} d(X_j, X_{j+1}) = \frac{1}{2} \sum_{j=0}^{n} \rho(X_j, X_{j+1}),$$

where $d(X_j, X_k)$ and $\rho(X_j, X_k)$ are the metrics determined by the relationships (4.1.5) and (4.1.3), respectively. Here, the values of indices are denoted modulo $n+1$, i.e., $X_{n+1} \equiv X_0$.

The signal $\xi(t)/X = \bigcup_\alpha X_\alpha$ with continuous structure $\{X_\alpha\}$ (with continuous time domain) in metric space Ω may be represented by a continuous closed line $l(\{X_\alpha\})$ situated upon a sphere $Sp(\mathbf{O}, R)$, which at the same time is a fragment of n-dimensional simplex $Sx(X)$ inscribed into this sphere $Sp(\mathbf{O}, R)$, and in series connects the vertices $\{X_j\}$ of a given simplex.

On the basis of the introduced notions one can distinguish two forms of structural diversity of the signal $\xi(t)/X = \bigcup_\alpha X_\alpha$, respectively, two measures of a collection of the elements of the signal $\xi(t)$, i.e., *overall quantity of information* and *relative quantity of information* introduced by the following definitions.

Definition 4.1.10. *Overall quantity of information $I[\xi(t)]$ contained in the signal $\xi(t)/X$ considered a collection of the elements $\{X_\alpha\}$ is an information quantity equal to the measure of their sum:*

$$I[\xi(t)] = m(X) = m(\bigcup_\alpha X_\alpha) = m(\sum_\alpha X_\alpha). \tag{4.1.6}$$

Evidently, this measure of structural diversity is identical to the quantity of absolute information I_X contained in the signal $\xi(t)/X$: $I[\xi(t)] = I_X$.

Definition 4.1.11. *Relative quantity of information $I_\Delta[\xi(t)]$ contained in the signal $\xi(t)$ owing to distinctions between the elements of a collection $\{X_\alpha\}$ is an information quantity equal to the measure of symmetric difference between these elements:*

$$I_\Delta[\xi(t)] = m(\underset{\alpha}{\Delta} X_\alpha), \tag{4.1.7}$$

where $\underset{\alpha}{\Delta} X_\alpha = \ldots \Delta X_\alpha \Delta X_\beta \ldots$ is symmetric difference of the elements of a collection $\{X_\alpha\}$ of the signal $\xi(t)/X$.

The presence of two measures of structural diversity of the signal $\xi(t)/X$ is stipulated by Stone's duality between generalized Boolean algebra $\mathbf{B}(\Omega)$ with signature $(+, \cdot, -, \mathbf{O})$ of the type $(2,2,2,0)$ and generalized Boolean ring $\mathbf{BR}(\Omega)$ with signature $(\Delta, \cdot, \mathbf{O})$ of the type $(2,2,0)$, that are isomorphic between each other [175], [176], [223]:

$$\mathbf{B}(\Omega) \overset{Iso}{\leftrightarrow} \mathbf{BR}(\Omega).$$

The overall quantity of information $I[\xi(t)]$ has a sense of an entire information quantity contained in the signal $\xi(t)/X$ considered a collection of the elements with an ordered structure. Relative quantity of information $I_\Delta[\xi(t)]$ has a sense of maximal information quantity that can be extracted from the signal $\xi(t)$ by the proper processing.

Consider the main relationships characterizing the introduced measures of a collection of the elements $m(\sum_j X_j)$ and $m(\underset{j}{\Delta} X_j)$ for the signal $\xi(t)/X$ with discrete structure $\{X_j\}$:

$$m(\sum_j X_j) = \sum_j m(X_j) - \sum_{0 \leq j < k \leq n} m(X_j X_k) +$$

$$+ \sum_{0 \leq j < k < l \leq n} m(X_j X_k X_l) - \ldots + (-1)^n m(\prod_{j=0}^n X_j); \tag{4.1.8}$$

4.1 Physical and Informational Signal Spaces

$$m(\underset{j}{\Delta} X_j) = m(\sum_j X_j) - m(\sum_{0 \le j < k \le n} X_j X_k) +$$

$$+ m(\sum_{0 \le j < k < l \le n} X_j X_k X_l) - \ldots + (-1)^n m(\prod_{j=0}^n X_j). \quad (4.1.9)$$

The identities (4.1.8) and (4.1.9) imply the double inequality:

$$m(\underset{j}{\Delta} X_j) \le m(\sum_j X_j) \le \sum_j m(X_j), \quad (4.1.10)$$

which implies the double informational inequality:

$$I_\Delta[\xi(t)] \le I[\xi(t)] \le \sum_j m(X_j). \quad (4.1.11)$$

If the signal $\xi(t)/X$ considered as a collection $\{X_j\}$ consists of the disjoint elements $(X_j \cdot X_k = \mathbf{O})$, then the inequality (4.1.11) transforms to the identity:

$$I_\Delta[\xi(t)] = I[\xi(t)] = \sum_j m(X_j). \quad (4.1.12)$$

For the signal $\xi(t)/X$ with discrete structure $\{X_j\}$, $j = 0, 1, \ldots, n$, the following relationship holds:

$$m(\sum_j X_j) = P(\{X_j\}) + m(\prod_j X_j), \quad (4.1.13)$$

where $P(\{X_j\}) = \frac{1}{2} \sum_{j=0}^n m(X_j \Delta X_{j+1}) = \sum_{j=0}^n d(X_j, X_{j+1})$ is the perimeter of a closed polygonal line $l(\{X_j\})$ that connects in series the ordered elements $\{X_j\}$ of the signal $\xi(t)/X$ in metric signal space Ω. Here, the values of indices are denoted modulo $n + 1$, i.e., $X_{n+1} \equiv X_0$.

The relationships (4.1.9) and (4.1.13) imply the equality:

$$m(\underset{j}{\Delta} X_j) = P(\{X_j\}) - P(\{X_j \cdot X_{j+1}\}) + P(\{X_j \cdot X_{j+1} \cdot X_{j+2}\}) - \ldots \quad (4.1.14)$$

$$+ (-1)^k P(\{\prod_{i=j}^{j+k} X_i\}) + \ldots + m(\prod_j X_j) \cdot \mathrm{mod}_2(n-1),$$

where $\mathrm{mod}_2(n-1) = \begin{cases} 1, & n = 2k, \ k \in \mathbf{N}; \\ 0, & n = 2k + 1, \end{cases}$

$P(\{X_j\}) = \sum_{j=0}^n d(X_j, X_{j+1})$ is the perimeter of a closed polygonal line $l(\{X_j\})$ that in series connects the ordered elements $\{X_j\}$ of the signal $\xi(t)/X$ in metric signal space Ω; $P(\{X_j \cdot X_{j+1}\}) = \sum_{j=0}^n d[(X_{j-1} \cdot X_j), (X_j \cdot X_{j+1})]$ is the perimeter of a closed polygonal line $l(\{X_{\Pi j}\})$ that connects in series the ordered elements $\{X_{\Pi j}\}$:

$X_{\Pi j} = X_j \cdot X_{j+1}$ of a set $X_\Pi = \bigcup_j X_{\Pi j}$ in metric space Ω; and $P(\{\prod_{i=j}^{j+k} X_i\}) = \sum_{j=0}^{n} d[\prod_{i=j}^{j+k} X_i, \prod_{i=j+1}^{j+k+1} X_i]$ is the perimeter of a closed polygonal line $l(\{X_{\Pi j, j+k}\})$ that connects in series the ordered elements $\{X_{\Pi j,k}\}$: $X_{\Pi j,k} = \prod_{i=j}^{j+k} X_i$ of a set $X_{\Pi k} = \bigcup_j X_{\Pi j,k}$ in metric space Ω.

The relationship (4.1.13) implies that overall quantity of information $I[\xi(t)]$ is an information quantity contained in the signal $\xi(t)/X$ due to the presence of metric distinctions between the elements of a collection $\{X_j\}$. This overall quantity of information is determined by perimeter $P(\{X_j\})$ of a closed polygonal line $l(\{X_j\})$ in metric space Ω and also by the quantity of mutual information between the elements of a collection defined by measure of the product of the elements $m(\prod_j X_j)$.

The relationship (4.1.14) implies that relative quantity of information $I_\Delta[\xi(t)]$ is an information quantity contained in the signal $\xi(t)/X$ due to the presence of metric distinctions between the elements of a collection $\{X_j\}$, as well as overall quantity of information. Unlike overall quantity of information $I[\xi(t)]$, relative quantity of information $I_\Delta[\xi(t)]$ is defined by perimeter $P(\{X_j\})$ of a closed polygonal line $l(\{X_j\})$ in metric space Ω, taking into account the influence of metric distinctions between the products $\prod_{i=j}^{j+k} X_i$ of the elements of a collection $\{X_j\}$.

For the signal $\xi(t)/X$ with continuous structure $\{X_\alpha\}$, the identity holds:

$$m(\sum_\alpha X_\alpha) = P(\{X_\alpha\}), \qquad (4.1.15)$$

where $P(\{X_\alpha\}) = \lim_{d\alpha \to 0} \frac{1}{2} \sum_{\alpha \in A} m(X_\alpha \Delta X_{\alpha + d\alpha})$ is a length of some line $l(\{X_\alpha\})$ (in general case, not a straight one) that connects in series the ordered elements $\{X_\alpha\}$ of the signal $\xi(t)/X$ in space Ω; $A = [\alpha_0, \alpha_\mathbf{I}]$ is an interval of definition of a parameter α; $\lim_{d\alpha \to 0} \frac{1}{2} m(X_{\alpha_\mathbf{I}} \Delta X_{\alpha_\mathbf{I} + d\alpha}) = \frac{1}{2} m(X_{\alpha_\mathbf{I}} \Delta X_{\alpha_0})$.

The identity (4.1.15) means that overall quantity of information $I[\xi(t)]$ contained in the signal $\xi(t)/X$ considered a collection of the elements $\{X_\alpha\}$ is numerically equal to the perimeter $P(\{X_\alpha\})$ of a closed polygonal line $l(\{X_\alpha\})$ that in series connects the elements $\{X_\alpha\}$ in metric space Ω with metric $d(X_\alpha, X_\beta) = \frac{1}{2} m(X_\alpha \Delta X_\beta)$.

For a signal $\xi(t)/X$ characterized by IDD in the form of δ-function or Heaviside step function $\frac{1}{a}\left[1(\tau + \frac{a}{2}) - 1(\tau - \frac{a}{2})\right]$, the measures of structural diversity (4.1.7) and (4.1.6) are equivalent:

$$m(\underset{\alpha}{\Delta} X_\alpha) = m(\sum_\alpha X_\alpha), \qquad (4.1.16)$$

and the relative quantity of information $I_\Delta[\xi(t)]$ contained in the signal $\xi(t)/X$ is equal to the overall quantity of information $I[\xi(t)]$:

$$I_\Delta[\xi(t)] = I[\xi(t)]. \qquad (4.1.17)$$

For a signal $\xi(t)/X$ characterized by IDD in the form of the other functions, the measure of structural diversity (4.1.7) is equal to a half of the measure (4.1.6):

$$m(\underset{\alpha}{\Delta} X_\alpha) = \frac{1}{2} m(\sum_\alpha X_\alpha), \qquad (4.1.18)$$

and correspondingly, relative quantity of information $I_\Delta[\xi(t)]$ contained in a signal $\xi(t)/X$ is equal to half of the overall quantity of information $I[\xi(t)]$:

$$I_\Delta[\xi(t)] = \frac{1}{2} I[\xi(t)]. \qquad (4.1.19)$$

Both measures of structural diversity of a signal $\xi(t)/X$ with discrete (continuous) structure, $m(\sum_j X_j)$ and $m(\underset{j}{\Delta} X_j)$ ($m(\sum_\alpha X_\alpha)$ and $m(\underset{\alpha}{\Delta} X_\alpha)$), are functions of perimeter $P(\{X_j\})$ ($P(\{X_\alpha\})$) of an ordered structure of a set X in the form of a closed polygonal line $l(\{X_j\})$ ($l(\{X_\alpha\})$) that in series connects the elements $\{X_j\}$ ($\{X_\alpha\}$) in metric space Ω with metric $d(X_j, X_k) = \frac{1}{2} m(X_j \Delta X_k)$. The overall quantity of information $I[\xi(t)]$ has a sense of an entire quantity of information contained in a signal $\xi(t)/X$ considered a collection of the elements with an ordered structure. Relative quantity of information $I_\Delta[\xi(t)]$ has a sense of an information quantity that can be extracted from a signal $\xi(t)/X$ by the proper processing.

The interests of information theory and signal processing theory require the formulation of the main relationships, characterizing informational interrelations between the signals and their elements, under all the transformations realized over processed signals. The next section is devoted to establishing such relationships.

4.2 Homomorphic Mappings in Signal Space Built upon Generalized Boolean Algebra

4.2.1 Homomorphic Mappings of Continuous Signal Into a Finite (Countable) Sample Set in Signal Space Built upon Generalized Boolean Algebra with a Measure

The sampling theorem is a cornerstone of signal processing theory that was known for a long time [238], [239], [52]. As a theoretical base for a transition from Hilbert signal space into finite-dimensional Euclidean space, the sampling theorem overcomes the main difficulties of research and signal processing problems, in an attempt to achieve a harmony between continual and discrete. Its conclusions and relationships have extremely important theoretical and applied significance; they are used to solve various signal processing theory problems. Nevertheless, the use of the sampling theorem, as an exact statement with respect to the real signals, and attempts to organize the technological methods of signal processing without losses of useful information contained in the signals, encounters a number of difficulties. Theoretical complications that arise during use of this theorem and its interpretations are well known from a large body of scientific literature ranging from narrow and specialized [240] to educational [241].

The basis of the classical formulation of the sampling theorem is the principle of equivalence. According to this principle, the transition from continuous signal to the result of its discretization (sampling or quantization in time) is realized. The principle of equivalent representation of continuous signal by a finite (countable) set of samples is based on continuous function expansion in a series of orthogonal functions (generalized Fourier series) in Hilbert space.

Disadvantages of this theorem and its interpretations were discussed in Chapter 1. In the most general sense, all disadvantages of classical sampling theorem relate to ignoring the informational properties of continuous signal that must be represented by a finite (countable) set of the samples, and on the other hand, to neglecting geometrical properties of signal space.

In this section, we consider possible variants of sampling theorem formulation with application to stochastic processes, taking into account the informational characteristics and properties of stochastic signals covered in Section 3.5.

Consider the sampling of continuous stochastic process (signal) on the basis of theoretical-set approach, whose essence is revealed below.

The mapping (3.5.1) of IDD $i_\xi(t_\alpha, t)$ of an arbitrary sample $\xi(t_\alpha)$ of stochastic process $\xi(t)$ into its ordinate set $X_\alpha - \varphi\colon \{i_\xi(t_\alpha, t)\} \to \{X_\alpha\}$ considers any stochastic process (signal) $\xi(t)$ with IDD $i_\xi(t_\alpha, t)$ in the space $\mathbf{\Omega}$ as a collection of statistically dependent samples $\{\xi(t_\alpha)\}$, and also as a subalgebra $\mathbf{B}(X)$ of generalized Boolean algebra $\mathbf{B}(\Omega)$ with a measure m and signature $(+, \cdot, -, \mathbf{O})$ of the type $(2, 2, 2, 0)$. The set of the elements X is characterized by continuous structure $\{X_\alpha\}$:

$$\xi(t) \to X = \bigcup_\alpha X_\alpha,$$

and a measure of a single element X_α possesses the normalization property $m(X_\alpha) = 1$.

Definition 4.2.1. By *homomorphism of stochastic process* $\xi(t)$ with IDD $i_\xi(t_\alpha, t)$, defined in the interval $T_\xi = [t_0, t_*]$, $t \in T_\xi$, into stochastic process $\eta(t')$ with IDD $i_\eta(t'_\beta, t')$, defined in the interval $T_\eta = [t'_0, t'_*]$, $t' \in T_\eta$, in the terms of the mapping (3.5.1):

$$\varphi\colon \{i_\xi(t_\alpha, t)\} \to \{X_\alpha\}, \quad \xi(t) \to X = \bigcup_\alpha X_\alpha;$$

$$\varphi\colon \{i_\eta(t'_\beta, t')\} \to \{Y_\beta\}, \quad \eta(t') \to Y = \bigcup_\alpha Y_\beta,$$

we mean the mapping $h\colon \xi(t) \to \eta(t')$, $X \to Y$ of generalized Boolean algebra $\mathbf{B}(X)$ into generalized Boolean algebra $\mathbf{B}(Y)$ preserving all its signature operations:

1. $X_{\alpha 1} + X_{\alpha 2} = Y_{\beta 1} + Y_{\beta 2}$ (addition)

2. $X_{\alpha 1} \cdot X_{\alpha 2} = Y_{\beta 1} \cdot Y_{\beta 2}$ (multiplication)
3. $X_{\alpha 1} - X_{\alpha 2} = Y_{\beta 1} - Y_{\beta 2}$ (difference/relative complement operation)
4. $\mathbf{O}_X \equiv \mathbf{O}_Y \equiv \mathbf{O}$ (identity of null element)

Definition 4.2.2. By *isomorphism of stochastic process* $\xi(t)$ with IDD $i_\xi(t_\alpha, t)$, defined in the interval $T_\xi = [t_0, t_*]$, $t \in T_\xi$, into stochastic process $\eta(t')$ with IDD $i_\eta(t'_\beta, t')$, defined in the interval $T_\eta = [t'_0, t'_*]$, $t' \in T_\eta$, in the terms of the mapping (3.5.1), we mean bijective homomorphism.

Stochastic processes $\xi(t)$ and $\eta(t')$ connected by homomorphism (isomorphism) are called *homomorphic (isomorphic)* in terms of the mappings (3.5.1).

As a result of sampling, the stochastic process (signal) $\xi(t)$ is represented by a finite (countable) set of the samples $\Xi = \{\xi(t_j)\}$. The last set may be associated with a set of the elements X' with discrete structure $\{X_j\}$:

$$\Xi = \{\xi(t_j)\} \to X' = \bigcup_j X_j,$$

and every element of a set X' with discrete structure is simultaneously the element of a set X with continuous structure.

Thus, discretization (sampling) $D: \xi(t) \to \{X_j\}$ of continuous stochastic process $\xi(t)$ with IDD $i_\xi(t_\alpha, t)$ may be considered a mapping of the set X with continuous structure $\{X_\alpha\}$, $X = \bigcup_\alpha X_\alpha$ into the set $X' = \bigcup_j X_j$ with discrete structure $\{X_j\}$, and every element of the set X' with discrete structure is simultaneously the element of a set X with continuous structure, and the distinct elements $X_\alpha, X_\beta \in X$, $X_\alpha \neq X_\beta$ of the set X are mapped into the distinct elements $X_j, X_k \in X'$, $X_j \neq X_k$ of the set X':

$$D: X \to X'; \qquad (4.2.1)$$
$$X_\alpha \to X_j, \ X_\beta \to X_k; \qquad (4.2.1a)$$
$$X_\alpha, X_\beta \in X, \ X_\alpha \neq X_\beta; \ X_j, X_k \in X', \ X_j \neq X_k.$$

The mapping D, possessing the property (4.2.1a), is injective. Homomorphism (4.2.1) maps generalized Boolean algebra $\mathbf{B}(X)$ with signature $(+, \cdot, -, \mathbf{O})$ of the type $(2, 2, 2, 0)$ into subalgebra $\mathbf{B}(X')$, preserving the operations:

1. $X_\alpha + X_\beta = X_j + X_k$ (addition)
2. $X_\alpha \cdot X_\beta = X_j \cdot X_k$ (multiplication)
3. $X_\alpha - X_\beta = X_j - X_k$ (difference/relative complement operation)
4. $\mathbf{O}_X \equiv \mathbf{O}_{X'} \equiv \mathbf{O}$ (identity of null element)

In a general case, discretization (sampling) of continuous stochastic process $\xi(t)$ is not an isomorphic mapping preserving both measures of structural diversity of a signal $I[\xi(t)]$ (4.1.6) and $I_\Delta[\xi(t)]$ (4.1.7):

$$I[\xi(t)] \neq I[\{\xi(t_j)\}], \quad I_\Delta[\xi(t)] \neq I_\Delta[\{\xi(t_j)\}]; \qquad (4.2.2a)$$

$$I(X) \neq I(X'), \quad I_\Delta(X) \neq I_\Delta(X'), \tag{4.2.2b}$$

where $I[\xi(t)]$ is overall quantity of information contained in continuous signal $\xi(t)$, $I[\xi(t)] = I(X)$; $I[\{\xi(t_j)\}]$ is overall quantity of information contained in a sampled signal $\{\xi(t_j)\}$, $I[\{\xi(t_j)\}] = I(X')$; $I_\Delta[\xi(t)]$ is relative quantity of information contained in a continuous signal $\xi(t)$, $I_\Delta[\xi(t)] = I_\Delta(X)$; $I_\Delta[\{\xi(t_j)\}]$ is relative quantity of information contained in a sampled signal $\{\xi(t_j)\}$, $I_\Delta[\{\xi(t_j)\}] = I_\Delta(X')$.

It should be noted that notations (4.2.2a) and (4.2.2b), here and below, are used to denote informational characteristics of continuous signal $\xi(t)$ and the result of its discretization $\{\xi(t_j)\}$, and also to denote equivalent informational characteristics of their images X, X': $\xi(t) \to X$, $\{\xi(t_j)\} \to X'$ considered in terms of general Boolean algebra $\mathbf{B}(X)$.

Since discretization D of a continuous signal (4.2.1), like an arbitrary homomorphism, in a general case, does not preserve measures of structural diversity $I[\xi(t)]$, $I_\Delta[\xi(t)]$ (4.2.2a) of a signal, it is impossible to formulate strictly the sampling theorem valid for arbitrary stochastic processes. So, the principle of equivalent representation on any formulation of the sampling theorem will be used depending on informational properties of the signals.

Consider the main informational relationships characterizing a representation of continuous stochastic process (signal) $\xi(t)$ by a finite set of samples in a bounded time interval $T_\xi = [0, T]$, $t \in T_\xi$.

The following informational inequalities hold under discretization $D : X \to X'$ (4.2.1):

$$I(X) \geq I(X'), \tag{4.2.3a}$$

$$I(X') \geq I_\Delta(X'). \tag{4.2.3b}$$

The first inequality (4.2.3a) is stipulated by the relationship $X' \subset X \Rightarrow m(X') \leq m(X)$. The second (4.2.3b) is stipulated by the relationship (4.1.10) between overall and relative quantity of information contained in the result of discretization $\{\xi(t_j)\}$ of the signal $\xi(t)$. The relationship between relative quantity of information $I_\Delta(X)$ in continuous signal $\xi(t)$, and relative quantity of information $I_\Delta(X')$ in the result of its discretization $\{\xi(t_j)\}$ requires comment. This relationship essentially depends on the kind of IDD of stochastic process $\xi(t)$. For continuous weakly (wide-sense) stationary stochastic process $\xi(t)$ with IDD $i_\xi(\tau)$ that is differentiable in the point $\tau = 0$, there exists such an interval of discretization Δt between the samples $\{\xi(t_j)\}$ that relative quantity of information $I_\Delta(X)$ in continuous signal $\xi(t)$, and relative quantity of information $I_\Delta(X')$ in the result of its discretization $\{\xi(t_j)\}$ are connected by the inequality:

$$I_\Delta(X) \leq I_\Delta(X'). \tag{4.2.4}$$

For continuous stochastic process (signal) $\xi(t)$ with IDD $i_\xi(\tau)$ that is non-differentiable in the point $\tau = 0$ on arbitrary values of discretization interval (sampling interval) Δt between the samples $\{\xi(t_j)\}$, relative quantity of information $I_\Delta(X)$ in continuous signal $\xi(t)$ and relative quantity of information $I_\Delta(X')$ in the result of its discretization $\{\xi(t_j)\}$ are connected by the relationship:

$$I_\Delta(X) \geq I_\Delta(X'). \tag{4.2.5}$$

4.2 Homomorphic Mappings in Signal Space Built upon Generalized Boolean Algebra

Overall quantity of information $I(X)$ contained in a continuous stochastic process (signal) $\xi(t)$ considered a set X with continuous structure $\{X_\alpha\}$ is equal to the sum of overall quantity of information $I(X')$ contained in a finite set of the samples $\Xi = \{\xi(t_j)\}/X'$ (i.e., the result of discretization of stochastic process (signal) $\xi(t)/$ a set X) and an information quantity I'_L existing due to a curvature of a structure of a set X in signal space Ω that is lost under its discretization:

$$I(X) = I(X') + I'_L. \tag{4.2.6}$$

Information quantity I'_L is called *information losses of the first genus*. For a set of the elements X of a stochastic process (signal) $\xi(t)$ with a continuous structure $\{X_\alpha\}$ in the space Ω, we can introduce a measure of a *curvature* of a structure $c(X)$, which characterizes a deflection of a locus of a structure $\{X_\alpha\}$ of a set X from a line:

$$c(A) = \frac{I(X) - I(X')}{I(X)}. \tag{4.2.7}$$

Evidently, the curvature $c(X)$ of a set of the elements X can be varied within $0 \leq c(X) \leq 1$. If for an arbitrary pair of the adjacent elements $X_j, X_{j+1} \in X'$ of a discrete set X' and the element $X_\alpha \in X$, $X_\alpha \in [X_j X_{j+1}, X_j + X_{j+1}]$, the metric identity holds:

$$d[X_j, X_{j+1}] = d[X_j, X_\alpha] + d[X_\alpha, X_{j+1}],$$

where $d[X_\alpha, X_\beta] = \frac{1}{2} m(X_\alpha \Delta X_\beta)$ is a metric in the space Ω, then the curvature $c(X)$ of a set of elements X with continuous structure is equal to zero.

Overall quantity of information $I(X')$ contained in a set X' with discrete structure $\{X_j\}$ is equal to the sum of relative quantity of information $I_\Delta(X')$ of this set and a quantity of redundant information I''_L contained in a set X' due to nonempty pairwise intersections (due to mutual information) between its elements:

$$I(X') = I_\Delta(X') + I''_L. \tag{4.2.8}$$

Information quantity I''_L is called *information losses of the second genus*. For the result of discretization of continuous stochastic process (signal) $\xi(t)$, i.e., a set of the elements X' with discrete structure $\{X_j\}$ of the space Ω, one can introduce a measure of informational redundancy $r(X')$, which characterizes a presence of informational interrelations between the elements of discrete structure $\{X_j\}$ of a set X' (or simply mutual information between them):

$$r(X') = \frac{I(X') - I_\Delta(X')}{I(X')}. \tag{4.2.9}$$

If for an arbitrary pair of the adjacent elements $X_j, X_{j+1} \in X'$, the measure of their product is equal to zero: $m(X_j X_{j+1}) = 0$, then a measure of informational redundancy $r(X')$ of a structure $\{X_j\}$ also is equal to zero: $r(X') = 0$.

Substituting the equality (4.2.8) into (4.2.6), we obtain the following relationship:

$$I(X) = I_\Delta(X') + I'_L + I''_L. \tag{4.2.10}$$

The meaning of the relationship (4.2.10) can be elucidated. One cannot extract more information quantity from the signal $\xi(t)$ (i.e., a collection of the elements X with continuous structure) than a value of relative quantity of information $I_\Delta(X')$ contained in this signal $\xi(t)$ due to diversity between the elements of its structure. Information losses I'_L and I''_L take place, on the one hand, in the consequence of curvature $c(X)$ of a structure $\{X_\alpha\}$ of signal $\xi(t)$ in metric space Ω, and on the other hand, due to some informational redundancy $r(X')$ of a discrete set X' (some mutual information between the samples $\{X_j\}$).

Discretization $D: \xi(t) \to \{\xi(t_j)\}$, according to the identity (4.2.10), has to be realized to provide maximum ratio of relative quantity of information $I_\Delta(X')$ contained in the sampled signal $\{\xi(t_j)\}$ to overall quantity of information $I(X)$ contained in continuous signal $\xi(t)$ in a bounded time interval $T_\xi = [0, T]$, $t \in T_\xi$:

$$\frac{I_\Delta(X')}{I(X)} \to \max, \tag{4.2.11}$$

or, equivalently, to provide minimum sum of information losses of the first I'_L and the second I''_L genera:

$$I'_L + I''_L \to \min.$$

On the base of criterion (4.2.11), the sampling theorem may be formulated for signals with various informational and probabilistic-statistical properties. We limit the consideration of possible variants of the sampling theorem to a stationary stochastic process (signal), whose IDD $i_\xi(t_\alpha, t)$ of an arbitrary sample $\xi(t_\alpha)$ is characterized by the property: $i_\xi(t_\alpha, t) \equiv i_\xi(\tau)$, $\tau = t - t_\alpha$.

Theorem 4.2.1. *Theorem on equivalent (from the standpoint of preserving overall quantity of information) representation of continuous stochastic process by a finite set of the samples. Continuous stationary stochastic process (signal) $\xi(t)$ with IDD $i_\xi(\tau)$ $(0 < i_\xi(0) < \infty)$ defined in the interval $T_\xi = [0, T]$, $t \in T_\xi$ can be equivalently represented by a finite set of the samples $\Xi = \{\xi(t_j)\}$ without any information losses if a set of the elements $X = \bigcup_\alpha X_\alpha$ of the signal $\xi(t)$ can be represented by a partition $X' = \bigcup_j X_j$ of its orthogonal elements $\{X_j\}$: $X_j \cdot X_k = \mathbf{O}$, $j \neq k$, so that for overall quantity of information $I[\{\xi(t_j)\}]$ in the sampled signal $\{\xi(t_j)\}$ and overall quantity of information $I[\xi(t)]$ in continuous signal $\xi(t)$, the equality holds:*

$$X \equiv X' \Rightarrow I[\xi(t)] = I[\{\xi(t_j)\}]. \tag{4.2.12}$$

This identity makes possible the extraction of the entire information quantity, contained in this process $\xi(t)$ in time interval $T_\xi = [0, T]$, from a finite set of the samples $\Xi = \{\xi(t_j)\}$ of stochastic process $\xi(t)$ without any information losses. Such representation of continuous stochastic process $\xi(t)$ by a finite set of samples $\Xi = \{\xi(t_j)\}$ in the interval $T_\xi = [0, T]$, $t \in T_\xi$, when the relationship (4.2.12) holds, is equivalent from the standpoint of preserving overall quantity of information.

The main condition of the Theorem 4.2.1, i.e., orthogonality of the elements of continuous structure of stochastic process (signal) $\xi(t)$, is provided if and only if its

IDD $i_\xi(\tau)$ has a uniform distribution in a bounded interval $[-\Delta/2, \Delta/2]$ and takes the form:

$$i_\xi(\tau) = \frac{1}{\Delta}[1(\tau + \frac{\Delta}{2}) - 1(\tau - \frac{\Delta}{2})],$$

where $1(t)$ is a Heaviside step function.

In this case, the identity holds between both measures of structural diversity of continuous stochastic process (signal) $\xi(t)$ and a set of the samples $\Xi = \{\xi(t_j)\}$ (the sets of the elements with continuous $\{X_\alpha\}$ and discrete $\{X_j\}$ structures, respectively), i.e., between the measures of the sets X and X':

$$m(X) = m(\sum_\alpha X_\alpha) \equiv m(\sum_j X_j) = m(X'),$$

and a measure of symmetric difference of a set $m(\underset{j}{\triangle} X_j)$:

$$m(\sum_\alpha X_\alpha) = m(\sum_j X_j) = m(\underset{j}{\triangle} X_j) = m(\underset{\alpha}{\triangle} X_\alpha),$$

that from an informational view is equivalent to the identity:

$$I(X) = I(X') = I_\Delta(X') = I_\Delta(X).$$

Fulfillment of these identities is provided by orthogonality of the elements $\{X_j\}$, so measure of a set X' with a discrete structure $\{X_j\}$ is equal to a measure of symmetric difference of the elements of this set:

$$m(\sum_j X_j) = m(\underset{j}{\triangle} X_j).$$

The overall quantity of information $I(X')$ of a set X' is equal to the relative quantity of information $I_\Delta(X')$ of this set: $I(X') \equiv I_\Delta(X')$. This means that under discrete representation of a stochastic process (signal) $\xi(t)$ possessing the mentioned property, the information losses of the first I'_L and the second I''_L genera are equal to zero: $I'_L = I''_L = 0$.

The presence of a measure of structural diversity $I_\Delta[\xi(t)]$ (relative quantity of information) of the signal $\xi(t)$, based on symmetric differences of the elements, allows us to formulate the following variant of the sampling theorem.

Theorem 4.2.2. *Theorem on equivalent (from the standpoint of preserving relative quantity of information) representation of continuous stochastic process by a finite set of the samples. A continuous stationary stochastic process (signal) $\xi(t)$ with IDD $i_\xi(\tau)$ $(0 < i_\xi(0) < \infty)$ defined in the interval $T_\xi = [0, T]$, $t \in T_\xi$ may be equivalently represented by a finite set of the samples $\Xi = \{\xi(t_j)\}$, if the equality between relative quantity of information $I_\Delta[\xi(t)]$ contained in stochastic process (signal)$\xi(t)$ in the interval $T_\xi = [0, T]$ and relative quantity of information $I_\Delta[\{\xi(t_j)\}]$ contained in a finite set of the samples $\Xi = \{\xi(t_j)\}$ holds:*

$$I_\Delta[\xi(t)] = I_\Delta[\{\xi(t_j)\}]; \ t, t_j \in T_\xi. \tag{4.2.13}$$

This identity provides the possibility of extracting the same relative quantity of information contained in signal $\xi(t)$ in the interval $T_\xi = [0, T]$ from a finite set of the samples $\Xi = \{\xi(t_j)\}$ of stochastic process (signal) $\xi(t)$. Such representation of continuous stochastic process $\xi(t)$ determined in the interval $T_\xi = [0, T]$ by a finite set of the samples $\Xi = \{\xi(t_j)\}$ under the condition (4.2.13) is equivalent from the standpoint of preserving relative quantity of information.

Theorem 4.2.2 has the following corollaries.

Corollary 4.2.1. *Continuous stationary stochastic process (signal) $\xi(t)$ with IDD $i_\xi(\tau)$ that has a uniform distribution in the bounded interval $[-\Delta/2, \Delta/2]$ of the form:*

$$i_\xi(\tau) = \frac{1}{\Delta}\left[1\left(\tau + \frac{\Delta}{2}\right) - 1\left(\tau - \frac{\Delta}{2}\right)\right], \qquad (4.2.14)$$

may be equivalently represented by a finite set of the samples $\Xi = \{\xi(t_j)\}$ if the discretization interval (the sampling interval) $\Delta t = t_{j+1} - t_j$ is chosen to be equal to $\Delta t = 1/i_\xi(0)$.

Corollary 4.2.2. *Continuous stationary stochastic process (signal) $\xi(t)$ with IDD $i_\xi(\tau)$ of the following form:*

$$i_\xi(\tau) = \begin{cases} \frac{1}{a}\left[1 - \frac{|\tau|}{a}\right], & |\tau| \leq a; \\ 0, & |\tau| > a, \end{cases}$$

may be equivalently represented by a finite set of the samples $\Xi = \{\xi(t_j)\}$ if the discretization interval (the sampling interval) $\Delta t = t_{j+1} - t_j$ is chosen within the interval $\Delta t \in]0, 2/i_\xi(0)]$.

In the cases cited in Corollaries 4.2.1 and 4.2.2 between measures of structural diversity of continuous stochastic process (signal) $\xi(t)$ and a set of the samples $\Xi = \{\xi(t_j)\}$ (sets of the elements with continuous $\{X_\alpha\}$ and discrete $\{X_j\}$ structures, respectively), namely between measures of symmetric difference of the sets X and X', the identity holds:

$$m(\underset{\alpha}{\Delta} X_\alpha) = m(\underset{j}{\Delta} X_j),$$

and from an informational view, is equivalent to the identity:

$$I_\Delta(X) = I_\Delta(X').$$

For stochastic processes with IDD $i_\xi(\tau)$ differentiable in the point $\tau = 0$, the sampling theorem may be formulated in the following way.

Theorem 4.2.3. *Theorem on equivalent (from the standpoint of maximal ratio (4.2.11)) representation of continuous stochastic process by a finite set of the samples. Continuous stationary stochastic process (signal) $\xi(t)$, defined in the interval $T_\xi = [0, T]$, $t \in T_\xi$, with IDD $i_\xi(\tau)$ ($0 < i_\xi(0) < \infty$) that is differentiable in the point $\tau = 0$, may be equivalently represented by a finite set of the samples $\Xi = \{\xi(t_j)\}$, if*

4.2 Homomorphic Mappings in Signal Space Built upon Generalized Boolean Algebra 141

a chosen value of discretization interval (sampling interval) Δt provides the maximal ratio of relative quantity of information $I_\Delta(X')$ contained in the sampled signal $\{\xi(t_j)\}$ to overall information quantity $I(X)$ contained in continuous signal $\xi(t)$ in its domain of definition $T_\xi = [0, T]$, $t \in T_\xi$:

$$\Delta t_{\text{opt}} = \arg\max_{\Delta t} \left[\frac{I_\Delta(X')}{I(X)} \right]. \qquad (4.2.15)$$

Representation of continuous signal defined by Theorem 4.2.3 is equivalent in the sense that it is impossible to extract from the initial signal $\xi(t)$ larger information quantity than $I_\Delta(X')$ contained in a collection of the samples $\{\xi(t_j)\}$ that follow through the interval defined by the relationship (4.2.15). Other variants of Theorems 4.2.1 through 4.2.3 can be found in [242].

Recommendations concerning a discretization (sampling) of continuous stochastic signals may be stated in less strict formulation that is closer to applied problems. Their variant is based upon an approximation of the real IDD $i_\xi(\tau)$ of the signal by the function (4.2.14) where $i_\xi(0) \cdot \Delta = 1$ and may be expressed by the following remark.

Remark 4.2.1. Stationary stochastic process (signal) $\xi(t)$ with IDD $i_\xi(\tau)$ ($0 < i_\xi(0) < \infty$) defined in the interval $T_\xi = [0, T]$, $t \in T_\xi$ may be represented by a finite set of the samples $\Xi = \{\xi(t_j)\}$ that follow through the interval:

$$\Delta t = t_{j+1} - t_j = 1/i_\xi(0). \qquad (4.2.16)$$

Graphs of dependences of the ratio $[I_\Delta(X')/I(X)](\Delta t)$ of relative quantity of information $I_\Delta[\{\xi(t_j)\}]$ contained in a sampled signal $\{\xi(t_j)\}$ to overall quantity of information $I[\xi(t)]$ contained in continuous signal on the discretization interval Δt are shown in Fig. 4.2.1. Graphs of dependences are plotted for stochastic processes with the following IDDs:

$$i_1(\tau) = \frac{1}{\sqrt{2\pi}a} \exp\left[-\frac{\tau^2}{2a^2}\right]; \quad i_2(\tau) = \frac{1}{\pi\left(1+\left(\frac{\tau}{b}\right)^2\right)};$$

$$i_3(\tau) = c\exp(-2c|\tau|); \quad i_4(\tau) = \begin{cases} \frac{1}{d}\left[1-\frac{|\tau|}{d}\right], & |\tau| \leq d; \\ 0, & |\tau| > d. \end{cases}$$

Analysis of the dependences of the ratio $\left[\frac{I_\Delta(X')}{I(X)}\right](\Delta t)$ shown in Fig. 4.2.1 shows certain features typical for discretization of continuous stochastic processes (signals).

1. For the plotted dependences, the limit of the ratio $[I_\Delta(X')/I(X)](\Delta t)$ on $\Delta t \to 0$ is equal to $1/2$:

$$\lim_{\Delta t \to 0} \left[\frac{I_\Delta(X')}{I(X)} \right] (\Delta t) = \frac{1}{2}.$$

This confirms the result (4.1.19) concerning the ratio of relative quantity of information $I_\Delta[\xi(t)]$ to overall quantity of information $I[\xi(t)]$ contained in a continuous signal $\xi(t)$ with IDD that differs from a uniform one:

$$I_\Delta[\xi(t)] = 0.5 I[\xi(t)].$$

2. For stochastic processes with IDDs $i_1(\tau)$ and $i_2(\tau)$ (IDD $i_\xi(\tau)$ is differentiable in the point $\tau = 0$), the dependence $[I_\Delta(X')/I(X)](\Delta t)$ has a pronounced maximum. This feature allows choosing the discretization interval Δt for practical applications, maximizing the ratio $I_\Delta(X')/I(X)$, according to Theorem 4.2.3 and Formula (4.2.15).

On $I_\Delta(X')/I(X) > 1/2$, the inequality $I_\Delta(X') > I_\Delta(X)$ holds. Under a discretization through the interval Δt, which is equal to $1/i_\xi(0)$, the deflection from maximum value $[I_\Delta(X')/I(X)]_{\max}$ for the plotted dependences does not exceed 10%:

$$\frac{|I_\Delta(X')/I(X) - [I_\Delta(X')/I(X)]_{\max}|}{[I_\Delta(X')/I(X)]_{\max}} \leq 0.1.$$

3. For stochastic processes with IDDs $i_3(\tau)$ and $i_4(\tau)$ (IDD $i_\xi(\tau)$ is nondifferentiable in the point $\tau = 0$), the function $[I_\Delta(X')/I(X)](\Delta t)$ is nonincreasing and has no maximum. In this case, the inequality $I_\Delta(X') \leq I_\Delta(X)$ holds. Discretization of the process with an arbitrary interval Δt is accompanied by the losses of relative quantity of information $I_\Delta[\xi(t)]$ that may be extracted from the signal.

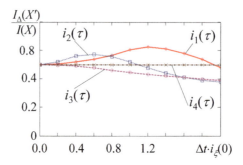

FIGURE 4.2.1 Graphs of dependences of ratio $I_\Delta(X')/I(X)$ on discretization interval Δt. Modified from [242], with permission

For most stochastic processes possessing the ability to carry information, the values $i_\xi(0)$ belong to the interval $[2\Delta f, 4\Delta f]$, where Δf is the real effective width of hyperspectral density $\sigma_\xi(\omega)$ equal to: $\Delta f = [\int_0^\infty \sigma_\xi(\omega)d\omega]/[2\pi\sigma_\xi(0)]$, so the values of the discretization (sampling) interval $\Delta t = 1/i_\xi(0)$ could change within the interval $[1/(4\Delta f), 1/(2\Delta f)]$.

Aforementioned variants of the sampling theorem are formulated for stationary stochastic processes possessing the ability to carry information (see Theorem 3.4.2)

and characterized by unbounded (in frequency domain) power spectral density. For such stochastic processes, the interval of discretization is entirely determined by IDD introduced by Definition 3.4.1, which in the case of Gaussian stochastic processes, is entirely determined by their normalized correlation function. The suggested approach to formulating the sampling theorem can be generalized for nonstationary processes. In this case, the discretization of continuous signal will be characterized by nonequidistant (in time domain) samples. Depending on IDD, one should distinguish two types of stochastic processes possessing the ability to carry information. Under discretization of stochastic processes with IDD, which is differentiable in the point $\tau = 0$, by processing a finite set of the samples $\{\xi(t_j)\}$, $j = 1, \ldots, n$ in the interval $T_\xi = [0, T]$ instead of processing the sample continuum $\{\xi(t_\alpha)\}$, $t_\alpha \in T_\xi$, one can obtain a quantitative gain in the use of information contained in the signal, according to the relationship (4.2.4). On the other hand, processing a finite set of the samples $\{\xi(t_j)\}$ of stochastic processes with IDD $i_\xi(\tau)$, which is nondifferentiable in the point $\tau = 0$, from an informational view, is inexpedient (see the relationship (4.2.5)). It should realize processing of such signals as a whole in continuum of the samples $\{\xi(t_\alpha)\}$, $t_\alpha \in T_\xi$.

4.2.2 Theorems on Isomorphism in Signal Space Built upon Generalized Boolean Algebra with a Measure

The mappings (3.5.1) and (3.5.26) that allow us to use generalized Boolean algebra with a measure for describing informational properties of stochastic processes (signals), create the possibility to overformulate the known results with application to the stochastic processes (the signals) in the forms of the following theorems and their corollaries.

Let $\varphi \colon \mathbf{\Gamma} \to \mathbf{\Omega}$ be morphism of physical signal space $\mathbf{\Gamma}$ into informational signal space $\mathbf{\Omega}$ defined by the relationship (4.1.1). Let also G be a group of automorphisms of generalized Boolean algebra $\mathbf{B}(\Omega)$ with a measure m, and besides, for an arbitrary mapping g of the group G:

$$g \colon \quad \xi(t) \to \xi'(t); \quad \xi(t), \xi'(t) \in \mathbf{\Gamma}; \quad g \in G,$$

where g is isomorphic mapping of stochastic process (signal) in the sense of Definition 4.2.2.

On the base of Theorem 2.3.3, one can formulate a similar theorem for stochastic processes (signals).

Theorem 4.2.4. *Theorem on isomorphic mapping preserving a measure of information quantity. If in metric signal space $\mathbf{\Omega}$, stochastic signal $\xi(t)$, considered a subalgebra $\mathbf{B}(X)$ of generalized Boolean algebra $\mathbf{B}(\Omega)$ with a measure m and signature $(+, \cdot, -, \mathbf{O})$ of the type $(2, 2, 2, 0)$, and characterized by IDD $i_\xi(t_\alpha, t)$, $(0 < i_\xi(t_\alpha, t) < \infty)$ in the interval $T_\xi = [t_0, t_*]$, $t \in T_\xi$, is isomorphic to stochastic signal $\xi'(t) \colon g \colon \xi(t) \to \xi'(t) \colon g \in G$ in the sense of Definition 4.2.2, then the mapping g is a measure preserving isomorphism.*

Corollary 4.2.3. *Isomorphic mapping g of stochastic process (signal) $\xi(t)$ into stochastic process (signal) $\xi'(t)$:*

$$g: \quad \xi(t) \to \xi'(t); \qquad (4.2.17)$$

$$\{i_\xi(t_\alpha, t)\} \to \{i'_\xi(t_\alpha, t)\};$$

$$\{X_\alpha\} \to \{X'_\alpha\}; \quad X = \bigcup_\alpha X_\alpha \to X' = \bigcup_\alpha X'_\alpha,$$

preserves the measures of all the binary operations between an arbitrary pair of the elements $X_{\alpha 1}, X_{\alpha 2} \in X$ of a set X of the corresponding pair of samples $\xi(t_{\alpha 1})$, $\xi(t_{\alpha 2})$ of the signal $\xi(t)/X = \bigcup_\alpha X_\alpha$ considered as subalgebra $\mathbf{B}(X)$ of generalized Boolean algebra $\mathbf{B}(\Omega)$ with signature $(+, \cdot, -, \mathbf{O})$ of the type $(2, 2, 2, 0)$:

$$m(X_{\alpha 1} + X_{\alpha 2}) = m(X'_{\alpha 1} + X'_{\alpha 2}); \qquad (4.2.18a)$$

$$m(X_{\alpha 1} \cdot X_{\alpha 2}) = m(X'_{\alpha 1} \cdot X'_{\alpha 2}); \qquad (4.2.18b)$$

$$m(X_{\alpha 1} - X_{\alpha 2}) = m(X'_{\alpha 1} - X'_{\alpha 2}); \qquad (4.2.18c)$$

$$m(X_{\alpha 1} \Delta X_{\alpha 2}) = m(X'_{\alpha 1} \Delta X'_{\alpha 2}). \qquad (4.2.18d)$$

Group of the identities (4.2.18) implies the following corollary.

Corollary 4.2.4. *Isomorphic mapping g (4.2.17) preserves all the sorts of information quantities (see Definitions 3.5.1 through 3.5.5) between arbitrary elements $X_{\alpha 1}, X_{\alpha 2} \in X$ of a set X of the corresponding samples $\xi(t_{\alpha 1})$, $\xi(t_{\alpha 2})$ of the signal $\xi(t)/X = \bigcup_\alpha X_\alpha$ considered a subalgebra $\mathbf{B}(X)$ of generalized Boolean algebra $\mathbf{B}(\Omega)$ with signature $(+, \cdot, -, \mathbf{O})$ of the type $(2, 2, 2, 0)$:*

$$I_{X_{\alpha 1} + X_{\alpha 2}} = I_{X'_{\alpha 1} + X'_{\alpha 2}}; \qquad (4.2.19a)$$

$$I_{X_{\alpha 1} \cdot X_{\alpha 2}} = I_{X'_{\alpha 1} \cdot X'_{\alpha 2}}; \qquad (4.2.19b)$$

$$I_{X_{\alpha 1} - X_{\alpha 2}} = I_{X'_{\alpha 1} - X'_{\alpha 2}}; \qquad (4.2.19c)$$

$$I_{X_{\alpha 1} \Delta X_{\alpha 2}} = I_{X'_{\alpha 1} \Delta X'_{\alpha 2}}. \qquad (4.2.19d)$$

Corollary 4.2.5. *Isomorphic mapping g (4.2.17) preserves overall quantity of information $I[\xi(t)]$ contained in stochastic process (signal) $\xi(t)$ considered a collection of the elements $X = \bigcup_\alpha X_\alpha$; each element corresponds to IDD $i_\xi(t_\alpha, t)$ of the sample $\xi(t_\alpha)$:*

$$I[\xi(t)] = m\left(\sum_\alpha X_\alpha\right) = m\left(\sum_\alpha X'_\alpha\right) = I[\xi'(t)].$$

Corollary 4.2.6. *Isomorphic mapping g (4.2.17) preserves relative quantity of information $I_\Delta[\xi(t)]$ contained in stochastic process (signal) $\xi(t)$ considered as a collection of the elements $X = \bigcup_\alpha X_\alpha$; each of them corresponds to IDD $i_\xi(t_\alpha, t)$ of the sample $\xi(t_\alpha)$:*

$$I_\Delta[\xi(t)] = m(\underset{\alpha}{\Delta} X_\alpha) = m(\underset{\alpha}{\Delta} X'_\alpha) = I_\Delta[\xi'(t)].$$

4.3 Homomorphic Mappings in Signal Space Built upon Generalized Boolean Algebra

The conclusions of Theorems 4.2.4 and 2.3.4 may be generalized upon an isomorphic mapping of stochastic processes $\xi(t)$, $\eta(t')$.

Theorem 4.2.5. *Theorem on isomorphic mapping of stochastic processes. Isomorphic mapping g (4.2.17) of a pair of stochastic processes (signals) $\xi(t)$, $\eta(t')$ — φ: $\xi(t)/X = \bigcup_\alpha X_\alpha$, $\eta(t')/Y = \bigcup_\beta Y_\beta$ considered a subalgebras $\mathbf{B}(X)$, $\mathbf{B}(Y)$ of generalized Boolean algebra $\mathbf{B}(\Omega)$ with a measure m and signature $(+, \cdot, -, \mathbf{O})$ of the type $(2, 2, 2, 0)$, into a pair of stochastic processes (signals) $\xi'(t)/X' = \bigcup_\alpha X'_\alpha$, $\eta'(t')/Y' = \bigcup_\beta Y'_\beta$, respectively:*

$$g: \quad \xi(t) \to \xi'(t), \quad \eta(t') \to \eta'(t'), \quad g \in G; \tag{4.2.20}$$

$$X' = g(X), \quad X = g^{-1}(X'); \quad Y' = g(Y), \quad Y = g^{-1}(Y'), \tag{4.2.20a}$$

preserves the measures of all the binary operations between the sets $X, Y \subset X \cup Y$, whose union $X \cup Y$ is also a subalgebra $\mathbf{B}(X + Y)$ of generalized Boolean algebra $\mathbf{B}(\Omega)$:

$$m(X + Y) = m(X' + Y');$$
$$m(X \cdot Y) = m(X' \cdot Y');$$
$$m(X - Y) = m(X' - Y');$$
$$m(X \Delta Y) = m(X' \Delta Y').$$

Corollary 4.2.7. *Isomorphic mapping g (4.2.20) preserves all the sorts of information quantities between the signals $\xi(t)/X = \bigcup_\alpha X_\alpha$, $\eta(t')/Y = \bigcup_\beta Y_\beta$ considered subalgebras $\mathbf{B}(X)$, $\mathbf{B}(Y)$ of generalized Boolean algebra $\mathbf{B}(\Omega)$ with a measure m and signature $(+, \cdot, -, \mathbf{O})$ of the type $(2, 2, 2, 0)$:*
Quantity of overall information:

$$I^+[\xi(t), \eta(t')] = I_{X+Y} = I_{X'+Y'} = I^+[\xi'(t), \eta'(t')]; \tag{4.2.21}$$

Quantity of mutual information:

$$I[\xi(t), \eta(t')] = I_{X \cdot Y} = I_{X' \cdot Y'} = I[\xi'(t), \eta'(t')]; \tag{4.2.22}$$

Quantity of particular relative information:

$$I^-[\xi(t), \eta(t')] = I_{X-Y} = I_{X'-Y'} = I^-[\xi'(t), \eta'(t')]; \tag{4.2.23}$$

Quantity of relative information:

$$I^\Delta[\xi(t), \eta(t')] = I_{X \Delta Y} = I_{X' \Delta Y'} = I^\Delta[\xi'(t), \eta'(t')]. \tag{4.2.24}$$

4.3 Features of Signal Interaction in Signal Spaces with Various Algebraic Properties

In most of the literature, the problems of signal processing in the presence of interference (noise) are formulated in terms of linear signal space $\mathcal{LS}$, where the result of interaction x of useful signal s and interference (noise) n is described by the operation of addition $x = s+n$ of an additive commutative group [148], [143], [122], [153]. Meanwhile, some authors generally state the problem of signal processing in the presence of interference (noise) with respect to the kind of interaction [159], [156], [243]: $x = F(s,n)$, where $F(s,n)$ is some deterministic function.

The results of synthesis of concrete signal processing algorithms and units are obtained on the base of criteria of optimality, usually assuming an additive (in terms of linear space) interaction x of signal s and interference (noise) n: $x = s+n$. As a rule, the question concerning optimality with respect to the kind of signal interaction is not investigated. How close to optimal is the model of such an additive signal interaction in the input of a processing unit with respect to its content? Could better quality indices of signal processing be achieved during interaction between the signals, if the signal space is not a linear? Could the problem of synthesis of optimal signal processing algorithm and unit be solved in the most general form on the assumption that the properties of signal space where the signal interaction $x = F(s,n)$ is realized are not defined? In other words, could we find an unknown function $x = F(s,n)$ that determines the kind of interaction between useful signal s and interference (noise) n, while solving some specific problems of synthesis of signal processing algorithms and units on the base of the known criteria of optimality?

If within such a formulation, the problem of synthesis has no solution, one can try to obtain a solution using various optimality criteria, first, to determine the kind of signal interaction $F(s,n)$ in the input of signal processing unit and second to realize a synthesis of optimal signal processing algorithm and unit.

If the algebraic properties of the signal space are defined, where the best signal interaction $F(s,n)$ of a corresponding kind takes place, then the requirement of determinacy of a function $F(s,n)$ used for the following synthesis of optimal signal processing algorithms will not seem to be too strict a confinement.

In order to answer these and related questions, it is necessary to provide a generalized approach concerning the analysis of informational relationships that take place during signal interaction in signal spaces with various algebraic properties.

An important circumstance should be emphasized here. Shannon's information theory does not assume the notion of information quantity contained in a signal in general to exist, so there is no sense of asking: "What is the quantity of information contained in a signal a?", inasmuch as the notion of information quantity has a sense here according to a pair of signals only [51]. In this case, the correct formulation of the question is: "How much information does the signal b contain concerning the signal a?" This means that classical information theory operates exclusively with the notion of mutual information which may be applied to a pair of signals only.

4.3 Features of Signal Interaction in Signal Spaces with Various Algebraic Properties 147

On the basis of classical information theory one can evaluate mutual information $I[x, a]$ contained in the signal x concerning the signal a, if, for instance, the signal x is the result of additive interaction of two signals a and b: $x = a + b$ in linear signal space $\mathcal{LS}$. However, it is impossible to evaluate quantity of information losses that take place during such an interaction. One cannot compare the sum $I_a + I_b$ of information quantities contained in the signals a and b, respectively, and an information quantity I_x contained in the signal x, which is a result of interaction of signals a and b. The reason is the same, inasmuch as the values I_x, I_a, I_b, corresponding to information quantities contained in each signal x, a, b separately, are not defined.

The goal of this section is to further develop the approach evaluating informational relationships between the signals covered in Sections 3.4, 3.5, 4.1, and 4.2, to carry out the comparative analysis of informational properties of signal spaces with various algebraic properties stipulated by signal interactions in these spaces.

There are two problems formulated and solved in this section. First, the main informational relationships between the signals before and after their interaction in signal spaces with various algebraic properties are determined. Second, depending on a quantitative content of these informational relationships, at a quality level, the types of interactions between the signals in signal spaces are distinguished. Inasmuch as this section provides a research transition from informational signal space to the physical signal spaces, we change the notation system based on the Greek alphabet accepted for the signals (stochastic processes) before. Here and below we will use the Latin alphabet to denote useful and interference signals while considering the questions of their processing that form the main content of Chapter 7.

4.3.1 Informational Paradox of Additive Signal Interaction in Linear Signal Space: Notion of Ideal Signal Interaction

The notions of physical and informational signal spaces, and also physical interactions and informational interrelations between the signals were introduced in Section 4.1.

Unlike informational interrelations between the signals that do not directly affect their physical interaction, the last impacts the informational relationships between the signals and the results of their interaction. Consider these features of physical signal interaction, citing as an example the additive interaction of two stationary statistically independent Gaussian stochastic processes $a(t)$ and $b(t)$ with zero expectations in linear (Hilbert) space $\mathcal{LS}$, $a(t), b(t) \in \mathcal{LS}$, based on the ideas stated in Sections 3.3 and 4.1:

$$x(t) = a(t) + b(t), t \in T. \qquad (4.3.1)$$

Let the centered stochastic processes (the signals) $a(t)$ and $b(t)$ be characterized by the variances D_a, D_b and identical normalized correlation functions $r_a(\tau) = r_b(\tau) = r(\tau) \geq 0$, respectively. The correlation function $R_x(\tau)$ of an additive mixture $x(t)$ (4.3.1) of these two signals is equal to:

$$R_x(\tau) = D_a r_a(\tau) + D_b r_b(\tau) = (D_a + D_b) r(\tau),$$

and its normalized correlation function $r_x(\tau)$ is identically equal to the normalized correlation function of the signals $a(t)$ and $b(t)$:

$$r_x(\tau) = r_a(\tau) = r_b(\tau) = r(\tau). \tag{4.3.2}$$

For Gaussian stochastic signals, the normalized correlation function defines its normalized function of statistical interrelationship (NFSI) introduced in Section 3.1. The equality (4.3.2), according to the formula (3.1.3), implies the identity of NFSIs $\psi_x(\tau)$, $\psi_a(\tau)$, $\psi_b(\tau)$ of the signals $x(t)$, $a(t)$, $b(t)$:

$$\psi_x(\tau) = \psi_a(\tau) = \psi_b(\tau), \tag{4.3.3}$$

which implies the identity of information distribution densities (IDDs) $i_x(\tau)$, $i_a(\tau)$, $i_b(\tau)$ of the signals $x(t)$, $a(t)$, $b(t)$ (see formula (3.4.3)):

$$i_x(\tau) = i_a(\tau) = i_b(\tau). \tag{4.3.4}$$

On the basis of Axiom 4.1.1 formulated in Section 4.1, depending on the sort of signature relations between the images A, B of the signals $a(t)$, $b(t)$ in informational signal space Ω built upon a generalized Boolean algebra $\mathbf{B}(\Omega)$ with a measure m and signature $(+, \cdot, -, \mathbf{O})$, $A \in \Omega$, $B \in \Omega$, the following main types of information quantity are defined: quantity of absolute information I_A, I_B, quantity of mutual information I_{AB}, and quantity of overall information I_{A+B} determined by the following relationships, respectively:

$$I_A = m(A), \quad I_B = m(B); \tag{4.3.5a}$$

$$I_{AB} = m(AB); \tag{4.3.5b}$$

$$I_{A+B} = m(A+B) = m(A) + m(B) - m(AB) = I_A + I_B - I_{AB}. \tag{4.3.5c}$$

The images A and B of the corresponding signals $a(t)$ and $b(t)$ are connected by the mapping (3.5.26a).

Identity (4.3.4), according to the formula (3.5.6b), means that, within the time interval T, the signals $x(t)$, $a(t)$, $b(t)$ carry the same quantity of absolute information:

$$I_X = I_A = I_B. \tag{4.3.6}$$

However, taking into account statistical independence of the signals $a(t)$ and $b(t)$ (i.e., $I_{AB} = 0$), it is easy to conclude that information contents carried by these signals are absolutely distinct despite quantitative equality $I_A = I_B$. Thus, the quantity of overall information I_{A+B} carried by the signals $a(t)$ and $b(t)$ together must equal the sum of I_A and I_B, and according to the relationships (4.3.5c) and (4.3.6), is equal to:

$$I_{A+B} = I_A + I_B = 2I_X. \tag{4.3.7}$$

Comparing the equations (4.3.6) and (4.3.7), it is natural to ask: why the quantity of overall information I_{A+B} carried by two independent Gaussian signals $a(t)$ and $b(t)$ with identical normalized correlation functions is twice as large as the quantity of absolute information I_X contained in additive mixture $x(t)$ of the signals $a(t)$ and $b(t)$, although the strict equality between them seems to be intuitively correct:

$$I_{A+B} = I_X?$$

Nevertheless, during additive interaction (4.3.1) of two stationary statistically independent signals $a(t)$, $b(t)$ with zero expectations and arbitrary NFSIs $\psi_a(\tau)$, $\psi_b(\tau)$ (and, respectively, IDDs $i_a(\tau)$, $i_b(\tau)$) in linear space $\mathcal{LS}$, the informational inequality always holds:

$$I_X < I_{A+B} = I_A + I_B. \tag{4.3.8}$$

We shall call this relationship *informational paradox* (informational inequality) of additive interaction between the signals in linear space, whose essence lies in nonequivalence ($I_{A+B} \neq I_X$, $I_{A+B} > I_X$) of both quantity of overall information $I_{A+B} = I_A + I_B$ contained in two statistically independent signals $a(t)$, $b(t)$ and quantity of absolute information I_X contained in their additive sum $x(t)$.

On the qualitative level, this paradox can be elucidated by the losses of information ΔI that accompany an additive interaction of statistically independent signals $a(t)$ and $b(t)$, and are stipulated by destroying the part of information contained in the signals:

$$\Delta I = I_{A+B} - I_X.$$

Example 4.3.1. In the interaction of two Gaussian statistically independent stochastic signals with identical normalized correlation functions $r_a(\tau) = r_b(\tau) = r(\tau)$ (i.e. $I_A = I_B = I_X$) and arbitrary variances D_a, D_b, the value of losses ΔI is equal to the sum of information quantities carried by both signals:

$$\Delta I = (I_A + I_B)/2 = I_A = I_B = I_X. \quad \triangledown$$

Thus, interaction of the signals in linear space $\mathcal{LS}$ is accompanied by losses of information contained in them.

We will distinguish the corresponding sorts of information quantity defined for informational $\boldsymbol{\Omega}$ and physical $\boldsymbol{\Gamma}$ signal spaces. For quantity of absolute information, quantity of mutual information, and quantity of overall information, defined in informational signal space $\boldsymbol{\Omega}$, we preserve the designations accepted in Section 4.1, respectively: I_A, I_B; I_{AB}; I_{A+B}. All these notions defined in physical signal space $\boldsymbol{\Gamma}$ with an interaction of the signals $a(t)$, $b(t)$ in the form of $x(t) = a(t) \oplus b(t)$, where $\oplus$ is some binary operation in $\boldsymbol{\Gamma}$, we shall denote as I_a, I_b; I_{ab}; I_{a+b}, respectively.

The problem of information losses during signal interaction in physical signal space $\boldsymbol{\Gamma}$ in general, and in particular, during their additive interaction (4.3.1) in linear space $\mathcal{LS}$, has, surely, nontrivial character. In physical signal spaces that differ from generalized Boolean algebra in their algebraic structure, the informational relationships between the interacting signals obtained in Section 4.1 for informational signal space $\boldsymbol{\Omega}$ cease to hold. This means that physical interaction of the signals negatively affects the value of some types of information quantities defined for physical signal space $\boldsymbol{\Gamma}$. In particular, quantity of overall information $I_{a \oplus b} \neq I_{A+B}$

and quantity of mutual information $I_{ab} \neq I_{AB}$ corresponding to a pair of interacting signals $a(t)$, $b(t)$ are changed. Meanwhile, the quantity of absolute information contained in each signal remains unchanged:

$$I_a = I_A, \quad I_b = I_B. \tag{4.3.9}$$

Thus, the notion of quantity of absolute information does not need refinement, inasmuch as it is defined irrespective of other signals of physical signal space; it is introduced exclusively with respect to a single signal.

Now, for physical signal space Γ with binary operation $\oplus$ defined on it, informational inequality (4.3.8) can be written in the form:

$$I_{a \oplus b} \leq I_{A+B} = I_A + I_B - I_{AB}, \tag{4.3.10}$$

where $I_{a \oplus b}$ is a quantity of overall information contained in a pair of signals $a(t)$ and $b(t)$ in physical space Γ that is equal to the quantity of absolute information $I_x = I_X$ contained in the result of their interaction $x(t) = a(t) \oplus b(t)$: $I_{a \oplus b} = I_x$; I_{A+B} is a quantity of overall information contained in a pair of signals $a(t)$ and $b(t)$ in informational signal space Ω; I_{AB} is a quantity of mutual information contained in a pair of signals $a(t)$ and $b(t)$ in informational signal space Ω; $I_a = I_A$, $I_b = I_B$ is a quantity of absolute information contained in every signal $a(t)$ and $b(t)$ separately. Unlike inequality (4.3.8), inequality (4.3.10) is not strict, assuming the existence of signal spaces, where the signal interaction occurs without information losses.

For any physical signal space Γ with arbitrary algebraic properties, there exists a pair of signals $a(t)$ and $b(t)$, for which informational inequality (4.3.10) turns to identity. Informational properties of such signals are introduced by the following definition.

Definition 4.3.1. Two signals $a(t)$ and $b(t)$ of physical signal space Γ, $a(t), b(t) \in \Gamma$ are called *identical in an informational sense*, if for them the inequality (4.3.10) turns to the identity:

$$I_{a \oplus b} = I_{A+B} = I_A = I_B,$$

so that the images A and B of the signals $a(t)$ and $b(t)$ in informational space Ω are identical: $A \equiv B$.

As follows from Definition 4.3.1, any signal $a(t)$ is identical to itself in an informational sense. For instance, in the case of additive interaction in linear space $\mathcal{LS}$, the signals $a(t)$ and $b(t)$ are identical in an informational sense, if they are connected by linear dependence: $a(t) = k \cdot b(t)$, where $k = \text{const}$.

According to formulated informational inequality (4.3.10), the following questions may be formulated.

1. Why do the losses of information take place during signal interaction in linear space $\mathcal{LS}$?

2. Is the signal interaction accompanied by the losses of information in all the spaces?

3. If the answer to the second question is affirmative, what are the signal spaces where such losses are minimal? If the answer to the second question is negative, what are the signal spaces where the signal interaction is possible without losses of information?

One can answer these questions, based on the following definition.

Definition 4.3.2. *Ideal interaction* $x(t) = a(t) \oplus b(t)$ between two statistically independent signals $a(t)$ and $b(t)$ in physical space Γ, where two binary operations are defined: addition $\oplus$ and multiplication $\otimes$, is a binary operation $\oplus$, which provides the quantities of overall information $I_{a \oplus b}$, I_{A+B}, contained in a pair of signals $a(t)$, $b(t)$ and defined for both physical Γ and informational Ω signal spaces, respectively, to be equivalent:

$$I_{a \oplus b} = I_{A+B}. \tag{4.3.11}$$

Here and below, the binary operations of addition $\oplus$ and multiplication $\otimes$ are understood as abstract operations over some algebraic structure. Find out, what are the algebraic properties of the physical signal space providing the identity (4.3.11) to hold. The answer to this question is given by the following theorem.

Theorem 4.3.1. *Let there are two binary operations of addition $\oplus$ and multiplication $\otimes$ defined in physical signal space Γ. Then for the identity (4.3.11) to hold, it is necessary and sufficient that physical signal space Γ be a generalized Boolean algebra* $\mathbf{B}(\Gamma)$ *with a measure* m_Γ.

Proof of necessity. Identity (4.3.11) may be written in more detailed form:

$$I_{a \oplus b} = I_x = I_X = I_{A+B} = I_A + I_B - I_{AB} = I_a + I_b - I_{ab}. \tag{4.3.12}$$

According to (4.3.9), the identities between the quantities of absolute information of the signals $a(t)$ and $b(t)$ defined for both physical Γ and informational Ω signal spaces, respectively, hold: $I_a = I_A$, $I_b = I_B$. Besides, the identity holds $I_{ab} = I_{AB}$ between the quantities of mutual information I_{ab} and I_{AB} defined for both physical Γ and informational Ω signal spaces, respectively. Then a measure m_Γ of the elements (the signals) $\{a(t), b(t), \ldots\}$ of physical space Γ is isomorphic to a measure m of informational signal space Ω: i.e., for every $c(t) \in \Gamma$, $\exists C \in \Omega$, $C = \varphi[c(t)]$: $m_\Gamma[c(t)] = m(C)$ [176]; and the mapping $\varphi \colon \Gamma \to \Omega$ defines isomorphism of the spaces Γ and Ω into each other: i.e., $\forall \varphi$, $\exists \varphi^{-1}$: $\varphi^{-1} \colon \Omega \to \Gamma$. Thus, physical signal space Γ, where the identity (4.3.11) holds, is the algebraic structure identical to informational signal space Ω, i.e., generalized Boolean algebra $\mathbf{B}(\Gamma)$ with a measure m_Γ and signature $(\oplus, \otimes, -, \mathbf{O})$. □

Proof of sufficiency. If physical signal space Γ is a generalized Boolean algebra $\mathbf{B}(\Gamma)$ with a measure m_Γ and signature $(\oplus, \otimes, -, \mathbf{O})$, then the mapping φ defined by the relationship (4.1.2) $\varphi \colon \Gamma \to \Omega$ is a homomorphism preserving all the signature operations, and the following equations hold:

$$\varphi[a(t) \oplus b(t)] = \varphi[a(t)] + \varphi[b(t)]; \tag{4.3.13a}$$

$$\varphi[a(t) \otimes b(t)] = \varphi[a(t)] \cdot \varphi[b(t)], \tag{4.3.13b}$$

where $x(t) = a(t) \oplus b(t)$, $\tilde{x}(t) = a(t) \otimes b(t)$; $\varphi[a(t)] = A$, $\varphi[b(t)] = B$, $\varphi[a(t) \oplus b(t)] = X$, $\varphi[a(t) \otimes b(t)] = \tilde{X}$; $a(t), b(t), x(t), \tilde{x}(t) \in \Gamma$; $A, B, X, \tilde{X} \in \Omega$.

According to (4.3.9), the identities between the quantities of absolute information of the signals $a(t), b(t), x(t), \tilde{x}(t)$ defined for both physical Γ and informational Ω signal spaces, respectively, hold: $I_a = I_A$, $I_b = I_B$, $I_x = I_X$, $I_{\tilde{x}} = I_{\tilde{X}}$. These identities define isomorphism of the measures m_Γ, m of both physical Γ and informational Ω signal spaces: $m_\Gamma[a(t)] = m(A)$, $m_\Gamma[b(t)] = m(B)$, $m_\Gamma[x(t)] = m(X)$, $m_\Gamma[\tilde{x}(t)] = m(\tilde{X})$. Then the mapping $\varphi\colon \Gamma \to \Omega$ is an isomorphism preserving a measure and $m_\Gamma[x(t)] = m(X) \Rightarrow I_x = I_X \Rightarrow I_{a \oplus b} = I_{A+B}$. □

Note that a measure $m_\Gamma[\tilde{x}(t)]$ of the signal $\tilde{x}(t) = a(t) \otimes b(t)$ gives a sense of the quantity of mutual information I_{ab} contained in both the signal $a(t)$ and the signal $b(t)$, which will be also denoted below as $I_{a \otimes b}$, indicating the relation of this measure to the binary operation $\otimes$ of the space Γ.

Thus, the main content of the Theorem 4.3.1 claims that during the interaction $x(t) = a(t) \oplus b(t)$ of the signals $a(t)$ and $b(t)$ in physical space Γ in the form of generalized Boolean algebra, the measures of the corresponding sorts of information quantities defined for both physical Γ and informational Ω signal spaces are isomorphic and the identities hold:

$$I_a = I_A,\ I_b = I_B,\ I_x = I_X,\ I_{\tilde{x}} = I_{\tilde{X}},\ I_{a \oplus b} = I_{A+B},\ I_{ab} = I_{a \otimes b} = I_{AB}.$$

The answers to the questions above are as follows.

1. During signal interaction in linear space $\mathcal{LS}$, the losses of information take place. The presence of such losses is explained by the fact that linear space $\mathcal{LS}$ is not isomorphic to general Boolean algebra with a measure.

2. Interaction of the signals is accompanied by losses of information that do not affect all the spaces. The exceptions are the spaces with the properties of a generalized Boolean algebra with a measure.

Unfortunately, Theorem 4.3.1 does not give concrete recommendations for obtaining the signal spaces with these useful informational properties. The requirement for the informational identity (4.3.11) to be valid for practical application may be too strict. In this case, it is enough to require the quantity of mutual information $I_{ax} = I_{a \otimes x}$, $I_{bx} = I_{b \otimes x}$, contained in the signals $a(t)$ and $b(t)$ and in the result of their interaction $x(t)$, to be identically equal to the quantities of absolute information I_a and I_b contained in these signals:

$$\begin{cases} I_{a \otimes x} = I_a; \\ I_{b \otimes x} = I_b. \end{cases}$$

On the basis of this approach, one may define a sort of interaction of the signals in physical signal space that differs from ideal interaction with respect to its informational properties. This formulation is closer to applied aspects of signal processing and is expanded in its algebraic interpretation.

Definition 4.3.3. *Quasi-ideal interaction* $x(t) = a(t) \oplus b(t)$ of two signals $a(t)$ and $b(t)$ in physical signal space Γ, where two binary operations of addition $\oplus$ and

4.3 Features of Signal Interaction in Signal Spaces with Various Algebraic Properties

multiplication $\otimes$ are defined, is a binary operation $\oplus$ providing that the quantity of mutual information $I_{ax} = I_{a\otimes x}$, $I_{bx} = I_{b\otimes x}$ contained in the signals $a(t)$ and $b(t)$, and in the result of their interaction $x(t)$ is equal to the quantity of absolute information I_a, I_b contained in these signals, respectively:

$$\begin{cases} I_{a\otimes x} = I_a; & (a) \\ I_{b\otimes x} = I_b; & (b) \\ x(t) = a(t) \oplus b(t). & (c) \end{cases} \quad (4.3.14)$$

Find out what are the algebraic properties of physical signal space Γ for the equation system (4.3.14) to hold. The answer to this question is given by the following theorem.

Theorem 4.3.2. *Let there be two binary operations of addition $\oplus$ and multiplication $\otimes$ defined in physical signal space Γ. Then for the equation system (4.3.14) to hold, it is necessary and sufficient that physical signal space Γ be a lattice with a measure m_Γ and operations of join $\oplus$ and meet $\otimes$.*

Proof. If physical signal space Γ is a lattice with operations of join $a(t) \oplus b(t)$ and meet $a(t) \otimes b(t)$, then the following relationships hold:

$$\begin{cases} a(t) \otimes x(t) = a(t); & (a) \\ b(t) \otimes x(t) = b(t); & (b) \\ x(t) = a(t) \oplus b(t), & (c) \end{cases} \quad (4.3.15)$$

where $a(t) \oplus b(t) = \sup_\Gamma \{a(t), b(t)\}$; $a(t) \otimes b(t) = \inf_\Gamma \{a(t), b(t)\}$.

Identities (4.3.15a) and (4.3.15b) define axioms of absorption of lattice [223], [221]. If physical signal space Γ is a lattice with a measure m_Γ, then the system (4.3.15) determines the following identities:

$$\begin{cases} m_\Gamma(a(t) \otimes x(t)) = m_\Gamma(a(t)); & (a) \\ m_\Gamma(b(t) \otimes x(t)) = m_\Gamma(b(t)); & (b) \\ x(t) = a(t) \oplus b(t), & (c) \end{cases} \quad (4.3.16)$$

where $m_\Gamma(a(t) \otimes x(t)) = I_{a\otimes x}$, $m_\Gamma(b(t) \otimes x(t)) = I_{b\otimes x}$, $m_\Gamma(a(t)) = I_a$, $m_\Gamma(b(t)) = I_b$.

Thus, for the identities (4.3.14a) and (4.3.14b) of the system (4.3.14) to hold, it is sufficient that physical signal space Γ be a lattice with a measure. □

Thus, for the identities (4.3.14a) and (4.3.14b) of the system (4.3.14) to hold, i.e., for quasi-ideal interaction in physical signal space Γ to take place, it is sufficient that physical signal space Γ be a lattice $\Gamma(\vee, \wedge)$ with operations of join and meet, respectively: $a(t) \vee b(t) = \sup_\Gamma \{a(t), b(t)\}$, $a(t) \wedge b(t) = \inf_\Gamma \{a(t), b(t)\}$.

Then for interaction of two signals $a(t)$, $b(t)$ in physical signal space Γ with lattice operations: $x(t) = a(t) \vee b(t)$ or $\tilde{x}(t) = a(t) \wedge b(t)$, the initial requirement (4.3.14) could be written in slightly extended form owing to duality of lattice operation properties with the help of two equation systems, respectively:

$$\begin{cases} I_{a\wedge x} = I_a; & (a) \\ I_{b\wedge x} = I_b; & (b) \\ x(t) = a(t) \vee b(t), & (c) \end{cases} \quad (4.3.17)$$

$$\begin{cases} I_{a \vee \tilde{x}} = I_a; & (a) \\ I_{b \vee \tilde{x}} = I_b; & (b) \\ \tilde{x}(t) = a(t) \wedge b(t). & (c) \end{cases} \quad (4.3.18)$$

It should be noted that in physical signal space Γ, where ideal interaction of the signals exists (i.e., the identity (4.3.11) holds), the relationships (4.3.14a) and (4.3.14b) unconditionally hold, and correspondingly there exists a quasi-ideal interaction of the signals. Physical signal space Γ, which is a generalized Boolean algebra $\mathbf{B}(\Gamma)$ with a measure m_Γ and signature $(\oplus, \otimes, -, \mathbf{O})$, is also a lattice of signature $(\oplus, \otimes)$ with operations of least upper bound and greatest lower bound (join and meet), respectively:

$$a(t) \oplus b(t) = \sup_{\Gamma}\{a(t), b(t)\}; \quad a(t) \otimes b(t) = \inf_{\Gamma}\{a(t), b(t)\}.$$

On the base of Definition 4.3.3, another variant of formulation of the notion of quasi-ideal interaction may be suggested that does not directly reflect informational properties of interaction of this kind, putting an accent on its algebraic properties only.

Definition 4.3.4. *Quasi-ideal interaction $x(t) = a(t) \oplus b(t)$ of useful $a(t)$ and interference $b(t)$ signals in physical signal space Γ, where two binary operations of addition $\oplus$ and multiplication $\otimes$ are defined, is a binary operation $\oplus$ forming a result of signal interaction $x(t)$ that allows extracting a completely known signal $a(t)$ from a mixture $x(t)$ without losses of information, so that the identity holds:*

$$\hat{a}(t) \otimes x(t) = a(t), \quad (4.3.19)$$

where $\hat{a}(t)$ is some deterministic function of the signals $a(t)$ and $x(t)$.

Find out what are the algebraic properties of physical signal space Γ for the identity (4.3.19) to hold. Also we must establish the kind of the function $\hat{a}(t)$ satisfying Equation (4.3.19). The answer to this question is given by the following theorem.

Theorem 4.3.3. *Let there be two binary operations of addition $\oplus$ and multiplication $\otimes$, which are defined in physical signal space Γ. Then for the identity (4.3.19) to hold, it is sufficient that physical signal space Γ is a lattice with signature $(\oplus, \otimes)$ and operations of join and meet, respectively:*

$$a(t) \oplus b(t) = \sup_{\Gamma}\{a(t), b(t)\}; \quad a(t) \otimes b(t) = \inf_{\Gamma}\{a(t), b(t)\}.$$

Proof. Definition 4.3.4 implies that in quasi-ideal interaction $x(t) = a(t) \oplus b(t)$ of a completely known useful signal $a(t)$ with interference (noise) signal $b(t)$ in physical signal space Γ, there exists a binary operation $\otimes$ that allows, with the help of the estimator $\hat{a}(t)$ of the signal $a(t)$ with completely known parameters ($\hat{a}(t) = a(t)$), obtaining (extracting) the useful signal $a(t)$ from the result of interaction $x(t)$ without information losses:

$$y(t) = \hat{a}(t) \otimes x(t) = a(t) \otimes (a(t) \oplus b(t)) = a(t). \quad (4.3.20)$$

4.3 Features of Signal Interaction in Signal Spaces with Various Algebraic Properties

The last interrelation determines the absorption property for a lattice with operations of join and meet, respectively: $a(t) \oplus b(t)$ and $a(t) \otimes b(t)$. This means that for quasi-ideal interaction of useful $a(t)$ and interference $b(t)$ signals defined above to exist in physical signal space $\boldsymbol{\Gamma}$, it is sufficient that signal space $\boldsymbol{\Gamma}$ be a lattice with signature $(\oplus, \otimes)$. $\square$

It should be noted that for physical signal space $\boldsymbol{\Gamma}$ with lattice properties and signature $(\oplus, \otimes)$, for dual interaction of the signals in the form of $\tilde{x}(t) = a(t) \otimes b(t)$, the relationship that is dual with respect to the equality (4.3.20), holds:

$$y(t) = \hat{a}(t) \oplus x(t) = a(t) \oplus (a(t) \otimes b(t)) = a(t). \tag{4.3.21}$$

The utility of the notion of quasi-ideal interaction is illustrated by the following examples.

Example 4.3.2. Consider a model of interaction of the signal $s_i(t)$ from the set of deterministic signals $S = \{s_i(t)\}$, $i = 1, \ldots, m$ and noise $n(t)$ in signal space $\boldsymbol{\Gamma}(\vee, \wedge)$ with lattice properties and operations of join $a(t) \vee b(t)$ and meet $a(t) \wedge b(t)$ $(a(t), b(t) \in \boldsymbol{\Gamma}(\vee, \wedge))$, respectively:

$$x(t) = s_i(t) \vee n(t), \; t \in T_s, \; i = 1, \ldots, m, \tag{4.3.22}$$

where $T_s = [t_0, t_0 + T]$ is the domain of definition of the signal $s_i(t)$; t_0 is a known time of signal arrival; T is a signal duration; $m \in \mathbf{N}$, $\mathbf{N}$ is the set of natural numbers.

We consider that the noise $n(t)$ is characterized by arbitrary probabilistic-statistical properties. In the presence of the signal $s_k(t)$, $s_k(t) \in S$, $1 \leq k \leq m$ in the observed process $x(t)$: $x(t) = s_k(t) \vee n(t)$, $t \in T_s$, to solve a classification problem for the signal $s_k(t)$ from a set of deterministic signals $S = \{s_i(t)\}$ in the presence of noise $n(t)$, it is necessary to form the estimator $\hat{s}_k(t)$, which is equal to the received signal $s_k(t)$, so that the structure-forming function of the optimal signal processing algorithm has the form $\hat{s}_k(t) \wedge x(t)$, and the result of optimal processing $y_k(t)$, according to the lattice absorption axiom, is identically equal to the received signal $s_k(t)$:

$$y_k(t) = \hat{s}_k(t) \wedge x(t) = s_k(t) \wedge [s_k(t) \vee n(t)] = s_k(t).$$

Thus, regardless of parametric and nonparametric prior uncertainty conditions and correspondingly, regardless of probabilistic-statistical properties of noise (interference), the optimal demodulator of deterministic signals in the signal space with lattice properties accurately classifies (demodulates) the signals from the given set $S = \{s_i(t)\}$, $i = 1, \ldots, m$. $\triangledown$

Example 4.3.3. Shannon's theorem on capacity of communication channel with additive white Gaussian noise implies a lower bound value of the ratio E_b/N_0 of the energy E_b that falls to one bit of information contained in the signal to the value of noise power spectral density N_0; this ratio is called an ultimate Shannon limit. This value E_b/N_0 equal to $\ln 2$ establishes the limit of errorless transmission of information; see, for instance, [164]. The former example implies that while solving the classification problem of deterministic signals in the presence of noise, the value E_b/N_0 and the probability of error receiving of the signal $s_i(t)$ from a set of deterministic signals $S = \{s_i(t)\}$, $i = 1, \ldots, m$ may be arbitrarily small. This, however, does not mean that the unbounded capacity of a communication channel with noise can be provided in such signal spaces. In particular, Chapter 5 will show that communication channel capacity, even in the absence of noise, is always a finite value. ▽

4.3.2 Informational Relationships Characterizing Signal Interaction in Signal Spaces with Various Algebraic Properties

Analysis of informational relationships taking place under different sorts of signal interactions in the signal spaces, i.e., spaces with various algebraic properties, has to be performed from unified positions. To provide such an approach, we use the following considerations.

The statement of Sections 3.2 and 3.3 that any pair of stochastic processes can be considered a partially ordered set may be generalized for a physical signal space as a whole. Any physical signal space $\mathbf{\Gamma}$ can be considered a partially ordered set. In every time instant $t \in T$ between two instantaneous values (samples) a_t, b_t of the signals $a(t), b(t) \in \mathbf{\Gamma}$, a relation of order $a_t \leq b_t$ (or $a_t \geq b_t$) is defined. Then the partially ordered set $\mathbf{\Gamma}$ is a lattice with operations of join and meet, respectively: $a_t \vee b_t = \sup_{\mathbf{\Gamma}}\{a_t, b_t\}$, $a_t \wedge b_t = \inf_{\mathbf{\Gamma}}\{a_t, b_t\}$, and if $a_t \leq b_t$, then $a_t \wedge b_t = a_t$ and $a_t \vee b_t = b_t$ [221]:

$$a_t \leq b_t \Leftrightarrow \begin{cases} a_t \wedge b_t = a_t; \\ a_t \vee b_t = b_t. \end{cases}$$

Thus, let a_t and b_t be the instantaneous values (samples) of stochastic signals $a(t)$ and $b(t)$ with symmetric (even) univariate probability density functions (PDFs) $p_a(u)$, $p_b(v)$: $p_a(u) = p_a(-u)$; $p_b(v) = p_b(-v)$. The quantity $\nu_{\mathbf{P}}(a_t, b_t)$, introduced by Definition 3.3.1 in Section 3.3, is called a probabilistic measure of statistical interrelationship (PMSI) between a pair of the samples a_t, b_t of stochastic signals $a(t)$ and $b(t)$ in physical signal space $\mathbf{\Gamma}$: $a(t), b(t) \in \mathbf{\Gamma}$:

$$\nu_{\mathbf{P}}(a_t, b_t) = 3 - 4\mathbf{P}[a_t \vee b_t > 0], \qquad (4.3.23)$$

where $\mathbf{P}[a_t \vee b_t > 0]$ is the probability that random variable $a_t \vee b_t$, which is equal to join of the samples a_t and b_t, takes a value greater than zero.

Within further consideration of informational relationships taking place during signal interaction in signal spaces with various algebraic properties, in the essential measure, the notion of PMSI and also the material of Section 3.3 will be used.

The notions of ideal and quasi-ideal interactions between the signals are introduced by Definitions 4.3.2, 4.3.3 respectively, and besides, it is specified that physical signal space $\mathbf{\Gamma}$, where these sorts of interactions take place, necessarily has to possess two binary operations, i.e., the operations of addition $\oplus$ and multiplication $\otimes$ appearing in definition of the notions of quantity of overall information $I_{a \oplus b}$ and quantity of mutual information $I_{ab} = I_{a \otimes b}$, respectively. The last ones are equal to the quantity of absolute information I_x, $I_{\tilde{x}}$ contained in the results of these binary operations $x(t) = a(t) \oplus b(t)$, $\tilde{x}(t) = a(t) \otimes b(t)$ between the signals $a(t)$, $b(t)$, respectively:

$$I_x = I_{a \oplus b}; \quad I_{\tilde{x}} = I_{a \otimes b}. \tag{4.3.24}$$

Note that two aforementioned binary operations (of addition $\oplus$ and multiplication $\otimes$) cannot be defined at the same time in any physical signal space $\mathbf{\Gamma}$. The exceptions are the signal spaces with group (semigroup) properties, where the only binary operation is defined over the signals, i.e., either addition $\oplus$ or multiplication $\otimes$.

There exists a theoretical problem which will be considered within physical signal space $\mathbf{\Gamma}$ with the properties of an additive group of linear space, where only one binary operation between the signals is defined, i.e., addition $\oplus$. By a method based on the relationships (4.3.24), we can define only a quantity of overall information $I_{a \oplus b}$ contained in the result of interaction $x(t) = a(t) \oplus b(t)$ in such a signal space. It is not possible to define similarly a quantity of mutual information contained in both signals $a(t)$ and $b(t)$.

In order to provide an approach to analyze informational relationships taking place in the signal space with various types interactions, it is necessary to define the quantity of mutual information contained in both signals $a(t)$ and $b(t)$ in such a way, that is acceptable for the signal spaces with minimal number of binary operations between the signals.

It is obvious that the minimal number of operations is equal to one; in this case, the only binary operation characterizes an interaction of the signals in physical signal space with group (semigroup) properties. On the other hand, a definition of a quantity I_{ab} should provide the introduced notion will not contradict the obtained results. The necessary approach to a formulation of the quantity of mutual information I_{ab} that is acceptable for the signal spaces with arbitrary algebraic properties is given by the following definition.

Definition 4.3.5. *For stationary and stationary coupled stochastic signals $a(t)$ and $b(t)$, which interact in physical signal space $\mathbf{\Gamma}$ with arbitrary algebraic properties, quantity of mutual information, contained in the signals $a(t)$ and $b(t)$, is a quantity I_{ab} equal to:*

$$I_{ab} = \nu_{ab} \min(I_a, I_b), \tag{4.3.25}$$

where $\nu_{ab} = \nu_{\mathbf{P}}(a_t, b_t)$ is PMSI between the samples a_t and b_t of stochastic signals $a(t)$ and $b(t)$ in physical signal space $\mathbf{\Gamma}$ determined by the relationship (4.3.23).

With help from Definition 4.3.5, for an arbitrary signal $a(t)$ in physical signal space $\mathbf{\Gamma}$ with arbitrary algebraic properties, we introduce a measure of information quantity m_Γ defined by the Equation (4.3.25):

$$m_\Gamma[a(t)] = I_{aa} = \nu_{aa} \min(I_a, I_a) = I_a.$$

As shown in Section 3.3, for physical signal space $\mathbf{\Gamma}$ with group properties, the notions of PMSI (see Definition 3.3.1) and normalized measure of statistical interrelationship (NMSI) (see Definition 3.2.2) coincide; as for signal space $\mathbf{\Gamma}$ with lattice properties, these notions have different content. In this section, the further analysis of informational relationships between the signals interacting in physical signal space $\mathbf{\Gamma}$ with various algebraic properties will be performed on the base of quantity of mutual information (4.3.25), and this quantity I_{ab} will be used on the base of PMSI ($\nu_{ab} = \nu_\mathbf{P}(a_t, b_t)$), although it can be defined through NMSI ($\nu_{ab} = \nu(a_t, b_t)$). So, for instance, in the next section, this notion will be defined and used on the basis of NMSI ($\nu_{ab} = \nu(a_t, b_t)$).

As noted at the beginning of the section, in applied problems of radiophysics and radioengeneering, an additive interaction of useful signal and interference (noise) is considered. In some special cases, a multiplicative signal interaction is the subject of interest [244]. These and other sorts of signal interactions in physical signal space that differ in their informational properties from the above kinds of interaction will refer to a large group of interactions with so-called usual informational properties with help of the following definition based on the notion of quantity of mutual information introduced by Definition 4.3.5.

Definition 4.3.6. *Usual interaction* of two signals $a(t)$ and $b(t)$ in physical signal space $\mathbf{\Gamma}$ with semigroup properties is a binary operation $\oplus$ providing the quantity of mutual information I_{ax}, I_{bx} contained in the distinct signals $a(t)$, $b(t)$ ($a(t) \neq b(t)$) and in the result of their interaction $x(t) = a(t) \oplus b(t)$ is less than the quantity of absolute information I_a, I_b contained in these signals, respectively:

$$\begin{cases} I_{ax} < I_a; & (a) \\ I_{bx} < I_b; & (b) \\ x(t) = a(t) \oplus b(t). & (c) \end{cases} \quad (4.3.26)$$

Recall, that all the definitions of information quantities listed in Section 4.1 are based upon an axiomatic statement, according to which, the sorts of information quantities are completely defined by the kind of binary operation between the images A and B of the signals $a(t)$ and $b(t)$ in informational signal space $\mathbf{\Omega}$ built upon generalized Boolean algebra $\mathbf{B(\Omega)}$ with a measure m and signature $(+, \cdot, -, \mathbf{O})$. Meanwhile, the quantity of mutual information is introduced differently in Definition 4.3.5, i.e., by PMSI and a quantity of absolute information contained in the interacting signals $a(t)$, $b(t)$. First, it adjusts to the corresponding notion introduced in Section 4.1. Second, it allows considering the features of informational relationships between the signals interacting in physical signal space $\mathbf{\Gamma}$ with arbitrary algebraic properties.

For ideal interaction $x(t) = a(t) \oplus b(t)$ between the signals $a(t)$ and $b(t)$ in physical signal space $\mathbf{\Gamma}$ with the properties of generalized Boolean algebra $\mathbf{B(\Gamma)}$ with a measure m_Γ and signature $(\oplus, \otimes, -, \mathbf{O})$, the following informational relationships hold:

$$\begin{cases} I_a + I_b - I_{ab} = I_x; & (a) \\ I_{ax} = I_a; & (b) \\ I_{bx} = I_b, & (c) \end{cases} \quad (4.3.27)$$

where $I_a = m_\Gamma[a(t)] = I_A = m(A)$, $I_b = m_\Gamma[b(t)] = I_B = m(B)$, $I_x = m_\Gamma[x(t)] = I_X = I_{A+B} = m(A+B)$, $I_{ab} = m_\Gamma[a(t) \otimes b(t)] = I_{AB} = m(AB)$; m is a measure of informational signal space Ω that is isomorphic to a measure m_Γ of physical signal space Γ.

Equality (4.3.27a) represents an informational identity of ideal interaction (4.3.11) and (4.3.12). Equalities (4.3.27b) and (4.3.27c) represent informational identities of quasi-ideal interaction (4.3.14a), (4.3.14b) that take place during ideal interaction of signals. The values of quantities of mutual information I_{ab}, I_{ax}, I_{bx} appearing in all the equations of the system (4.3.27) should be considered in the context of Definition 4.3.5. The identities (4.3.27b) and (4.3.27c) are the consequence of lattice absorption axioms (4.3.15a) and (4.3.15b). Based on the equations in (4.3.27), the following system of equations can be obtained:

$$\begin{cases} \dfrac{I_{ax}}{I_x} + \dfrac{I_{bx}}{I_x} - \dfrac{I_{ab}}{I_x} = 1; & (a) \\ I_{ax} = I_a; & (b) \\ I_{bx} = I_b. & (c) \end{cases} \quad (4.3.28)$$

For quasi-ideal interaction $x(t) = a(t) \oplus b(t)$ between the signals $a(t)$ and $b(t)$ in physical signal space Γ with lattice properties with a measure m_Γ, the following main informational relationships hold:

$$\begin{cases} I_a + I_b - I_{ab} > I_x; & (a) \\ I_{ax} = I_a; & (b) \\ I_{bx} = I_b, & (c) \end{cases} \quad (4.3.29)$$

where $I_a = m_\Gamma[a(t)] = I_A = m(A)$, $I_b = m_\Gamma[b(t)] = I_B = m(B)$, $I_x = m_\Gamma[x(t)] = I_X = m(X)$, $I_{ab} = \nu_{ab}\min(I_a, I_b)$; m is a measure of informational signal space Ω; m_Γ is a measure of physical signal space Γ with lattice properties.

Here, the inequality (4.3.29a) represents the informational inequality (4.3.10), and the equalities (4.3.29b) and (4.3.29c) represent informational identities of quasi-ideal interaction (4.3.14a) and (4.3.14b). The quantities of mutual information I_{ab}, I_{ax}, I_{bx}, appearing in all the equations of the system (4.3.29), should be considered in the sense of Definition 4.3.5. The identities (4.3.29b) and (4.3.29c) follow from the results (3.3.9a) and (3.3.9b) of Theorem 3.3.7.

On the basis of relationships of the system (4.3.29), the following system of relationships can be obtained:

$$\begin{cases} \dfrac{I_{ax}}{I_x} + \dfrac{I_{bx}}{I_x} - \dfrac{I_{ab}}{I_x} > 1; & (a) \\ I_{ax} = I_a; & (b) \\ I_{bx} = I_b. & (c) \end{cases} \quad (4.3.30)$$

For interaction $x(t) = a(t) \oplus b(t)$ of a pair of stochastic signals $a(t)$ and $b(t)$ in physical signal space Γ with group properties (i.e., for additive interaction $x(t) = a(t) + b(t)$ in linear signal space $\mathcal{LS}$), the following main informational relationships hold:

$$\begin{cases} \nu_\mathbf{P}(a_t, x_t) + \nu_\mathbf{P}(b_t, x_t) - \nu_\mathbf{P}(a_t, b_t) = 1; & (a) \\ \nu_\mathbf{P}(a_t, x_t) < 1; & (b) \\ \nu_\mathbf{P}(b_t, x_t) < 1, & (c) \end{cases} \quad (4.3.31)$$

where $\nu_{\mathbf{P}}(a_t, x_t)$, $\nu_{\mathbf{P}}(b_t, x_t)$, $\nu_{\mathbf{P}}(a_t, b_t)$ are PMSIs of the corresponding pairs of their samples a_t, x_t; b_t, x_t; a_t, b_t of stochastic signals $a(t)$, $b(t)$, $x(t)$.

The equality (4.3.31a) represents the result (3.3.7) of the Theorem 3.3.6. The inequalities (4.3.31b), (4.3.31c) are the consequences of the inequalities (4.3.26a), (4.3.26b) of Definition 4.3.6. Considering the relationship (4.3.25) of Definition 4.3.5, the system (4.3.31) can be overwritten in the following form:

$$\begin{cases} \dfrac{I_{ax}}{I_a} + \dfrac{I_{bx}}{I_b} - \dfrac{I_{ab}}{\min[I_a, I_b]} = 1; & (a) \\ I_{ax} < I_a; & (b) \\ I_{bx} < I_b, & (c) \\ \dfrac{I_a}{I_x} + \dfrac{I_b}{I_x} - \dfrac{I_{ab}}{I_x} > 1, & (d) \end{cases} \quad (4.3.32)$$

where $I_a = I_A = m(A)$, $I_b = I_B = m(B)$, $I_x = I_X = m(X)$; $I_{ab} = \nu_{ab} \min(I_a, I_b)$, $I_{ax} = \nu_{ax} \min(I_a, I_x)$, $I_{bx} = \nu_{bx} \min(I_b, I_x)$; m is a measure of informational signal space Ω.

The equality (4.3.32a) is the consequence of joint fulfillment of the relationships (4.3.31a) and (4.3.25). The inequalities (4.3.32b), (4.3.32c) are the consequences of the inequalities (4.3.26a), (4.3.26b) of Definition 4.3.6. Inequality (4.3.32d) represents the informational inequality (4.3.10).

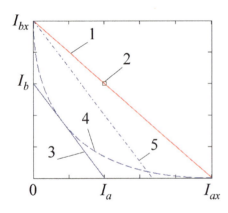

FIGURE 4.3.1 Dependences $I_{bx}(I_{ax})$ of quantity of mutual information I_{bx} on quantity of mutual information I_{ax}

Fig. 4.3.1 illustrates the dependences $I_{bx}(I_{ax})$ of quantity of mutual information I_{bx} contained in the signal $x(t)$ with respect to the signal $b(t)$ on quantity of mutual information I_{ax}, contained in the signal $x(t)$ with respect to the signal $a(t)$: line 1 corresponds to ideal interaction; point 2 is the quasi-ideal interaction; line 3 illustrates the usual interaction of statistically independent signals (i.e., on $I_{ab} = 0$) in physical signal space $\mathbf{\Gamma}$ with group properties (in linear signal space $\mathcal{LS}$); curve 4 is the usual interaction of statistically independent signals in physical signal space $\mathbf{\Gamma}$ with group properties (in linear signal space $\mathcal{LS}$) on $I_{bx}(I_{ax}) \le \sup\limits_{I_a+I_b=\text{const}} [I_{bx}(I_{ax})]|_{I_{ab}=0}$; line 5 is the usual interaction of statistically dependent signals in physical signal space $\mathbf{\Gamma}$ with group properties (in linear signal space $\mathcal{LS}$) built on the relationships (4.3.28), (4.3.30), and (4.3.32), respectively.

For the ideal interaction $x(t) = a(t) \oplus b(t)$ between the signals $a(t)$, $b(t)$ in physical signal space $\mathbf{\Gamma}$ with properties of generalized Boolean algebra $\mathbf{B}(\mathbf{\Gamma})$ with signature $(\oplus, \otimes, -, \mathbf{O})$, the dependence 1 is determined by the line equation (4.3.28a), and $I_{bx} = 0$ on $I_{ax} = I_a + I_b$, and $I_{ax} = 0$ on $I_{bx} = I_a + I_b$. The dependence 1 determines the upper bound of possible values of quantities of mutual information I_{ax} and I_{bx}

of interacting signals $a(t)$ and $b(t)$. No sorts of signal interactions allow achieving more high performance informational relationships on the fixed sum $I_a + I_b$=const.

For quasi-ideal interaction $x(t) = a(t) \oplus b(t)$ of a pair of signals $a(t)$ and $b(t)$ in physical signal space Γ with lattice properties, the dependence 2 is determined by the point $I_{ax} = I_a$, $I_{bx} = I_b$ that corresponds to the identities (4.3.29b) and (4.3.29c).

For interaction $x(t) = a(t) \oplus b(t)$ between statistically independent signals $a(t)$ and $b(t)$ in physical signal space Γ with group properties (for additive interaction $x(t) = a(t) + b(t)$ in linear signal space $\mathcal{LS}$), the dependence 3 is determined by the line equation (4.3.32a) passing through the points $(I_a, 0)$ and $(0, I_b)$ correspondingly, so that for independent signals $a(t)$ and $b(t)$, quantity of mutual information is equal to zero: I_{ab}=0. The dependence 3 determines the lower bound of possible values of quantities of mutual information I_{ax} and I_{bx} of interacting signals $a(t)$ and $b(t)$. Whatever the signal interactions, one cannot achieve worse informational relationships on the fixed sum $I_a + I_b$=const.

The dependence 4 $\sup_{I_a+I_b=\text{const}} [I_{bx}(I_{ax})]|_{I_{ab}=0}$ determines the upper bound of possible values of quantities of mutual information I_{ax} and I_{bx} of the signals $a(t)$ and $b(t)$ additively interacting in physical signal space Γ with group properties (in linear space $\mathcal{LS}$). Whatever the signal interactions, one cannot achieve higher performance informational relationships on the fixed sum $I_a + I_b$=const. This function is determined by the relationship:

$$\sup_{I_a+I_b=\text{const}} [I_{bx}(I_{ax})]|_{I_{ab}=0} = \left(\sqrt{I_a + I_b} - \sqrt{I_{ax}}\right)^2. \quad (4.3.33)$$

For the interaction $x(t) = a(t) \oplus b(t)$ between statistically dependent signals $a(t)$ and $b(t)$ in physical signal space Γ with group properties (for additive interaction $x(t) = a(t) + b(t)$ in linear signal space $\mathcal{LS}$), the dependence 5 is determined by the line equation (4.3.32a) passing through the points $(kI_a, 0)$ and $(0, kI_b)$, where k is equal to:

$$k = 1 + [I_{ab}/\min(I_a, I_b)], \quad (4.3.34)$$

and for the dependent signals $a(t)$ and $b(t)$, the quantity of mutual information I_{ab} is bounded by the quantity $I_{ab} \leq [\min(I_a, I_b)]^2/\max(I_a, I_b)$.

Dependence 5 determines the upper bound of possible values of quantities of mutual information I_{ax} and I_{bx} of statistically dependent signals $a(t)$ and $b(t)$ interacting in physical signal space Γ with group properties (in linear signal space $\mathcal{LS}$). Whatever the signal interactions, one cannot achieve more high performance informational relationships on the fixed sum I_a+I_b=const. If the interacting signals $a(t)$ and $b(t)$ are identical in informational sense (see Definition 4.3.1), i.e., they are characterized by the same information quantity $I_a = I_b$, so that information contained in these signals has the same content, then the value of coefficient k determined by relationship (4.3.34) becomes equal to 2: $k = 2$. In this (and only in this) case, the dependences 5 and 1 coincide.

Thus, the locus of the lines, that characterize the dependence (4.3.32a) for interaction $x(t) = a(t) \oplus b(t)$ of the signals $a(t)$ and $b(t)$ in physical signal space Γ

with group properties (in linear signal space $\mathcal{LS}$), is between the curves 1 and 3 in Fig. 4.3.1.

In summary, one can make the following conclusions.

1. As a rule, in all physical signal spaces, the signal interaction is accompanied by the losses of some quantity of overall information. Exceptions are the signal spaces with the properties of generalized Boolean algebra. In such spaces, the signal interaction is characterized by preservation of quantity of overall information contained in a pair of interacting signals, so that the informational identity (4.3.12) holds.

2. During a signal interaction in physical signal space that is non-isomorphic to generalized Boolean algebra with a measure, there inevitably appear the losses of some quantity of overall information. On the qualitative level, these losses are described by informational inequality (4.3.10). The interaction of the signals in physical signal space $\boldsymbol{\Gamma}$ with group properties (and also in linear space $\mathcal{LS}$) is always accompanied with losses of some part of overall information.

3. As a rule, in all physical signal spaces, the signal interaction is accompanied by the losses of some quantity of mutual information between the signals before and after their interactions. Exceptions are the signal spaces with lattice properties. The signal interactions in the spaces of such type are characterized by preservation of quantity of mutual information contained in one of the interacting signals ($a(t)$ or $b(t)$), and in the result of their interaction $x(t) = a(t) \oplus b(t)$, and the informational identities (4.3.14a) and (4.3.14b) hold.

4. During a signal interaction in physical signal space that is non-isomorphic to lattice, there inevitably appear the losses of some quantity of mutual information contained in one of the interacting signals ($a(t)$ or $b(t)$) and in the result of their interaction $x(t) = a(t) \oplus b(t)$. On the qualitative level, these losses are described by informational inequalities (4.3.24a), (4.3.24b). The signal interaction in physical signal space $\boldsymbol{\Gamma}$ with group properties (and also in linear space $\mathcal{LS}$) is always accompanied by the losses of some quantity of mutual information contained in one of the interacting signals and in the result of their interaction.

5. Depending on qualitative relations between quantity of overall information $I_{a \oplus b}$, quantity of mutual information I_{ab}, and quantities of absolute information I_a and I_b contained in the interacting signals $a(t)$ and $b(t)$, we distinguish the following types of interactions between the signals in physical signal space: ideal interaction, quasi-ideal interaction, and usual interaction introduced by Definitions 4.3.2, 4.3.3, and 4.3.6, respectively.

6. During ideal interaction of the signals in physical signal space with the properties of generalized Boolean algebra, the informational relationships (4.3.28) hold.

7. During quasi-ideal interaction of the signals in physical signal space with lattice properties, the informational relationships (4.3.30) hold.

8. During the signal interaction in physical signal space $\boldsymbol{\Gamma}$ with group properties (in linear signal space $\mathcal{LS}$), the informational relationships (4.3.32) hold.

9. Dependences 1 and 3 in Fig. 4.3.1 defined by the equalities (4.3.28a) and (4.3.32a) establish the corresponding upper and the lower theoretical bounds that confine the informational relationships of the signals interacting in signal spaces with various algebraic properties.

10. Informational signal space is an abstract notion that allows us to evaluate possible informational relationships between the signals interacting in real physical signal space. If we will discover a physical medium in which the signal interactions will have the properties of ideal interactions defined here, then highest possible quality indices of signal processing will be achieved. Unlike signal spaces admitting ideal interaction of signals, physical signal space with lattice properties can be realized now. The prerequisites for developing such signal spaces and their relations with linear signal spaces were defined by Birkhoff [221, Sections XIII.3 and XIII.4].

11. Unfortunately, the quantity equal to $1 - \nu_{\mathbf{P}}(a_t, b_t)$ that acts as a metric in a linear signal space is not a metric in a signal space with lattice properties and does not allow adequate evaluation of the informational relationships describing signal interactions in signal spaces with various algebraic properties based on unified positions. This statement arises from the definition of quantity of mutual information (4.3.25) based on the probabilistic measure of statistical interrelationship (PMSI). One main property of a measure of information quantity is providing signal space metrization that cannot be realized based on quantity of mutual information (4.3.25). To construct a metric signal space with concrete algebraic properties, we use the normalized measure of statistical interrelationship (NMSI) introduced in Section 3.2. That approach is the subject of Section 4.4.

4.4 Metric and Informational Relationships between Signals Interacting in Signal Space

Results of comparative analysis on the main informational relationships describing the signal interactions in both linear signal space and the signal space with lattice properties discussed in Section 4.3 require clarification concerning metric and informational properties of such physical signal spaces. As noted in Section 4.3, the physical interaction $x(t) = a(t) \oplus b(t)$ of two signals $a(t)$ and $b(t)$ in physical space $\boldsymbol{\Gamma}$ exert negative influences on values of some types of information quantities defined for physical signal space $\boldsymbol{\Gamma}$. The quantities of overall information $I_{a \oplus b} \neq I_{A+B}$

and mutual information $I_{ab} \neq I_{AB}$, corresponding to a pair of interacting signals $a(t)$ and $b(t)$ with respect to the quantities I_{A+B} and I_{AB} defining these notions in informational signal space Ω, are changed.

The only way to solve the problem of defining informational relationships between signals interacting in a physical signal is introducing a metric into it. Otherwise, one cannot establish how close or far the signals are from each other. In other words, introducing a metric in physical signal space provides adequacy of analysis of informational relationships of the signals. A newly introduced metric must satisfy a critical property: it must conform to the measure of information quantity introduced earlier.

The goals of this section are (1) metrization of physical signal space on the basis of a normalized metric $\mu(a_t, b_t)$ (3.2.1) between the samples a_t, b_t of stochastic signals $a(t)$, $b(t) \in \boldsymbol{\Gamma}$ and (2) definition of the main types of information quantities that characterize signal interactions in physical signal spaces with various informational properties.

A key notion of this section is a metric $\mu(a_t, b_t)$ between the samples a_t, b_t of stochastic signals $a(t)$, $b(t) \in \boldsymbol{\Gamma}$ interacting in physical signal space $\boldsymbol{\Gamma}$ with the properties of either a group $\boldsymbol{\Gamma}(+)$ or a lattice $\boldsymbol{\Gamma}(\vee, \wedge)$ of L-group $\boldsymbol{\Gamma}(+, \vee, \wedge)$.

Preserving the introduced notation system for the signals interacting in physical space $\boldsymbol{\Gamma}$, we consider an additive interaction of two stationary stochastic signals $a(t)$ and $b(t)$ in a group $\boldsymbol{\Gamma}(+)$ of L-group $\boldsymbol{\Gamma}(+, \vee, \wedge)$, $a(t), b(t) \in \boldsymbol{\Gamma}(+)$, $t \in T$:

$$x(t) = a(t) + b(t), \qquad (4.4.1)$$

and also their interaction in the signal space with lattice properties $\boldsymbol{\Gamma}(\vee, \wedge)$ of L-group $\boldsymbol{\Gamma}(+, \vee, \wedge)$ with operations of join and meet, respectively: $a(t) \vee b(t) = \sup_{\boldsymbol{\Gamma}}\{a(t), b(t)\}$, $a(t) \wedge b(t) = \sup_{\boldsymbol{\Gamma}}\{a(t), b(t)\}$, $a(t), b(t) \in \boldsymbol{\Gamma}(\vee, \wedge)$, $t \in T$:

$$x(t) = a(t) \vee b(t), \quad \tilde{x}(t) = a(t) \wedge b(t). \qquad (4.4.2)$$

On the basis of Axiom 4.1.1 formulated in Section 4.1, depending on the signature relations between the images A and B of the signals $a(t)$ and $b(t)$ in informational signal space Ω built upon generalized Boolean algebra $\mathbf{B}(\Omega)$ with a measure m and signature $(+, \cdot, -, \mathbf{O})$, $A \in \Omega$, $B \in \Omega$, the following sorts of information quantity are defined: quantities of absolute information I_A and I_B, quantity of mutual information I_{AB}, and quantity of overall information I_{A+B} defined by the following relationships:

$$I_A = m(A), \quad I_B = m(B); \qquad (4.4.3a)$$

$$I_{AB} = m(AB); \qquad (4.4.3b)$$

$$I_{A+B} = m(A+B) = I_A + I_B - I_{AB}, \qquad (4.4.3c)$$

so that the images A and B of the corresponding signals $a(t)$ and $b(t)$ are defined by their information distribution densities and mappings (3.5.26).

Theorems 4.4.1 through 4.4.3 formulated below allow us to realize metric transformation of physical signal space $\boldsymbol{\Gamma} - (\boldsymbol{\Gamma}, \mu)$ with a metric $\mu(a_t, b_t)$ (3.2.1) equal to:

$$\mu(a_t, b_t) = 1 - \nu(a_t, b_t), \qquad (4.4.4)$$

4.4 Metric and Informational Relationships between Signals Interacting in Signal Space

where $\nu(a_t, b_t)$ is a normalized measure of stochastic interrelation (NMSI) between the samples a_t and b_t of stochastic signals $a(t), b(t) \in \Gamma$ introduced by Definition 3.2.2.

Taking into account the peculiarities of the main informational relationships between the signals interacting in the signal space with lattice properties, which are cited in the previous section, we define a metric in physical signal space of this type on the basis of the metric (4.4.4), supposing the interacting signals to be stationary and stationary coupled, i.e., $\nu(a_t, b_t) = \nu_{ab} = $const, $t \in T$.

Theorem 4.4.1. *For stationary and stationary coupled stochastic signals $a(t), b(t) \in \Gamma$ in physical signal space Γ that are characterized by the quantity of absolute information I_a and I_b, respectively, and also by NMSI $\nu(a_t, b_t) = \nu_{ab}$ between the samples a_t and b_t, the function d_{ab} determined by the equation:*

$$d_{ab} = I_a + I_b - 2\nu_{ab}\min[I_a, I_b], \qquad (4.4.5)$$

is metric.

Before proving the theorem, we transform Equation (4.4.5) to the form that is convenient for the following reasoning, using the relationship (4.4.4) and the identity [221, Section XIII.4;(22)]:

$$\min[I_a, I_b] = 0.5(I_a + I_b - |I_a - I_b|), \qquad (4.4.6)$$

with whose help the function d_{ab} may be written in the equivalent form:

$$d_{ab} = \mu(a_t, b_t)(I_a + I_b) + (1 - \mu(a_t, b_t))|I_a - I_b|. \qquad (4.4.7)$$

We write the expression (4.4.7) in the form of the function of three variables:

$$d_{ab} = F(d_u, \mu_u^+, \mu_u^\Delta) = d_u \mu_u^+ + (1 - d_u)\mu_u^\Delta, \qquad (4.4.8)$$

where $d_u = \mu(a_t, b_t)$, $\mu_u^+ = \mu_{ab}^+ = I_a + I_b$, $\mu_u^\Delta = \mu_{ab}^\Delta = |I_a - I_b|$.

Similarly, we introduce the following designations for the corresponding functions d_{bc}, d_{ca} between the pairs of signals $b(t), c(t)$ and $c(t), a(t)$:

$$d_{bc} = F(d_v, \mu_v^+, \mu_v^\Delta) = d_v \mu_v^+ + (1 - d_v)\mu_v^\Delta, \qquad (4.4.9)$$

where $d_v = \mu(b_t, c_t)$, $\mu_v^+ = \mu_{bc}^+ = I_b + I_c$, $\mu_v^\Delta = \mu_{bc}^\Delta = |I_b - I_c|$;

$$d_{ca} = F(d_w, \mu_w^+, \mu_w^\Delta) = d_w \mu_w^+ + (1 - d_w)\mu_w^\Delta, \qquad (4.4.10)$$

where $d_w = \mu(c_t, a_t)$, $\mu_w^+ = \mu_{ca}^+ = I_c + I_a$, $\mu_w^\Delta = \mu_{ca}^\Delta = |I_c - I_a|$.

It should be noted that for the values $d_u, \mu_u^+, \mu_u^\Delta$; $d_v, \mu_v^+, \mu_v^\Delta$; $d_w, \mu_w^+, \mu_w^\Delta$, the inequalities hold:

$$d_u \leq d_v + d_w; \quad \mu_u^+ \leq \mu_v^+ + \mu_w^+; \quad \mu_u^\Delta \leq \mu_v^\Delta + \mu_w^\Delta, \qquad (4.4.11)$$

$$\mu_u^\Delta \leq \mu_u^+; \quad \mu_v^\Delta \leq \mu_v^+; \quad \mu_w^\Delta \leq \mu_w^+. \qquad (4.4.12)$$

Further we shall denote:

$$F(x_{u,v,w}) = F(d_{u,v,w}, \mu^+_{u,v,w}, \mu^\Delta_{u,v,w}), \qquad (4.4.13)$$

where $x_{u,v,w}$ is a variable denoting one of three variables of the function (4.4.13), under a condition in which another two are the constants:

$$\begin{cases} x_{u,v,w} = d_{u,v,w}; \\ \mu^+_{u,v,w} = \text{const}; \\ \mu^\Delta_{u,v,w} = \text{const}, \end{cases} \text{or} \begin{cases} d_{u,v,w} = \text{const}; \\ x_{u,v,w} = \mu^+_{u,v,w}; \\ \mu^\Delta_{u,v,w} = \text{const}, \end{cases} \text{or} \begin{cases} d_{u,v,w} = \text{const}; \\ \mu^+_{u,v,w} = \text{const}; \\ x_{u,v,w} = \mu^\Delta_{u,v,w}. \end{cases} \qquad (4.4.14)$$

For the next proof, the generalization of lemma 9.0.2 in [245] will be used. It defines the sufficient conditions for the function (4.4.13) to make a space to remain pseudometric.

Lemma 4.4.1. *Let $F(d_u, \mu^+_u, \mu^\Delta_u)$ be a monotonic, nondecreasing, convex upward function such that $F(d_u = 0, \mu^+_u, \mu^\Delta_u) = 0$, $F(d_u, \mu^+_u = 0, \mu^\Delta_u) \geq 0$, $F(d_u, \mu^+_u, \mu^\Delta_u = 0) \geq 0$, which is defined by the relationship (4.4.8). Then if (Γ, d_u) is a pseudometric space, then $(\Gamma, F(d_u, \mu^+_u, \mu^\Delta_u))$ is also a pseudometric space.*

Proof of lemma. Fulfillment of the condition $F(d_u = 0, \mu^+_u, \mu^\Delta_u) = 0$ of the function (4.4.8) is obvious, inasmuch as condition $d_u = \mu(a_t, b_t) = 0$ implies the identity $a \equiv b$, thus, $I_a = I_b$. Taking into account the designations introduced earlier, to prove a triangle inequality, it is enough to show, that if x_u, x_v, x_w are nonnegative values such that $x_u \leq x_v + x_w$, then $F(x_u) \leq F(x_v) + F(x_w)$. It is obvious, that for nondecreasing function $F(x_{u,v,w})$, the inequality holds:

$$F(x_u) \leq F(x_v + x_w). \qquad (4.4.15)$$

For nondecreasing convex upward function $F(x_{u,v,w})$, under the condition that $F(0) \geq 0$ for $\forall x: 0 \leq x \leq x_v + x_w$, the relation holds:

$$F(x)/x \geq F(x_v + x_w)/(x_v + x_w), \qquad (4.4.16)$$

where x and $x_{v,w}$ are the variables denoting one of three in function (4.4.13), according to accepted designations (4.4.14).

The inequality (4.4.16) implies the inequalities:

$$F(x_v) \geq x_v F(x_v + x_w)/(x_v + x_w); \qquad (4.4.17a)$$

$$F(x_w) \geq x_w F(x_v + x_w)/(x_v + x_w). \qquad (4.4.17b)$$

Taking into account the inequality (4.4.15), summing the left and the right parts of the inequalities (4.4.17a) and (4.4.17b), we obtain the triangle inequality being proved:

$$F(x_v) + F(x_w) \geq F(x_v + x_w) \geq F(x_u).$$

The symmetry property of pseudometric (4.4.5) and (4.4.7) is obvious and follows from the symmetry of the functions $\mu(a_t, b_t)$, $\mu^+_{ab} = I_a + I_b$, $\mu^\Delta_{ab} = |I_a - I_b|$. □

Proof of theorem. According to Lemma 4.4.1, the function $d_{ab} = F(d_u, \mu_u^+, \mu_u^\Delta)$ (4.4.8) (and also the function (4.4.7)) is pseudometric. We prove the identity property of a metric. Equality $d_{ab} = 0$ implies the equalities $\mu(a_t, b_t) = 0$ and $I_a = I_b$, and the equality $\mu(a_t, b_t) = 0$ implies the identity $a \equiv b$. Conversely, the identity $a \equiv b$ implies the equalities $\mu(a_t, b_t) = 0$ and $I_a = I_b$ that implies the equality $d_{ab} = 0$. □

After introducing the metric d_{ab} in physical signal space Γ, which is determined by the relationship (4.4.5), one should define the main sorts of information quantities.

The notions of the quantity of absolute information in physical and informational signal spaces, as shown in the previous section, completely coincide, inasmuch as physical interaction of the signals has no influence upon this sort of information quantity carried by the signals. The notion of quantity of absolute information for physical signal spaces with arbitrary properties remains without changes within the initial variant contained in the Definition 4.1.4.

Unlike the quantity of absolute information, the other sorts of information quantities introduced by Definitions 4.1.5, 4.1.6, and 4.1.8 for informational signal space Ω need their content clarifying with respect to physical signal space Γ. Such notions are introduced by Definitions 4.4.1 through 4.4.3.

Definition 4.4.1. *Quantity of mutual information I_{ab} contained in a pair of stochastic signals $a(t), b(t) \in \Gamma$, so that each is characterized by quantity of absolute information I_a and I_b, respectively, with NMSI $\nu(a_t, b_t) = \nu_{ab}$ between the samples a_t and b_t, is an information quantity equal to:*

$$I_{ab} = \nu_{ab} \min[I_a, I_b]. \tag{4.4.18}$$

Definition 4.4.1 establishes a measure of information quantity contained simultaneously in two distinct signals $a(t) \neq b(t)$ and also a measure of information quantity for two identical signals $a(t) \equiv b(t)$. This essential feature of the quantity of mutual information is below in Remark 4.4.1.

The list below shows (without a proof) the main properties of quantity of mutual information I_{ab} of stationary coupled stochastic signals $a(t)$ and $b(t)$ that interact in physical signal space Γ.

1. $I_{ab} \leq I_a$, $I_{ab} \leq I_b$ (boundedness)
2. $I_{ab} \geq 0$ (nonnegativity)
3. $I_{ab} = I_{ba}$ (symmetry)
4. $I_{aa} = I_a$, $I_{bb} = I_b$ (idempotency)
5. $I_{ab} = 0$ if and only if the stochastic signals $a(t)$ and $b(t)$ are statistically independent ($\nu(a_t, b_t) = 0$)

Definition 4.4.2. *Quantity of relative information I_{ab}^Δ contained in a pair of stochastic signals $a(t), b(t) \in \Gamma$, so that each is characterized by quantity of absolute information I_a and I_b, respectively, with NMSI $\nu(a_t, b_t) = \nu_{ab}$ between the samples a_t and b_t, is an information quantity equal to d_{ab} (4.4.5):*

$$I_{ab}^\Delta = d_{ab} = I_a + I_b - 2\nu_{ab} \min[I_a, I_b]. \tag{4.4.19}$$

Definition 4.4.3. *Quantity of overall information I_{ab}^+ contained in a pair of stochastic signals $a(t), b(t) \in \Gamma$, so that each is characterized by quantity of absolute information I_a and I_b, respectively, with NMSI $\nu(a_t, b_t) = \nu_{ab}$ between the samples a_t and b_t, is an information quantity equal to:*

$$I_{ab}^+ = I_a + I_b - \nu_{ab} \min[I_a, I_b]. \tag{4.4.20}$$

Remark 4.4.1. Quantity of absolute information I_a contained in the signal $a(t) \in \Gamma$ of physical signal space Γ is identically equal to the quantity of mutual information I_{aa} and quantity of overall information I_{aa}^+ contained in the signal $a(t)$ interacting with itself in the signal space with corresponding algebraic properties:

$$I_a = I_{aa} = I_{aa}^+. \tag{4.4.21}$$

Proof of the identity (4.4.21) follows from the relations (4.4.18) and (4.4.20), under the condition that $\nu_{aa} = 1$.

Thus, quantity of mutual information I_{ab} between the signals establishes a measure of information quantity that corresponds to a measure for informational signal space introduced in Section 4.1.

Remark 4.4.2. Quantity of overall information I_{ab}^+ contained in a pair of stochastic signals $a(t), b(t) \in \Gamma$ interacting in physical signal space Γ is equal to the sum of the quantity of mutual information I_{ab} and the quantity of relative information $I_{ab}^\triangle$ and is bounded below by $\max[I_a, I_b]$:

$$I_{ab}^+ = I_{ab} + I_{ab}^\triangle \geq \max[I_a, I_b]. \tag{4.4.22}$$

Proof of the inequality (4.4.22) follows directly from joint fulfillment of triplets of the relationships (4.4.18), (4.4.19), (4.4.20), and proof of the identity (4.4.22) follows from the definition (4.4.20) by realizing identical transformations:

$$I_{ab}^+ = I_a + I_b - \nu_{ab} \min[I_a, I_b] =$$
$$= \max[I_a, I_b] + \min[I_a, I_b] - \nu_{ab} \min[I_a, I_b] =$$
$$= \max[I_a, I_b] + (1 - \nu_{ab}) \min[I_a, I_b] \geq \max[I_a, I_b]. \quad \square$$

Remark 4.4.3. Quantity of absolute information I_x ($I_{\tilde{x}}$) contained in the result of interaction $x(t) = a(t) \vee b(t)$ ($\tilde{x}(t) = a(t) \wedge b(t)$) between the stochastic signals $a(t), b(t) \in \Gamma$ in physical signal space $\Gamma(\vee, \wedge)$ with lattice properties is bounded below by the value of linear combination between the quantities of mutual information of a pair of signals $a(t)$ and $b(t)$ and the result of their interaction $x(t)$, and also by $\min[I_a, I_b]$:

$$I_x \geq I_{ax} + I_{bx} - I_{ab} \geq \min[I_a, I_b]; \tag{4.4.23a}$$

$$I_{\tilde{x}} \geq I_{a\tilde{x}} + I_{b\tilde{x}} - I_{ab} \geq \min[I_a, I_b]. \tag{4.4.23b}$$

Proof of remark. Write a triangle inequality d_{ab}: $d_{ab} \leq d_{ax} + d_{xb}$, substituting in

4.4 Metric and Informational Relationships between Signals Interacting in Signal Space

both parts the relationship between metric d_{ab} and quantity of relative information $I_{ab}^{\triangle}$ (4.4.19) that implies the inequality:

$$I_x \geq I_{ax} + I_{bx} - I_{ab} \geq \nu_{ax} I_a + \nu_{bx} I_b - \nu_{ab} \min[I_a, I_b] \geq \\ \geq (\nu_{ax} + \nu_{bx} - \nu_{ab}) \min[I_a, I_b]. \quad (4.4.24)$$

Substituting into the last inequality the value of linear combination $\nu_{ax} + \nu_{bx} - \nu_{ab} = 1$ defined by the relationship (3.2.33a), we obtain the initial inequality (4.4.23a). Proof of the inequality (4.4.23b) is similar. □

Remark 4.4.4. Quantity of absolute information I_x contained in the result of interaction $x(t) = a(t) + b(t)$ of a pair of stochastic signals $a(t), b(t) \in \mathbf{\Gamma}$ in physical signal space $\mathbf{\Gamma}(+)$ with group properties, is bounded below by the value of a linear combination between the quantities of mutual information of a pair of signals $a(t)$ and $b(t)$ and the result of their interaction $x(t)$, and also by $\min[I_a, I_b]$:

$$I_x \geq I_{ax} + I_{bx} - I_{ab} \geq \min[I_a, I_b]. \quad (4.4.25)$$

The proof of remark is the same as the previous one, except that the value of linear combination $\nu_{ax} + \nu_{bx} - \nu_{ab} = 1$ is defined by the relationship (3.2.18).

As accepted for informational signal space $\mathbf{\Omega}$, similarly, for physical signal space $\mathbf{\Gamma}$ as a unit of information quantity measurement, we take the quantity of absolute information $I[\xi(t_\alpha)]$ contained in a single element $\xi(t_\alpha)$ of stochastic signal $\xi(t)$, and according to the relationship (4.1.4), it is equal to:

$$I[\xi(t_\alpha)] = m(X_\alpha)|_{t_\alpha \in T_\xi} = \int_{T_\xi} i_\xi(t_\alpha; t) dt = 1 \text{abit}, \quad (4.4.26)$$

where T_ξ is a domain of definition of the signal $\xi(t)$ (in the form of a discrete set or continuum); $i_\xi(t_\alpha; t)$ is an information distribution density (IDD) of the signal $\xi(t)$.

In the physical signal space $\mathbf{\Gamma}$, the unit of information quantity is introduced by the definition that corresponds to Definition 4.1.9.

Definition 4.4.4. In physical signal space $\mathbf{\Gamma}$, *unit of information quantity* is the quantity of absolute information, which is contained in a single element (a sample) $\xi(t_\alpha)$ of stochastic process (signal) $\xi(t)$ considered a subalgebra $\mathbf{B}(X)$ of generalized Boolean algebra $\mathbf{B}(\Omega)$ with a measure m, and is called *absolute unit* or abit.

Below we discuss informational relationships that characterize interactions of the signals $a(t)$ and $b(t)$ in physical signal space $\mathbf{\Gamma}$, $a(t), b(t) \in \mathbf{\Gamma}$:

$$x(t) = a(t) \oplus b(t),$$

where $\oplus$ is a binary operation of a group $\mathbf{\Gamma}(+)$ or a lattice $\mathbf{\Gamma}(\vee, \wedge)$.

Consider a group of identities that characterize informational relations for an arbitrary pair of stochastic signals $a(t)$, $b(t) \in \mathbf{\Gamma}$ interacting in physical signal space $\mathbf{\Gamma}$ with properties of a group $\mathbf{\Gamma}(+)$ and a lattice $\mathbf{\Gamma}(\vee, \wedge)$:

$$I_{ab} = \nu_{ab} \min[I_a, I_b]; \tag{4.4.27a}$$

$$I_{ab}^{\Delta} = I_a + I_b - 2I_{ab}; \tag{4.4.27b}$$

$$I_{ab}^{+} = I_a + I_b - I_{ab}; \tag{4.4.27c}$$

$$I_{ab}^{+} = I_{ab} + I_{ab}^{\Delta}; \tag{4.4.27d}$$

$$I_{ax} = \nu_{ax} \min[I_a, I_x] = \nu_{ax} I_a \leq I_a; \tag{4.4.27e}$$

$$I_{ax}^{\Delta} = I_a + I_x - 2\nu_{ax} \min[I_a, I_x] = I_x + (\mu(a_t, x_t) - \nu_{ax}) I_a; \tag{4.4.27f}$$

$$I_{ax}^{+} = I_a + I_x - \nu_{ax} \min[I_a, I_x] = I_x + \mu(a_t, x_t) I_a. \tag{4.4.27g}$$

Consider also a group of inequalities that characterize informational relations for an arbitrary pair of stochastic signals $a(t)$, $b(t) \in \mathbf{\Gamma}$ interacting in physical signal space $\mathbf{\Gamma}$ with properties of a group $\mathbf{\Gamma}(+)$ and a lattice $\mathbf{\Gamma}(\vee, \wedge)$:

$$I_a + I_b \geq I_{ab}^{+} \geq I_{ab}^{\Delta}; \tag{4.4.28a}$$

$$I_a + I_b \geq I_{ab}^{+} \geq I_{ab}; \tag{4.4.28b}$$

$$I_a + I_b > I_x \geq I_{ax} + I_{bx} - I_{ab} \geq \min[I_a, I_b]; \tag{4.4.28c}$$

$$I_a + I_b > I_{ab}^{+} \geq \max[I_a, I_b]. \tag{4.4.28d}$$

For an arbitrary pair of stochastic signals $a(t)$, $b(t) \in \mathbf{\Gamma}$ interacting in physical signal space $\mathbf{\Gamma}(\vee, \wedge)$ with lattice properties: $x(t) = a(t) \vee b(t)$, $\tilde{x}(t) = a(t) \wedge b(t)$, as follows from the relation (3.2.38b), the identities hold:

$$\begin{cases} I_{ax} = I_a/2; \\ I_{a\tilde{x}} = I_a/2; \\ I_{ax} + I_{a\tilde{x}} = I_a, \end{cases} \tag{4.4.29a}$$

$$\begin{cases} I_{bx} = I_b/2; \\ I_{b\tilde{x}} = I_b/2; \\ I_{bx} + I_{b\tilde{x}} = I_b, \end{cases} \tag{4.4.29b}$$

and if the signals $a(t)$ and $b(t)$ are statistically independent, then the identities that directly follows from the relation (3.2.39b) hold:

$$I_{ab} = I_{x\tilde{x}} = 0. \tag{4.4.30}$$

The invariance property of the quantities I_{ab}, I_{ab}^{Δ}, I_{ab}^{+} introduced by Definitions 4.4.1 through 4.4.3 under bijection mappings of stochastic signals $a(t)$ and $b(t)$ is defined by the following theorem.

4.4 Metric and Informational Relationships between Signals Interacting in Signal Space

Theorem 4.4.2. *For a pair of stationary stochastic signals $a(t)$ and $b(t)$ with even univariate PDFs in L-group $\Gamma(+, \vee, \wedge)$: $a(t), b(t) \in \Gamma$, $t \in T$, the quantities I_{ab}, I_{ab}^{Δ}, I_{ab}^{+} introduced by Definitions 4.4.1 through 4.4.3 are invariants of a group H of continuous mappings $\{h_{\alpha,\beta}\}$, $h_{\alpha,\beta} \in H$; $\alpha, \beta \in A$ of stochastic signals preserving neutral element (zero) 0 of a group $\Gamma(+)$ of L-group $\Gamma(+, \vee, \wedge)$: $h_{\alpha,\beta}(0) = 0$:*

$$I_{ab} = I_{a'b'}, I_{ab}^{\Delta} = I_{a'b'}^{\Delta}, I_{ab}^{+} = I_{a'b'}^{+}; \qquad (4.4.31)$$

$$h_{\alpha} : a(t) \to a'(t), \quad h_{\beta} : b(t) \to b'(t); \qquad (4.4.31a)$$

$$h_{\alpha}^{-1} : a'(t) \to a(t), \quad h_{\beta}^{-1} : b'(t) \to b(t), \qquad (4.4.31b)$$

where $I_{a'b'}$, $I_{a'b'}^{\Delta}$, $I_{a'b'}^{+}$ are the quantities of mutual, relative, and overall information between a pair of signals $a'(t)$ and $b'(t)$, that are the results of mappings (4.4.31a) of the signals $a(t)$ and $b(t)$, respectively.

Proof. Corollary 3.5.3 of the Theorem 3.1.1. (3.5.15) (and also Corollary 4.2.5 of the Theorem 4.2.4) implies invariance property of the quantity of absolute information I_a and I_b contained in the signals $a(t)$ and $b(t)$, respectively:

$$I_a = I_{a'}, \quad I_b = I_{b'}. \qquad (4.4.32)$$

Joint fulfillment of the equalities (4.4.18), (4.4.32), and (3.2.11) implies the identity that determines invariance of the quantity of mutual information I_{ab}:

$$I_{ab} = I_{a'b'}, \qquad (4.4.33)$$

while joint fulfillment of the identities (4.4.32), (4.4.33), (4.4.27b) and (4.4.27c) implies the identity that determines invariance of the quantity of relative information I_{ab}^{Δ}:

$$I_{ab}^{\Delta} = I_{a'b'}^{\Delta}, \qquad (4.4.34)$$

and also implies the identity that determines the invariance of the quantity of overall information I_{ab}^{+}:

$$I_{ab}^{+} = I_{a'b'}^{+}. \qquad (4.4.35)$$

□

On the basis of Theorem 4.4.1 defining a metric d_{ab} (4.4.5) for the stationary and stationary coupled stochastic signals $a(t), b(t') \in \Gamma$ in physical signal space Γ, one can also define a metric $d_{ab}(t, t')$ between a pair of the samples $a_t = a(t)$, $b'_t = b(t')$ for nonstationary signals:

$$d_{ab}(t, t') = I_a + I_b - 2\nu(a_t, b_{t'}) \min[I_a, I_b], \quad t, t' \in T, \qquad (4.4.36)$$

where I_a and I_b is the quantity of absolute information contained in the signals $a(t)$ and $b(t')$; $\nu(a_t, b_{t'}) = 1 - \mu(a_t, b_{t'})$ is NMSI of a pair of the samples $a_t = a(t)$, $b'_t = b(t')$ of the signals $a(t), b(t')$; $\mu(a_t, b_{t'})$ is a normalized metric between the samples $a_t = a(t)$, $b'_t = b(t')$ of the signals $a(t)$ and $b(t')$, which is defined by relationship (3.2.1).

Such a possibility is provided by the following theorem.

Theorem 4.4.3. *For nonstationary stochastic signals $a(t), b(t') \in \Gamma$ interacting in physical signal space Γ in such a way that for an arbitrary pair of the samples $a_t = a(t)$, $b'_t = b(t')$, the condition $\nu_{ab} = \sup\limits_{t,t' \in T} \nu(a_t, b_{t'})$ holds, the function ρ_{ab} determined by the relationship:*

$$\rho_{ab} = \inf_{t,t' \in T} d_{ab}(t,t') = I_a + I_b - 2\nu_{ab}\min[I_a, I_b], \quad (4.4.37)$$

is a metric.

Proof. Two stochastic signals $a(t), b(t') \in \Gamma$ could be considered as the corresponding nonintersecting sets of the samples $A = \{a_t\}$, $B = \{b_{t'}\}$ with a distance between the samples (4.4.36). Then the distance ρ_{ab} between the sets $A = \{a_t\}$, $B = \{b_{t'}\}$ of the samples (between the signals $a(t), b(t')$) is determined by the identity [246, Section IV.1;(1)]:

$$\rho_{ab} = \inf_{t,t' \in T} d_{ab}(t,t') = I_a + I_b - 2\sup_{t,t' \in T}\nu(a_t, b_{t'})\min[I_a, I_b]. \quad (4.4.38)$$

Under the condition $\nu_{ab} = \sup\limits_{t,t' \in T} \nu(a_t, b_{t'})$, from the equality (4.4.38) we obtain the initial relationship. $\square$

The condition $\nu_{ab} = \sup\limits_{t,t' \in T} \nu(a_t, b_{t'})$ appearing in Theorem 4.4.3 considers the fact when the closest, i.e., statistically most dependent samples of two signals $a(t)$ and $b(t')$ interact at the same time instant t. When the signals $a(t)$ and $b(t')$ are statistically independent, this condition is always valid.

The results reported in this section allow us to draw the following conclusions.

1. The metric (4.4.5) introduced in physical signal space provide the adequacy of the following analysis of informational relationships between the signals interacting in physical signal space with both group and lattice properties.

2. Quantity of mutual information introduced by Definition 4.4.1 establishes a measure of information quantity, which completely corresponds to a measure of information quantity for informational signal space introduced earlier in Section 4.1.

3. The obtained informational relationships create the basis for the analysis of quality indices (possibilities) of signal processing in signal spaces with both group and lattice properties.

5
Communication Channel Capacity

The intermediate inferences on informational signal space shown in Chapter 4 allow us to obtain the final results that are important for both signal processing theory and information theory and, in particular, determine the regularities and capacities of discrete and continuous noiseless communication channels.

In the most general case, the signal $s(t)$ carrying information can be represented in the following form [247], [248]:

$$s(t) = M[c(t), u(t)],$$

where $M[*, *]$ is a modulating function; $c(t)$ is a *carrier signal* (or simply a carrier); $u(t)$ is a transmitted message.

Modulation is changing some parameter of a carrier signal according to a transmitted message. A variable (under a modulation) parameter is called an informational one. Naturally, to extract the transmitted message $u(t)$ at the receiving side, the modulating function $M[*, *]$ has to be a one-to-one relation:

$$u(t) = M^{-1}[c(t), s(t)],$$

where $M^{-1}[*, *]$ is a function that is an inverse with respect to the initial $M[*, *]$.

To transmit information over some distance, the systems use the signals possessing the ability to be propagated as electromagnetic, hydroacoustic, and other oscillations within the proper physical medium separating a sender and an addressee. The wide class of such signals is described by harmonic functions. The transmitted information must be included in high-frequency oscillation $\cos \omega_0 t$ called a carrier:

$$s(t) = A(t)\cos(\omega_0 t + \varphi(t)) = A(t)\cos(\Phi(t)),$$

where the amplitude $A(t)$ and/or phase $\varphi(t)$ are changed according to the transmitted message $u(t)$.

Depending on which parameter of a carrier signal is changed, we distinguish amplitude, frequency, and phase modulations. Modulation of a carrier signal by a discrete message $u(t) = \{u_j(t)\}$, $j = 1, 2, \ldots, n$; $u_j(t) \in \{u_i\}$; $i = 1, 2, \ldots, q$; $q = 2^k$; $k \in \mathbf{N}$ is called *keying*, and the signal $s(t)$ is a manipulated one. In the following section, the informational characteristics of discrete (binary ($q = 2$) and m-ary ($q = 2^k$; $k > 1$)) and continuous signals are considered.

5.1 Information Quantity Carried by Discrete and Continuous Signals

5.1.1 Information Quantity Carried by Binary Signals

A large class of binary signals can be represented in the following form [115]:

$$s(t) = u(t)c_1(t) + [1 - u(t)]c_2(t), \tag{5.1.1}$$

where $u(t)$ is a binary informational message taking two values only $\{u_1, u_2\}$, $u_1 \neq u_2$, $u_{1,2} \in \{0, 1\}$; $c_1(t)$ and $c_2(t)$ are some narrowband deterministic signals.

In particular, the signals with amplitude-shift, frequency-shift, and phase-shift keying can be associated with this class.

Let $u(t)$ be a discrete stochastic process that may accept one of two values u_1 and u_2 at arbitrary instant t with the same probabilities $p_1 = \mathbf{P}\{u(t) = u_1\} = 0.5$, $p_2 = \mathbf{P}\{u(t) = u_2\} = 0.5$, so that the change of the states (the values) is possible at the fixed instants t_j:

$$t_j = \Delta \pm j\tau_0,$$

where τ_0 is a duration of an elementary signal; $j = 0, 1, 2, \ldots$ is an integer nonnegative number; Δ is a random variable that does not depend on $u(t)$ and is uniformly distributed in the interval $[0, \tau_0]$.

If the probabilities of state transition $\mathbf{P}\{u_1 \to u_2\}$, $\mathbf{P}\{u_2 \to u_1\}$ and probabilities of state preservation $\mathbf{P}\{u_1 \to u_1\}$, $\mathbf{P}\{u_2 \to u_2\}$ are assumed to be equal:

$$\mathbf{P}\{u_1 \to u_2\} = \mathbf{P}\{u_2 \to u_1\} = p_c; \quad \mathbf{P}\{u_1 \to u_1\} = \mathbf{P}\{u_2 \to u_2\} = p_s = 1 - p_c,$$

then the transition matrix of one step probabilities of a transition $\mathbf{\Pi}_1$ from one state to another is equal to [249]:

$$\mathbf{\Pi}_1 = \begin{bmatrix} p_s & p_c \\ p_c & p_s \end{bmatrix},$$

and the matrix of probabilities of transition $\mathbf{\Pi}_n$ from one state to another for n steps is equal to [115], [249]:

$$\mathbf{\Pi}_n = \begin{bmatrix} \Pi_{11} & \Pi_{12} \\ \Pi_{21} & \Pi_{22} \end{bmatrix},$$

where $\Pi_{ij} = 0.5(1 + (-1)^{i+j}\Pi)$; $\Pi = (p_s - p_c)^{\left|\left[\frac{\tau}{\tau_0}\right]\right|}\left(1 - 2p_c\left(\frac{|\tau|}{\tau_0} - \left|\left[\frac{\tau}{\tau_0}\right]\right|\right)\right)$; $n = \left[\frac{\tau}{\tau_0}\right]$ is an integer part of ratio $\frac{\tau}{\tau_0}$; $\tau = t_j - t_k$ is a time difference between two arbitrary samples $u(t_j)$ and $u(t_k)$ of discrete stochastic process (message) $u(t)$.

The joint bivariate probability density function (PDF) $p(x_1, x_2; \tau)$ of two samples $u(t_j)$ and $u(t_k)$ of discrete stochastic process $u(t)$ is determined by the expression:

$$p(x_1, x_2; \tau) = 0.5\Pi_{11}\delta(x_1 - u_1)\delta(x_2 - u_1) + 0.5\Pi_{12}\delta(x_1 - u_1)\delta(x_2 - u_2) +$$
$$+ 0.5\Pi_{21}\delta(x_1 - u_2)\delta(x_2 - u_1) + 0.5\Pi_{22}\delta(x_1 - u_2)\delta(x_2 - u_2), \tag{5.1.2}$$

5.1 Information Quantity Carried by Discrete and Continuous Signals

and univariate PDFs of the samples $u(t_j)$ and $u(t_k)$ of discrete stochastic process $u(t)$ are equal to:

$$p(x_1) = 0.5\delta(x_1 - u_1) + 0.5\delta(x_1 - u_2); \qquad (5.1.3a)$$

$$p(x_2) = 0.5\delta(x_2 - u_1) + 0.5\delta(x_2 - u_2), \qquad (5.1.3b)$$

where $\delta(x)$ is Dirac delta function.

Substituting the relationships (5.1.2), (5.1.3a), and (5.1.3b) into the formula (3.1.2), we obtain the normalized function of statistical interrelationship (NFSI) $\psi_u(\tau)$ of stochastic process $u(t)$ which for arbitrary values of state transition probability p_c, is defined by the expression:

$$\psi_u(\tau) = |1 - 2p_c|^{\left|\left[\frac{\tau}{\tau_0}\right]\right|} \left(1 - (1 - |1 - 2p_c|) \cdot \left(\frac{|\tau|}{\tau_0} - \left|\left[\frac{\tau}{\tau_0}\right]\right|\right)\right). \qquad (5.1.4)$$

From the relationships (5.1.2) and (5.1.3) one can also find the normalized autocorrelation function (ACF) $r_u(\tau)$ of stochastic process $u(t)$, which, on arbitrary values of state transition probabilities p_c, is determined by the relationship:

$$r_u(\tau) = (1 - 2p_c)^{\left|\left[\frac{\tau}{\tau_0}\right]\right|} \left(1 - 2p_c \left(\frac{|\tau|}{\tau_0} - \left|\left[\frac{\tau}{\tau_0}\right]\right|\right)\right). \qquad (5.1.5)$$

For stochastic process $u(t)$ with state transition probability p_c taking the values in the interval $[0; 0.5]$, the normalized ACF $r_u(\tau)$ is a strictly positive function, and NFSI $\psi_u(\tau)$ is identically equal to it: $\psi_u(\tau) = r_u(\tau) \geq 0$ (see Fig. 5.1.1(a)).

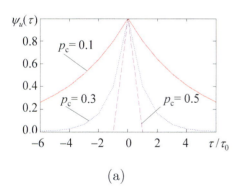
(a)

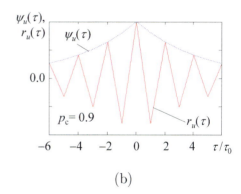
(b)

FIGURE 5.1.1 NFSI $\psi_u(\tau)$ and normalized ACF $r_u(\tau)$ of stochastic process $u(t)$ with state transition probability p_c taking values in intervals (a) $[0; 0.5]$, (b) $]0.5; 1]$

For stochastic processes $u(t)$ with state transition probability p_c taking the values in the interval $]0.5; 1]$, normalized ACF $r_u(\tau)$ has an oscillated character possessing both positive and negative values.

An example of a relation between normalized ACF and NFSI for this case is shown in the Fig. 5.1.1(b).

In the case of statistical independence of the symbols $\{u_j(t)\}$ of a message $u(t)$ ($p_c = p_s = 1/2$), normalized ACF $r_u(\tau)$ and NFSI $\psi_u(\tau)$ are determined by the expression:

$$\psi_u(\tau) = r_u(\tau) = \begin{cases} 1 - \dfrac{|\tau|}{\tau_0}, & |\tau| \leq \tau_0; \\ 0, & |\tau| > \tau_0. \end{cases}$$

It should be noted that for stochastic processes with state transition probability $p_c = P$ taking the values in the interval $[0; 0.5]$, NFSI $\psi_u(\tau)|_P$ is identically equal to NFSI $\psi_u(\tau)|_{1-P}$ of stochastic processes with state transition probability $p_c = 1-P$ taking the values in the interval $[0.5; 1]$:

$$\psi_u(\tau)|_P = \psi_u(\tau)|_{1-P} = \psi_u(\tau). \tag{5.1.6}$$

According to the coupling equation between information distribution density (IDD) and NFSI (3.4.3), IDD $i_u(\tau)$ of stochastic process $u(t)$ is determined by the expression (see Fig. 5.1.2):

$$i_u(\tau) = (1 - |1 - 2p_c|) \cdot |1 - 2p_c|^{\left|\left[\frac{2\tau}{\tau_0}\right]\right|}. \tag{5.1.7}$$

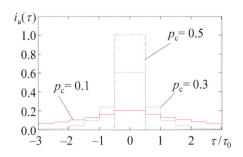

FIGURE 5.1.2 IDD $i_u(\tau)$ of stochastic process $u(t)$ that is determined by Equation (5.1.7)

In the case of statistical independence of the symbols $\{u_j(t)\}$ of a message $u(t)$ ($p_c = p_s = 1/2$), IDD $i_u(\tau)$ is determined by the expression:

$$i_u(\tau) = [1(\tau + \tau_0/2) - 1(\tau - \tau_0/2)]/\tau_0.$$

Quantity of absolute information $I_u(T)$ carried by stochastic process $u(t)$ in the interval $t \in [0; T]$, according to (3.5.7), is equal to:

$$I_u(T) = T i_u(0), \tag{5.1.8}$$

where $i_u(0)$ is the maximum value of IDD $i_u(\tau)$ of stochastic process $u(t)$ on $\tau = 0$.

There are the sequences with state transition probability $p_c = 1/2$ among all binary random sequences that carry the largest information quantity $I_u(T)$ during the time interval T, inasmuch as in this case, there is no statistical interrelation between the elements of the sequence:

$$\sup_{p_c} I_u(T) = T \cdot \sup_{p_c} i_u(0) = T/\tau_0 = n; \tag{5.1.9}$$

5.1 Information Quantity Carried by Discrete and Continuous Signals

$$\arg\sup_{p_c} I_u(T) = 1/2,$$

where n is a number of binary symbols of the sequence $u(t)$.

Conversely, some sequences among all binary random sequences carry the least information quantity $I_u(T)$ if their state transition and state preservation probabilities are equal to $p_c = 0$, $p_s = 1$ ($p_c = 1$, $p_s = 0$), respectively:

$$\inf_{p_c} I_u(T) = T \cdot \inf_{p_c} i_u(0) = 0. \qquad (5.1.10)$$

According to the expression (5.1.10), binary sequences in the form of a constant ($p_c = 0$, $u(t) = \{\ldots u_1, u_1, u_1, \ldots\}$ or $\{\ldots u_2, u_2, u_2, \ldots\}$) and also in the form of a meander ($p_c = 1$, $u(t) = \{\ldots u_1, u_2, u_1, u_2, u_1, u_2, \ldots\}$) do not carry any information.

The mathematical apparatus for the construction of the stated fundamentals of signal processing theory and information theory is Boolean algebra with a measure. As a unit of measurement of information quantity, we accept the quantity of absolute information $I[s(t_\alpha)]$ contained in a single element $s(t_\alpha)$ of stochastic process $s(t)$ with a domain of definition T_s (in the form of discrete set or continuum) and IDD $i(t_\alpha; t)$ which, according to the relationship (3.5.3), is equal to:

$$I[s(t_\alpha)] = \int_{T_s} i(t_\alpha; t) dt = 1 \text{ abit}.$$

The unit of the quantity of absolute information contained in a single element of stochastic process, according to Definition 3.5.6, is called *absolute unit* or *abit*. For newly introduced unit of measurement of information quantity, it is necessary to establish interrelation with known units of measurement of information quantity that take their origins from the logarithmic measure of Hartley [50].

Let $\xi(t_j)$ be a discrete random sequence with independent elements, which is defined upon a finite set of instants $T_\xi = \{t_j\}$; $j = 1, 2, \ldots, n$, and the random variable $\xi_j = \xi(t_j)$ equiprobably takes the values from the set $\Xi = \{x_i\}$; $i = 1, 2, \ldots, q$. Then the information quantity $I_\xi[n, q]$ contained in discrete random sequence $\xi(t_j)$ is determined by the known relationship:

$$I_\xi[n, q] = n \log q. \qquad (5.1.11)$$

For a logarithmic measure to base 2, information quantity $I_\xi[n, q]$ is equal to $n \log_2 q$ binary units (bit).

According to the formula (5.1.9), information quantity $I_u[n, q]$, contained in a binary sequence $u(t)$ ($q = 2$) with equiprobable states $\{u_1, u_2\}$ ($\mathbf{P}\{u_1\} = \mathbf{P}\{u_2\} = 0.5$) and state transition probability $p_c = 1/2$, is equal to:

$$I_u[n, q] = n \log_2 q = n, \text{ (bit)} \qquad (5.1.12)$$

that corresponds to the result determined by formula (5.1.9).

Thus, for binary sequence $u(t)$ containing n elements with two equiprobable symbols (states $\{u_1, u_2\}$) ($\mathbf{P}\{u_1\} = \mathbf{P}\{u_2\} = 0.5$) and state transition probability $p_c = 1/2$, the quantity of absolute information $I_u(T)$ measured in abits is equal to information quantity $I_u[n, 2]$ measured in bits:

$$I_u(T) \text{ (abit)} = I_u[n, 2] = n. \text{ (bit)} \qquad (5.1.13)$$

Despite the equality (5.1.13), a quantity of absolute information $I_u(T)$ measured in abits is not equivalent to information quantity $I_u[n, 2]$ measured in bits, inasmuch as the first one is measured based on IDD, and the second is measured on the base of entropy. The difference between these measures of information quantity will be elucidated more thoroughly below. Now we consider informational characteristics of binary signals $s(t)$ formed on the base of binary messages $u(t)$.

The signal representation (5.1.1) admits the possibility of unique extraction (demodulation) of initial message $u(t)$ from the signal $s(t)$ using information concerning the carrier signals $c_{1,2}(t)$. Thus, for a message $u(t)$, informational signal $s(t)$, and carrier signal $c_{1,2}(t)$, there exists some one-to-one modulating function $M[*, *]$ that the following relation holds:

$$s(t) = M[u(t), c_{1,2}(t)]; \quad u(t) = M^{-1}[s(t), c_{1,2}(t)], \qquad (5.1.14)$$

where $M^{-1}[*, *]$ is a function that is inverse with respect to initial one.

According to Theorem 3.1.1, under the isomorphic mapping of the processes $u(t)$ and $s(t)$ into each other:

$$u(t) \underset{M^{-1}}{\overset{M}{\rightleftarrows}} s(t), \qquad (5.1.15)$$

NFSI is preserved:

$$\psi_u(t_j, t_k) = \psi_s(t'_j, t'_k), \qquad (5.1.16)$$

and the overall quantity of information $I[u(t)]$ ($I[s(t)]$) contained in stochastic process $u(t)$ ($s(t)$), is preserved too:

$$I[u(t)] = I[s(t)]. \qquad (5.1.17)$$

If the mapping (5.1.14) preserves domain $T_u = [t_0; t_0 + T] = T_s$ of stochastic process $u(t)$ ($s(t)$), then the identity holds:

$$\psi_u(t_j, t_k) = \psi_s(t_j, t_k). \qquad (5.1.18)$$

Let M_{AM}, M_{PM}, M_{FM} be the functions describing the transformation (5.1.15) of a binary message $u(t)$ into a binary signal $s(t)$ under amplitude-shift $s_{AM}(t)$, phase-shift $s_{PM}(t)$, and frequency-shift $s_{FM}(t)$ keying of binary message $u(t) = \{u_j(t)\}$, $j = 1, 2, \ldots, n; u_j(t) \in \{u_i\}; i = 1, 2$, respectively:

$$M_{AM}: u(t) \to \{s_{AMj}(t)\} = s_{AM}(t);$$

$$M_{PM}: u(t) \to \{s_{PMj}(t)\} = s_{PM}(t);$$

$$M_{FM}: u(t) \to \{s_{FMj}(t)\} = s_{FM}(t).$$

Assume the processes $s_{AM}(t)$, $s_{PM}(t)$, $s_{FM}(t)$ are stationary. Then, according to Theorem 3.1.1, the relationships, which characterize the identities between single characteristics of the signal $s(t)$ and a message $u(t)$, hold:
Identity between NFSIs:

$$\psi_{sAM}(\tau) = \psi_{sPM}(\tau) = \psi_{sFM}(\tau) = \psi_u(\tau);$$

Identity between IDDs:

$$i_{sAM}(\tau) = i_{sPM}(\tau) = i_{sFM}(\tau) = i_u(\tau);$$

Identity between HSDs:

$$\sigma_{sAM}(\omega) = \sigma_{sPM}(\omega) = \sigma_{sFM}(\omega) = \sigma_u(\omega);$$

Identity of overall quantity of information:

$$I[u(t)] = I[s_{AM}(t)] = I[s_{PM}(t)] = I[s_{FM}(t)].$$

5.1.2 Information Quantity Carried by m-ary Signals

At arbitrary instant t, let each symbol $u_j(t)$ of discrete stochastic process $u(t) = \{u_j(t)\}$, $j = 0, 1, 2, \ldots$ takes one of q values from the set $\{u_i\}$, $i = 1, 2, \ldots, q$; $q = 2^k$; $k \in \mathbf{N}$ with the same probabilities $p_i = \mathbf{P}\{u(t) = u_i\} = 1/q$, so that state transition is possible at the fixed instants t_j only:

$$t_j = \Delta \pm j\tau_0,$$

where τ_0 is duration of an elementary signal; $j = 0, 1, 2, \ldots$ is an integer nonnegative number; Δ is a random variable that does not depend on $u(t)$ and is uniformly distributed in the interval $[0, \tau_0]$.

If state transition probabilities $\mathbf{P}\{u_l \to u_m\}$ $(1 \le l \le q; 1 \le m \le q)$ and state preservation probabilities $\mathbf{P}\{u_l \to u_l\}$ are, respectively, equal to:

$$\mathbf{P}\{u_l \to u_m\} = p_{lm},$$

then transition matrix of one-step probabilities of a transition $\mathbf{\Pi}_1$ from one state to another is equal to:

$$\mathbf{\Pi}_1 = \begin{bmatrix} p_{11} & p_{12} & \cdots & p_{1q} \\ p_{21} & p_{22} & \cdots & p_{2q} \\ \cdots & \cdots & \cdots & \cdots \\ p_{q1} & p_{q2} & \cdots & p_{qq} \end{bmatrix},$$

$$\sum_{m=1}^{q} p_{lm} = 1.$$

In general, it is extremely difficult to obtain the expressions for normalized ACF $r_u(\tau)$ and NFSI $\psi_u(\tau)$ of a discrete random sequence $u(t)$ with arbitrary state transition probabilities p_{lm}, $(l \neq m)$ and cardinality $\text{Card}\{u_i\} = q$ of a set $\{u_i\}$ of values. Nevertheless, if we consider the case when state transition probabilities $\mathbf{P}\{u_l \to u_m\}$, $(l \neq m)$ and state preservation probabilities $\mathbf{P}\{u_l \to u_l\}$ are equal to:

$$\mathbf{P}\{u_l \to u_m\} = p_c; \quad \mathbf{P}\{u_l \to u_l\} = p_s,$$

then the transition matrix of one-step probabilities of a transition $\mathbf{\Pi}_1$ from one state to another is equal to:

$$\mathbf{\Pi}_1 = \begin{bmatrix} p_s & p_c & \cdots & p_c \\ p_c & p_s & \cdots & p_c \\ \cdots & \cdots & \cdots & \cdots \\ p_c & p_c & \cdots & p_s \end{bmatrix},$$

where $p_c \in]0; 1/(q-1)]$; $p_s \in [0; (q-1)p_c]$, and the transition matrix of probabilities of transition $\mathbf{\Pi}_n$ from one state to another for n steps is equal to [115], [249]:

$$\mathbf{\Pi}_n = \begin{bmatrix} \Pi_{11} & \Pi_{12} & \cdots & \Pi_{1q} \\ \Pi_{21} & \Pi_{22} & \cdots & \Pi_{2q} \\ \cdots & \cdots & \cdots & \cdots \\ \Pi_{q1} & \Pi_{q2} & \cdots & \Pi_{qq} \end{bmatrix},$$

where $\Pi_{ij} = (1 + (-1)^{i+j}\Pi)/q$; $\Pi = (1 - qp_c)^{\left|\left[\frac{\tau}{\tau_0}\right]\right|}\left(1 - qp_c\left(\frac{|\tau|}{\tau_0} - \left|\left[\frac{\tau}{\tau_0}\right]\right|\right)\right)$; $n = \left[\frac{\tau}{\tau_0}\right]$ is an integer part of the ratio $\frac{\tau}{\tau_0}$; $q = 2^k$; $k \in \mathbf{N}$; $\tau = t_j - t_k$ is a time difference between two arbitrary samples $u(t_j)$, $u(t_k)$ of discrete stochastic process $u(t)$.

Then joint bivariate PDF $p(x_1, x_2; \tau)$ of two samples $u(t_j)$, $u(t_k)$ of discrete stochastic process $u(t)$ is determined by the expression:

$$p(x_1, x_2; \tau) = \frac{1}{q}\sum_{i,j=1}^{q} \Pi_{ij}\delta(x_1 - u_i)\delta(x_2 - u_j), \tag{5.1.19}$$

and univariate PDFs $p(x_1)$, $p(x_2)$ of the samples $u(t_j)$ and $u(t_k)$ of discrete stochastic process $u(t)$ are, respectively, equal to:

$$p(x_1) = \frac{1}{q}\sum_{i=1}^{q}\delta(x_1 - u_i); \quad p(x_2) = \frac{1}{q}\sum_{i=1}^{q}\delta(x_2 - u_i). \tag{5.1.20}$$

Substituting the relationships (5.1.19) and (5.1.20) into the formula (3.1.2), we obtain NFSI $\psi_u(\tau)$ of stochastic process $u(t)$, which, on arbitrary values of state transition probabilities p_c, is determined by the expression:

$$\psi_u(\tau) = |1 - q \cdot p_c|^{\left|\left[\frac{\tau}{\tau_0}\right]\right|}\left(1 - (1 - |1 - q \cdot p_c|) \cdot \left(\frac{|\tau|}{\tau_0} - \left|\left[\frac{\tau}{\tau_0}\right]\right|\right)\right). \tag{5.1.21}$$

From the relationships (5.1.19) and (5.1.20), we can also find the normalized ACF

5.1 Information Quantity Carried by Discrete and Continuous Signals

$r_u(\tau)$ of stochastic process $u(t)$, which, on arbitrary values of state transition probabilities p_c, is determined by the expression:

$$r_u(\tau) = (1 - q \cdot p_c)^{\left|\left[\frac{\tau}{\tau_0}\right]\right|} \left(1 - q \cdot p_c \cdot \left(\frac{|\tau|}{\tau_0} - \left|\left[\frac{\tau}{\tau_0}\right]\right|\right)\right), \tag{5.1.22}$$

where $\left[\frac{\tau}{\tau_0}\right]$ is an integer part of the ratio $\frac{\tau}{\tau_0}$.

For discrete stochastic process $u(t)$ with state transition probability p_c taking the values in the interval $[0; 1/q]$, normalized ACF $r_u(\tau)$ is a strictly positive function, and NFSI $\psi_u(\tau)$ is identically equal to it: $\psi_u(\tau) = r_u(\tau) \geq 0$. Conversely, for stochastic process $u(t)$ with state transition probability p_c taking the values in the interval $]1/q; 1/(q-1)]$, normalized ACF $r_u(\tau)$ is an oscillating function, taking both positive and negative values. The examples of NFSIs for multipositional sequence $u(t)$ with state transition probability p_c equal to $p_c = 1/(q-1)$ on $q = 8; 64$ are represented in Fig. 5.1.3.

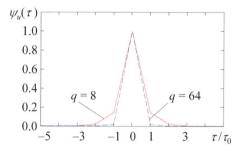

FIGURE 5.1.3 NFSI of multipositional sequence

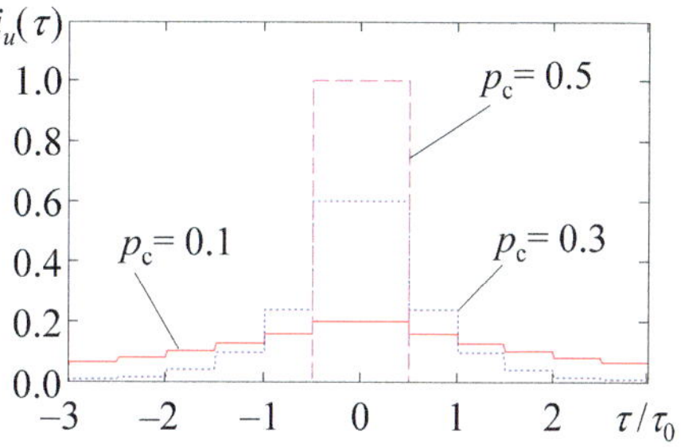

FIGURE 5.1.4 IDD of multipositional sequence

In the case of statistical independence of the symbols $\{u_j(t)\}$ of a message $u(t)$ ($p_{lm} = p_c = p_s = 1/q$), normalized ACF $r_u(\tau)$ and NFSI $\psi_u(\tau)$ are determined by the relationship:

$$\psi_u(\tau) = r_u(\tau) = \begin{cases} 1 - \frac{|\tau|}{\tau_0}, & |\tau| \leq \tau_0; \\ 0, & |\tau| > \tau_0. \end{cases}$$

According to the coupling equation (3.4.3) between IDD and NFSI (5.1.21), IDD $i_u(\tau)$ of stochastic process $u(t)$ is determined by the expression:

$$i_u(\tau) = (1 - |1 - q \cdot p_c|) \cdot |1 - q \cdot p_c|^{\left|\left[\frac{2\tau}{\tau_0}\right]\right|}. \tag{5.1.23}$$

The examples of IDDs for multipositional sequence $u(t)$ with state transition probability p_c equal to $p_c = 1/(q-1)$ on $q = 8; 64$ are represented in Fig. 5.1.4. In the case of statistical independence of the symbols $\{u_j(t)\}$ of a message $u(t)$ ($p_{lm} = 1/q$), IDD $i_u(\tau)$ is defined by the expression:

$$i_u(\tau) = [1(\tau + \tau_0/2) - 1(\tau - \tau_0/2)]/\tau_0,$$

where $1(t)$ is Heaviside step function.

Quantity of absolute information $I_u(T)$ carried by stochastic process $u(t)$ in the interval $t \in T_u = [t_0, t_0 + T]$, according to (5.1.8), is equal to:

$$I_u(T) = T i_u(0) = n \text{ (abit)}, \tag{5.1.24}$$

where $i_u(0)$ is maximum value of IDD $i_u(\tau)$ of stochastic process $u(t)$ on $\tau = 0$; n is a number of symbols of a sequence.

Comparing the obtained result (5.1.24) with formula (5.1.9), one may conclude that the quantity of absolute information $I_u(T)$, contained in a discrete message $u(t)$ being the sequence of statistically independent symbols $\{u_j(t)\}$, whose every symbol $u_j(t)$ takes one of q values from the set u_i, is equal to the number n of symbols of the sequence:

$$I_u(T) = n \text{ (abit)},$$

i.e., quantity of absolute information (in abits) carried by a discrete sequence does not depend on cardinality $\text{Card}\{u_i\} = q$ of the set of values $\{u_i\}$ of sequence symbols. This result can be elucidated by the fact that abit characterizes the information quantity contained in a single element $u_j(t)$ of statistical collection and it is defined by a measure $m(U_j)$ of ordinate set $\varphi[i(t_j, t)]$ of IDD $i(t_j, t)$ of this element:

$$m(U_j) = \int_{T_u} i(t_j, t) dt = 1,$$

where $U_j = \varphi[i(t_j, t)] = \{(t, y) : t \in T_u, 0 \leq y \leq i(t_j, t)\}$.

Thus, independently of cardinality $\text{Card}\{u_i\} = q$ of a set of the values $\{u_i\}$ of the sequence symbols $\{u_j(t)\}$, information quantity (in abits) contained in a discrete message in the case of statistical independence of the symbols is determined by their number n. It is fair to ask: if an equality between the information quantity $I_u(T)$, contained in discrete random sequence $\{u_j(t)\}$ and measured in abits, and the information quantity $I_u[n, q]$, measured in bits, holds with respect to binary sequences only ($q = 2$) (5.1.13):

$$I_u(T) = I_u[n, q] = n, \tag{5.1.25}$$

is introducing a new measure and its application with respect to discrete random sequences necessary and reasonable? In the case of arbitrary discrete sequence with $q > 2$, the equality (5.1.25) does not hold:

$$I_u(T) \neq I_u[n, q].$$

Moreover, discrete random sequences $\{u_j(t)\}$, $\{w_j(t)\}$ with the same length n (there is no statistical dependence between the symbols) and distinct cardinalities $\text{Card}\{u_i\}$, $\text{Card}\{w_k\}$ of the sets of the values $\{u_i\}$, $\{w_k\}$ ($i = 1, 2, \ldots, q_u$; $k = 1, 2, \ldots, q_w$; $q_u \neq q_w$) contain the same quantity of absolute information $I_u(T)$, $I_w(T)$ measured in abits:

$$I_u(T) = I_w(T),$$

5.1 Information Quantity Carried by Discrete and Continuous Signals

whereas information quantities $I_u[n, q_u]$, $I_w[n, q_w]$ measured in bits are not equal to each other:

$$I_u[n, q_u] \neq I_w[n, q_w], \quad q_u \neq q_w.$$

Thus, for a newly introduced unit of information quantity measurement, it is necessary to establish the interrelation with known units based on logarithmic measure of Hartley [50]. It may be realized by the definition of a new notion.

Definition 5.1.1. Discrete random sequence $\xi(t) = \{\xi_j(t)\}$ is defined on a finite set $T_\xi = \bigcup_j T_j$ of time intervals $T_j =]t_{j-1}, t_j]$:

$$t_j = \Delta + j\tau_0,$$

where τ_0 is a duration of an elementary signal; $j = 1, 2, \ldots, n$ is an integer positive number; Δ is a random variable that does not depend on $\xi(t)$ and is uniformly distributed in the segment $[0, \tau_0]$, is called *generalized discrete random sequence*, if the element of the sequence $\xi_j(t)$, $t \in T_j$ equiprobably takes the values from the set $\{z_i\}$, $i = 1, 2, \ldots, q$; $q = 2^k$; $k \in \mathbf{N}$, so that, in general case, $\{z_i\}$ are statistically independent random variables and each of them has symmetric (with respect to c_i) PDF $p_i(c_i, z)$:

$$p_i(c_i, z) = p_i(c_i, 2c_i - z),$$

where c_i is a center of symmetry of the function $p_i(c_i, z)$.

Thus, a generalized discrete random sequence $\xi(t)$ consists of the elements $\{\xi_j(t)\}$, so that each of them may be considered a stochastic process $\xi_j(t)$ defined in the interval $T_j =]t_{j-1}, t_j]$, and an arbitrary pair of elements $\xi_j(t)$, $t \in T_j$; $\xi_k(t)$, $t \in T_k$ is a pair of statistically dependent stochastic processes.

For PDF $p_i(c_i, z)$ of random variable z_i with domain D_z, we denote its ordinate set Z_i by $\varphi[p_i(c_i, z)]$:

$$Z_i = \varphi[p_i(c_i, z)] = \{(z, p) : z \in D_z, 0 \leq p \leq p_i(c_i, z)\}. \tag{5.1.26}$$

It is not difficult to notice that for an arbitrary pair of random variables z_i and z_l from the set of values $\{z_i\}$ of the element $\xi_j(t)$ of generalized discrete random sequence $\xi(t)$ with PDFs $p_i(c_i, z)$ and $p_l(c_l, z)$, respectively, the metric identity holds:

$$d_{il} = \frac{1}{2} \int_{D_z} |p_i(c_i, z) - p_l(c_l, z)| dz = \frac{1}{2} m[Z_i \Delta Z_l], \tag{5.1.27}$$

where d_{il} is a metric between PDF $p_i(c_i, z)$ of random variable z_i and PDF $p_l(c_l, z)$ of random variable z_l; $m[Z_i \Delta Z_l]$ is a measure of symmetric difference of the sets Z_i and Z_l; $Z_i = \varphi[p_i(c_i, z)]$; $Z_l = \varphi[p_l(c_l, z)]$.

The mapping (5.1.26) transforms PDFs $p_i(c_i, z)$, $p_l(c_l, z)$ into corresponding equivalent sets Z_i, Z_l and defines isometric mapping of the function space of PDFs $\{p_i(c_i, z)\} - (P, d)$ with metric d_{il} (5.1.27) into the set space $\{Z_i\} - (Z, d^*)$ with metric $d_{il}^* = \frac{1}{2} m[Z_i \Delta Z_l]$.

The normalization property of PDF provides normalization of a measure $m(Z_i)$ of random variable z_i from a set of the values $\{u_i\}$:

$$m(Z_i) = \int_{D_z} p_i(c_i, z) dz = 1.$$

The mapping (5.1.26) permits considering a set of values $\{z_i\}$ of generalized discrete random sequence $\xi(t)$ as a collection of sets $\{Z_i\}$:

$$Z = \bigcup_{i=1}^{q} Z_i,$$

so that a measure of a single element Z_i is normalized: $m(Z_i) = 1$.

By a measure $m(Z)$ we mean a quantity of absolute information $I[\{z_i\}]$, which is contained in a set of values $\{z_i\}$ of a symbol $\xi_j(t)$ of generalized discrete random sequence $\xi(t)$, as in a collection of the elements $\{Z_i\}$:

$$I[\{z_i\}] = m[\bigcup_{i=1}^{q} Z_i].$$

The information quantity $I_\xi(n,q)$, contained in generalized discrete random sequence $\xi(t)$ and measured in bits, can be evaluated on the base of logarithmic measure of Hartley [50] and two measures of quantity of absolute information $I[\xi(t)]$, $I[\{z_i\}]$ contained in this sequence and measured in abits:

$$I_\xi(n,q) = I[\xi(t)] \cdot \log_2 I[\{z_i\}] \text{ (bit)}, \tag{5.1.28}$$

where $I[\xi(t)] = m(\bigcup_{j=1}^{n} X_j) = T \cdot i_u(0)$ (abit) is a quantity of absolute information contained in generalized discrete random sequence $\xi(t)$, as in a collection of the elements X_j; $X_j = \varphi[i(t_j, t)] = \{(t, y) : t \in T, 0 \leq y \leq i(t_j, t)\}$ is an ordinate set of IDD $i(t_j, t)$ of a symbol $\xi_j(t)$ of a sequence $\xi(t)$; $I[\{z_i\}] = m[\bigcup_{i=1}^{q} Z_i]$ (abit) is a quantity of absolute information contained in a set of values $\{z_i\}$ of a symbol $\xi_j(t)$ of a sequence $\xi(t)$; $Z_i = \varphi[p_i(c_i, z)] = \{(z, p) : z \in D_z, 0 \leq p \leq p_i(c_i, z)\}$ is an ordinate set of PDF $p_i(c_i, z)$ of random values $\{z_i\}$ forming a set of values of a symbol $\xi_j(t)$ of a sequence $\xi(t)$.

Example 5.1.1. Let $\xi(t)$ be generalized discrete random sequence with statistically independent elements $\{\xi_j(t)\}$, so that the element of a sequence $\xi_j = \xi_j(t)$, $t \in T_j$ equiprobably takes the values from a set $\{z_i\}$, $i = 1, 2, \ldots, q$; and $\{z_i\}$ are statistically independent random variables with PDF $p_i(c_i, z) = \delta(z - c_i)$, $c_i \neq c_l$, $i \neq l$.

The information quantity $I_\xi(n,q)$ contained in generalized discrete random sequence $\xi(t)$ and measured in bits, according to (5.1.28), is equal to:

$$I_\xi(n,q) = n \log_2 q \text{ (bit)}, \tag{5.1.29}$$

inasmuch as $I[\xi(t)] = n$ (abit) (see (5.1.24)), and $I[\{z_i\}] = m[\bigcup_{i=1}^{q} Z_i] = q$ (abit). $\triangledown$

A random sequence in the output of a message source is an example of practical application of a generalized discrete random sequence of this type in the absence of various destabilizing and interfering factors influencing symbol generation.

Example 5.1.2. Let $\xi(t)$ be a generalized discrete random sequence with statistically independent elements $\{\xi_j(t)\}$, so that the element of a sequence $\xi_j(t)$, $t \in T_j$ equiprobably takes the values from a set $\{z_i\}$, $i = 1, 2, \ldots, q$; and $\{z_i\}$ are statistically independent random variables with PDF $p_i(c_i, z)$:

$$p_i(c_i, z) = \begin{cases} \frac{1}{b}\left(1 - \frac{|z - c_i|}{b}\right), & |z - c_i| \leq b; \\ 0, & |z - c_i| > b, \end{cases}$$

where $c_i = b \cdot i$ is a center of symmetry of PDF $p_i(c_i, z)$; b is a scale parameter of PDF $p_i(c_i, z)$.

The information quantity $I_\xi(n, q)$ contained in generalized discrete random sequence $\xi(t)$ and measured in bits, according to (5.1.28), is equal to:

$$I_\xi(n, q) = n \log_2(3q/4) < n \log_2 q \text{ (bit)},$$

inasmuch as $I[\xi(t)] = n$ (abit), and $I[\{z_i\}] = m[\bigcup_{i=1}^{q} Z_i] = 3q/4$ (abit). $\triangledown$

A random sequence in the outputs of transmitting and receiving sets, under influence of various destabilizing and/or interfering factors during generation and receiving message symbols, respectively, is an example of practical application of a generalized discrete random sequence of this type ($p_i(c_i, z) \neq \delta(z - c_i)$).

Example 5.1.3. Let $\xi(t)$ be a generalized discrete random sequence with statistically independent elements $\{\xi_j(t)\}$, so that the element of a sequence $\xi_j = \xi_j(t)$, $t \in T_j$ equiprobably takes the values from a set $\{z_i\}$, $i = 1, 2, \ldots, q$; and $\{z_i\}$ are statistically independent random variables with PDF $p_i(c_i, z) = \delta(z - c_i)$, $c_i \neq c_l$, $i \neq l$, and IDD $i_\xi(\tau)$ of a sequence $\xi(t)$ is determined by the expression (5.1.23):

$$i_u(\tau) = (1 - |1 - q \cdot p_c|) \cdot |1 - q \cdot p_c|^{\left\lfloor \left\lfloor \frac{2\tau}{\tau_0} \right\rfloor \right\rfloor},$$

where $p_c = 0.5/q$ is the state transition probability $\mathbf{P}\{u_l \to u_m\}$ ($l \neq m$).

The information quantity $I_\xi(n, q)$ contained in a generalized discrete random sequence $\xi(t)$ and measured in bits, according to (5.1.28), is equal to:

$$I_\xi(n, q) = T i_\xi(0) \log_2 q = 0.5n \log_2 q \text{ (bit)},$$

inasmuch as $I[\xi(t)] = T i_\xi(0) = 0.5n$ (abit), and $I[\{z_i\}] = m[\bigcup_{i=1}^{q} Z_i] = q$ (abit). $\triangledown$

The result (5.1.29) obtained first by Hartley [50] corresponds to the informational characteristic of an ideal message source. From a formal standpoint, for a finite time interval $T_\xi = [t_0, t_0 + T]$, an ideal message source is able to produce information quantity $I_\xi(n, q)$ equal to:

$$I_\xi(n, q) = n \log_2 q \text{ (bit)}, \qquad (5.1.30)$$

where $n = T/\tau_0$ is a number of sequence symbols in the interval T_ξ; τ_0 is a duration of a sequence symbol.

Information quantity $I_\xi(1, q)$ corresponding to one symbol $\xi_j(t)$ of a multiposition discrete random sequence $\{\xi_j(t)\}$ is evaluated by the formula:

$$I_\xi(1, q) = \log_2 q \text{ (bit)}. \qquad (5.1.31)$$

According to (5.1.30), a multiposition discrete random sequence with an arbitrarily large code base (alphabet size) q permits generating and transmitting an arbitrarily large information quantity $I_\xi(n, q)$ for a finite time interval:

$$\lim_{q \to \infty} I_\xi(n, q) = \lim_{q \to \infty} (n \log_2 q) = \infty. \qquad (5.1.32)$$

Then information quantity $I_\xi(1, q)$, transmitted by a single symbol of multiposition discrete random sequence with an arbitrarily large code base q, is an infinitely large quantity:

$$\lim_{q \to \infty} I_\xi(1, q) = \infty. \qquad (5.1.33)$$

This feature of multiposition discrete random sequences was reported by Wiener in [193]: "One single voltage measured to an accuracy of one part in ten trillion could convey all the information in the *Encyclopedia Britannica*, if the circuit noise did not hold us to an accuracy of measurement of perhaps one part in ten thousand". According to Wiener's view, the only restriction influencing the information transmission process with the use of multiposition discrete random sequences is interfering action of noise. However, above all, there exist fundamental properties of the signals that confine severely information quantity transmitted through the communication channels but neither noise nor interference. We shall show that excepting noise and interference, another confining condition does not permit transmitting infinitely large information quantity $I_\xi(n, q) \to \infty$ for a finite time interval $T_\xi = [t_0, t_0 + T]$.

Let $u(t) = \{u_j(t)\}$ be discrete random sequences with statistically independent symbols, where each symbol $u_j(t)$ at an arbitrary instant t may take one of q values from a set $\{u_i\}$, $i = 1, 2, \ldots, q$ with the same probabilities $p_i = \mathbf{P}\{u_j(t) = u_i\} = 1/q$, so that the state (value) transition is possible at the fixed instants t_j:

$$t_j = \Delta \pm j\tau_0,$$

where τ_0 is a duration of an elementary signal; $j = 1, 2, \ldots, n$ is an integer positive number; Δ is some deterministic quantity.

Then the following theorem holds.

5.1 Information Quantity Carried by Discrete and Continuous Signals

Theorem 5.1.1. *Information quantity $I_u[n,q]$, carried by discrete sequence $u(t)$ with n symbols and code base q, is defined by Hartley's measure but does not exceed $n \log_2 n$ bit:*

$$I_u[n,q] = \begin{cases} n \cdot \log_2 q, & q < n; \\ n \cdot \log_2 n, & q \geq n. \end{cases} \tag{5.1.34}$$

Proof. Obviously, the fact that a part of identity in (5.1.34), corresponding to the case when a strict inequality $q < n$ holds, does not require a proof. So we consider the proof of identity when $q \geq n$. Let f be a function that realizes a bijection mapping of discrete message $u(t)$ in the form of a discrete random sequence into a discrete sequence $w(t)$:

$$u(t) \underset{f^{-1}}{\overset{f}{\rightleftarrows}} w(t), \tag{5.1.35}$$

and the mapping f is such that each symbol $u_j(t)$ of initial message $u(t) = \{u_j(t)\}$, $j = 1, 2, \ldots, n$, accepting one of q values of a set $\{u_i\}$, $i = 1, 2, \ldots, q$, (Card$\{u_i\} = q$), is mapped into corresponding symbol $w_l(t)$ of multipositional discrete sequence $w(t) = \{w_l(t)\}$, $l = 1, 2, \ldots, n$, which can accept one of n values of a set $\{w_l\}$, (Card$\{w_l\} = n$):

$$f : u_j(t) \to w_l(t). \tag{5.1.36}$$

The possibility of such mapping f exists because the maximum number of distinct symbols of initial sequence $u(t) = \{u_j(t)\}$, $j = 1, 2, \ldots, n$ and the maximum number of distinct symbols of the sequence $w(t) = \{w_l(t)\}$, $l = 1, 2, \ldots, n$ are equal to the number of sequence symbols n. Under bijection mapping f (5.1.35), the initial discrete random sequence $u(t)$ with n symbols and code base q transforms into discrete random sequence $w(t)$ with n symbols and code base n. Then, according to the main axiom of signal processing theory 2.3.1, information quantity $I_w[n,n]$ contained in discrete random sequence $w(t)$ is equal to information quantity $I_u[n,q]$ contained in the initial discrete random sequence $u(t)$:

$$I_u[n,q] = I_w[n,n] = n \log_2[\min(q,n)] \text{ (bit)}. \tag{5.1.37}$$

□

Equality (5.1.37) can be interpreted in the following way: discrete random sequence $u(t) = \{u_j(t)\}$, $j = 1, 2, \ldots, n$, whose every symbol $u_j(t)$ at an arbitrary instant equiprobably accepts the values from a set $\{u_i\}$, $i = 1, 2, \ldots, q$, contains the information quantity $I[u(t)]$ defined by Hartley's logarithmic measure (5.1.34) but not greater than $n \log_2 n$ bit. Consider another proof of equality (5.1.34).

Proof. Consider the mapping h of a set of symbols $U_x = \{u_j(t)\}$ of discrete random sequence $u(t)$ onto a set of values $U_z = \{u_i\}$:

$$h : U_x = \{u_j(t)\} \to U_z = \{u_i\}. \tag{5.1.38}$$

The mapping h has a property, according to which, for each element u_i from a set of values $U_z = \{u_i\}$ there exists at least one element $u_j(t)$ from a set of symbols

U_x, $u_j(t) \in U_x = \{u_j(t)\}$, i.e., h is a surjective mapping of a set $U_x = \{u_j(t)\}$ onto a set $U_z = \{u_i\}$. This implies, that for an arbitrary sequence $u(t) = \{u_j(t)\}$, $j = 1, 2, \ldots, n$, a cardinality $\text{Card}U_z$ of a set $U_z = \{u_i\}$ does not exceed a cardinality $\text{Card}U_x$ of a set $U_x = \{u_j(t)\}$:

$$\text{Card}U_z \leq \text{Card}U_x, \tag{5.1.39}$$

where $\text{Card}U_x = n$.

Cardinality $\text{Card}U$ of a set U of all possible mappings of the elements $\{u_j(t)\}$ of a set U_x onto a set $U_z = \{u_i\}$ is equal to:

$$\text{Card}U = (\text{Card}U_z)^{\text{Card}U_x}. \tag{5.1.40}$$

Finding the logarithm of both parts of (5.1.40), we obtain Hartley's logarithmic measure defining information quantity $I_u[n, q]$ carried by discrete random sequence $u(t)$:

$$I_u[n, q] = \log_2(\text{Card}U) = \text{Card}U_x \log_2(\text{Card}U_z). \tag{5.1.41}$$

Applying the inequality (5.1.39) to the relationship (5.1.41) and replacing a cardinality $\text{Card}U_x$ of a set U_x with n, we obtain the upper bound of information quantity $I_u[n, q]$ contained in a discrete random sequence:

$$I_u[n, q] = \text{Card}U_x \log_2(\text{Card}U_z) \leq \text{Card}U_x \log_2(\text{Card}U_x);$$

$$I_u[n, q] \leq n \log_2 n. \tag{5.1.42}$$

$\square$

Theorem 5.1.1, taken with the relationship (5.1.34) and inequality (5.1.42), permits us to draw some conclusions.

1. Upper bound of information quantity $I_u[n, q]$, that can be transmitted by a discrete random sequence $u(t)$ with a length of n symbols and code base q, is determined by a number of symbols n of the sequence only, and does not depend on code base (alphabet size) q.

2. If a discrete random sequence $u(t)$ is defined on a finite time interval $T_u = [t_0, t_0 + T]$, then the information quantity $I_u[n, q]$ contained in this sequence is always a finite quantity:

$$I_u[n, q] \leq \frac{T}{\tau_0} \log_2 \frac{T}{\tau_0} \text{ (bit)}, \tag{5.1.43}$$

where τ_0 is a duration of sequence symbol.

3. There exists an exclusively theoretical possibility of transmitting an infinitely large information quantity I_j by a single symbol $u_j(t)$ of discrete message $u(t)$, that can be realized only on the basis of the sequence of an infinitely large length $n \to \infty$:

$$\lim_{n \to \infty} I_j = \lim_{n \to \infty} \log_2 n = \infty \text{ (bit)}.$$

5.1 Information Quantity Carried by Discrete and Continuous Signals

4. Generally, it is impossible to transmit an arbitrary information quantity by a single elementary signal $u_j(t)$ considered outside statistical collection $\{u_j(t)\}$. Here, the reason cannot be elucidated by the absence of necessary technical means and ill effect of interference (noise). The right answer is stipulated by a fundamental property of information, i.e., its ability to exist exclusively within a statistical collection of structural elements of its carrier (the signal).

5. If $\xi(t) = \{\xi_j(t)\}$ is a generalized discrete random sequence defined on a finite set $T_\xi = \bigcup_j T_j$ of time intervals $T_j =]t_{j-1}, t_j]$:

$$t_j = \Delta + j\tau_0,$$

where τ_0 is a duration of an elementary signal; $j = 1, 2, \ldots, n$ is an integer positive number; Δ is a random variable that does not depend on $\xi(t)$ and is uniformly distributed in a segment $[0, \tau_0]$; and the element of the sequence $\xi_j(t)$, $t \in T_j$ equiprobably accepts the values from a set $\{z_i\}$, $i = 1, 2, \ldots, q$; $q = 2^k$; $k \in \mathbf{N}$, so that, in general case, $\{z_i\}$ are statistically dependent random variables, each of them has PDF $p_i(c_i, z)$ that is symmetric with respect to c_i:

$$p_i(c_i, z) = p_i(c_i, 2c_i - z),$$

where c_i is a center of symmetry of PDF $p_i(c_i, z)$, then the following statements hold.

(a) Information quantity $I_\xi(n, q)$ carried by generalized discrete random sequence $\xi(t)$ is determined by the relationships (5.1.28) and (5.1.34):

$$I_\xi(n, q) = \begin{cases} I[\xi(t)] \cdot \log_2 I[\{z_i\}], & q < n; \\ I[\xi(t)] \cdot \log_2 I[\xi(t)], & q \geq n, \end{cases} \text{(bit)} \qquad (5.1.44)$$

where $I[\xi(t)] = m(\bigcup_{j=1}^{n} X_j) = T \cdot i_u(0)$ (abit) is a quantity of absolute information contained in generalized discrete random sequence $\xi(t)$ considered as a collection of the elements X_j; $X_j = \varphi[i(t_j, t)] = \{(t, y) : t \in T, 0 \leq y \leq i(t_j, t)\}$ is an ordinate set of IDD $i(t_j, t)$ of the symbol $\xi_j(t)$ of the sequence $\xi(t)$; $I[\{z_i\}] = m[\bigcup_{i=1}^{q} Z_i]$ (abit) is a quantity of absolute information contained in a set of values $\{z_i\}$ of the symbol $\xi_j(t)$ of the sequence $\xi(t)$; $Z_i = \varphi[p_i(c_i, z)] = \{(z, p) : z \in D_z, 0 \leq p \leq p_i(c_i, z)\}$ is an ordinate set of PDF $p_i(c_i, z)$ of random variables $\{z_i\}$ forming a set of values of the symbol $\xi_j(t)$ of the sequence $\xi(t)$.

The relationship (5.1.44) means that if cardinality Card$\{z_i\}$ of a set of values $\{z_i\}$, $i = 1, 2, \ldots, q$ of generalized discrete random sequence

$\xi(t) = \xi_j(t)$, $j = 1, 2, \ldots, n$ is less than cardinality of a set of the symbols $\mathrm{Card}\{\xi_j(t)\}$:

$$\mathrm{Card}\{z_i\} = q < \mathrm{Card}\{\xi_j(t)\} = n,$$

the information quantity $I_\xi(n,q)$ carried by this sequence is equal to:

$$I_\xi(n,q) = I[\xi(t)] \cdot \log_2 I[\{z_i\}]. \text{ (bit)}$$

If a cardinality $\mathrm{Card}\{z_i\}$ of a set of values $\{z_i\}$, $i = 1, 2, \ldots, q$ of generalized discrete random sequence $\xi(t) = \xi_j(t)$, $j = 1, 2, \ldots, n$ is greater than or equal to a cardinality of a set of the symbols $\mathrm{Card}\{\xi_j(t)\}$:

$$\mathrm{Card}\{z_i\} = q \geq \mathrm{Card}\{\xi_j(t)\} = n,$$

then the information quantity $I_\xi(n,q)$ carried by this sequence is equal to:

$$I_\xi(n,q) = I[\xi(t)] \cdot \log_2 I[\xi(t)] \text{ (bit)}.$$

(b) Information quantity $I_\xi(n,q)$ carried by a generalized discrete random sequence $\xi(t)$ cannot exceed some upper bound:

$$I_\xi(n,q) \leq I[\xi(t)] \cdot \log_2 I[\xi(t)] \text{ (bit)}, \qquad (5.1.45)$$

where $I[\xi(t)] = m(\bigcup_{j=1}^{n} X_j)$ is quantity of absolute information contained in generalized discrete random sequence $\xi(t)$ considered a collection of the elements $\{X_j\}$. Thus, for generalized discrete random sequences $\xi(t)$ with relatively large cardinality of a set of values $\mathrm{Card}\{z_i\} = q > n$, information quantity $I_\xi(n,q)$ carried by this sequence and measured in bits (5.1.45) is entirely determined by quantity of absolute information $I[\xi(t)]$ measured in abits.

5.1.3 Information Quantity Carried by Continuous Signals

As shown in the previous subsection, information quantity $I_\xi(n,q)$ contained in a generalized discrete random sequence $\xi(t)$ and measured in bits can be evaluated on the base of Hartley's logarithmic measure and two measures of quantity of absolute information $I[\xi(t)]$ and $I[\{z_i\}]$ contained in this sequence and measured in abits (see Formula (5.1.28)):

$$I_\xi(n,q) = I[\xi(t)] \cdot \log_2 I[\{z_i\}] \text{ (bit)},$$

where $I[\xi(t)] = m(\bigcup_{j=1}^{n} X_j)$ (abit) is a quantity of absolute information contained in a generalized discrete random sequence $\xi(t)$ considered a collection of the elements $\{X_j\}$; $I[\{z_i\}] = m[\bigcup_{i=1}^{q} Z_i]$ (abit) is a quantity of absolute information contained in a set of the values $\{z_i\}$ of the symbols $\xi_j(t)$ of the sequence $\xi(t)$.

5.1 Information Quantity Carried by Discrete and Continuous Signals

The result (5.1.28) may be generalized, introducing, as the variables, the abstract measures M_T, M_X of a set of the symbols $\{\xi_j(t)\}$, $j = 1, 2, \ldots, n$ and a set of the values $\{z_i\}$, $i = 1, 2, \ldots, q$, respectively:

$$I_\xi = M_T \log_2 M_X. \text{ (bit)} \qquad (5.1.46)$$

As to the information quantity I_ξ carried by a usual discrete random sequence, the measures M_T and M_X are determined by the cardinalities $\text{Card}\{\xi_j(t)\} = n$, $\text{Card}\{z_i\} = q$ of the sets $\{\xi_j(t)\}$, $\{z_i\}$, respectively:

$$M_T = \text{Card}\{\xi_j(t)\} = n; \quad M_X = \text{Card}\{z_i\} = q.$$

Regarding information quantity I_ξ carried by a generalized discrete random sequence, the measures M_T and M_X are determined by the quantities of absolute information $I[\xi(t)]$ and $I[\{z_i\}]$ of the sets $\{\xi_j(t)\}$ and $\{z_i\}$, respectively:

$$M_T = I[\xi(t)] = m(\bigcup_{j=1}^{n} X_j); \quad M_X = I[\{z_i\}] = m[\bigcup_{i=1}^{q} Z_i].$$

The result of application of Hartley's generalized logarithmic measure (5.1.46) with respect to continuous stochastic processes (to stochastic functions or signals) to evaluate the information quantity they contain and measured in bits is included in the following theorem.

Theorem 5.1.2. *Overall quantity of information I_ξ contained in continuous stochastic signal $\xi(t)$ and evaluated by logarithmic measure (5.1.46) is determined by the expression:*

$$I_\xi = I[\xi(t)] \cdot \log_2 I[\xi(t)] \text{ (bit)}.$$

Proof. Continuous stochastic function $\xi(t)$, defined in the interval $T_\xi = [t_0, t_0 + T]$, can be considered a mapping of a set of the values of argument $T_\xi = \{t_\alpha\}$ (domain of definition of the function) into a set of the values of the function $\Xi = \{x_\alpha\}$ (codomain of the function): $x_\alpha = \xi(t_\alpha)$, x_α is a random variable. For any $t_\alpha, t_\beta \in T$, the equality $\xi(t_\alpha) = \xi(t_\beta)$ implies the equality $t_\alpha = t_\beta$, i.e., stochastic function $\xi(t)$ is an injective mapping [233]. For continuous random variables $x_\alpha = \xi(t_\alpha)$, $x_\beta = \xi(t_\beta)$, probability $\mathbf{P}[x_\beta = x/x_\alpha = x]$ that random variable x_β takes the same value x taken by random variable x_α is equal to zero $\mathbf{P}[x_\beta = x/x_\alpha = x] = 0$, if $t_\alpha \neq t_\beta$. On the other hand, for any $x_\alpha \in \Xi$, there exists at least one element $t_\alpha \in T_\xi$, such that $\xi(t_\alpha) = x_\alpha$, i.e., stochastic function is a surjective mapping [233]. Being simultaneously both injective and surjective mapping, stochastic function $\xi(t)$ is also bijection mapping.

Each element t_α from a set T_ξ (every sample $\xi(t_\alpha)$ of stochastic process $\xi(t)$) is associated with a function called IDD $i(t_\alpha; t)$, which is associated with its ordinate set X_α with normalized measure $m(X_\alpha) = 1$ by the mapping (3.5.1). Hence, the set of values of argument $T_\xi = \{t_\alpha\}$ (the set of samples $\{\xi(t_\alpha)\}$) is associated with a set $\bigcup_\alpha X_\alpha$ with a measure $m(\sum_\alpha X_\alpha)$ called overall quantity of information $I[\xi(t)]$, which is contained in a stochastic function $\xi(t)$ (see Section 4.1). The stochastic function $\xi(t)$ is a one-to-one correspondence between the sets $T_\xi = \{t_\alpha\}$ and $\Xi = \{x_\alpha\}$. This means that both the set of values of an argument $T_\xi = \{t_\alpha\}$ and the set of values of the function $\Xi = \{x_\alpha\}$ can be associated with a set $\bigcup_\alpha X_\alpha$ with a measure $m(\sum_\alpha X_\alpha)$. This implies that a measure M_T of the set of the values of argument of the function T_ξ is equal to a measure M_X of the set of the values of the function Ξ and is equal to a measure $m(\sum_\alpha X_\alpha)$ called overall quantity of information $I[\xi(t)]$: $M_T = M_X = I[\xi(t)] = m(\sum_\alpha X_\alpha)$ (see Definition 3.5.4), which is equal to the quantity of absolute information (see Definition 4.1.4).

Thus, for a continuous stochastic signal, a measure M_T of a domain of definition $T_\xi = \{t_\alpha\}$ and a measure M_X of a codomain $\Xi = \{x_\alpha\}$ are identical and equal to the overall quantity of information $I[\xi(t)]$ contained in a stochastic signal in the interval of its existence:
$$M_T = M_X = I[\xi(t)].$$

Thus, the overall quantity of information $I[\xi(t)]$ contained in a stochastic signal $\xi(t)$ and evaluated by logarithmic measure (5.1.46) is determined by the expression:

$$I_\xi = I[\xi(t)] \log_2 I[\xi(t)] \text{ (bit)}. \qquad (5.1.47)$$

$\square$

It should be reminded here that overall quantity of information $I[\xi(t)]$, which is included in the formula (5.1.47), is measured in abits.

Theorem 5.1.2. has a corollary.

Corollary 5.1.1. *Relative quantity of information $I_{\xi,\Delta}$ contained in stochastic signal $\xi(t)$ and evaluated by logarithmic measure (5.1.46) is defined by the following expression:*

$$I_{\xi,\Delta} = I_\Delta[\xi(t)] \log_2 I_\Delta[\xi(t)] \text{ (bit)}. \qquad (5.1.48)$$

Relative quantity of information $I_{\xi,\Delta}$ incoming into the formula (5.1.47) is measured in abits and connected with overall quantity of information $I[\xi(t)]$ by the relationship (4.1.19).

It should be noted that overall quantity of information I_ξ contained in continuous stochastic signal $\xi(t)$ and evaluated by logarithmic measure (5.1.47) does not possess the property of additivity, unlike overall quantity of information $I[\xi(t)]$ measured in abits. Let a collection $\bigcup_{k=1}^{K} \xi_k(t)$, $t \in T_k$ form a partition of a stochastic process $\xi(t)$:

$$\xi(t) = \bigcup_{k=1}^{K} \xi_k(t), \ t \in T_k; \qquad (5.1.49)$$

$$T_k \cap T_l = \emptyset, \ k \neq l, \ \bigcup_{k=1}^{K} T_k = T_\xi, \ T_\xi = [t_0, t_0 + T];$$

5.1 Information Quantity Carried by Discrete and Continuous Signals

$$\xi_k(t) = \xi(t), \ t \in T_k,$$

where $T_\xi = [t_0, t_0 + T]$ is a domain of definition of $\xi(t)$.

Let $I[\xi_k(t)]$ be an overall quantity of information contained in an elementary signal $\xi_k(t)$ and measured in abits, and $I_{\xi,k}$ be an overall quantity of information contained in an elementary signal $\xi_k(t)$ and evaluated by logarithmic measure:

$$I_{\xi,k} = I[\xi_k(t)] \log_2 I[\xi_k(t)] \ \text{(bit)}. \tag{5.1.50}$$

Then the identity holds:

$$I[\xi(t)] = \sum_{k=1}^{K} I[\xi_k(t)]. \tag{5.1.51}$$

Substituting the equality (5.1.51) into the expression (5.1.47), we obtain the following relationship:

$$I_\xi = \left(\sum_{k=1}^{K} I[\xi_k(t)]\right) \log_2 \left(\sum_{k=1}^{K} I[\xi_k(t)]\right) =$$
$$= \sum_{k=1}^{K} \left(I[\xi_k(t)] \cdot \log_2 \left(\sum_{k=1}^{K} I[\xi_k(t)]\right)\right). \tag{5.1.52}$$

Using the identity (5.1.52) and the inequality:

$$\sum_{k=1}^{K} \left(I[\xi_k(t)] \cdot \log_2 \left(\sum_{k=1}^{K} I[\xi_k(t)]\right)\right) > \sum_{k=1}^{K} (I[\xi_k(t)] \cdot \log_2 I[\xi_k(t)]),$$

we obtain the following relationship:

$$I_\xi > \sum_{k=1}^{K} I_{\xi,k}. \tag{5.1.53}$$

Thus, the relationship (5.1.53) implies that overall quantity of information I_ξ contained in continuous stochastic signal $\xi(t)$ and evaluated by logarithmic measure (5.1.47), does not possess the property of additivity, and this overall quantity of information is always greater than the sum of information quantities contained in separate parts $\xi_k(t)$ of an entire stochastic process (see (5.1.49)).

There are a lot of practical applications in which it is necessary to operate with information quantity contained in an unknown nonstochastic signal (or an unknown nonrandom parameter of this signal). An evaluation of this quantity is given by the following theorem.

Theorem 5.1.3. *Information quantity $I_s(T_\infty)$ contained in an unknown nonstochastic signal $s(t)$ defined in an infinite time interval $t \in T_\infty$ by some continuous function:*

$$s(t) = f(\lambda, t), \ t \in T_\infty =]-\infty, \infty[,$$

where λ is an unknown nonrandom parameter is equal to 1 abit:

$$I_s(T_\infty) = 1 \ \text{(abit)}.$$

Corollary 5.1.2. *Information quantity $I_\lambda(T_\infty)$ contained in an unknown nonrandom parameter $\lambda(t)$, which is a constant in an infinite time interval $t \in T_\infty$:*

$$\lambda(t) = \lambda = \text{const}, \ t \in T_\infty =]-\infty, \infty[,$$

is equal to 1 abit:

$$I_\lambda(T_\infty) = 1 \ (\text{abit}).$$

Corollary 5.1.3. *Information quantity $I_s(T_0)$ contained in an unknown non-stochastic signal $s(t)$ defined in a bounded open time interval $t \in T_0$ by some continuous function:*

$$s(t) = f(\lambda, t), \ t \in T_0 =]t_0, t_0 + T[,$$

where λ is an unknown nonrandom parameter, is equal to 1 abit:

$$I_s(T_0) = 1 \ (\text{abit}).$$

Corollary 5.1.4. *Information quantity $I_\lambda(T_0)$ contained in an unknown nonrandom parameter $\lambda(t)$, which is a constant in a finite time interval $t \in T_0$:*

$$\lambda(t) = \lambda = \text{const}, \ t \in T_0 =]t_0, t_0 + T[,$$

is equal to 1 abit:

$$I_\lambda(T_0) = 1 \ (\text{abit}).$$

Proof of Theorem 5.1.3 and Corollaries 5.1.2 through 5.1.4. According to a condition of the theorem, arbitrary samples $s(t')$ and $s(t'')$, $t'' = t' + \Delta$ of the signal $s(t)$ are connected by one-to-one transformation:

$$\begin{cases} s(t'') = f(\lambda, t' + \Delta) = s(t' + \Delta); \\ s(t') = f(\lambda, t'' - \Delta) = s(t'' - \Delta). \end{cases}$$

The NFSI $\psi_s(t', t'')$ of the signal $s(t)$ (3.1.2) between its arbitrary samples $s(t')$ and $s(t'')$ is equal to one:

$$\psi_s(t', t'') = 1.$$

This identity holds if and only if an arbitrary sample $s(t')$ of the signal $s(t)$ is characterized by IDD $i_s(t', t)$ of the following form:

$$i_s(t', t) = \lim_{a \to \infty} \left\{ \frac{1}{a} \left[1\left(t - (t' - \frac{a}{2})\right) - 1\left(t - (t' + \frac{a}{2})\right) \right] \right\},$$

where $1(t)$ is Heaviside step function.

According to the formula (3.5.6), overall quantity of information $I[\xi(t)]$ contained in the signal $\xi(t)$ with IDD $i_\xi(t_\alpha, t)$ in the interval $T_\xi = [t_0, t_0 + T_x]$ is equal to:

$$I[\xi(t)] = \int_{T_\xi} \sup_{t_\alpha \in T_\xi} [i_\xi(t_\alpha, t)] dt.$$

5.1 Information Quantity Carried by Discrete and Continuous Signals

Thus, information quantity $I_s(T_\infty)$ contained in an unknown nonstochastic signal $s(t)$ defined in infinite time interval $t \in T_\infty$ is equal to:

$$I_s(T_\infty) = \int_{T_\infty} \sup_{t' \in T_\infty} [i_s(t', t)] dt = 1 \text{ (abit)}.$$

The Corollary 5.1.2 is proved similarly.

To prove the Corollary 5.1.3, consider the one-to-one mapping g of the signal $s(t) = f(\lambda, t)$, defined on an infinite time interval $t \in T_\infty =\,]-\infty, \infty[$, into the signal $s^*(t) = f^*(\lambda, t)$ defined on an open time interval $t \in T_0 =\,]t_0, t_0 + T[$:

$$s(t) \underset{g^{-1}}{\overset{g}{\rightleftarrows}} s^*(t).$$

According to Corollary 4.2.5 of Theorem 4.2.4, such a mapping g preserves overall quantity of information $I_s(T_\infty)$ contained in the signal $s(t)$:

$$I_s(T_\infty) = I_s(T_0) = 1 \text{ (abit)}.$$

The Corollary 5.1.4 is proved similarly.

Thus, information quantity contained in an unknown nonstochastic signal $s(t) = f(\lambda, t)$ and its constant parameter λ in a bounded open time interval $t \in T_0 =\,]t_0, t_0 + T[$ is equal to 1 abit:

$$I_s(T_0) = I_\lambda(T_0) = 1 \text{ (abit)}. \qquad \square$$

Hereinafter this statement will be expanded upon the signals (and their parameters) defined on the closed interval $[t_0, t_0 + T]$.

Thus, for continuous stochastic signal $\xi(t)$ characterized by the partition (5.1.49), the following statements hold.

1. Overall quantity of information $I[\xi(t)]$ measured in abits possesses the property of additivity (5.1.51).

2. Conversely, overall quantity of information I_ξ contained in continuous stochastic signal $\xi(t)$ and evaluated by logarithmic measure (5.1.47), does not possess the property of additivity. This quantity I_ξ is always greater than the sum of the quantities $I_{\xi,k}$ characterizing overall quantity of information contained in the elements $\{\xi_k(t)\}$ of a partition of the signal $\xi(t)$ (5.1.53).

3. Information quantity contained in an unknown nonstochastic signal $s(t) = f(\lambda, t)$ and its constant parameter λ in a finite time interval is equal to 1 abit:

$$I_s(T_0) = I_\lambda(T_0) = 1 \text{ (abit)}.$$

As for information quantity carried by continuous stochastic processes, one should summarize the following facts. First, information quantity (both overall and relative) carried by continuous stochastic process $\xi(t)$ in a bounded time interval

$T_\xi = [t_0, t_0 + T]$ independently of the units is always a finite quantity. The exceptions are stochastic processes with IDD in the form of δ-function, and, in particular, delta-correlated Gaussian processes. Second, overall quantity of information I_ξ contained in a continuous stochastic signal $\xi(t)$ and evaluated by logarithmic measure (5.1.47), does not possess the property of additivity, unlike overall quantity of information $I[\xi(t)]$, which is measured in abits. In this case, the inequality (5.1.53) confirms Aristotle's pronouncement that "the whole is always greater than the sum of its parts".

5.2 Capacity of Noiseless Communication Channels

Using the formalized descriptions of signals as the elements of metric space with special properties and introduced measures of information quantities, this section considers informational characteristics of communication channels to establish the most general regularities defining information transmitting in the absence of noise (interference). Informational characteristics of communication channels under noise (interference) influence will be investigated in Chapter 6.

Channels and signals considered within informational systems are characterized by diversity of their physical natures and properties. To determine the general regularities, it is necessary to abstract from their concrete physical content and to operate with general notions.

An *information transmitting and receiving channel* (communication channel or, simply, channel), in a wide sense, is a collection of transmitting, receiving, and processing units (sets) and also physical medium providing information transmission from one point of space to another by signals, their receiving, and processing to extract useful information contained in signals [247], [248]. In information theory, one should distinguish continuous and discrete channels. Continuous (discrete) channels are intended for transmitting continuous (discrete) messages, respectively. Real channels are complicated inertial nonlinear objects, whose characteristics change in a random manner over time. To analyze such channels, special mathematical models are used. They differ in the complexity level and adequacy degree. Gaussian channels are the types of real channel models built under the following suppositions [247], [248], [70]:

1. The main physical parameters of a channel are its known nonrandom quantities.
2. Channel passband is bounded by some upper bound frequency F.
3. A stochastic Gaussian signals with bounded power, normal distribution of instantaneous values, and uniform power spectral density are transmitted over a channel within a bounded channel passband.
4. A channel contains an additive Gaussian quasi-white noise with bounded power and uniform power spectral density bounded by a channel passband.

5.2 Capacity of Noiseless Communication Channels

5. There are no statistical relations between a signal and noise.

This model needs more detail concerning a boundedness of a channel passband. An ideal filter with rectangular passband is physically unrealizable inasmuch as it does not satisfy the causality principle [235]. The need to apply an ideal filter as a theoretical model of a channel was stipulated by sampling theorem requiring the boundedness of a signal band by some value F. Here we shall consider physically feasible channels operating with "effective bandwidth of a channel" and "effective bandwidth of signal power spectral density".

If the ill effects of noise and/or interference in channels can be ignored, then analyzing channels involves a model in the form of an ideal channel, also called a noiseless channel. When a high level of information transmission quality is required and noise (interference) influence in a channel can not be neglected, a more complicated model is used known as a noisy channel (channel with noise/interference).

Capacity is the main informational characteristic of a channel. In the most general case, noiseless channel capacity C means maximum of information quantity $I(s)$ that can be transmitted over the channel by the signal $s(t)$ (discrete or continuous) for a signal duration T:

$$C = \max_s \frac{I(s)}{T}.$$

Earlier we introduced two measures of information quantity carried by a signal, i.e., overall quantity of information $I(s)$ and relative quantity of information $I_\Delta(s)$. According to this approach, one should distinguish channel capacity C evaluated by overall quantity of information $I(s)$ and channel capacity C_Δ evaluated by relative quantity of information $I_\Delta(s)$. By denoting channel capacities C (by o.q.i.) and C_Δ (by r.q.i.), we mean the capacities evaluated by these measures of information quantity.

Definition 5.2.1. By *noiseless channel capacity C (by o.q.i.)* we mean maximum of information quantity $I(s)$ can be transmitted through the channel by the signal $s(t)$ with duration T:

$$C = \max_s \frac{I(s)}{T} \text{ (abit/s).} \tag{5.2.1}$$

Substituting into Definition (5.2.1) a value of overall quantity of information $I(s)$, which, according to the formula (5.1.8), is expressed over information distribution density (IDD) $i_s(\tau)$ of the signal $s(t)$, we obtain one more variant of definition of noiseless channel capacity C:

$$C = \max_s \frac{T \cdot i_s(0)}{T} = \max_s i_s(0) \text{ (abit/s).} \tag{5.2.2}$$

Definition 5.2.2. By *noiseless channel capacity C_Δ (by r.q.i.)* we mean maximum of information quantity $I_\Delta(s)$ that can be extracted from the signal $s(t)$ under its transmitting through a noiseless channel for a signal duration T:

$$C_\Delta = \max_s \frac{I_\Delta(s)}{T} \text{ (abit/s).} \tag{5.2.3}$$

In all the variants of the capacity Definitions (5.2.1) through (5.2.3), the choice of the best and the most appropriate signal $s(t)$ for this channel is realized over all the possible signals from signal space. In this sense, the channel is considered to be matched with the signal $s(t)$ if its capacity evaluated by overall (relative) quantity of information is equal to the ratio of overall (relative) quantity of information contained in the signal $s(t)$ to its duration T:

$$C = \frac{I(s)}{T} = \frac{T \cdot i_s(0)}{T} = i_s(0) \text{ (abit/s)}; \qquad (5.2.4a)$$

$$C_\Delta = \frac{I_\Delta(s)}{T} \text{ (abit/s)}, \qquad (5.2.4b)$$

where $i_s(0)$ is a value of IDD $i_s(\tau)$ in the point $\tau = 0$.

As shown in Section 4.1, for the signals characterized by IDD in the form of δ-function or in the form of the difference of two Heaviside step functions $\frac{1}{a}\left[1(\tau + \frac{a}{2}) - 1(\tau - \frac{a}{2})\right]$, relative quantity of information $I_\Delta(s)$ contained in the signal $s(t)$ is equal to overall quantity of information $I(s)$ (see Formula (4.1.17)):

$$I_\Delta(s) = I(s).$$

Meanwhile, for the signals characterized by IDD in the form of other functions, relative quantity of information $I_\Delta(s)$ contained in the signal $s(t)$ is equal to a half of overall quantity of information $I(s)$ (see Formula (4.1.19)):

$$I_\Delta(s) = \frac{1}{2}I(s).$$

Taking into account the interrelation between relative quantity of information $I_\Delta(s)$ and overall quantity of information $I(s)$ contained in the signal $s(t)$, the relationship between noiseless channel capacity evaluated by relative quantity of information and overall quantity of information is defined by IDD of the signal $s(t)$.

As for the signals characterized by IDD in the form of δ-function or in the form of the difference of two Heaviside step functions $\frac{1}{a}\left[1(\tau + \frac{a}{2}) - 1(\tau - \frac{a}{2})\right]$, noiseless channel capacity C_Δ (by r.q.i.) is equal to a capacity C (by o.q.i.):

$$C_\Delta = C. \qquad (5.2.5)$$

As for the signals characterized by IDD in the form of other functions, noiseless channel capacity (by r.q.i.) C_Δ is equal to a half of capacity (by o.q.i.) C:

$$C_\Delta = \frac{1}{2}C. \qquad (5.2.6)$$

Using the relationships obtained in Section 5.1, we evaluate the capacities of both discrete and continuous noiseless channels, which characterize information quantities carried by both discrete and continuous stochastic signals respectively.

5.2 Capacity of Noiseless Communication Channels

5.2.1 Discrete Noiseless Channel Capacity

According to Theorem 5.1.1, discrete random sequence $u(t) = \{u_j(t)\}$, $j = 1, 2, \ldots, n$ with statistically independent symbols in which each symbol $u_j(t)$ equiprobably takes the values from the set $\{u_i\}$, $i = 1, 2, \ldots, q$ at arbitrary instant $t \in T_u = [t_0, t_0 + T]$ contains information quantity $I[u(t)] = I_u[n, q]$ determined by Hartley's logarithmic measure $I_u[n, q] = n \log_2 q$ (bit), but it is always less than or equal to $n \log_2 n$ bit (see (5.1.34)):

$$I_u[n,q] = \begin{cases} n \cdot \log_2 q, & q < n; \\ n \cdot \log_2 n, & q \geq n. \end{cases} \text{(bit)}$$

We shall assume that a discrete random sequence $u(t)$ is matched with a channel. Then a discrete noiseless channel capacity (by o.q.i.) C (bit/s) measured in bit/s is equal to the ratio of overall quantity of information $I_u[n, q]$ contained in a signal $u(t)$ to its duration T:

$$C = \frac{I_u[n,q]}{T} = \begin{cases} \dfrac{1}{\tau_0} \cdot \log_2 q, & q < n; \\ \dfrac{1}{\tau_0} \cdot \log_2 n, & q \geq n, \end{cases} \quad (5.2.7)$$

where τ_0 is a duration of elementary signal $u_j(t)$.

Consider the case in which discrete random sequence $u(t) = \{u_j(t)\}$ is characterized by a statistical interrelation between the symbols with IDD $i_u(\tau)$, and is matched with a channel. The discrete noiseless channel capacity (by o.q.i.) C (bit/s) measured in bit/s, according to the relationship (5.1.45), is bounded by the quantity:

$$C \, (\text{bit/s}) \leq i_u(0) \log_2[i_u(0)T]. \quad (5.2.8)$$

The relationships (5.2.7) and (5.2.8) imply that discrete noiseless channel capacity (by o.q.i.) C (bit/s) measured in bit/s is always a finite quantity under the condition that a duration T of a sequence is a bounded value: $T < \infty$.

Let a discrete random sequence $u(t)$ with statistically independent symbols $\{u_j(t)\}$ characterized by IDD $i_u(\tau)$ in the form of the difference of two Heaviside step functions $\frac{1}{\tau_0}\left[1(\tau + \frac{\tau_0}{2}) - 1(\tau - \frac{\tau_0}{2})\right]$ be matched with a channel. Then the discrete noiseless channel capacity (by r.q.i.) C_Δ (bit/s) is equal to channel capacity (by o.q.i.) C (bit/s):

$$C_\Delta \, (\text{bit/s}) = C \, (\text{bit/s}), \quad (5.2.9)$$

and the last, according to the relationships (5.2.7) and (5.1.40), is determined by the expression:

$$C \, (\text{bit/s}) = \begin{cases} C \, (\text{abit/s}) \cdot \log_2 q, & q < n; \\ C \, (\text{abit/s}) \cdot \log_2 n, & q \geq n, \end{cases} \quad (5.2.10)$$

where C (abit/s) $= i_u(0) = 1/\tau_0$.

5.2.2 Continuous Noiseless Channel Capacity

According to Theorem 5.1.2, overall quantity of information I_ξ contained in continuous signal $\xi(t)$, $t \in T_\xi = [t_0, t_0 + T]$ and evaluated by logarithmic measure is determined by the expression (5.1.47):

$$I_\xi = I[\xi(t)] \log_2 I[\xi(t)] \text{ (bit)}.$$

where $I[\xi(t)] = m(\bigcup_\alpha X_\alpha)$ (abit) is overall quantity of information contained in continuous signal $\xi(t)$ considered a collection of the elements $\{X_\alpha\}$.

We assume that the continuous stochastic signal is matched with a channel and vice versa. The continuous noiseless channel capacity (by o.q.i.) C (bit/s) measured in bit/s is equal to the ratio of overall quantity of information I_ξ contained in the signal $\xi(t)$ to its duration T:

$$C \text{ (bit/s)} = \frac{I[\xi(t)]}{T} \log_2 I[\xi(t)] = C \text{ (abit/s)} \cdot \log_2 I[\xi(t)] =$$

$$= C \text{ (abit/s)} \cdot \log_2[C \text{ (abit/s)}T], \qquad (5.2.11)$$

where C (abit/s) is continuous noiseless channel capacity (by o.q.i.) measured in abit/s.

The expression (5.2.11) defines ultimate information quantity I_ξ that can be transmitted by the signal $\xi(t)$ over the channel for a time equal to a signal duration T.

The relative quantity of information $I_{\xi,\Delta}$ contained in continuous stochastic signal $\xi(t)$ and evaluated by logarithmic measure is determined by Expression (5.1.48):

$$I_{\xi,\Delta} = I_\Delta[\xi(t)] \log_2 I_\Delta[\xi(t)] \text{ (bit)}.$$

Continuous noiseless channel capacity (by r.q.i.) C_Δ (bit/s) measured in bit/s is equal to the ratio of relative quantity of information $I_{\xi,\Delta}$ contained in the signal $\xi(t)$ to its duration T:

$$C_\Delta \text{ (bit/s)} = \frac{I_\Delta[\xi(t)]}{T} \log_2 I_\Delta[\xi(t)] = C_\Delta \text{ (abit/s)} \cdot \log_2 I_\Delta[\xi(t)] =$$

$$= C_\Delta \text{ (abit/s)} \cdot \log_2[C_\Delta \text{ (abit/s)}T], \qquad (5.2.12)$$

where C_Δ (abit/s) is a continuous noiseless channel capacity (by r.q.i.) measured in abit/s.

The expression (5.2.12) defines ultimate quantity of useful information $I_{\xi,\Delta}$ that can be extracted from the transmitted signal $\xi(t)$ for a time equal to a signal duration T.

The formulas converting continuous noiseless channel capacity from unit measured in abit/s to unit measured in bit/s (5.2.11), (5.2.12) have the following features. First, continuous noiseless channel capacity, evaluated by overall (relative) quantity of information C (bit/s) (C_Δ (bit/s)) and measured in bit/s, unlike channel capacity C (abit/s) (C_Δ (abit/s)) measured in abit/s, depends on IDD $i_\xi(\tau)$, and also on overall (relative) quantity of information $I[\xi(t)]$ ($I_\Delta[\xi(t)]$) contained in the signal $\xi(t)$. Second, continuous noiseless channel capacity, evaluated by overall (relative) quantity of information $C(\text{bit/s})$ ($C_\Delta(\text{bit/s})$) and measured in bit/s, as well as channel capacity C (abit/s) (C_Δ (abit/s)) measured in abit/s, is always a finite quantity under the condition that duration T of continuous signal is a bounded value: $T < \infty$. Third, continuous noiseless channel capacity, evaluated by overall (relative) quantity of information C (bit/s) (C_Δ (bit/s)) and measured in bit/s, is uniquely determined only when a signal duration T is known. Fourth, (5.2.11) (and (5.2.12)) imply that under any signal-to-noise ratio and by means of any signal processing units, one cannot transmit (extract) greater information quantity (quantity of useful information) for a signal duration T than the quantities determined by these relationships, respectively.

Thus, when talking about a continuous noiseless channel capacity, one should take into account the following considerations. On the base of introduced notions of quantities of absolute, mutual, and relative information along with overall and relative quantities of information, a continuous noiseless channel capacity can be defined uniquely. Conversely, applying logarithmic measure of information quantity, a continuous noiseless channel capacity can be uniquely defined exclusively for a fixed time, for instance, for a signal duration T, or for a time unit (e.g., a second). In the last case, the formulas converting continuous noiseless channel capacity evaluated by overall (relative) quantity of information (5.2.11), (5.2.12) may be written as follows:

$$C\,(\text{bit/s}) = C\,(\text{abit/s}) \cdot \log_2[C\,(\text{abit/s}) \cdot 1\,\text{s}]; \qquad (5.2.13\text{a})$$

$$C_\Delta\,(\text{bit/s}) = C_\Delta\,(\text{abit/s}) \cdot \log_2[C_\Delta\,(\text{abit/s}) \cdot 1\,\text{s}]. \qquad (5.2.13\text{b})$$

Continuous noiseless channel capacity (by o.q.i.) C (bit/s) (5.2.13a) measured in bit/s defines maximum information quantity that can be transmitted through a noiseless channel for one second. Continuous noiseless channel capacity (by r.q.i.) C_Δ (bit/s) (5.2.13b) measured in bit/s defines maximum quantity of useful information that can be extracted from a signal transmitted through a noiseless channel for one second.

5.2.3 Evaluation of Noiseless Channel Capacity

5.2.3.1 Evaluation of Capacity of Noiseless Channel Matched with Stochastic Stationary Signal with Uniform Information Distribution Density

We can determine a capacity of noiseless channel matched with a stochastic stationary signal $s(t)$ characterized by a uniform (rectangular) IDD $i_s(\tau)$:

$$i_s(\tau) = \frac{1}{a}\left[1(\tau + \frac{a}{2}) - 1(\tau - \frac{a}{2})\right], \qquad (5.2.14)$$

where $1(\tau)$ is the Heaviside step function; a is a scale parameter.

There is a multipositional discrete random sequence $s(t) = \{s_j(t)\}$, $j = 1, 2, \ldots, n$ with statistically independent symbols $\{s_j(t)\}$ that is an example of such a signal, the duration τ_0 of a symbol $s_j(t)$ of this sequence is equal to a appearing in the formula: $\tau_0 = a$.

Normalized function of statistical interrelationship (NFSI) $\psi_s(\tau)$ of a stationary signal $s(t)$ is completely defined by IDD $i_s(\tau)$ by the relationship (3.4.2):

$$\psi_s(\tau) = \begin{cases} 1 - \dfrac{|\tau|}{a}, & |\tau| \leq a; \\ 0, & |\tau| > a. \end{cases} \qquad (5.2.15)$$

Hyperspectral density (HSD) $\sigma_s(\omega)$ of the signal $s(t)$, according to the formula (3.1.11), is connected with NFSI $\psi_s(\tau)$ over Fourier transform:

$$\sigma_s(\omega) = \mathcal{F}[\psi_s(\tau)] \frac{1}{2\pi} \int_{-\infty}^{\infty} \psi_s(\tau) \exp[-i\omega\tau] d\tau. \qquad (5.2.16)$$

Substituting the value of NFSI $\psi_s(\tau)$ (5.2.15) into (5.2.16), we obtain the expression for HSD $\sigma_s(\omega)$ of the signal $s(t)$:

$$\sigma_s(\omega) = \frac{a}{2\pi} \left[\frac{\sin(a\omega/2)}{a\omega/2} \right]^2. \qquad (5.2.17)$$

Effective width $\Delta\omega$ of HSD $\sigma_s(\omega)$ is equal to:

$$\Delta\omega = \frac{1}{\sigma_s(0)} \int_0^{\infty} \sigma_s(\omega) d\omega = \frac{1}{2\sigma_s(0)} = \frac{\pi}{a}. \qquad (5.2.18)$$

Noiseless channel capacity (by o.q.i.) C, under the condition that the channel is matched with a stochastic signal with a rectangular IDD, according to the formula (5.2.4a), is determined by maximum value $i_s(0)$ of IDD $i_s(\tau)$:

$$C = i_s(0) = \frac{1}{a} = 2\Delta f \text{ (abit/s)}, \qquad (5.2.19)$$

where $\Delta f = \Delta\omega/(2\pi) = 1/(2a) = 0.5 i_s(0)$ is a real effective width of HSD $\sigma_s(\omega)$.

Noiseless channel capacity (by r.q.i.) C_Δ, under the condition that the channel is matched with a stochastic signal with a rectangular IDD, according to the formula (5.2.5), is equal to channel capacity (by o.q.i.) C:

$$C_\Delta = C = i_s(0) = 2\Delta f \text{ (abit/s)}. \qquad (5.2.20)$$

On the base of a continuous noiseless channel capacity evaluated by overall (relative) quantity of information C (abit/s) (C_Δ (abit/s)) (5.2.19) and (5.2.20), we shall determine a channel capacity C (bit/s) (C_Δ (bit/s)) measured in bit/s on the assumption that a duration T of the signal $s(t)$ is equal to one second. In this case, according to the formulas (5.2.13), the channel capacities C (bit/s) (C_Δ (bit/s)) are equal to:

$$C \text{ (bit/s)} = 2\Delta f \cdot \log_2[2\Delta f \cdot 1\,\text{s}]; \qquad (5.2.21a)$$

$$C_\Delta \text{ (bit/s)} = 2\Delta f \cdot \log_2[2\Delta f \cdot 1\,\text{s}], \qquad (5.2.21b)$$

where $\Delta f = 0.5 i_s(0)$ is a real effective width of HSD $\sigma_s(\omega)$ of the signal $s(t)$.

5.2 Capacity of Noiseless Communication Channels

Example 5.2.1. For a channel matched with a signal characterized by IDD $i_s(\tau)$ in the form (5.2.14) with real effective width of HSD $\Delta f = 3.4$ kHz, channel capacities C (bit/s) (C_Δ (bit/s)), determined by the formulas (5.2.21a) and (5.2.21b), are equal to each other: C (bit/s)$=C_\Delta$ (bit/s)$= 87$ kbit/s. The value C (bit/s) determines a maximum information quantity that can be transmitted over this channel by the signal $s(t)$ for one second. The value C_Δ (bit/s) determines a maximum quantity of useful information that can be extracted from the signal $s(t)$ for one second. For any channel, as follows from Formulas (5.2.21), both these values are potentially achievable and are the finite values under a finite value of a real effective width of HSD Δf of the signal $s(t)$. It should be noted that C (bit/s) and C_Δ (bit/s) are not identical quantities, and their possible equality is determined by features of IDD $i_s(\tau)$ (5.2.14) of the signal $s(t)$. The expression (5.2.21a) (the expression (5.2.21b)) implies that under any signal-to-noise ratio in a channel and by means of any signal processing units, one cannot transmit (extract) more information quantity (quantity of useful information) for a second than the quantity determined by this formulas. $\triangledown$

As for Gaussian channels, one should note the following consideration. Squared module of frequency response function (amplitude-frequency characteristic) $K(\omega)$ of a channel, matched with Gaussian signal $s(t)$, is equal to its power spectral density $S(\omega)$:

$$|K(\omega)|^2 = S(\omega). \tag{5.2.22}$$

Despite the nonlinear dependence between NFSI $\psi_s(\tau)$ and normalized autocorrelation function $r_s(\tau)$ of the signal $s(t)$, which is defined by the relationship (3.1.3), one may consider the power spectral density $S(\omega)$ to be approximately proportional to HSD $\sigma_s(\omega)$:

$$S(\omega) \approx A\sigma_s(\omega), \tag{5.2.23}$$

where A is a coefficient of proportionality providing the normalization property of HSD: $\int_{-\infty}^{\infty} \sigma_\xi(\omega)d\omega = 1$ (see Section 3.1).

The effective width $\Delta\omega_{\text{eff}}$ of a squared module of a frequency response function $K(\omega)$ of a channel, matched with Gaussian signal $s(t)$, is approximately equal to an effective width $\Delta\omega$ of HSD $\sigma_s(\omega)$:

$$\Delta\omega_{\text{eff}} = \frac{1}{|K(0)|^2} \int_0^\infty |K(\omega)|^2 d\omega \approx \Delta\omega. \tag{5.2.24}$$

Real effective width F_{eff} of a squared module of frequency response function $K(\omega)$ of a channel matched with a Gaussian signal $s(t)$ is connected with effective width $\Delta\omega_{\text{eff}}$ by the known relationship and is approximately equal to real effective width of HSD Δf:

$$F_{\text{eff}} = \Delta\omega_{\text{eff}}/2\pi \approx \Delta f. \tag{5.2.25}$$

According to the last relationship, the formulas (5.2.21) for a channel matched with a Gaussian signal $s(t)$ may be written in the following form:

$$C\,(\text{bit/s}) \approx 2F_{\text{eff}} \cdot \log_2[2F_{\text{eff}} \cdot 1\,\text{s}]; \qquad (5.2.26\text{a})$$

$$C_\Delta\,(\text{bit/s}) \approx 2F_{\text{eff}} \cdot \log_2[2F_{\text{eff}} \cdot 1\,\text{s}]. \qquad (5.2.26\text{b})$$

5.2.3.2 Evaluation of Capacity of Noiseless Channel Matched with Stochastic Stationary Signal with Laplacian Information Distribution Density

We now determine a capacity of a noiseless channel matched with a stochastic stationary signal $s(t)$ characterized by Laplacian IDD $i_s(\tau)$:

$$i_s(\tau) = b\exp(-2b|\tau|), \qquad (5.2.27)$$

where b is a scale parameter.

NFSI $\psi_s(\tau)$ of a stationary signal $s(t)$ is completely defined by IDD $i_s(\tau)$ over the relationship (3.4.2):

$$\psi_s(\tau) = \exp(-b|\tau|). \qquad (5.2.28)$$

HSD $\sigma_s(\omega)$ of a signal $s(t)$, according to formula (3.1.11), is connected with NFSI $\psi_s(\tau)$ by Fourier transform and is equal to:

$$\sigma_s(\omega) = \mathcal{F}[\psi_s(\tau)] = \frac{1}{2\pi}\int_{-\infty}^{\infty}\psi_s(\tau)\exp[-i\omega\tau]d\tau = \frac{b}{\pi(b^2+\omega^2)}. \qquad (5.2.29)$$

Effective width $\Delta\omega$ of HSD $\sigma_s(\omega)$ is equal to:

$$\Delta\omega = \frac{1}{\sigma_s(0)}\int_0^{\infty}\sigma_s(\omega)d\omega = \frac{1}{2\sigma_s(0)} = \frac{\pi\cdot b}{2}. \qquad (5.2.30)$$

Noiseless channel capacity (by o.q.i.) C, under the condition that the channel is matched with a stationary signal with Laplacian IDD, according to the formula (5.2.4a), is determined by maximum value $i_s(0)$ of IDD $i_s(\tau)$:

$$C = i_s(0) = b = 4\Delta f\,(\text{abit/s}), \qquad (5.2.31)$$

where $\Delta f = \Delta\omega/(2\pi) = b/4 = i_s(0)/4$ is the real effective width of HSD $\sigma_s(\omega)$.

Noiseless channel capacity (by r.q.i.) C_Δ, under the condition that the channel is matched with a stationary signal with Laplacian IDD, according to the formula (5.2.6), is equal to half of noiseless channel capacity over overall quantity of information C:

$$C_\Delta = \frac{1}{2}C = \frac{1}{2}i_s(0) = 2\Delta f\,(\text{abit/s}). \qquad (5.2.32)$$

On the base of noiseless channel capacity evaluated by overall (relative) quantity of information C (abit/s) (C_Δ (abit/s)) (5.2.31) and (5.2.32), we determine channel capacities C (bit/s) and C_Δ (bit/s) measured in bit/s, on the assumption that duration T of a signal $s(t)$ is equal to one second. In this case, according to the formulas (5.2.13), channel capacities C (bit/s), C_Δ (bit/s) are equal to:

$$C\,(\text{bit/s}) = 4\Delta f \cdot \log_2[4\Delta f \cdot 1\,\text{s}]; \tag{5.2.33a}$$

$$C_\Delta\,(\text{bit/s}) = 2\Delta f \cdot \log_2[2\Delta f \cdot 1\,\text{s}], \tag{5.2.33b}$$

where $\Delta f = i_s(0)$ is real effective width of HSD $\sigma_s(\omega)$ of a signal $s(t)$.

Example 5.2.2. For a channel matched with a signal characterized by IDD $i_s(\tau)$ in the form (5.2.27) with the real effective width of HSD $\Delta f = 3.4$ kHz, noiseless channel capacities C (bit/s) (C_Δ (bit/s)) determined by the formulas (5.2.33a) and (5.2.33b) are, respectively, equal to C (bit/s)$=187$ kbit/s, C_Δ (bit/s)$= 87$ kbit/s. The quantity C (bit/s) determines maximum of information quantity, which can be transmitted over such a channel by a signal $s(t)$ for one second, and the quantity C_Δ (bit/s) determines maximum quantity of useful information, which can be extracted from a signal $s(t)$ for one second. Both values with respect to any channel, as follows from the formulas (5.2.33), are potentially achievable and are the finite values under a finite value of real effective width of HSD Δf of a signal $s(t)$. Thus, the expression (5.2.33a) (the expression (5.2.33b)) implies that under any signal-to-noise ratio and by means of any signal processing units one cannot transmit (extract) greater information quantity (quantity of useful information) for a second than the quantity determined by this formulas, respectively. ▽

According to the relationship (5.2.25), the formulas (5.2.33) for a noiseless channel matched with Gaussian signal $s(t)$ may be written as follows:

$$C\,(\text{bit/s}) \approx 4F_\text{eff} \cdot \log_2[4F_\text{eff} \cdot 1\,\text{s}]; \tag{5.2.34a}$$

$$C_\Delta\,(\text{bit/s}) \approx 2F_\text{eff} \cdot \log_2[2F_\text{eff} \cdot 1\,\text{s}], \tag{5.2.34b}$$

where F_eff is real effective width of a squared module of frequency-response function of the channel matched with a Gaussian signal $s(t)$.

Comparing the expressions (5.2.21) and (5.2.33) determining the capacities C (bit/s) (C_Δ (bit/s)) for two distinct channels matched with signals characterized by the same real effective width of HSD Δf, one can draw the following conclusions. First, a channel matched with a signal characterized by Laplacian IDD (5.2.27), permits transmitting for one second more than double information quantity the channel matched with a signal characterized by rectangular IDD (5.2.14). This conclusion questions: which one from a collection of channels, under a fixed value of real effective width of HSD $\Delta f=$const, permits transmission of greater information quantity for a time unit. Second, in spite of marked feature, one can extract the same quantity of useful information for a time unit from the signal characterized by Laplacian IDD (5.2.27) that can be extracted from the signal characterized by rectangular IDD (5.2.14), under the condition that the values of real effective width of HSD Δf of both signals, with which these channels are matched, are the same. This conclusion allows us to formulate a question: which one from a collection of channels, under a fixed value of real effective width of HSD $\Delta f=$const, permits extraction of greater quantity of useful information from a transmitted signal for a time unit.

Among all the physically feasible channels with fixed values of real effective width of HSD Δf=const, only a channel matched with stochastic stationary signal characterized by Laplacian IDD in the form of (5.2.27) can transmit the largest information quantity per a time unit equal to $i_s(0) = 4\Delta f$ (abit/s). From this signal transmitted through such a channel, one can extract the largest quantity of useful information per a time unit equal to $0.5i_s(0) = 2\Delta f$ (abit/s). The last property belongs also to all the discrete random sequences (both binary and multipositional) with statistically independent symbols with rectangular IDDs in the form (5.2.14).

Investigation of noisy channel capacity is essentially based on the general relationships characterizing the efficiency of signal processing under noise (interference) background, so the approaches to evaluation of capacity of the channels with noise (interference), will be considered in Chapter 6.

6

Quality Indices of Signal Processing in Metric Spaces with L-group Properties

In electronic systems that transmit, receive, and extract information, signal processing is always accompanied with interactions between useful signals and interference (noise). The presence of interference (noise) in the input of a signal processing unit will prevent accurate reproduction of the useful signal or transmitted message in the output of the signal processing unit because of information losses.

A signal processing unit may be considered optimal if the losses of information are minimal. Solving signal processing problems requires various criteria and quality indices.

The conditions of interactions of useful signals and interference (noise) require optimal signal processing algorithms to ensure minimal information losses. The level of information losses defines the potential quality indices of signal processing.

Generally, the evaluation of potential quality indices of signal processing and also synthesis and analysis of signal processing algorithms are fundamental problems of signal processing theory. In contrast to synthesis, the establishment of potential quality indices of signal processing is not based on any possible criteria of optimality. Potential quality indices of signal processing that determine the upper bound of efficiency of solving signal processing problems are not defined by the structure of signal processing unit. They follow from informational properties of the signal space where signal processing is realized.

Naturally, under certain conditions of interactions between useful signals and interference (noise), the algorithms of signal processing in the presence of interference (noise) obtained via the synthesis based on an optimality criterion cannot provide quality indices of signal processing that are better than those of potential ones.

The main results obtained in Chapter 4 with respect to physical signal space create the basis for achieving the final results that are important for signal processing theory which deals with establishment of potential quality indices of signal processing in signal spaces with various algebraic properties. These final results also are important for information theory with respect to the relationships defining the capacity of discrete and continuous communication channels, taking into account the influence of interference (noise) there.

We will formulate *the main five signal processing problems*, excluding the problem of signal recognition, that are covered in the known works on statistical signal processing, statistical radio engineering, and statistical communication theory [159], [163], [155]. We will extend their content upon the signal spaces with lattice properties, so that consideration of these problems is not confined within signal spaces with group properties (within the linear spaces).

6.1 Formulation of Main Signal Processing Problems

Let a useful signal $s(t)$ interact with noise (interference) $n(t)$ in physical signal space $\boldsymbol{\Gamma}$ with the properties of L-group $\boldsymbol{\Gamma}\left(+,\vee,\wedge\right)$:

$$x(t) = s(t) \oplus n(t), \ t \in T_s, \qquad (6.1.1)$$

where $\oplus$ is one of three binary operations of L-group $\boldsymbol{\Gamma}\left(+,\vee,\wedge\right)$; $T_s = [t_0, t_0 + T]$ is a domain of definition of a useful signal $s(t)$; T is a duration of useful signal $s(t)$.

Probabilistic-statistical properties of a stochastic signal $s(t)$ and noise (interference) $n(t)$ are known to some extent.

Useful signal $s(t)$, as a rule, is a deterministic one-to-one function $M[*,*]$ of a *carrier signal* $c(t)$ and vector of informational parameters $\boldsymbol{\lambda}(t)$ changing on a domain of definition T_s of a signal $s(t)$:

$$s(t) = M[c(t), \boldsymbol{\lambda}(t)], \qquad (6.1.2)$$

where $\boldsymbol{\lambda}(t) = [\lambda_1(t), \ldots, \lambda_m(t)]$, $\boldsymbol{\lambda}(t) \in \Lambda$ is a vector of informational parameters of a signal $s(t)$.

Evaluation of an *estimator* $\hat{s}(t+\tau)$ of a signal $s(t)$ as a functional $F_{\hat{s}}[x(t)]$ of an observed realization $x(t)$, $t \in T_s$ on $\tau = 0$ is called a signal *filtering (extraction)* problem, and on $\tau < 0$ it is called a signal *interpolation (smoothing)* problem $s(t)$ [159], [163], [155]:

$$\hat{s}(t+\tau) = F_{\hat{s}}[x(t)], \ t \in T_s. \qquad (6.1.3)$$

Taking into account the one-to-one property of modulating function $M[*,*]$, the problems of evaluation of the estimators $\hat{s}(t)$ and $\hat{\boldsymbol{\lambda}}(t)$ are equivalent in a sense that by having an estimator $\hat{s}(t)$ of a signal $s(t)$, it is easy to obtain an *estimator* $\hat{\boldsymbol{\lambda}}(t)$ of a parameter $\boldsymbol{\lambda}(t)$ and vice versa:

$$\hat{\boldsymbol{\lambda}}(t) = M^{-1}[c(t), \hat{s}(t)]; \qquad (6.1.4a)$$

$$\hat{s}(t) = M[c(t), \hat{\boldsymbol{\lambda}}(t)]. \qquad (6.1.4b)$$

If an estimated vector parameter $\boldsymbol{\lambda}(t)$ does not change on a domain of definition T_s of a signal $s(t)$, the problem of *parameter filtering* $\boldsymbol{\lambda}(t)$ is equivalent to the problem of *estimation*, which is formulated in the following way.

Let a useful signal $s(t)$ be a deterministic one-to-one function $M[*,*]$ of a carrier signal $c(t)$ and informational parameter $\boldsymbol{\lambda}$ that remains permanent on a domain of definition T_s of a signal $s(t)$:

$$s(t) = M[c(t), \boldsymbol{\lambda}], \qquad (6.1.5)$$

where $\boldsymbol{\lambda} = [\lambda_1, \ldots, \lambda_m]$, $\boldsymbol{\lambda} \in \boldsymbol{\Lambda}$ is a vector of unknown informational parameters of a signal $s(t)$, $\lambda_1 = \text{const}, \ldots, \lambda_m = \text{const}$.

Then the problem of *estimation* of an unknown vector parameter $\boldsymbol{\lambda} = [\lambda_1, \ldots, \lambda_m]$ of a signal $s(t)$ under influence of noise (interference) $n(t)$ lies in forming (according to a certain criterion) an estimator vector $\hat{\boldsymbol{\lambda}}$ in the form of a vector functional $F_{\hat{\boldsymbol{\lambda}}}[x(t)]$ of an observed realization $x(t)$:

$$\hat{\boldsymbol{\lambda}} = F_{\hat{\boldsymbol{\lambda}}}[x(t)], \ t \in T_s, \ \hat{\boldsymbol{\lambda}} \in \boldsymbol{\Lambda}. \tag{6.1.6}$$

Undoubtedly, the problem of signal extraction (filtering) is more general and complex than the problem of parameter estimation [163], [155], inasmuch as an estimator $\hat{\boldsymbol{\lambda}}$ of parameter $\boldsymbol{\lambda}$ of a signal $s(t)$ may be obtained from the solution of a filtering problem (6.1.3) on the base of the relationship (6.1.4a):

$$\hat{\boldsymbol{\lambda}} = M^{-1}[c(t), \hat{s}(t)]; \tag{6.1.7a}$$

$$\hat{s}(t) = F_{\hat{s}}[x(t)]. \tag{6.1.7b}$$

Consider now a model of interaction between a signal $s_i(t)$ from a signal set $S = \{s_i(t)\}$, $i = 1, \ldots, m$ and interference (noise) $n(t)$ in a signal space with the properties of L-group $\boldsymbol{\Gamma}(+, \vee, \wedge)$:

$$x(t) = s_i(t) \oplus n(t), \ t \in T_s, \ i = 1, \ldots, m, \tag{6.1.8}$$

where $\oplus$ is some binary operation of L-group $\boldsymbol{\Gamma}(+, \vee, \wedge)$; $T_s = [t_0, t_0 + T]$ is a domain of definition of a signal $s_i(t)$; t_0 is a time of signal arrival; T is a signal duration; $m \in \mathbf{N}$, $\mathbf{N}$ is the set of natural numbers.

The problem of *classification* of the signals in the presence of interference (noise) lies in making a decision, using some criterion, distinguishing which one from a set of the signals $S = \{s_i(t)\}$, $i = 1, \ldots, m$ is contained in the observed process $x(t)$.

Detection of the signal $s(t)$ under influence of interference (noise) $n(t)$ is a case of the problem of binary classification of two signals on $S = \{s_1(t) = 0, s_2(t) = s(t)\}$, $m = 2$.

Consider a model of interaction between a signal $s_i(t, \lambda)$ from a signal set $S = \{s_i(t, \lambda)\}$, $i = 1, \ldots, m$ and interference (noise) $n(t)$ in a signal space with the properties of L-group $\boldsymbol{\Gamma}(+, \vee, \wedge)$:

$$x(t) = \left(\bigoplus_{i=1}^{m} \theta_i s_i(t; \lambda)\right) \oplus n(t), \ t \in T_{\text{obs}}, \ i = 1, \ldots, m, \tag{6.1.9}$$

where $\oplus$ is some binary operation of L-group $\boldsymbol{\Gamma}(+, \vee, \wedge)$; T_{obs} is an observation interval of the signals $s_i(t, \lambda)$; θ_i is a random parameter taking the values from the set $\{0, 1\}$: $\theta_i \in \{0, 1\}$; λ is an unknown nonrandom scalar parameter of a signal $s_i(t, \lambda)$ that takes the values on a set Λ: $\lambda \in \Lambda$; $m \in \mathbf{N}$, $\mathbf{N}$ is the set of natural numbers.

The problem of signal *resolution-classification* under interference (noise) background lies in making a decision, using some criterion, distinguishing which one of signal combinations from a set $S = \{s_i(t, \lambda)\}$, $i = 1, \ldots, m$ is contained in the observed process $x(t)$. If the signals from a set $S = \{s_i(t, \lambda)\}$, $i = 1, \ldots, m$ are such that the formulated problem has the solutions on any of 2^m signal combinations, then one can claim that the signals $\{s_i(t, \lambda)\}$ are resolvable in a parameter λ.

It is obvious that the formulated problem of signal resolution-classification is equivalent to the problem of signal classification from a set $S' = \{s_k(t, \lambda)\}$, $k = 1, \ldots, 2^m$.

Within the process of resolution-classification, one can consider *resolution-detection* of a useful signal $s_i(t, \lambda)$ in the presence of interference (noise) $n(t)$ and other interfering signals $\{s_j(t, \lambda)\}$, $j \neq i$, $j = 1, \ldots, m$. Thus, under resolution-detection, the problem of signal processing is the establishment of the presence of the i-th signal in the observed process $x(t)$ (6.1.9), i.e., its separate detection.

If the type of signal combination (6.1.9) from a set $S = \{s_i(t, \lambda)\}$ is established and information extraction lies in estimation of their parameters taken separately, then the problem of signal processing turns to *resolution-estimation*.

Thus, two from five formulated (above) main signal processing problems are the most general, i.e., filtering (extraction) of the signals and/or their parameters under interference (noise) background, and also signal classification (or multiple-alternative detection) in the presence of interference (noise), that is repeatedly stressed in the works on statistical communication theory and statistical radio engineering [163], [155].

Within every section of this chapter, we establish the potential quality indices of signal processing that define the efficiency upper bound of solving each main signal processing problem formulated here.

6.2 Quality Indices of Signal Filtering in Metric Spaces with L-group Properties

Let useful signal $s(t)$ interact with interference (noise) $n(t)$ in physical signal space Γ with the properties of L-group $\Gamma(+, \vee, \wedge)$:

$$x(t) = s(t) \oplus n(t),\ t \in T_s, \quad (6.2.1)$$

where $\oplus$ is some binary operation of L-group $\Gamma(+, \vee, \wedge)$; $T_s = [t_0, t_0 + T]$ is a domain of definition of a useful signal $s(t)$; T is a duration of useful signal $s(t)$.

Useful signal $s(t)$ is a deterministic one-to-one function $M[*, *]$ of a carrier signal $c(t)$ and informational parameter $\lambda(t)$ changing on a domain of definition T_s of a signal $s(t)$:

$$s(t) = M[c(t), \lambda(t)].$$

Then, by *filtering (extraction)* of a signal $s(t)$ (or its parameter $\lambda(t)$) in the presence of interference (noise) $n(t)$, we mean the formation of an estimator $\hat{s}(t)$ of a signal $s(t)$ (or estimator $\hat{\lambda}(t)$ of its parameter $\lambda(t)$) [163], [155]. Taking into account the one-to-one property of modulating function $M[*, *]$, the problems of formation of

6.2 Quality Indices of Signal Filtering in Metric Spaces with L-group Properties

the estimators $\hat{s}(t)$ and $\hat{\lambda}(t)$ are equivalent within the sense, that an estimator $\hat{s}(t)$ of a signal $s(t)$ makes it easy to obtain an estimator $\hat{\lambda}(t)$ of a parameter $\lambda(t)$ and vice versa (see the relationships (6.1.4)).

Generally, any estimator $\hat{s}(t)$ of useful signal $s(t)$ (or estimator $\hat{\lambda}(t)$ of its parameter $\lambda(t)$) is some function $f_{\hat{s}}[x(t)]$ ($f_{\hat{\lambda}}[x(t)]$) of an observed process $x(t)$, which is a result of interaction between a signal $s(t)$ and interference (noise) $n(t)$:

$$\hat{s}(t) = f_{\hat{s}}[x(t)]; \tag{6.2.2a}$$

$$\hat{\lambda}(t) = f_{\hat{\lambda}}[x(t)]. \tag{6.2.2b}$$

Under given probabilistic-statistical properties of the signal and interference (noise), and a certain type of interaction between them, the quality of the estimator $\hat{s}(t)$ ($\hat{\lambda}$) of a signal $s(t)$ (or its parameter λ) is defined by the kind of the processing function $f_{\hat{s}}$ ($f_{\hat{\lambda}}$).

Meanwhile, one can claim with confidence that a quality of the best (optimal with respect to a chosen criterion) estimators will be defined by probabilistic-statistical properties of useful signal and interference (noise) and also by the type of their interaction. Thus, to establish the limiting values (as a rule, the upper bounds) of quality indices of estimation of signals and their parameters, it is not necessary to know the concrete kind of processing function $f_{\hat{s}}$ ($f_{\hat{\lambda}}$).

In this section, we do not formulate the problem of establishing processing function (estimating function) $f_{\hat{s}}$ ($f_{\hat{\lambda}}$) which provides the best estimator of useful signal $s(t)$ (or its parameter λ); the next chapter is intended to solve this problem.

This section has two goals: (1) explaining the relationships determining the potential possibilities for signal extraction (filtering) under interference (noise) conditions in linear signal space and signal space with lattice properties based on analyzing simple informational relationships between the processed signals and (2) specifying the results of evaluating the capacity of a noiseless communication channel with respect to the channels operating in the presence of interference (noise).

We again consider the interaction of stochastic signals $a(t)$ and $b(t)$ in physical signal space $\mathbf{\Gamma}$: $a(t), b(t) \in \mathbf{\Gamma}$, $t \in T_s$ with properties of L-group $\mathbf{\Gamma}(+, \vee, \wedge)$, bearing in mind that the result of their interaction $v(t)$ is described by some binary operation of L-group $\mathbf{\Gamma}(+, \vee, \wedge)$:

$$v(t) = a(t) \oplus b(t), \ t \in T_s.$$

For stochastic signals $a(t)$ and $b(t)$ interacting in signal space $\mathbf{\Gamma}$: $a(t), b(t) \in \mathbf{\Gamma}$, by Definition 4.4.1, we introduced I_{ab} called the quantity of mutual information and equal to:

$$I_{ab} = \nu_{ab} \min[I_a, I_b], \tag{6.2.3}$$

where $\nu_{ab} = \nu(a_t, b_t)$ is a normalized measure of statistical relationship (NMSI) of the samples a_t, b_t of stochastic signals $a(t)$, $b(t)$; I_a and I_b represent the quantity of absolute information contained in the signals $a(t)$ and $b(t)$, respectively.

The relationship (6.2.3) permits establishing the correspondence between quantity of mutual information I_{sx} contained in the result of interaction $x(t)$ of useful signal $s(t)$ and interference (noise) $n(t)$ (6.2.1), and quantity of absolute information

I_s contained in useful signal $s(t)$ in physical signal space $\mathbf{\Gamma}$ with arbitrary algebraic properties.

With respect to useful signal $s(t)$ and the result of interaction $x(t)$ (6.2.1), the relationship (6.2.3) takes the form:

$$I_{sx} = \nu_{sx} \min[I_s, I_x], \qquad (6.2.4)$$

where I_{sx} is the quantity of mutual information contained in the result of interaction $x(t)$ concerning a useful signal $s(t)$; $\nu_{sx} = \nu(s_t, x_t)$ is NMSI of the samples s_t, x_t of stochastic signals $s(t)$, $x(t)$; I_s, I_x is quantity of absolute information contained in useful signal $s(t)$ and observed process $x(t)$, respectively.

On the base of Theorem 3.2.14 stated for metric signal space $(\mathbf{\Gamma}, \mu)$ with metric μ (3.2.1), one can formulate the similar theorem for an estimator $\hat{s}(t)$ of useful signal $s(t)$, whose interaction with interference (noise) $n(t)$ in physical signal space is described by some binary operation of L-group $\mathbf{\Gamma}(+, \vee, \wedge)$.

Theorem 6.2.1. *Metric inequality of signal processing.* In metric signal space $(\mathbf{\Gamma}, \mu)$ with metric μ (3.2.1), while forming the estimator $\hat{s}(t) = f_{\hat{s}}[x(t)]$, $t \in T_s$ of useful signal $s(t)$ by processing of the result of interaction $x_t = s_t \oplus n_t$ (1) between the samples s_t and n_t of stochastic signals $s(t)$, $n(t)$, $t \in T_s$ defined by a binary operation of L-group $\mathbf{\Gamma}(+, \vee, \wedge)$, the metric inequality holds:

$$\mu_{sx} = \mu(s_t, x_t) \leq \mu_{s\hat{s}} = \mu(s_t, \hat{s}_t), \qquad (6.2.5)$$

and inequality (6.2.5) turns into identity if and only if the mapping $f_{\hat{s}}$ belongs to a group $H = \{h_\alpha\}$ of continuous mappings $f_{\hat{s}} \in H$ preserving null (identity) element 0 of a group $\mathbf{\Gamma}(+)$: $h_\alpha(0) = 0$, $h_\alpha \in H$.

Thus, Theorem 6.2.1 implies the expression determining a lower bound of metric $\mu_{s\hat{s}}$ between useful signal $s(t)$ and its estimator $\hat{s}(t)$:

$$\mu_{sx} = \inf_{s,n,\hat{s} \in \mathbf{\Gamma}} \mu_{s\hat{s}} \leq \mu_{s\hat{s}}. \qquad (6.2.6)$$

The sense of the expression (6.2.6) lies in the fact that no signal processing in physical signal space $\mathbf{\Gamma}$ with arbitrary algebraic properties permit achieving a value of metric $\mu_{s\hat{s}}$ between useful signal $s(t)$ and its estimator $\hat{s}(t)$, that is less than established by the relationship (6.2.6) (see Fig. 6.2.1).

To characterize the quality of the estimator $\hat{s}(t)$ (6.2.2a) of useful signal $s(t)$ obtained as a result of its filtering (extraction) in the presence of interference (noise), the following definition is introduced.

Definition 6.2.1. By *quality index of estimator* of a signal $s(t)$, while solving the problem of its filtering (extraction) under interference (noise) background within the framework of the model (6.2.1), we mean the NMSI $\nu_{s\hat{s}} = \nu(s_t, \hat{s}_t)$ between useful signal $s(t)$ and its estimator $\hat{s}(t)$.

Taking into account the known relation between metric and NMSI (3.2.7), as a corollary of Theorem 6.2.1, one can formulate independent theorems for NMSIs characterizing the same pairs of the signals: $s(t)$, $x(t)$ and $s(t)$, $\hat{s}(t)$.

6.2 Quality Indices of Signal Filtering in Metric Spaces with L-group Properties

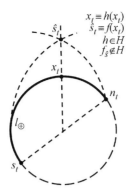

FIGURE 6.2.1 Metric relationships between signals elucidating Theorem 6.2.1

Theorem 6.2.2. *Signal processing inequality.* In metric signal space (Γ, μ) with metric μ (3.2.1), while forming the estimator $\hat{s}(t) = f_{\hat{s}}[x(t)]$, $t \in T_s$ of useful signal $s(t)$ by processing of the result of interaction $x_t = s_t \oplus n_t$ (6.2.1) between the samples s_t, n_t of stochastic signals $s(t)$, $n(t)$, $t \in T_s$ defined by a binary operation of L-group $\Gamma(+, \vee, \wedge)$, the inequality holds:

$$\nu_{s\hat{s}} = \nu(s_t, \hat{s}_t) \le \nu_{sx} = \nu(s_t, x_t), \tag{6.2.7}$$

and inequality (6.2.7) turns into identity if and only if the mapping $f_{\hat{s}}$ belongs to a group $H = \{h_\alpha\}$ of continuous mappings $f_{\hat{s}} \in H$ preserving null (identity) element 0 of a group $\Gamma(+)$: $h_\alpha(0) = 0$, $h_\alpha \in H$.

According to Theorem 6.2.2, the relationship (6.2.7) determines maximal part of information quantity, which contains in the result of interaction $x(t)$ with respect to useful signal $s(t)$ and can be extracted with a useful signal $s(t)$ processed in the presence of interference (noise) $n(t)$ by signal filtering, i.e., while forming the estimator $\hat{s}(t)$ of a useful signal $s(t)$ in a physical signal space.

The relationship (6.2.7) determines the potential possibilities of extraction (filtering) of a useful signal $s(t)$ from the mixture $x(t)$, i.e., it determines the quality of the estimator $\hat{s}(t)$ of the useful signal $s(t)$, thus, establishing the upper bound of NMSI $\nu_{s\hat{s}} = \nu(s_t, \hat{s}_t)$ between a useful signal $s(t)$ and its estimator $\hat{s}(t)$:

$$\nu_{s\hat{s}} \le \sup_{s,n,\hat{s} \in \Gamma} \nu_{s\hat{s}} = \nu_{sx}. \tag{6.2.8}$$

The sense of the relationship (6.2.8) lies in the fact that no signal processing in physical signal space Γ with arbitrary algebraic properties can provide quality index of a useful signal estimator determined by NMSI $\nu_{s\hat{s}}$ between a useful signal $s(t)$ and its estimator $\hat{s}(t)$ that is greater than the quantity of NMSI ν_{sx} between a useful signal $s(t)$ and the result of its interaction $x(t)$ with interference (noise) $n(t)$.

Theorem 6.2.3. *Informational inequality of signal processing.* In metric signal space (Γ, μ) with metric μ (3.2.1), while forming the estimator $\hat{s}(t) = f_{\hat{s}}[x(t)]$, $t \in T_s$ of useful signal $s(t)$ by processing of the result of interaction $x_t = s_t \oplus n_t$ (1) between the samples s_t, n_t of stationary stochastic signals $s(t)$, $n(t)$, $t \in T_s$ determined by a binary operation of L-group $\Gamma(+, \vee, \wedge)$, between the quantities of mutual information I_{sx}, $I_{s\hat{s}}$ contained in the pairs of signals $s(t)$, $x(t)$ and $s(t)$, $\hat{s}(t)$, respectively, the inequality holds:

$$I_{s\hat{s}} \leq I_{sx}, \qquad (6.2.9)$$

and inequality (6.2.9) turns into identity if and only if the mapping $f_{\hat{s}}$ belongs to a group $H = \{h_\alpha\}$ of continuous mappings $f_{\hat{s}} \in H$ preserving null (identity) element 0 of the group $\Gamma(+)$: $h_\alpha(0) = 0$, $h_\alpha \in H$.

Proof. Multiplying the left and right parts of inequality (6.2.7) by the left and right parts of obvious inequality $\min(I_s, I_{\hat{s}}) \leq \min(I_s, I_x)$, respectively, we obtain:

$$\nu_{s\hat{s}} \min(I_s, I_{\hat{s}}) \leq \nu_{sx} \min(I_s, I_x) \Rightarrow I_{s\hat{s}} \leq I_{sx}.$$

$\square$

According to Theorem 6.2.3, the relationship (6.2.9) determines maximum quantity of information I_{sx}, which is contained in the result of interaction $x(t)$ concerning useful signal $s(t)$, and can be extracted with a useful signal $s(t)$ processed in the presence of interference (noise) $n(t)$ by signal filtering, i.e., by forming the estimator $\hat{s}(t)$ of a useful signal $s(t)$ in a physical signal space. Besides, the relationship (6.2.9) determines the potential possibilities of useful signal extraction (filtering) $s(t)$ from a mixture $x(t)$ (6.2.1), i.e., determines informational quality of the estimator $\hat{s}(t)$ of a useful signal $s(t)$, thus establishing the upper bound of quantity of mutual information $I_{s\hat{s}}$ between a useful signal $s(t)$ and its estimator $\hat{s}(t)$:

$$I_{s\hat{s}} \leq \sup_{s, n, \hat{s} \in \Gamma} I_{s\hat{s}} = I_{sx}. \qquad (6.2.10)$$

The sense of the relationship (6.2.10) lies in the fact that no signal processing in physical signal space Γ with arbitrary algebraic properties can provide the quantity of mutual information $I_{s\hat{s}}$ between a useful signal $s(t)$ and its estimator $\hat{s}(t)$ that is greater than the value of quantity of mutual information I_{sx} between the useful signal $s(t)$ and the result of its interaction $x(t)$ with interference (noise) $n(t)$.

In the case of additive interaction between a statistically independent Gaussian useful signal $s(t)$ and interference (noise) $n(t)$ in the form of white Gaussian noise (WGN) in linear signal space $\Gamma(+)$:

$$x(t) = s(t) + n(t), \qquad (6.2.11)$$

according to Theorem 3.2.3, the following relationships hold:

$$\nu_{sx} = \frac{2}{\pi} \arcsin[\rho_{sx}]; \qquad (6.2.12)$$

$$I_{sx} = \nu_{sx} \min[I_s, I_x] = \frac{2}{\pi} \arcsin[\rho_{sx}] I_s; \qquad (6.2.13)$$

$$\rho_{sx} = q/\sqrt{1+q^2}, \qquad (6.2.14)$$

where $\nu_{sx} = \frac{2}{\pi}\arcsin[\rho_{sx}]$ is NMSI between the samples s_t and x_t of Gaussian signals $s(t)$ and $x(t)$; ρ_{sx} is the correlation coefficient between the samples s_t and x_t of Gaussian signals $s(t)$ and $x(t)$; $q^2 = S_0/N_0 = D_s/D_n$ is the ratio of the energy of Gaussian useful signal to power spectral density of interference (noise) $n(t)$; $D_s = R(0) = \frac{1}{2\pi}\int_{-\infty}^{\infty} S(\omega)d\omega$ is a variance of useful signal $s(t)$; $S_0 = S(0) = \int_{-\infty}^{\infty} R(\tau)d\tau$; $D_n = N_0(\int_{-\infty}^{\infty} S(\omega)d\omega)/(2\pi S_0)$ is a variance of interference (noise) $n(t)$ brought to the energy of useful signal $s(t)$; $S(\omega)$ is power spectral density of useful signal $s(t)$; N_0 is power spectral density of interference (noise) $n(t)$ in the form of WGN.

In the case of interactions between a useful signal $s(t)$ and interference (noise) $n(t)$ in signal space $\Gamma(\vee, \wedge)$ with lattice properties:

$$x(t) = s(t) \vee n(t); \qquad (6.2.15a)$$
$$\tilde{x}(t) = s(t) \wedge n(t), \qquad (6.2.15b)$$

according to the identities (3.2.38b) and (6.2.4), the following relationships determining the values of the quantities of mutual information and also their sum hold:

$$I_{sx} = 0.5 I_s; \qquad (6.2.16a)$$
$$I_{s\tilde{x}} = 0.5 I_s; \qquad (6.2.16b)$$
$$I_{sx} + I_{s\tilde{x}} = I_s. \qquad (6.2.16c)$$

The relationships (6.2.13), (6.2.16a), and (6.2.16b) determine information quantity contained in the results of interactions (6.2.11), (6.2.15a), and (6.2.15b) concerning the useful signal $s(t)$ that can be extracted by useful signal filtering, i.e., by forming the estimator $\hat{s}(t)$ in linear signal space $\Gamma(+)$ and in the signal space $\Gamma(\vee, \wedge)$ with lattice properties, respectively. The relationships (6.2.13) and (6.2.16) determine the potential possibilities of signal filtering from the received mixture in linear signal space and in signal space with lattice properties, respectively. The quantity of mutual information (6.2.13), (6.2.16a), and (6.2.16b), according to the relationship (6.2.9) of informational inequality of signal processing represents itself the information quantity contained in the optimal (according to criterion of maximum quantity of extracted useful information) estimator $\hat{s}(t)$ of useful signal $s(t)$ obtained by processing the observed stochastic process $x(t)$. The relationships (6.2.16a) and (6.2.16b) imply that in signal space $\Gamma(\vee, \wedge)$ with lattice properties there exists a possibility of useful signal extraction (filtering) from a mixture with permanent quality that does not depend on energetic relations of useful and interference signals. Optimal estimators $\hat{s}_a(t) = f_a[x(t)]$, $\hat{s}_b(t) = f_b[\tilde{x}(t)]$ of useful signal $s(t)$ in the form of (6.2.2a) obtained as a result of proper processing (filtering) of the received mixture $x(t)$ (6.2.15a) and $\tilde{x}(t)$ (6.2.15b) in signal space $\Gamma(\vee, \wedge)$ with lattice properties contain exactly half of the quantity of absolute information I_s contained in useful signal $s(t)$.

Due to its importance, the relationship (6.2.16c) demands a single elucidation. It claims that under the proper use of the results of simultaneous processing of observations in the form of join (6.2.15a) and meet (6.2.15b) of the lattice $\Gamma(\vee, \wedge)$, one can avoid the losses of information contained in useful signals. It is elucidated by the statistical independence of the processes $x(t)$ (6.2.15a) and $\tilde{x}(t)$ (6.2.15b) owing to fulfillment of the relationship (3.2.39b); each of them, according to the identities (6.2.16a) and (6.2.16b), contains half the information quantity contained in useful signal $s(t)$. On a qualitative level, this information has distinct content because of statistical independence of the processes $x(t)$, $\tilde{x}(t)$, when taken together (but not in the sum) they contain all information of a useful signal, not only on a quantitative level, but on a qualitative one. While processing the signal $y(t) = x(t) \bigcup \tilde{x}(t)$ (i.e., under simultaneous processing of the signals $x(t)$ and $\tilde{x}(t)$ taken separately), according to Theorem 6.2.3 and the relationships (6.2.16), the inequality holds:

$$I_{s\hat{s}} \leq I_{sy} = I_{sx} + I_{s\tilde{x}} = I_s,$$

and implies that the upper bound of quality index $\nu_{s\hat{s}}$ of an estimator $\hat{s}(t)$ of a signal $s(t)$, while solving the problem of its filtering (extraction) in signal space $\Gamma(\vee, \wedge)$ with lattice properties is equal to 1: $\sup \nu_{s\hat{s}} = 1$.

The last relation means that the possibilities of signal filtering (extraction) under interference (noise) background in signal space $\Gamma(\vee, \wedge)$ with lattice properties are not bounded by the conditions of parametric and nonparametric prior uncertainties. There exists (at least, theoretically) possibility of signal processing without losses of information contained in the processed signals $x(t)$ and $\tilde{x}(t)$.

In linear signal space $\Gamma(+)$, as follows from (6.2.13), the potential possibilities of extraction of useful signal $s(t)$ while forming the estimator $\hat{s}(t)$, as against the signal space $\Gamma(\vee, \wedge)$ with lattice properties, are essentially bounded by the energetic relations between interacting useful and interference signals, as illustrated with the following example.

Example 6.2.1. Consider a useful Gaussian signal $s(t)$ with power spectral density $S(\omega)$:

$$S(\omega) = \frac{S_0}{1 + (\omega T)^2} \tag{6.2.17}$$

and interference (noise) $n(t)$ in the form of white Gaussian noise with power spectral density (PSD) N_0 additively interact with each other in linear signal space $\Gamma(+)$.

The minimal variance of filtering error (fluctuation error of filtering) D_ε is determined by the value [159, (2.122)]:

$$D_\varepsilon = \mathbf{M}\{[s(t) - \hat{s}(t)]^2\} = \frac{S_0}{T(\sqrt{1 + q^2} + 1)}, \tag{6.2.18}$$

where $\mathbf{M}(*)$ is a symbol of mathematical expectation, and the minimal relative filtering error (relative fluctuation error of filtering) δ_ε is, respectively, equal to:

$$\delta_\varepsilon = D_\varepsilon/(2D_s) = \frac{2}{\sqrt{1 + q^2} + 1}, \tag{6.2.19}$$

where D_s is a signal variance equal to $D_s = S_0/(4T)$; T is a constant of time; $q^2 = S_0/N_0$ is a ratio of signal energy to noise PSD.

6.2 Quality Indices of Signal Filtering in Metric Spaces with L-group Properties

The relationships (6.2.18) and (6.2.19) imply that correlation coefficient $\rho_{s\hat{s}}$ between useful signal $s(t)$ and its estimator $\hat{s}(t)$ is determined by the quantity:

$$\rho_{s\hat{s}} = 1 - \delta_\varepsilon = \frac{\sqrt{1+q^2}-1}{\sqrt{1+q^2}+1}. \tag{6.2.20}$$

The relationship (6.2.7) implies that, between NMSI ν_{sx} of the samples s_t, x_t of Gaussian signals $s(t)$, $x(t)$ and NMSI $\nu_{s\hat{s}}$ of the samples s_t, $\hat{s}_t$ of Gaussian signals $s(t)$, $\hat{s}(t)$, the following inequality holds:

$$\nu_{s\hat{s}} = \frac{2}{\pi}\arcsin[\rho_{s\hat{s}}] \le \nu_{sx} = \frac{2}{\pi}\arcsin[\rho_{sx}]. \tag{6.2.21}$$

Substituting the values of NMSIs for Gaussian signals determined by the equality (3.2.12) and also the values of correlation coefficients from the relations (6.2.20) and (6.2.14) into the inequality (6.2.21), we obtain the inequality:

$$\nu_{s\hat{s}}(q) = \frac{2}{\pi}\arcsin\left(\frac{\sqrt{1+q^2}-1}{\sqrt{1+q^2}+1}\right) \le \nu_{sx}(q) = \frac{2}{\pi}\arcsin\left(\frac{q}{\sqrt{1+q^2}}\right). \tag{6.2.22}$$

The relationships (3.2.51a), (6.2.22), and (3.2.7) imply that metric μ_{sx} between the samples s_t, x_t of Gaussian signals $s(t)$, $x(t)$ and metric $\mu_{s\hat{s}}$ between the samples s_t, $\hat{s}_t$ of Gaussian signals $s(t)$, $\hat{s}(t)$ are connected by the inequality:

$$\mu_{s\hat{s}}(q) = 1 - \frac{2}{\pi}\arcsin\left(\frac{\sqrt{1+q^2}-1}{\sqrt{1+q^2}+1}\right) \ge$$
$$\ge \mu_{sx}(q) = 1 - \frac{2}{\pi}\arcsin\left(\frac{q}{\sqrt{1+q^2}}\right). \quad \triangledown \tag{6.2.23}$$

The graphs of dependences of NMSIs $\nu_{s\hat{s}}(q)$, $\nu_{sx}(q)$ and metrics $\mu_{s\hat{s}}(q)$, $\mu_{sx}(q)$ on a signal-to-noise ratio $q^2 = S_0/N_0$ are shown in Figs. 6.2.2 and 6.2.3, respectively.

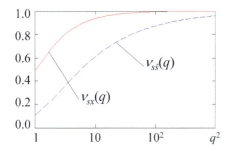

FIGURE 6.2.2 Dependences of NMSIs $\nu_{s\hat{s}}(q)$, $\nu_{sx}(q)$ on signal-to-noise ratio $q^2 = S_0/N_0$

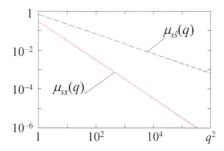

FIGURE 6.2.3 Dependences of metrics $\mu_{s\hat{s}}(q)$, $\mu_{sx}(q)$ on signal-to-noise ratio $q^2 = S_0/N_0$

The relationship (6.2.22) (see Fig. 6.2.2) implies that NMSI dependence $\nu_{sx}(q)$ determines the upper bound of NMSI $\nu_{s\hat{s}}(q)$ between useful signal $s(t)$ and its estimator $\hat{s}(t)$, which cannot be exceeded by any methods of Gaussian signal processing in linear signal space $\Gamma(+)$ and does not depend on the kinds of useful signal modulation. Similarly, the relationship (6.2.23) determines the lower bound $\mu_{sx}(q)$ of metric $\mu_{s\hat{s}}(q)$ between useful signal $s(t)$ and its estimator $\hat{s}(t)$, which cannot be smaller while using methods of Gaussian signal processing in linear space.

Generally, the properties of the estimators of the signals extracted under interference (noise) background are not included directly into a subject of information theory. At the same time, consideration of informational properties of signal estimators in informational relationships between processed signals is of interest, first, to establish constraint relationships for quality indices of signal extraction (filtering) in the presence of interference (noise), and second, to determine the capacity of communication channels with interference (noise) that operate in linear signal spaces and in signal spaces with lattice properties. The questions of evaluation of noiseless channel capacity are considered within Section 5.2.

6.3 Quality Indices of Unknown Nonrandom Parameter Estimation

This section discusses comparative characteristics of unknown nonrandom parameter estimators in sample space with group properties and sample space with lattice properties. Algebraic structures unifying the properties of a group and a lattice are called L-groups. In existing algebraic literature, L-groups have been known for a long time and are well investigated [221], [233]. *Sample space* $\mathcal{L}(\mathcal{X}, \mathcal{B}_\mathcal{X}; +, \vee, \wedge)$ *with L-group properties* is defined as a probabilistic space $(\mathcal{X}, \mathcal{B}_\mathcal{X})$ in which the axioms of distributive lattice $\mathcal{L}(\mathcal{X}; \vee, \wedge)$ with operations of join and meet hold: $a \vee b = \sup_\mathcal{L}\{a, b\}$, $a \wedge b = \inf_\mathcal{L}\{a, b\}$; $a, b \in \mathcal{L}(\mathcal{X}; \vee, \wedge)$, and also the axioms of a group $\mathcal{L}(\mathcal{X}; +)$ with operation of addition hold.

In a wide range of literature on mathematical statistics, the attention is devoted to estimator properties in sample space, where interaction of some deterministic function $f(\lambda)$ of unknown nonrandom parameter λ with estimation errors (measurement errors) $\{N_i\}$ is realized on the base of operation of addition of additive commutative group $X_{+,i} = f(\lambda) + N_i$ [250], [231], [251]. The goals of this section are, first, introducing quality indices of estimators based on metric relationships between random variables (statistics), and second, consideration of the main qualitative and quantitative characteristics of some estimators in sample space with L-group properties.

We now perform comparative analysis of characteristics of unknown nonrandom parameter estimators for the models of a direct estimation (measurement) in sample space $\mathcal{L}(\mathcal{X}, \mathcal{B}_\mathcal{X}; +, \vee, \wedge)$ with group and lattice properties of L-group $\mathcal{L}(\mathcal{X}; +, \vee, \wedge)$, respectively:

$$X_{+,i} = \lambda + N_i; \qquad (6.3.1a)$$

$$X_{\vee,i} = \lambda \vee N_i; \qquad (6.3.1b)$$

$$X_{\wedge,i} = \lambda \wedge N_i. \qquad (6.3.1c)$$

where $\{N_i\}$ are independent estimation (measurement) errors that are represented by the sample $N = (N_1, \ldots, N_n)$, $N_i \in N$, $N \in \mathcal{L}(\mathcal{X}, \mathcal{B}_\mathcal{X}; +, \vee, \wedge)$ with distribution from a distribution class with symmetric probability density function (PDF) $p_N(z) = p_N(-z)$; $\{X_{+,i}\}, \{X_{\vee,i}\}, \{X_{\wedge,i}\}$ are the results of estimation (measurement) represented by the sample $X_+ = (X_{+,1}, \ldots, X_{+,n})$, $X_\vee = (X_{\vee,1}, \ldots, X_{\vee,n})$, $X_\wedge = (X_{\wedge,1}, \ldots, X_{\wedge,n})$; $X_{+,i} \in X_+$, $X_{\vee,i} \in X_\vee$, $X_{\wedge,i} \in X_\wedge$, respectively: $X_+, X_\vee, X_\wedge \in \mathcal{L}(\mathcal{X}, \mathcal{B}_\mathcal{X}; +, \vee, \wedge)$; $+, \vee, \wedge$ are operations of addition, join, and meet of sample space $\mathcal{L}(\mathcal{X}, \mathcal{B}_\mathcal{X}; +, \vee, \wedge)$ with properties of L-group $\mathcal{L}(\mathcal{X}; +, \vee, \wedge)$, respectively; $i = 1, \ldots, n$ is an index of the elements of statistical collections $\{N_i\}, \{X_{+,i}\}, \{X_{\vee,i}\}, \{X_{\wedge,i}\}$; n represents size of the samples $N = (N_1, \ldots, N_n)$, $X_+ = (X_{+,1}, \ldots, X_{+,n})$, $X_\vee = (X_{\vee,1}, \ldots, X_{\vee,n})$, $X_\wedge = (X_{\wedge,1}, \ldots, X_{\wedge,n})$.

For the model (6.3.1a), the estimator $\hat{\lambda}_{n,+}$, which is a sample mean, is a uniformly minimum variance unbiased estimator [250], [251]:

$$\hat{\lambda}_{n,+} = \frac{1}{n} \sum_{i=1}^n X_{+,i}. \qquad (6.3.2)$$

As the estimator $\hat{\lambda}_{n,\wedge}$ of a parameter λ for the model (6.3.1b), we consider meet of lattice $\mathcal{L}(\mathcal{X}; +, \vee, \wedge)$:

$$\hat{\lambda}_{n,\wedge} = \bigwedge_{i=1}^n X_{\vee,i}, \qquad (6.3.3)$$

where $\bigwedge_{i=1}^n X_{\vee,i} = \inf_{X_\vee}\{X_{\vee,i}\}$ is the least element of the sample $X_\vee = (X_{\vee,1}, \ldots, X_{\vee,n})$.

As the estimator $\hat{\lambda}_{n,\vee}$ of a parameter λ for the model (6.3.1c), we take the join of lattice $\mathcal{L}(\mathcal{X}; +, \vee, \wedge)$:

$$\hat{\lambda}_{n,\vee} = \bigvee_{i=1}^n X_{\wedge,i}, \qquad (6.3.4)$$

where $\bigvee_{i=1}^n X_{\wedge,i} = \sup_{X_\wedge}\{X_{\wedge,i}\}$ is the largest element of the sample $X_\wedge = (X_{\wedge,1}, \ldots, X_{\wedge,n})$.

Of course, there are no doubts concerning optimality of the estimator $\hat{\lambda}_{n,+}$, at least within Gaussian distribution of estimation (measurement) errors. At the same time, it should be noted that the questions about the estimators $\hat{\lambda}_{n,\wedge}$ (6.3.3), $\hat{\lambda}_{n,\vee}$ (6.3.4) on the basis of optimality criteria, are considered in Chapter 7.

For normally distributed errors of estimation (measurement) $\{N_i\}$, cumulative distribution function (CDF) $F_{\hat{\lambda}n,+}(z)$ of the estimator $\hat{\lambda}_{n,+}$ is determined by the formula:

$$F_{\hat{\lambda}n,+}(z) = \int_{-\infty}^z p_{\hat{\lambda}n,+}(x)dx, \qquad (6.3.5)$$

where $p_{\hat{\lambda}n,+}(x) = (2\pi D_{\hat{\lambda}n,+})^{-1/2} \exp\left(-(x-\lambda)^2/2D_{\hat{\lambda}n,+}\right)$ is the PDF of the estimator $\hat{\lambda}_{n,+}$; $D_{\hat{\lambda}n,+} = D/n$ is a variance of the estimator $\hat{\lambda}_{n,+}$; D is a variance of estimation (measurement) errors $\{N_i\}$; λ is an estimated parameter.

Consider the expressions for CDFs $F_{\hat{\lambda}n,\wedge}(z)$, $F_{\hat{\lambda}n,\vee}(z)$ of the estimators $\hat{\lambda}_{n,\wedge}$ (6.3.3) and $\hat{\lambda}_{n,\vee}$ (6.3.4) and also for CDFs $F_{X_{\vee,i}}(z)$, $F_{X_{\wedge,i}}(z)$ of estimation (measurement) results $X_{\vee,i}$ (6.3.1b) and $X_{\wedge,i}$ (6.3.1c), respectively, supposing that CDF $F_N(z)$ of estimation (measurement) errors $\{N_i\}$ is an arbitrary one. The relationships [115, (3.2.87)] and [115, (3.2.82)] imply that CDFs $F_{X_{\vee,i}}(z)$, $F_{X_{\wedge,i}}(z)$ are, respectively, equal to:

$$F_{X_{\vee,i}}(z) = F_\lambda(z)F_N(z); \tag{6.3.6a}$$

$$F_{X_{\wedge,i}}(z) = F_\lambda(z) + F_N(z) - F_\lambda(z)F_N(z), \tag{6.3.6b}$$

where $F_\lambda(z) = 1(z - \lambda)$ is CDF of unknown nonrandom parameter λ; $1(z)$ is Heaviside step function; $F_N(z)$ is the CDF of estimation (measurement) errors $\{N_i\}$.

The relationships [252, (2.1.2)] and [252, (2.1.1)] imply that CDFs $F_{\hat{\lambda}n,\wedge}(z)$, $F_{\hat{\lambda}n,\vee}(z)$ of the estimators $\hat{\lambda}_{n,\wedge}$ (6.3.3) and $\hat{\lambda}_{n,\vee}$ (6.3.4) are, respectively, equal to:

$$F_{\hat{\lambda}n,\wedge}(z) = [1 - (1 - F_N(z))^n]F_\lambda(z); \tag{6.3.7a}$$

$$F_{\hat{\lambda}n,\vee}(z) = F_{X_{\wedge,i}}^n(z). \tag{6.3.7b}$$

Generalizing the relationships (6.3.6), (6.3.7), the expressions for CDFs $F_{X_{\vee,i}}(z)$, $F_{X_{\wedge,i}}(z)$ of estimation (measurement) results $X_{\vee,i}$ (6.3.1b), $X_{\wedge,i}$ (6.3.1c) and CDFs $F_{\hat{\lambda}n,\wedge}(z)$, $F_{\hat{\lambda}n,\vee}(z)$ of the estimators $\hat{\lambda}_{n,\wedge}$ (6.3.3) and $\hat{\lambda}_{n,\vee}$ (6.3.4) are, respectively:

$$F_{X_{\vee,i}}(z) = \inf_z[F_\lambda(z), F_N(z)]; \tag{6.3.8a}$$

$$F_{X_{\wedge,i}}(z) = \sup_z[F_\lambda(z), F_N(z)]; \tag{6.3.8b}$$

$$F_{\hat{\lambda}n,\wedge}(z) = \inf_z[F_\lambda(z), (1 - (1 - F_N(z))^n)]; \tag{6.3.8c}$$

$$F_{\hat{\lambda}n,\vee}(z) = \sup_z[F_\lambda(z), F_N^n(z)], \tag{6.3.8d}$$

where $F_\lambda(z) = 1(z - \lambda)$ is the CDF of unknown nonrandom parameter λ; $1(z)$ is Heaviside step function; $F_N(z)$ is the CDF of estimation (measurement) errors $\{N_i\}$; n represents size of the samples $N = (N_1, \ldots, N_n)$, $X_\vee = (X_{\vee,1}, \ldots, X_{\vee,n})$, $X_\wedge = (X_{\wedge,1}, \ldots, X_{\wedge,n})$.

For standard normal distribution of estimation (measurement) errors $\{N_i\}$ with the variance $D = 1$, the graphs of CDFs (6.3.8a,b) are shown in the Fig. 6.3.1(a), and the graphs of CDFs of the estimators (6.3.8c,d) are shown in the Fig. 6.3.1(b).

In the graphs, CDFs $F_{X_{\wedge,i}}(z)$, $F_{\hat{\lambda}n,\vee}(z)$ are shown by dotted lines, and CDFs $F_{X_{\vee,i}}(z)$, $F_{\hat{\lambda}n,\wedge}(z)$ are shown by solid lines. As follows from the relationships (6.3.8a,b) and (6.3.8c,d), CDFs $F_{X_{\vee,i}}(z)$, $F_{X_{\wedge,i}}(z)$, $F_{\hat{\lambda}n,\wedge}(z)$, $F_{\hat{\lambda}n,\vee}(z)$ are continuous on a union $D_{<\lambda} \cup D_{>\lambda}$ of open intervals $D_{<\lambda} =]-\infty, \lambda[$, $D_{>\lambda} =]\lambda, \infty[$, i.e.,

6.3 Quality Indices of Unknown Nonrandom Parameter Estimation

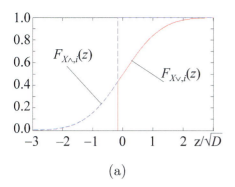
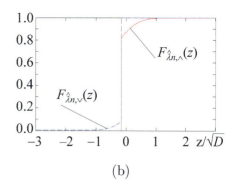

FIGURE 6.3.1 CDFs: (a) CDFs $F_{X_{\vee,i}}(z)$ (6.3.8a) and $F_{X_{\wedge,i}}(z)$ (6.3.8b) of estimation (measurement) results $X_{\vee,i}$ (6.3.1b), $X_{\wedge,i}$ (6.3.1c); (b) CDFs $F_{\hat{\lambda}_{n,\wedge}}(z)$ (6.3.8c) and $F_{\hat{\lambda}_{n,\vee}}(z)$ (6.3.8d) of estimators $\hat{\lambda}_{n,\wedge}$ (6.3.3) and $\hat{\lambda}_{n,\vee}$ (6.3.4)

there is a unilateral continuity, and $z = \lambda$ is a discontinuity point of the first kind. The limit on left is not equal to the limit on right of the CDF in this point:

$$\lim_{z \to \lambda - 0} F_{X_{\vee,i}}(z) = F_{X_{\vee,i}}(\lambda - 0), \quad \lim_{z \to \lambda + 0} F_{X_{\vee,i}}(z) = F_{X_{\vee,i}}(\lambda + 0); \quad (6.3.9a)$$

$$\lim_{z \to \lambda - 0} F_{X_{\wedge,i}}(z) = F_{X_{\wedge,i}}(\lambda - 0), \quad \lim_{z \to \lambda + 0} F_{X_{\wedge,i}}(z) = F_{X_{\wedge,i}}(\lambda + 0); \quad (6.3.9b)$$

$$\lim_{z \to \lambda - 0} F_{\hat{\lambda}_{n,\wedge}}(z) = F_{\hat{\lambda}_{n,\wedge}}(\lambda - 0), \quad \lim_{z \to \lambda + 0} F_{\hat{\lambda}_{n,\wedge}}(z) = F_{\hat{\lambda}_{n,\wedge}}(\lambda + 0); \quad (6.3.9c)$$

$$\lim_{z \to \lambda - 0} F_{\hat{\lambda}_{n,\vee}}(z) = F_{\hat{\lambda}_{n,\vee}}(\lambda - 0), \quad \lim_{z \to \lambda + 0} F_{\hat{\lambda}_{n,\vee}}(z) = F_{\hat{\lambda}_{n,\vee}}(\lambda + 0). \quad (6.3.9d)$$

To determine a quality of the estimators (6.3.2), (6.3.3), and (6.3.4) in sample space with L-group properties $\mathcal{L}(\mathcal{X}, \mathcal{B}_\mathcal{X}; +, \vee, \wedge)$, we introduce the functions $\mu(\hat{\lambda}_n, X_i)$ (3.2.1), $\mu'(\hat{\lambda}_n, X_i)$ (3.2.1a), which characterize distinctions between one of the estimators $\hat{\lambda}_n$ (6.3.2), (6.3.3), and (6.3.4) of a parameter λ and the estimation (measurement) result $X_i = \lambda \oplus N_i$, where $\oplus$ is one of the operations of L-group $\mathcal{L}(\mathcal{X}; +, \vee, \wedge)$ (6.3.1a), (6.3.1b), and (6.3.1c), respectively:

$$\mu(\hat{\lambda}_n, X_i) = 2(\mathbf{P}[\hat{\lambda}_n \vee X_i > \lambda] - \mathbf{P}[\hat{\lambda}_n \wedge X_i > \lambda]); \quad (6.3.10a)$$

$$\mu'(\hat{\lambda}_n, X_i) = 2(\mathbf{P}[\hat{\lambda}_n \wedge X_i < \lambda] - \mathbf{P}[\hat{\lambda}_n \vee X_i < \lambda]). \quad (6.3.10b)$$

Theorem 3.2.1 establishes that, for random variables with special properties, the functions defined by the relationships (3.2.1), (3.2.1a) are metrics. Meanwhile, establishing this fact for the estimators (6.3.2), (6.3.3), and (6.3.4) requires a separate proof contained in two following theorems.

Theorem 6.3.1. *The functions $\mu(\hat{\lambda}_{n,+}, X_{+,i})$, $\mu(\hat{\lambda}_{n,\wedge}, X_{\vee,i})$ determined by the expression (6.3.10a) between the estimators $\hat{\lambda}_{n,+}$ (6.3.2), $\hat{\lambda}_{n,\wedge}$ (6.3.3) and the estimation (measurement) results $X_{+,i}$ (6.3.1a), $X_{\vee,i}$ (6.3.1b), respectively, are metrics.*

Proof. We use general designations to denote one of the estimators (6.3.2), (6.3.3) of a parameter λ as $\hat{\lambda}_n$ and denote the estimation (measurement) result as X_i, $X_i = \lambda \oplus N_i$, where $\oplus$ is one of operations of L-group $\mathcal{L}(\mathcal{X}; +, \vee, \wedge)$ (6.3.1a) and (6.3.1b), respectively. Consider the probabilities $\mathbf{P}[\hat{\lambda}_n \vee X_i > \lambda]$, $\mathbf{P}[\hat{\lambda}_n \wedge X_i > \lambda]$ that, according to the formulas [115, (3.2.80)] and [115, (3.2.85)], are equal to:

$$\mathbf{P}[\hat{\lambda}_n \wedge X_i > \lambda] = 1 - [F_{\hat{\lambda}}(\lambda + 0) + F_{X_i}(\lambda + 0) - F_{\hat{\lambda}, X_i}(\lambda + 0, \lambda + 0)]; \quad (6.3.11a)$$

$$\mathbf{P}[\hat{\lambda}_n \vee X_i > \lambda] = 1 - F_{\hat{\lambda}, X_i}(\lambda + 0, \lambda + 0). \quad (6.3.11b)$$

where $F_{\hat{\lambda}, X_i}(z_1, z_2)$, $F_{\hat{\lambda}}(z_1)$, $F_{X_i}(z_2)$ are joint and univariate CDFs of the estimator $\hat{\lambda}_n$ and the measurement result X_i, respectively, and $F_{\hat{\lambda}, X_i}(\lambda+0, \lambda+0) \neq 1$, $F_{\hat{\lambda}}(\lambda+0) \neq 1$, $F_{X_i}(\lambda+0) \neq 1$.

Then the function $\mathbf{P}[\hat{\lambda}_n > \lambda] = 1 - F_{\hat{\lambda}}(\lambda+0)$, $\mathbf{P}[X_i > \lambda] = 1 - F_{X_i}(\lambda+0)$ upon a lattice $\mathcal{L}(\mathcal{X}; \vee, \wedge)$ is a valuation, inasmuch as the relationships (6.3.11a,b) imply the identity [221, Section X.1 (V1)]:

$$\mathbf{P}[\hat{\lambda}_n > \lambda] + \mathbf{P}[X_i > \lambda] = \mathbf{P}[\hat{\lambda}_n \vee X_i > \lambda] + \mathbf{P}[\hat{\lambda}_n \wedge X_i > \lambda]. \quad (6.3.12)$$

Valuation $\mathbf{P}[\hat{\lambda}_n > \lambda]$, $\mathbf{P}[X_i > \lambda]$ is isotonic, inasmuch as the implication holds [221, Section X.1 (V2)]:

$$\hat{\lambda}_n \geq \hat{\lambda}'_n \Rightarrow \mathbf{P}[\hat{\lambda}_n > \lambda] \geq \mathbf{P}[\hat{\lambda}'_n > \lambda]; \quad (6.3.13a)$$

$$X_i \geq X'_i \Rightarrow \mathbf{P}[X_i > \lambda] \geq \mathbf{P}[X'_i > \lambda]. \quad (6.3.13b)$$

Joint fulfillment of the relationships (6.3.12) and (6.3.13a,b), according to Theorem 6.3.1 [221, Section X.1], implies the quantity (6.3.10a) that is equal to:

$$\mu(\hat{\lambda}_n, X_i) = 2(\mathbf{P}[\hat{\lambda}_n \vee X_i > \lambda] - \mathbf{P}[\hat{\lambda}_n \wedge X_i > \lambda]) =$$
$$= 2[F_{\hat{\lambda}}(\lambda + 0) + F_{X_i}(\lambda + 0)] - 4F_{\hat{\lambda}, X_i}(\lambda + 0, \lambda + 0), \quad (6.3.14)$$

is metric. □

For the quantity $\mu'(\hat{\lambda}_n, X_i)$ determined by the relationship (6.3.10b), one can formulate a theorem that is analogous to Theorem 6.3.1; and to prove it, the following lemma may be useful.

Lemma 6.3.1. *The functions $\mu(\hat{\lambda}_{n,\wedge}, X_{\vee,i})$, $\mu(\hat{\lambda}_{n,\vee}, X_{\wedge,i})$; $\mu'(\hat{\lambda}_{n,\wedge}, X_{\vee,i})$, $\mu'(\hat{\lambda}_{n,\vee}, X_{\wedge,i})$, determined by the expressions (6.3.10a,b) between the estimators $\hat{\lambda}_{n,\wedge}$ (6.3.3), $\hat{\lambda}_{n,\vee}$ (6.3.4) and measurement results $X_{\vee,i}$ (6.3.1b), $X_{\wedge,i}$ (6.3.1c), respectively, are equal to:*

$$\mu(\hat{\lambda}_{n,\wedge}, X_{\vee,i}) = 2[F_{\hat{\lambda}_{n,\wedge}}(\lambda + 0) - F_{X_{\vee,i}}(\lambda + 0)]; \quad (6.3.15a)$$

$$\mu(\hat{\lambda}_{n,\vee}, X_{\wedge,i}) = 2[F_{X_{\wedge,i}}(\lambda + 0) - F_{\hat{\lambda}_{n,\vee}}(\lambda + 0)]; \quad (6.3.15b)$$

$$\mu'(\hat{\lambda}_{n,\wedge}, X_{\vee,i}) = 2[F_{\hat{\lambda}_{n,\wedge}}(\lambda - 0) - F_{X_{\vee,i}}(\lambda - 0)]; \quad (6.3.15c)$$

$$\mu'(\hat{\lambda}_{n,\vee}, X_{\wedge,i}) = 2[F_{X_{\wedge,i}}(\lambda - 0) - F_{\hat{\lambda}_{n,\vee}}(\lambda - 0)]. \quad (6.3.15d)$$

6.3 Quality Indices of Unknown Nonrandom Parameter Estimation

Proof. Determine the values of functions $\mu(\hat{\lambda}_{n,\wedge}, X_{\vee,i})$, $\mu'(\hat{\lambda}_{n,\wedge}, X_{\vee,i})$ between the estimator $\hat{\lambda}_{n,\wedge}$ (6.3.3) and the estimation (measurement) results $X_{\vee,i}$ (6.3.1b). Join $\hat{\lambda}_{n,\wedge} \vee X_{\vee,i}$, which appears in the initial formulas (6.3.10a,b), according to the definition of the estimator $\hat{\lambda}_{n,\wedge}$ (6.3.3), and also according to the lattice absorption property, is equal to the estimation (measurement) result $X_{\vee,i}$:

$$\hat{\lambda}_{n,\wedge} \vee X_{\vee,i} = [(\bigwedge_{j \neq i} X_{\vee,j}) \wedge X_{\vee,i}] \vee X_{\vee,i} = X_{\vee,i}. \tag{6.3.16}$$

Meet $\hat{\lambda}_{n,\wedge} \wedge X_{\vee,i}$, which appears in the initial formulas (6.3.10a,b) according to the idempotency property of lattice, is equal to the estimator $\hat{\lambda}_{n,\wedge}$:

$$\hat{\lambda}_{n,\wedge} \wedge X_{\vee,i} = [(\bigwedge_{j \neq i} X_{\vee,j}) \wedge X_{\vee,i}] \wedge X_{\vee,i} = \hat{\lambda}_{n,\wedge}. \tag{6.3.17}$$

Substituting the values of join and meet (6.3.16) and (6.3.17) into the initial formulas (6.3.10a,b), we obtain the values of the functions $\mu(\hat{\lambda}_{n,\wedge}, X_{\vee,i})$, $\mu'(\hat{\lambda}_{n,\wedge}, X_{\vee,i})$ between the estimator $\hat{\lambda}_{n,\wedge}$ and the estimation (measurement) result $X_{\vee,i}$:

$$\mu(\hat{\lambda}_{n,\wedge}, X_{\vee,i}) = 2(\mathbf{P}[X_{\vee,i} > \lambda] - \mathbf{P}[\hat{\lambda}_{n,\wedge} > \lambda]) =$$
$$= 2[F_{\hat{\lambda}_{n,\wedge}}(\lambda + 0) - F_{X_{\vee,i}}(\lambda + 0)]; \tag{6.3.18a}$$

$$\mu'(\hat{\lambda}_{n,\wedge}, X_{\vee,i}) = 2(\mathbf{P}[\hat{\lambda}_{n,\wedge} < \lambda] - \mathbf{P}[X_{\vee,i} < \lambda]) =$$
$$= 2[F_{\hat{\lambda}_{n,\wedge}}(\lambda - 0) - F_{X_{\vee,i}}(\lambda - 0)], \tag{6.3.18b}$$

where

$$\mathbf{P}[X_{\vee,i} > \lambda] = 1 - F_{X_{\vee,i}}(\lambda + 0); \tag{6.3.19a}$$

$$\mathbf{P}[\hat{\lambda}_{n,\wedge} > \lambda] = 1 - F_{\hat{\lambda}_{n,\wedge}}(\lambda + 0); \tag{6.3.19b}$$

$$\mathbf{P}[X_{\vee,i} < \lambda] = 1 - F_{X_{\vee,i}}(\lambda - 0); \tag{6.3.19c}$$

$$\mathbf{P}[\hat{\lambda}_{n,\wedge} < \lambda] = 1 - F_{\hat{\lambda}_{n,\wedge}}(\lambda - 0), \tag{6.3.19d}$$

$F_{\hat{\lambda}_{n,\wedge}}(z) = \inf_{z}[F_{\lambda}(z), (1 - (1 - F_N(z))^n)]$ is the CDF of the estimator $\hat{\lambda}_{n,\wedge}$ determined by the expression (6.3.8c); $F_{X_{\vee,i}}(z) = \inf_{z}[F_{\lambda}(z), F_N(z)]$ is the CDF of random variable $X_{\vee,i}$ (6.3.1b) determined by the expression (6.3.8a).

Similarly, we obtain the values of the functions $\mu(\hat{\lambda}_{n,\vee}, X_{\wedge,i})$, $\mu'(\hat{\lambda}_{n,\vee}, X_{\wedge,i})$ determined by the expressions (6.3.10a) and (6.3.10b) between the estimator $\hat{\lambda}_{n,\vee}$ (6.3.4) and the estimation (measurement) result $X_{\wedge,i}$ (6.3.1c). In this case, join $\hat{\lambda}_{n,\vee} \vee X_{\wedge,i}$, which appears in the initial formulas (6.3.10a) and (6.3.10b), according to the definition of the estimator $\hat{\lambda}_{n,\vee}$ (6.3.4), and also according to the lattice idempotency property, is equal to the estimator $\hat{\lambda}_{n,\vee}$:

$$\hat{\lambda}_{n,\vee} \vee X_{\wedge,i} = [(\bigvee_{j \neq i} X_{\wedge,j}) \vee X_{\wedge,i}] \vee X_{\wedge,i} = \hat{\lambda}_{n,\vee}. \tag{6.3.20}$$

Meet $\hat{\lambda}_{n,\vee} \wedge X_{\wedge,i}$, which appears in the initial formulas (6.3.10a) and (6.3.10b), according to the lattice absorption property, is equal to the estimation (measurement) result $X_{\wedge,i}$:

$$\hat{\lambda}_{n,\vee} \wedge X_{\wedge,i} = [(\bigvee_{j\neq i} X_{\wedge,j}) \vee X_{\wedge,i}] \wedge X_{\wedge,i} = X_{\wedge,i}. \quad (6.3.21)$$

Substituting the values of join and meet (6.3.20) and (6.3.21) into the initial formulas (6.3.10a,b), we obtain the values of the functions $\mu(\hat{\lambda}_{n,\vee}, X_{\wedge,i})$, $\mu'(\hat{\lambda}_{n,\vee}, X_{\wedge,i})$ between the estimator $\hat{\lambda}_{n,\vee}$ and the estimation (measurement) result $X_{\wedge,i}$:

$$\mu(\hat{\lambda}_{n,\vee}, X_{\wedge,i}) = 2(\mathbf{P}[\hat{\lambda}_{n,\vee} > \lambda] - \mathbf{P}[X_{\wedge,i} > \lambda]) =$$
$$= 2[F_{X_{\wedge,i}}(\lambda + 0) - F_{\hat{\lambda}_{n,\vee}}(\lambda + 0)]; \quad (6.3.22a)$$

$$\mu'(\hat{\lambda}_{n,\vee}, X_{\wedge,i}) = 2(\mathbf{P}[X_{\wedge,i} < \lambda] - \mathbf{P}[\hat{\lambda}_{n,\vee} < \lambda]) =$$
$$= 2[F_{X_{\wedge,i}}(\lambda - 0) - F_{\hat{\lambda}_{n,\vee}}(\lambda - 0)], \quad (6.3.22b)$$

where

$$\mathbf{P}[X_{\wedge,i} > \lambda] = 1 - F_{X_{\wedge,i}}(\lambda + 0); \quad (6.3.23a)$$

$$\mathbf{P}[\hat{\lambda}_{n,\vee} > \lambda] = 1 - F_{\hat{\lambda}_{n,\vee}}(\lambda + 0); \quad (6.3.23b)$$

$$\mathbf{P}[X_{\wedge,i} < \lambda] = 1 - F_{X_{\wedge,i}}(\lambda - 0); \quad (6.3.23c)$$

$$\mathbf{P}[\hat{\lambda}_{n,\vee} < \lambda] = 1 - F_{\hat{\lambda}_{n,\vee}}(\lambda - 0), \quad (6.3.23d)$$

$F_{\hat{\lambda}_{n,\vee}}(z) = \sup_z [F_\lambda(z), F_N^n(z)]$ is the CDF of the estimator $\lambda_{n,\vee}$ determined by the expression (6.3.8d); $F_{X_{\wedge,i}}(z) = \sup_z [F_\lambda(z), F_N(z)]$ is the CDF of random variable $X_{\wedge,i}$ (6.3.1c) determined by the expression (6.3.8b). $\square$

Corollary 6.3.1. *The quantities $\mu(\hat{\lambda}_{n,\vee}, X_{\wedge,i})$, $\mu'(\hat{\lambda}_{n,\wedge}, X_{\vee,i})$ determined by the expressions (6.3.22a), (6.3.18b) are, respectively, equal to zero $\forall \lambda \in]-\infty, \infty[$:*

$$\mu(\hat{\lambda}_{n,\vee}, X_{\wedge,i}) = 0; \quad (6.3.24a)$$

$$\mu'(\hat{\lambda}_{n,\wedge}, X_{\vee,i}) = 0. \quad (6.3.24b)$$

The proof of the corollary is provided by the direct substitution of the values of CDFs equal to:

$$F_{\hat{\lambda}_{n,\vee}}(\lambda + 0) = F_{X_{\wedge,i}}(\lambda + 0) = 1; \quad (6.3.25a)$$

$$F_{\hat{\lambda}_{n,\wedge}}(\lambda - 0) = F_{X_{\vee,i}}(\lambda - 0) = 0, \quad (6.3.25b)$$

into the relationships (6.3.22a) and (6.3.18b), respectively.

Theorem 6.3.2. *The functions $\mu'(\hat{\lambda}_{n,+}, X_{+,i})$, $\mu'(\hat{\lambda}_{n,\vee}, X_{\wedge,i})$, determined by the expression (6.3.10b) between the estimators $\hat{\lambda}_{n,+}$ (6.3.2), $\hat{\lambda}_{n,\vee}$ (6.3.4) and the estimation (measurement) results $X_{+,i}$ (6.3.1a), $X_{\wedge,i}$ (6.3.1c), respectively, are metrics, and for $\forall \lambda \in]-\infty, \infty[$, the identities hold:*
Within the group $\mathcal{L}(\mathcal{X};+)$:

$$\mu'(\hat{\lambda}_{n,+}, X_{+,i}) = \mu(\hat{\lambda}_{n,+}, X_{+,i}); \quad (6.3.26a)$$

6.3 Quality Indices of Unknown Nonrandom Parameter Estimation

Within the lattice $\mathcal{L}(\mathcal{X}; \vee, \wedge)$:

$$\mu'(\hat{\lambda}_{n,\vee}, X_{\wedge,i})|_{\lambda=-\lambda'} = \mu(\hat{\lambda}_{n,\wedge}, X_{\vee,i})|_{\lambda=\lambda'}. \tag{6.3.26b}$$

Proof. We first prove the identity (6.3.26a) for the group $\mathcal{L}(\mathcal{X}; +)$, and then the identity (6.3.26b) for the lattice $\mathcal{L}(\mathcal{X}; \vee, \wedge)$.

According to the identity (6.3.12), metric $\mu(\hat{\lambda}_{n,+}, X_{+,i})$ is equal to:

$$\mu(\hat{\lambda}_{n,+}, X_{+,i}) = 2(\mathbf{P}[\hat{\lambda}_{n,+} \vee X_{+,i} > \lambda] - \mathbf{P}[\hat{\lambda}_{n,+} \wedge X_{+,i} > \lambda]) =$$
$$= 2[F_{\hat{\lambda}+}(\lambda+0) + F_{X_{+,i}}(\lambda+0)] - 4F_{\hat{\lambda}+,X_i}(\lambda+0, \lambda+0), \tag{6.3.27}$$

where $F_{\hat{\lambda}+,X_i}(z_1, z_2)$ and $F_{\hat{\lambda}+}(z_1)$, $F_{X_{+,i}}(z_2)$ are joint and univariate CDFs of the estimator $\hat{\lambda}_{n,+}$ and the estimation (measurement) result $X_{+,i}$ respectively.

Similarly, according to the formulas [115, (3.2.80)] and [115, (3.2.85)], the probabilities $\mathbf{P}[\hat{\lambda}_{n,+} \wedge X_{+,i} < \lambda]$, $\mathbf{P}[\hat{\lambda}_{n,+} \vee X_{+,i} < \lambda]$ are equal to:

$$\mathbf{P}[\hat{\lambda}_{n,+} \wedge X_{+,i} < \lambda] = F_{\hat{\lambda}+}(\lambda-0) + F_{X_{+,i}}(\lambda-0) - F_{\hat{\lambda}+,X_i}(\lambda-0, \lambda-0);$$

$$\mathbf{P}[\hat{\lambda}_{n,+} \vee X_{+,i} < \lambda] = F_{\hat{\lambda}+,X_i}(\lambda-0, \lambda-0),$$

so the function $\mu'(\hat{\lambda}_{n,+}, X_{+,i})$ is equal to:

$$\mu'(\hat{\lambda}_{n,+}, X_{+,i}) = 2(\mathbf{P}[\hat{\lambda}_{n,+} \wedge X_{+,i} < \lambda] - \mathbf{P}[\hat{\lambda}_{n,+} \vee X_{+,i} < \lambda]) =$$
$$= 2[F_{\hat{\lambda}+}(\lambda-0) + F_{X_{+,i}}(\lambda-0)] - 4F_{\hat{\lambda}+,X_i}(\lambda-0, \lambda-0). \tag{6.3.28}$$

Due to the continuity of CDFs $F_{\hat{\lambda}+,X_i}(z_1, z_2)$, $F_{\hat{\lambda}+}(z_1)$, $F_{X_{+,i}}(z_2)$, the identities hold:

$$F_{\hat{\lambda}+}(\lambda-0) = F_{\hat{\lambda}+}(\lambda+0) = F_{\hat{\lambda}+}(\lambda); \tag{6.3.29a}$$

$$F_{X_{+,i}}(\lambda-0) = F_{X_{+,i}}(\lambda+0) = F_{X_{+,i}}(\lambda); \tag{6.3.29b}$$

$$F_{\hat{\lambda}+,X_i}(\lambda-0, \lambda-0) = F_{\hat{\lambda}+,X_i}(\lambda+0, \lambda+0) = F_{\hat{\lambda}+,X_i}(\lambda, \lambda). \tag{6.3.29c}$$

Joint fulfillment of the equalities (6.3.29) implies the identity between the relationships (6.3.27) and (6.3.28):

$$\mu'(\hat{\lambda}_{n,+}, X_{+,i}) = \mu(\hat{\lambda}_{n,+}, X_{+,i}),$$

thus, the function $\mu'(\hat{\lambda}_{n,+}, X_{+,i})$, being the same as $\mu(\hat{\lambda}_{n,+}, X_{+,i})$, is metric. The initial statement (the identity (6.3.26a)) of the Theorem 6.3.2 is proved.

We prove the second statement of the Theorem 6.3.2. For CDF $F_N(z)$ of measurement errors $\{N_i\}$ and CDF $F_\lambda(z)$ of unknown nonrandom parameter λ that are symmetrical with respect to medians, the identity holds [253, (7.1)]:

$$F_N(-z) = 1 - F_N(z); \tag{6.3.30a}$$

$$F_\lambda(-z)|_{\lambda=-\lambda'} = 1 - F_\lambda(z)|_{\lambda=\lambda'}. \tag{6.3.30b}$$

Joint fulfillment of the relationships (6.3.30a,b), (6.3.8a), and (6.3.8b) implies the identities determining the symmetry of a pair of distributions $F_{X_{\vee,i}}(z)$, $F_{X_{\wedge,i}}(z)$ of measurement results [253, (7.1)]:

$$F_{X_{\vee,i}}(-z)|_{\lambda=-\lambda'} = 1 - F_{X_{\wedge,i}}(z)|_{\lambda=\lambda'} ; \quad (6.3.31a)$$

$$F_{X_{\wedge,i}}(-z)|_{\lambda=-\lambda'} = 1 - F_{X_{\vee,i}}(z)|_{\lambda=\lambda'} . \quad (6.3.31b)$$

Similarly, joint fulfillment of the relationships (6.3.30a,b), (6.3.8c), and (6.3.8d) implies the identities determining the symmetry of a pair of distributions $F_{\hat{\lambda}n,\wedge}(z)$, $F_{\hat{\lambda}n,\vee}(z)$ of the estimators $\hat{\lambda}_{n,\wedge}$, $\hat{\lambda}_{n,\vee}$ [253, (7.2)]:

$$F_{\hat{\lambda}n,\wedge}(-z)|_{\lambda=-\lambda'} = 1 - F_{\hat{\lambda}n,\vee}(z)|_{\lambda=\lambda'} ; \quad (6.3.32a)$$

$$F_{\hat{\lambda}n,\vee}(-z)|_{\lambda=-\lambda'} = 1 - F_{\hat{\lambda}n,\wedge}(z)|_{\lambda=\lambda'} . \quad (6.3.32b)$$

Joint fulfillment of the relationships (6.3.18a), (6.3.31a), (6.3.32a), and the relationships (6.3.22b), (6.3.31b), (6.3.32b), implies the identity (6.3.26b), so, the function $\mu'(\hat{\lambda}_{n,\vee}, X_{\wedge,i})$, as well as $\mu(\hat{\lambda}_{n,\wedge}, X_{\vee,i})$, is metric. □

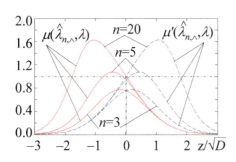

FIGURE 6.3.2 Graphs of metrics $\mu(\hat{\lambda}_{n,\wedge}, \lambda)$ and $\mu'(\hat{\lambda}_{n,\vee}, \lambda)$ depending on parameter λ; $n = 3, 5, 20$

Graphs of metrics $\mu(\hat{\lambda}_{n,\wedge}, X_{\vee,i}) = \mu(\hat{\lambda}_{n,\wedge}, \lambda)$, $\mu'(\hat{\lambda}_{n,\vee}, X_{\wedge,i}) = \mu'(\hat{\lambda}_{n,\vee}, \lambda)$, depending on parameter λ for the size n of the sample of the measurement results $\{X_{\vee,i}\}$ (6.3.1b), $\{X_{\wedge,i}\}$ (6.3.1c) that are equal to $n = 3, 5, 20$, under standard normal distribution of the measurement errors $\{N_i\}$, are shown in Fig. 6.3.2. The parts of graphs $\mu(\hat{\lambda}_{n,\wedge}, \lambda)$, $\mu'(\hat{\lambda}_{n,\vee}, \lambda)$ passing above 1.0 are conditionally called the parcels of "superefficiency" of the estimators. Further, for adequate and impartial description of estimator properties, the intervals where the metrics are less than or equal to 1.0 are conditionally called "operating" intervals. Operating interval for the metric $\mu(\hat{\lambda}_{n,\wedge}, \lambda)$ is the domain where parameter λ is positively defined: $\lambda \in [0, \infty[$, and for the metric $\mu'(\hat{\lambda}_{n,\vee}, \lambda)$, operating interval is the domain, where parameter λ is negatively defined: $\lambda \in]-\infty, 0]$.

The metrics $\mu(\hat{\lambda}_n, X_i)$ (6.3.10a) and $\mu'(\hat{\lambda}_n, X_i)$ (6.3.10b) between one of the estimators $\hat{\lambda}_n$ (6.3.2), (6.3.3), (6.3.4) of a parameter λ and the measurement result $X_i = \lambda \oplus N_i$, where $\oplus$ is one of operations of L-group $\mathcal{L}(\mathcal{X}; +, \vee, \wedge)$ (6.3.1a) through (6.3.1c), respectively, were introduced in order to characterize the quality of these estimators (6.3.2) through (6.3.4) in sample space $\mathcal{L}(\mathcal{X}, \mathcal{B}_\mathcal{X}; +, \vee, \wedge)$ with the properties of L-group $\mathcal{L}(\mathcal{X}; +, \vee, \wedge)$. Let us define the notion of quality of the estimator $\hat{\lambda}_n$ of an unknown nonrandom parameter λ in sample space $\mathcal{L}(\mathcal{X}, \mathcal{B}_\mathcal{X}; +, \vee, \wedge)$ with L-group properties.

6.3 Quality Indices of Unknown Nonrandom Parameter Estimation

Definition 6.3.1. In sample space $\mathcal{L}(\mathcal{X}, \mathcal{B}_\mathcal{X}; +, \vee, \wedge)$ with L-group properties, by a *quality index of estimator* $\hat{\lambda}_n$ we mean the quantity $q\{\hat{\lambda}_n\}$ equal to metric (6.3.10a) and (6.3.10b) between one of the estimators $\hat{\lambda}_n - \hat{\lambda}_{n,+}$ (6.3.2), $\hat{\lambda}_{n,\wedge}$ (6.3.3), and $\hat{\lambda}_{n,\vee}$ (6.3.4) of a parameter λ and the estimation (measurement) results $X_i = \lambda \oplus N_i - X_{+,i}$ (6.3.1a), $X_{\vee,i}$ (6.3.1b), $X_{\wedge,i}$ (6.3.1c), respectively, where $\oplus$ is one of operations of L-group $\mathcal{L}(\mathcal{X}; +, \vee, \wedge)$:
Within the group $\mathcal{L}(\mathcal{X}; +)$:

$$q\{\hat{\lambda}_{n,+}\} = \mu(\hat{\lambda}_{n,+}, X_{+,i}), \text{ on } \lambda \in]-\infty, \infty[; \quad (6.3.33a)$$

Within the lattice $\mathcal{L}(\mathcal{X}; \vee, \wedge)$:

$$q\{\hat{\lambda}_{n,\wedge}\} = \mu(\hat{\lambda}_{n,\wedge}, X_{\vee,i}), \text{ on } \lambda \in [0, \infty[; \quad (6.3.33b)$$

$$q\{\hat{\lambda}_{n,\vee}\} = \mu'(\hat{\lambda}_{n,\vee}, X_{\wedge,i}), \text{ on } \lambda \in]-\infty, 0]. \quad (6.3.33c)$$

For normally distributed estimation (measurement) errors $\{N_i\}$, joint CDF $F_{\hat{\lambda}_+, X_i}(z_1, z_2)$ is invariant with respect to a shift of parameter λ, so, according to the identity [115, (3.3.22)], it is equal to:

$$F_{\hat{\lambda}_+, X_i}(\lambda, \lambda) = F_{\hat{\lambda}_+, X_i}(0, 0)|_{\lambda=0} =$$

$$= \left(1 + \frac{2}{\pi} \arcsin(\rho_{\hat{\lambda}, X})\right)/4, \quad (6.3.34)$$

where $\rho_{\hat{\lambda}, X}$ is a correlation coefficient between the estimator $\hat{\lambda}_{n,+}$ (6.3.2) and the result of interaction $X_{+,i}$ (6.3.1a) equal to $\rho_{\hat{\lambda}, X} = 1/\sqrt{n}$; n is a size of the sample $N = (N_1, \ldots, N_n)$.

Then, substituting the values of joint CDF (6.3.34) and the values of univariate CDFs $F_{\hat{\lambda}_{n,+}}(\lambda) = 0.5$, $F_{X_{+,i}}(\lambda) = 0.5$ into the formula (6.3.27), we obtain the value of metric $\mu(\hat{\lambda}_{n,+}, X_{+,i})$ for the case of interaction between a parameter λ and the estimation (measurement) errors $\{N_i\}$ within the model (6.3.1a) under normal distribution of the last ones:

$$\mu(\hat{\lambda}_{n,+}, X_{+,i}) = 1 - \frac{2}{\pi} \arcsin\left(\frac{1}{\sqrt{n}}\right), \quad (6.3.35)$$

where n represents size of the samples $N = (N_1, \ldots, N_n)$, $X_+ = (X_{+,1}, \ldots, X_{+,n})$.

According to the Definition 6.3.1, the quality index $q\{\hat{\lambda}_{n,+}\}$ of the estimator $\hat{\lambda}_{n,+}$ (6.3.2) of an unknown nonrandom parameter λ, on $\lambda \in]-\infty, \infty[$, is equal to:

$$q\{\hat{\lambda}_{n,+}\} = \mu(\hat{\lambda}_{n,+}, X_{+,i}) = 1 - \frac{2}{\pi} \arcsin\left(\frac{1}{\sqrt{n}}\right). \quad (6.3.36)$$

As shown by the graphs in Fig. 6.3.2, the dependences $\mu(\hat{\lambda}_{n,\wedge}, X_{\vee,i})$ (6.3.18a), $\mu'(\hat{\lambda}_{n,\vee}, X_{\wedge,i})$ (6.3.22b), under an arbitrary symmetric distribution $F_N(z; \sigma)$ of the estimation (measurement) errors $\{N_i\}$ with scale parameter σ and the property

(6.3.30a) will be determined by the relationship λ/σ between an estimated parameter λ and scale parameter σ. In this connection, we obtain the dependences of quality index $q\{\hat{\lambda}_n\}$ of estimation of an unknown nonrandom parameter λ determined by the relationships (6.3.33b) and (6.3.33c) in large quantities of estimation (measurement) errors $\{N_i\}$ when the module of the relationship λ/σ is much less than 1: $|\lambda/\sigma| \ll 1$. Substituting the value of CDF $F_N(0) = 0.5$ of the estimation (measurement) errors $\{N_i\}$ with the property (6.3.30a) and also the value of CDF $F_\lambda(z) = 1(z - \lambda)|_{\lambda=0}$ of an unknown nonrandom parameter λ into the initial formulas for calculation of CDFs of the estimation (measurement) results (6.3.8a,b) and (6.3.8c,d), we obtain:

$$F_{X_\vee,i}(\lambda+0) = 0.5, \quad F_{\hat{\lambda}n,\wedge}(\lambda+0) = 1 - 2^{-n}; \quad (6.3.37a)$$

$$F_{X_\wedge,i}(\lambda-0) = 0.5, \quad F_{\hat{\lambda}n,\vee}(\lambda-0) = 2^{-n}. \quad (6.3.37b)$$

Substituting the pairs of values of CDFs (6.3.37a) into the expression (6.3.18a), and the pairs of values of CDFs (6.3.37b) into the expression (6.3.22b), we obtain the values of quality indices $q\{\hat{\lambda}_{n,\wedge}\}$, $q\{\hat{\lambda}_{n,\vee}\}$ of the estimators $\hat{\lambda}_{n,\wedge}$ (6.3.3), $\hat{\lambda}_{n,\vee}$ (6.3.4) of an unknown nonrandom parameter λ under an arbitrary symmetric distribution $F_N(z;\sigma)$ of the estimation (measurement) errors $\{N_i\}$ and on $|\lambda/\sigma| \to 0$:

$$q\{\hat{\lambda}_{n,\wedge}\} = \mu(\hat{\lambda}_{n,\wedge}, X_{\vee,i}) = 1 - 2^{-(n-1)}; \quad (6.3.38a)$$

$$q\{\hat{\lambda}_{n,\vee}\} = \mu'(\hat{\lambda}_{n,\vee}, X_{\wedge,i}) = 1 - 2^{-(n-1)}. \quad (6.3.38b)$$

The graphs of the dependences $q\{\hat{\lambda}_{n,+}\}$ (6.3.36), $q\{\hat{\lambda}_{n,\wedge}\}$ (6.3.38a) and $1 - \frac{1}{n}$ on a size of samples n are shown in Fig. 6.3.3(a). The graphs of dependences $1 - q\{\hat{\lambda}_{n,+}\}$, $1 - q\{\hat{\lambda}_{n,\wedge}\}$, $1/n$ on a size of samples n are shown in Fig. 6.3.3(b) in logarithmic scale.

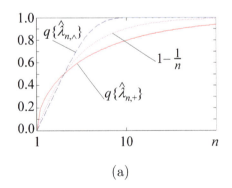
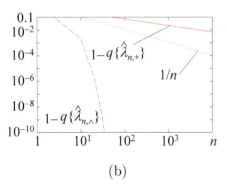

FIGURE 6.3.3 Dependences on size of samples n: (a) $q\{\hat{\lambda}_{n,+}\}$ (6.3.36), $q\{\hat{\lambda}_{n,\wedge}\}$ (6.3.38a), and $1 - \frac{1}{n}$; (b) $1 - q\{\hat{\lambda}_{n,+}\}$, $1 - q\{\hat{\lambda}_{n,\wedge}\}$, $1/n$

As seen in the graphs, the quality indices (6.3.38a,b) of the estimators in the sample space with properties of the lattice $\mathcal{L}(\mathcal{X}; \vee, \wedge)$ tend to 1 exponentially, whereas the quality index (6.3.36) of the estimator in the sample space with properties of the group $\mathcal{L}(\mathcal{X};+)$ tends to 1 more slowly than inversely proportional dependence $1 - \frac{1}{n}$. From an informational view, information, contained in the estimation (measurement) results on a qualitative level is used better while processing the estimation (measurement) results in the sample space with lattice properties, than in sample space with group properties. We should note that while considering information contained in the estimation (measurement) results (in the sample with independent elements), we mean information in an unknown nonrandom parameter λ and also information contained in the sample of the independent estimation (measurement) results $N = (N_1, \ldots, N_n)$, whose use provides achieving estimator quality index determined by the sample size n or, in other words, by the quantity of absolute information contained in this sample and measured in abits. Under an arbitrary symmetric distribution $F_N(z;\sigma)$ of the estimation (measurement) results $\{N_i\}$ and rather small ratio λ/σ of estimated parameter λ to scale parameter σ $|\lambda/\sigma| \to 0$, one can consider that quality indices $q\{\hat{\lambda}_{n,\wedge}\}, q\{\hat{\lambda}_{n,\vee}\}$ (6.3.38a,b) of the estimators $\hat{\lambda}_{n,\wedge}$ (6.3.3), $\hat{\lambda}_{n,\vee}$ (6.3.4) in sample space with properties of the lattice $\mathcal{L}(\mathcal{X};\vee,\wedge)$ are characterized by invariance property with respect to the conditions of nonparametric prior uncertainty. This property is not inherent to the estimator $\hat{\lambda}_{n,+}$ (6.3.2), whose quality is determined by CDF $F_N(z;\sigma)$ of the estimation (measurement) errors $\{N_i\}$ [250], [251].

6.4 Quality Indices of Classification and Detection of Deterministic Signals

6.4.1 Quality Indices of Classification of Deterministic Signals in Metric Spaces with *L*-group Properties

The classification of the signals with discrete informational parameters that do not change their values on a signal domain of definition plays a key role in information transmitting and receiving. In this subsection we determine quality of signal classification in such spaces and also establish a connection between quality of classification of the signals with discrete parameters and discrete communication channel capacity.

Consider a general model of interaction between the signal $s_i(t)$ from a set of deterministic signals $S = \{s_i(t)\}$, $i = 1, \ldots, m$ and interference (noise) $n(t)$ in signal space with properties of L-group $\mathbf{\Gamma}(+, \vee, \wedge)$:

$$x(t) = s_i(t) \oplus n(t), \ t \in T_s, \ i = 1, \ldots, m, \qquad (6.4.1)$$

where $\oplus$ is some binary operation of L-group $\mathbf{\Gamma}(+, \vee, \wedge)$; $T_s = [t_0, t_0 + T]$ is domain of definition of the signal $s_i(t)$; t_0 is a known time of arrival of the signal $s_i(t)$; T is a duration of the signal $s_i(t)$; $m \in \mathbf{N}$, $\mathbf{N}$ is the set of natural numbers.

Let the signals from a set $S = \{s_i(t)\}$, $i = 1, \ldots, m$ be characterized by the same energy $E_i = \int\limits_{t \in T_s} s_i^2(t)\mathrm{d}t = E$ and cross-correlation coefficients $r_{ik} = \int\limits_{t \in T_s} s_i(t)s_k(t)\mathrm{d}t/E$.

During signal classification, we should distinguish which one of m useful signals from a set $S = \{s_i(t)\}$, $i = 1,\ldots,m$ is received in the input of receiver at some moment of time.

Generally, quality of classification (as well as quality of multiple-alternative detection) of signals is defined by probability of a signal receiving error taking into account corresponding types of discrete modulation of informational parameters of a signal [163], [155], [149]. This probability of error is some function of a distance between the signals in Hilbert signal space [149, Section 4.2;(36)]. However, for the case of classification of $m > 2$ signals from a set $S = \{s_i(t)\}$, an exact evaluation of error probability is problematic even under normal distribution of interference (noise) in the input of a processing unit (see, for instance, [155, (2.4.43)], [254, Section 4.2;(59)]), so, the upper bound of error probability is often used for its evaluation (see [155, (2.4.46)], [254, Section 4.2;(65)]). In a given subsection, quality index of classification of deterministic signals is identified not by signal receiving error probability, but by a quality index of the estimator $\hat{s}_k(t)$ of a received signal $s_k(t)$ from a set $S = \{s_i(t)\}$, $i = 1,\ldots,m$, $s_k(t) \in S$ based on metric relationships between the signals (in this case, between the signal $s_k(t)$ and its estimator $\hat{s}_k(t)$).

This section has two goals: (1) definition of quality indices of classification and detection of deterministic signals on the base of metric (3.2.1), and (2) the establishment of the most general qualitative and quantitative regularities in signal classification (detection) in linear signal space and signal space with lattice properties. As a model of signal space featuring the properties of these spaces, we use the signal space with the properties of L-group $\mathbf{\Gamma}(+, \vee, \wedge)$.

Further, on the base of metric (3.2.1), we shall define and evaluate the quality indices of signal classification in signal space with properties of a group $\mathbf{\Gamma}(+)$ and a lattice $\mathbf{\Gamma}(\vee, \wedge)$ of L-group $\mathbf{\Gamma}(+, \vee, \wedge)$.

Within the framework of the general model (6.4.1), we consider the case of additive interaction between the signal $s_i(t)$ from a set of deterministic signals $S = \{s_i(t)\}$, $i = 1,\ldots,m$ and interference (noise) $n(t)$ in signal space with the properties of the group $\mathbf{\Gamma}(+)$ of L-group $\mathbf{\Gamma}(+, \vee, \wedge)$:

$$x(t) = s_i(t) + n(t),\ t \in T_s,\ i = 1,\ldots,m, \qquad (6.4.2)$$

where $T_s = [t_0, t_0 + T]$ is a domain of definition of the signal $s_i(t)$; t_0 is a known time of arrival of the signal $s_i(t)$; T is a duration of the signal $s_i(t)$; $m \in \mathbf{N}$, $\mathbf{N}$ is the set of natural numbers.

Suppose the interference $n(t)$ is white Gaussian noise with power spectral density N_0.

While solving the problem of signal classification in linear space, i.e., when interaction equation (6.4.2) holds, the sufficient statistics $y_i(t)$ in the i-th processing channel is a scalar product $(x(t), s_i(t))$ of the signals $x(t)$, $s_i(t)$ in Hilbert space

6.4 Quality Indices of Classification and Detection of Deterministic Signals

equal to correlation integral $\int\limits_{t \in T_s} x(t)s_i(t)\mathrm{d}t$:

$$y_i(t) = \hat{s}_i(t) = (x(t), s_i(t)) = \int\limits_{t \in T_s} x(t)s_i(t)\mathrm{d}t, \qquad (6.4.3)$$

where $y_i(t)$ is the estimator of energetic parameter of the signal $s_i(t)$ (further, simply the estimator $\hat{s}_i(t)$ of the signal $s_i(t)$) in the i-th processing channel.

In the case of the presence of the signal $s_k(t)$ in the observed process $x(t) = s_k(t) + n(t)$, the problem of signal classification is solved by maximization of sufficient statistics $(x(t), s_i(t))$ (6.4.3):

$$\arg\max_{i \in I; s_i(t) \in S} [\int\limits_{t \in T_s} x(t)s_i(t)\mathrm{d}t]\,|_{x(t)=s_k(t)+n(t)} = \hat{k},$$

where $\hat{k}$ is the estimator of channel number corresponding to the number of a received signal $s_k(t)$ from a set $S = \{s_i(t)\}$.

Correlation integral $\int\limits_{t \in T_s} x(t)s_i(t)\mathrm{d}t$ in the i-th processing channel at the instant $t = t_0 + T$ takes a mean value equal to the product of cross-correlation coefficient r_{ik} between the signals $s_i(t)$, $s_k(t)$ and signal energy E:

$$\mathbf{M}\{y_i(t)\} = \mathbf{M}\{\int\limits_{t \in T_s} x(t)s_i(t)\mathrm{d}t\}\,|_{x(t)=s_k(t)+n(t)} = r_{ik}E, \qquad (6.4.4)$$

so that in the k-th processing channel $(i = k)$ at the instant $t = t_0 + T$, it takes a mean value equal to the energy E of the signal $s_k(t)$:

$$\mathbf{M}\{y_i(t)\} = \mathbf{M}\{\int\limits_{t \in T_s} x(t)s_i(t)\mathrm{d}t\}\,|_{i=k} = \int\limits_{t \in T_s} s_k(t)s_k(t)\mathrm{d}t = E, \qquad (6.4.5)$$

where $y_i(t)$ is the estimator $\hat{s}_i(t)$ of the signal $s_i(t)$ in the i-th processing channel in the presence of the signal $s_k(t)$ in the observed process $x(t)$; $\mathbf{M}(*)$ is the symbol of mathematical expectation.

In the output of the i-th processing channel, the probability density function (PDF) $p_{\hat{s}_i}(y)$ of the estimator $y_i(t) = \hat{s}_i(t)$ of the signal $s_i(t)$, at the instant $t = t_0 + T$ in the presence of the signal $s_k(t)$ in the observed process $x(t)$, is determined by the expression:

$$p_{\hat{s}_i}(y) = (2\pi D)^{-1/2} \exp\left\{\frac{(y - r_{ik}E)^2}{2D}\right\}, \qquad (6.4.6)$$

where $D = EN_0$ is a noise variance in the output of the i-th processing channel; E is the energy of the signal $s_i(t)$; N_0 is a power spectral density of interference (noise) in the input of the processing unit; r_{ik} is a cross-correlation coefficient between the signals $s_i(t)$, $s_k(t)$.

Consider the estimator $y_k(t) = \hat{s}_k(t)$ of the signal $s_k(t)$ in the k-th processing channel in signal presence in the observed process $x(t) = s_k(t) + n(t)$:

$$y_k(t) = \hat{s}_k(t)\big|_{x(t)=s_k(t)+n(t)}, \quad t = t_0 + T, \tag{6.4.7}$$

and also consider the estimator $y_k(t)\big|_{n(t)=0} = \hat{s}_k(t)\big|_{n(t)=0}$ of the signal $s_k(t)$ in the k-th processing channel in the absence of interference (noise) ($n(t) = 0$) in the observed process $x(t) = s_k(t)$:

$$y_k(t)\big|_{n(t)=0} = \hat{s}_k(t)\big|_{x(t)=s_k(t)}, \quad t = t_0 + T. \tag{6.4.8}$$

To characterize the quality of the estimator $y_k(t) = \hat{s}_k(t)$ of the signal $s_k(t)$ while solving the problem of classification of the signals from a set $S = \{s_i(t)\}$, $i = 1, \ldots, m$ that additively interact with interference (noise) $n(t)$ (6.4.2) in the group $\Gamma(+)$ of L-group $\Gamma(+, \vee, \wedge)$, we introduce the function $\mu(y_t, y_{t,0})$ that is analogous to metric (3.2.1) and characterizes the difference between the estimator $y_k(t)$ (6.4.7) of the signal $s_k(t)$ in the k-th processing channel in signal presence in the observed process $x(t) = s_k(t) + n(t)$ and the estimator $y_k(t)\big|_{n(t)=0}$ (6.4.8) of the signal $s_k(t)$ in the absence of interference (noise):

$$\mu(y_t, y_{t,0}) = 2[\mathbf{P}(y_t \vee y_{t,0} > h) - \mathbf{P}(y_t \wedge y_{t,0} > h)]. \tag{6.4.9}$$

In the equation, y_t is a sample of the estimator $y_k(t)$ (6.4.7) of the signal $s_k(t)$ in the output of the k-th processing channel at the instant $t = t_0 + T$ in the presence of the signal $s_k(t)$ in the observed process $x(t)$; $y_{t,0}$ is a sample of the estimator $y_k(t)\big|_{n(t)=0}$ (6.4.8) of the signal $s_k(t)$ in the output of the k-th processing channel at the instant $t = t_0 + T$ in the absence of interference (noise) ($n(t) = 0$) in the observed process $x(t)$, equal to the energy E of the signal: $y_{t,0} = E$; h is some threshold level $h < E$ determined by an average of two mathematical expectations of the processes in the outputs of the i-th (6.4.4) and the k-th (6.4.5) processing channels:

$$h = (r_{ik}E + E)/2, \tag{6.4.10}$$

r_{ik} is a cross-correlation coefficient between the signals $s_i(t)$ and $s_k(t)$.

Theorem 3.2.1 states that for random variables with special properties, the function defined by the relationship (3.2.1) is a metric. Meanwhile, the establishment of this fact for the function (6.4.9) requires a separate proof stated in the following theorem.

Theorem 6.4.1. *For a pair of samples $y_{t,0}$ and y_t of stochastic processes $y_k(t)\big|_{n(t)=0}$, $y_k(t)$ in the output of the unit of optimal classification of the signals from a set $S = \{s_i(t)\}$, $i = 1, \ldots, m$, which additively interact with interference (noise) $n(t)$ (6.4.2) in a group $\Gamma(+)$ of L-group $\Gamma(+, \vee, \wedge)$, the function $\mu(y_t, y_{t,0})$ defined by the relationship (6.4.9) is metric.*

Proof. Consider the probabilities $\mathbf{P}(y_t \vee y_{t,0} > h)$, $\mathbf{P}(y_t \wedge y_{t,0} > h)$ in formula (6.4.9). Joint fulfillment of the equality $y_{t,0} = E$ and the inequality $h < E$ implies that these probabilities are equal to:

$$\mathbf{P}(y_t \vee y_{t,0} > h) = \mathbf{P}(y_{t,0} > h) = 1; \tag{6.4.11a}$$

6.4 Quality Indices of Classification and Detection of Deterministic Signals

$$\mathbf{P}(y_t \wedge y_{t,0} > h) = \mathbf{P}(y_t > h) = 1 - F_{\hat{s}_k}(h), \tag{6.4.11b}$$

where $F_{\hat{s}_k}(y)$ is the cumulative distribution function (CDF) of the estimator $y_k(t) = \hat{s}_k(t)$ in the output of the k-th processing channel at the instant $t = t_0 + T$, which, according to the PDF (6.4.6), is equal to:

$$F_{\hat{s}_k}(y) = (2\pi D)^{-1/2} \int_{-\infty}^{y} \exp\left\{-\frac{(x-E)^2}{2D}\right\} dx. \tag{6.4.12}$$

The identities (6.4.11) imply the equality:

$$\mathbf{P}(y_t > h) + \mathbf{P}(y_{t,0} > h) = \mathbf{P}(y_t \vee y_{t,0} > h) + \mathbf{P}(y_t \wedge y_{t,0} > h). \tag{6.4.13}$$

The probability $\mathbf{P}(y_t > h)$ is a valuation [221, Section X.1 (V1)]. Besides, probability $\mathbf{P}(y_t > h)$ is isotonic inasmuch as the implication holds [221, Section X.1 (V2)]:

$$y_t \geq y'_t \Rightarrow \mathbf{P}(y_t > h) \geq \mathbf{P}(y'_t > h). \tag{6.4.14}$$

Joint fulfillment of the relationships (6.4.13), (6.4.14), according to Theorem 6.4.1 [221, Section X.1], implies that the quantity $\mu(y_t, y_{t,0})$ (6.4.9) is metric. □

Substituting the formulas (6.4.11) into the expression (6.4.9), we obtain a value for metric $\mu(y_t, y_{t,0})$ between the signal $s_k(t)$ and its estimator $y_k(t) = \hat{s}_k(t)$ while solving the problem of classification of the signals from a set $S = \{s_i(t)\}$, $i = 1, \ldots, m$ in the group $\mathbf{\Gamma}(+)$ of L-group $\mathbf{\Gamma}(+, \vee, \wedge)$:

$$\mu_{s\hat{s}} = \mu(y_t, y_{t,0}) = 2F_{\hat{s}_k}(h) = 2\left[1 - \Phi\left(\frac{E-h}{\sqrt{D}}\right)\right], \tag{6.4.15}$$

where $F_{\hat{s}_k}(y)$ is the CDF of the estimator $y_k(t) = \hat{s}_k(t)$ in the output of the k-th processing channel at the instant $t = t_0 + T$, which is determined by the formula (6.4.12); $\Phi(z) = (2\pi)^{-1/2} \int_{-\infty}^{z} \exp\left\{-\frac{x^2}{2}\right\} dx$ is a probability integral; $D = EN_0$ is a variance of noise in the output of the i-th processing channel; E is an energy of the signals $s_i(t)$ from a set $S = \{s_i(t)\}$, $i = 1, \ldots, m$; h is a threshold level defined by the relationship (6.4.10); N_0 is a power spectral density of interference (noise) in the input of a processing unit.

Thus, the expression (6.4.15) defines metric between the signal $s_k(t)$ and its estimator $y_k(t) = \hat{s}_k(t)$, which can be considered as a measure of their closeness.

Definition 3.2.2 establishes the relationship (3.2.7) between normalized measure of statistical interrelationship (NMSI) of a pair of samples and its metric (3.2.1). Similarly, the following definition establishes the connection between quality index of signal classification and metrics (6.4.9).

Definition 6.4.1. By *quality index of estimator* of the signals $\nu_{s\hat{s}}$ *while solving the problem of their classification* from a set $S = \{s_i(t)\}$, $i = 1, \ldots, m$ we mean NMSI $\nu(y_t, y_{t,0})$ between the samples $y_{t,0}$, y_t of stochastic processes $y_k(t)|_{n(t)=0}$, $y_k(t)$ in the output of a signal classification unit connected with metric $\mu(y_t, y_{t,0})$ (6.4.9) by the following relationship:

$$\nu_{s\hat{s}} = \nu(y_t, y_{t,0}) = 1 - \mu(y_t, y_{t,0}). \tag{6.4.16}$$

We now determine a value of quality index of an estimator of the signals (6.4.16) for the signals with arbitrary correlation properties from a set $S = \{s_i(t)\}$, $i = 1, \ldots, m$ under their classification in signal space with the properties of the group $\Gamma(+)$ of L-group $\Gamma(+, \vee, \wedge)$.

Substituting a value of correlation coefficient r_{ik} determining, according to the formula (6.4.10), the value of a threshold $h = (r_{ik}E + E)/2$, into the expression (6.4.15), and using simultaneously the coupling equation (6.4.16) between NMSI and metric, we obtain the dependence of the quality index of the estimator of the signals $\nu_{s\hat{s}}(q^2)$ on signal-to-noise ratio $q^2 = E/N_0$ for the signals with arbitrary correlation properties:

$$\nu_{s\hat{s}}(q^2) = 2\Phi\left(\sqrt{q^2(1 - r_{ik})/2}\right) - 1, \qquad (6.4.17)$$

where r_{ik} is a cross-correlation coefficient between the signals $s_i(t)$ and $s_k(t)$.

Consider the values of the quality index of the estimator of the signals (6.4.17) for the orthogonal ($r_{ik} = 0$) and opposite ($r_{ik} = -1$) signals from a set $S = \{s_i(t)\}$, $i = 1, \ldots, m$ under their classification in signal space with the properties of the group $\Gamma(+)$ of L-group $\Gamma(+, \vee, \wedge)$.

Based on the general formula (6.4.17), we obtain the dependences of quality indices of the estimators $\nu_{s\hat{s}}(q^2)$ on signal-to-noise ratio for orthogonal and opposite signals, respectively:

$$\nu_{s\hat{s}}(q^2)|_\perp = 2\Phi\left(\sqrt{q^2/4}\right) - 1, \quad s_i(t) \perp s_k(t), \; r_{ik} = 0; \qquad (6.4.18a)$$

$$\nu_{s\hat{s}}(q^2)|_- = 2\Phi\left(\sqrt{q^2}\right) - 1, \quad s_1(t) = -s_2(t), \; r_{12} = -1, \qquad (6.4.18b)$$

where $q^2 = E/N_0$ is signal-to-noise ratio.

The dependences $\nu_{s\hat{s}}(q^2)|_\perp$ (6.4.18a), $\nu_{s\hat{s}}(q^2)|_-$ (6.4.18b) are represented in Fig. 6.4.1 and are shown by the solid line and the dot line, respectively. Based on the formulas (6.4.18a), (6.4.18b), the use of opposite signals provides a gain in signal energy as against orthogonal signals, that corresponds rather well to known results [163], [155], [149]. Here, fourfold gain in signal energy elucidated by metric (6.4.15) is a function of square of Euclidean metric $\int_{t \in T_s}(s_i(t) - s_k(t))^2 dt$ between the signals $s_i(t)$ and $s_k(t)$ in Hilbert space.

Consider the model of interaction between the signal $s_i(t)$ from a set of deterministic signals $S = \{s_i(t)\}$, $i = 1, \ldots, m$ and interference (noise) $n(t)$ in signal space with the properties of the lattice $\Gamma(\vee, \wedge)$ of L-group $\Gamma(+, \vee, \wedge)$ with operations of join $s_i(t) \vee n(t)$ and meet $s_i(t) \wedge n(t)$, respectively:

$$x(t) = s_i(t) \vee n(t); \qquad (6.4.19a)$$

$$\tilde{x}(t) = s_i(t) \wedge n(t), \qquad (6.4.19b)$$

where $t \in T_s$; $T_s = [t_0, t_0 + T]$ is a domain of definition of the signal $s_i(t)$; t_0 is a known time of arrival of the signal $s_i(t)$; T is a duration of the signal $s_i(t)$; $m \in \mathbf{N}$, $\mathbf{N}$ is the set of natural numbers.

6.4 Quality Indices of Classification and Detection of Deterministic Signals

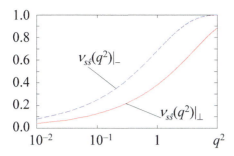

FIGURE 6.4.1 Dependences $\nu_{s\hat{s}}(q^2)|_{\perp}$ per Equation (6.4.18a); $\nu_{s\hat{s}}(q^2)|_{-}$ per Equation (6.4.18b)

Let the signals from a set $S = \{s_i(t)\}$, $i = 1, \ldots, m$ be characterized by the same energy $E_i = \int_{t \in T_s} s_i^2(t) \mathrm{d}t = E$ and cross-correlation coefficients $r_{ik} = \int_{t \in T_s} s_i(t) s_k(t) \mathrm{d}t / E$. Also suppose that interference (noise) $n(t)$ is characterized by arbitrary probabilistic-statistical properties.

Consider the estimators $y_k(t)$ and $\tilde{y}_k(t)$ of the signal $s_k(t)$ in the k-th processing channel of the classification unit in the presence of the signal $s_k(t)$ in the observed processes $x(t)$ (6.4.19a) and $\tilde{x}(t)$ (6.4.19b), which are determined by the following relationships:

$$y_k(t) = y_k(t)\big|_{x(t) = s_k(t) \vee n(t)} = s_k(t) \wedge x(t); \tag{6.4.20a}$$

$$\tilde{y}_k(t) = \tilde{y}_k(t)\big|_{\tilde{x}(t) = s_k(t) \wedge n(t)} = s_k(t) \vee \tilde{x}(t), \tag{6.4.20b}$$

and also consider the similar estimators $y_k(t)\big|_{n(t)=0}$, $\tilde{y}_k(t)\big|_{n(t)=0}$ of the signal $s_k(t)$ in the k-th processing channel in the absence of interference (noise) ($n(t) = 0$) in the observed processes $x(t)$ (6.4.19a) and $\tilde{x}(t)$ (6.4.19b):

$$y_k(t)\big|_{n(t)=0} = s_k(t) \wedge x(t)\big|_{x(t) = s_k(t) \vee 0}; \tag{6.4.21a}$$

$$\tilde{y}_k(t)\big|_{n(t)=0} = s_k(t) \vee \tilde{x}(t)\big|_{\tilde{x}(t) = s_k(t) \wedge 0}. \tag{6.4.21b}$$

We next determine the values of the function (6.4.9), which, as for the lattice $\Gamma(\vee, \wedge)$, is also metric, along with determining a quality index of the estimator of the signals (6.4.16) while solving the problem of signal classification from a set $S = \{s_i(t)\}$, $i = 1, \ldots, m$ in signal space with the properties of the lattice $\Gamma(\vee, \wedge)$ of L-group $\Gamma(+, \vee, \wedge)$:

$$\mu_{s\hat{s}} = \mu(y_t, y_{t,0}) = 2[\mathbf{P}(y_t \vee y_{t,0} > h) - \mathbf{P}(y_t \wedge y_{t,0} > h)]; \tag{6.4.22a}$$

$$\tilde{\mu}_{s\hat{s}} = \mu(\tilde{y}_t, \tilde{y}_{t,0}) = 2[\mathbf{P}(\tilde{y}_t \vee \tilde{y}_{t,0} > h) - \mathbf{P}(\tilde{y}_t \wedge \tilde{y}_{t,0} > h)], \tag{6.4.22b}$$

where y_t and $\tilde{y}_t$ are the samples of the estimators $y_k(t)$ (6.4.20a) and $\tilde{y}_k(t)$ (6.4.20b) of the signal $s_k(t)$ in the output of the k-th processing channel at the instant $t \in T_s$ in the presence of the signal $s_k(t)$ in the observed processes $x(t)$, $\tilde{x}(t)$; $y_{t,0}$ and $\tilde{y}_{t,0}$ are the samples of the estimators $y_k(t)\big|_{n(t)=0}$ (6.4.21a) and $\tilde{y}_k(t)\big|_{n(t)=0}$ (6.4.21b) of the signal $s_k(t)$ in the output of the k-th processing channel at the instant $t \in T_s$ in the absence of interference (noise) ($n(t) = 0$) in the observed processes $x(t)$, $\tilde{x}(t)$; h is some threshold level determined by energetic and correlation relations between the signals from a set $S = \{s_i(t)\}$, $i = 1, \ldots, m$.

Absorption axiom of a lattice $\Gamma(\vee, \wedge)$ of L-group $\Gamma(+, \vee, \wedge)$ contained in the third part of each of four following multilink identities implies that the estimators $y_k(t)$ (6.4.20a), $\tilde{y}_k(t)$ (6.4.20), and the estimators $y_k(t)\big|_{n(t)=0}$ (6.4.21a), $\tilde{y}_k(t)\big|_{n(t)=0}$ (6.4.21b) are identically equal to the received signal $s_k(t)$:

$$y_k(t) = s_k(t) \wedge x(t) = s_k(t) \wedge [s_k(t) \vee n(t)] = s_k(t); \qquad (6.4.23a)$$

$$\tilde{y}_k(t) = s_k(t) \vee \tilde{x}(t) = s_k(t) \vee [s_k(t) \wedge n(t)] = s_k(t); \qquad (6.4.23b)$$

$$y_k(t)\big|_{n(t)=0} = s_k(t) \wedge x(t) = s_k(t) \wedge [s_k(t) \vee 0] = s_k(t); \qquad (6.4.24a)$$

$$\tilde{y}_k(t)\big|_{n(t)=0} = s_k(t) \vee \tilde{x}(t) = s_k(t) \vee [s_k(t) \wedge 0] = s_k(t). \qquad (6.4.24b)$$

The obtained relationships imply that the samples y_t and $\tilde{y}_t$ of the estimators $y_k(t)$ (6.4.20a) and $\tilde{y}_k(t)$ (6.4.20b), and the samples $y_{t,0}$ and $\tilde{y}_{t,0}$ of the estimators $y_k(t)\big|_{n(t)=0}$ (6.4.21a) and $\tilde{y}_k(t)\big|_{n(t)=0}$ (6.4.21b) are identically equal to the received signal $s_k(t)$:

$$y_t = y_k(t) = s_k(t); \qquad (6.4.25a)$$

$$\tilde{y}_t = \tilde{y}_k(t) = s_k(t); \qquad (6.4.25b)$$

$$y_{t,0} = y_k(t)\big|_{n(t)=0} = s_k(t); \qquad (6.4.26a)$$

$$\tilde{y}_{t,0} = \tilde{y}_k(t)\big|_{n(t)=0} = s_k(t). \qquad (6.4.26b)$$

Substituting the obtained values of the samples y_t (6.4.25a) and $y_{t,0}$ (6.4.26a) into the relationship (6.4.22a), and also the values of the samples $\tilde{y}_t$ (6.4.25b) and $\tilde{y}_{t,0}$ (6.4.26b) into the relationship (6.4.22b), we note that the values of metrics $\mu(y_t, y_{t,0})$, $\mu(\tilde{y}_t, \tilde{y}_{t,0})$ between the signal $s_k(t)$ and its estimators $y_k(t)$ and $\tilde{y}_k(t)$, while solving the problem of classification of the signals from a set $S = \{s_i(t)\}$, $i = 1, \ldots, m$ in signal space with the properties of the lattice $\Gamma(\vee, \wedge)$ of L-group $\Gamma(+, \vee, \wedge)$, are identically equal to zero:

$$\mu_{s\hat{s}} = \mu(y_t, y_{t,0}) = 0; \qquad (6.4.27a)$$

$$\tilde{\mu}_{s\hat{s}} = \mu(\tilde{y}_t, \tilde{y}_{t,0}) = 0. \qquad (6.4.27b)$$

From the relationships (6.4.27) and the coupling equation (6.4.16), we obtain that quality indices of the estimators $\nu_{s\hat{s}}$, $\tilde{\nu}_{s\hat{s}}$ of the signals, while solving the problem of their classification from a set $S = \{s_i(t)\}$, $i = 1, \ldots, m$ in signal space with the properties of the lattice $\Gamma(\vee, \wedge)$ of L-group $\Gamma(+, \vee, \wedge)$, take absolute values equal to 1:

$$\nu_{s\hat{s}} = 1 - \mu(y_t, y_{t,0}) = 1; \qquad (6.4.28a)$$

$$\tilde{\nu}_{s\hat{s}} = 1 - \mu(\tilde{y}_t, \tilde{y}_{t,0}) = 1. \qquad (6.4.28b)$$

The relationships (6.4.28) imply very important conclusions on solving the problem of classification of the signals from a set $S = \{s_i(t)\}$, $i = 1, \ldots, m$ in signal space with the properties of the lattice $\boldsymbol{\Gamma}(\vee, \wedge)$ of L-group $\boldsymbol{\Gamma}(+, \vee, \wedge)$. There exists a possibility to provide absolute quality indices of the estimators of the signals $\nu_{s\hat{s}}$, $\tilde{\nu}_{s\hat{s}}$. This fact creates the necessary conditions to extract deterministic signals in the presence of interference (noise) without information losses. This is uncharacteristic for solving the same problem within the signal space with the properties of the group $\boldsymbol{\Gamma}(+)$ of L-group $\boldsymbol{\Gamma}(+, \vee, \wedge)$ (particularly, in linear signal space). Another advantage for solving the problem of classification of deterministic signals from a set $S = \{s_i(t)\}$, $i = 1, \ldots, m$ in signal space with the properties of the lattice $\boldsymbol{\Gamma}(\vee, \wedge)$ of L-group $\boldsymbol{\Gamma}(+, \vee, \wedge)$ is the invariance property of both metrics (6.4.27) and quality indices of the estimators of the signals (6.4.28) with respect to the conditions of parametric and nonparametric prior uncertainty. The quality indices of the estimators of the signals in the signal space with lattice properties (6.4.28) compared to the quality indices of the estimators of the signals in linear signal space (6.4.18), do not depend on signal-to-noise ratio and on interference (noise) distribution in the input of processing unit. The problem of synthesis of optimal algorithm of signal classification in signal space with lattice properties demands additional research that is the subject of consideration of the following chapter.

Generally, the results obtained in this subsection, will be used later to determine the capacity of discrete communication channels functioning in the signal space with the properties of the group $\boldsymbol{\Gamma}(+)$ and the lattice $\boldsymbol{\Gamma}(\vee, \wedge)$ of L-group $\boldsymbol{\Gamma}(+, \vee, \wedge)$.

6.4.2 Quality Indices of Deterministic Signal Detection in Metric Spaces with L-group Properties

The problem of signal binary detection can be considered as classification of a pair of signals, if one is assumed to be equal to zero. For signal detection in signal space with the properties of L-group $\boldsymbol{\Gamma}(+, \vee, \wedge)$ we will use the results from the previous subsection.

Generally, the quality of signal detection is determined by conditional probabilities of false alarm F and correct detection D (or false dismissal $1 - D$) [163], [155], [149]. Here, however, quality indices of signal detection will not be identified with conditional probabilities F, D, but will be shown within the problem of signal classification in signal space with the properties of L-group $\boldsymbol{\Gamma}(+, \vee, \wedge)$ and determined by the quality index of an estimator of the detected signal based on the metric relationships between the signals (in this case, between the signal $s(t)$ and its estimator $\hat{s}(t)$).

Consider the general model of interaction between deterministic signal $s(t)$ and interference (noise) $n(t)$ in signal space with the properties of L-group $\boldsymbol{\Gamma}(+, \vee, \wedge)$:

$$x(t) = \theta s(t) \oplus n(t), \ t \in T_s, \quad (6.4.29)$$

where $\oplus$ is some binary operation of L-group $\boldsymbol{\Gamma}(+, \vee, \wedge)$; θ is an unknown nonrandom parameter that can take only two values: $\theta \in \{0, 1\}$: $\theta = 0$ (signal is absent) or $\theta = 1$ (signal is present); T_s is domain of definition of the signal $s(t)$, $T_s = [t_0, t_1]$;

t_0 is known time of arrival of the signal $s(t)$; t_1 is known time of signal ending; $T = t_1 - t_0$ is a duration of the signal $s(t)$.

Within the general model (6.4.29), consider the additive interaction of deterministic signal and interference (noise) in signal space with the properties of the group $\Gamma(+)$ of L-group $\Gamma(+,\vee,\wedge)$:

$$x(t) = \theta s(t) + n(t), \ t \in T_s, \ \theta \in \{0,1\}. \tag{6.4.30}$$

Suppose that interference $n(t)$ is white Gaussian noise with a power spectral density N_0.

While solving the problem of signal detection in linear signal space, i.e., when the interaction equation (6.4.30) holds, the sufficient statistics $y(t)$ is a scalar product $(x(t), s(t))$ of the signals $x(t)$ and $s(t)$ in Hilbert space equal to the correlation integral $\int\limits_{t \in T_s} x(t)s(t)dt$:

$$y(t) = (x(t), s(t)) = \int\limits_{t \in T_s} x(t)s(t)dt, \tag{6.4.31}$$

where $y(t) = \hat{s}(t)$ is an estimator of an energetic parameter of the signal $s(t)$ (simply the estimator $\hat{s}(t)$ of the signal $s(t)$).

Signal detection problem can be solved by maximization of sufficient statistics $y(t)$ (6.4.31) on the domain of definition of the signal T_s and its comparison with a threshold l_0:

$$y(t) = \int\limits_{t \in T_s} x(t)s(t)dt \to \max_{T_s} y(t) \gtrless_{d_0}^{d_1} l_0,$$

where d_1 and d_0 are the decisions concerning the presence and the absence of the signal $s(t)$ in the observed process $x(t)$; $d_1 : \hat{\theta} = 1$, $d_0 : \hat{\theta} = 0$, respectively; $\hat{\theta}$ is an estimate of parameter θ, $\theta \in \{0,1\}$.

In the presence ($\theta = 1$) of the signal $s(t)$ in the observed process $x(t)$ (6.4.30), at the instant $t = t_0 + T$, correlation integral $\int\limits_{t \in T_s} x(t)s(t)dt$ takes an average value equal to energy E of the signal $s(t)$:

$$\mathbf{M}\{y(t)\} = \mathbf{M}\{\int\limits_{t \in T_s} x(t)s(t)dt\}|_{\theta=1} = \int\limits_{t \in T_s} s(t)s(t)dt = E, \tag{6.4.32}$$

where $y(t)$ is the estimator of the signal $s(t)$ in the output of detector; $\mathbf{M}(*)$ is a symbol of mathematical expectation, and in the absence ($\theta = 0$) of the signal $s(t)$ in the observed process $x(t)$ (6.4.30), at the instant $t = t_0 + T$, the correlation integral $\int\limits_{t \in T_s} x(t)s(t)dt$ takes an average value equal to zero:

$$\mathbf{M}\{y(t)\} = \mathbf{M}\{\int\limits_{t \in T_s} x(t)s(t)dt\}|_{\theta=0} = 0. \tag{6.4.33}$$

6.4 Quality Indices of Classification and Detection of Deterministic Signals

In the output of detector, PDF $p_{\hat{s}}(y)\,|_{\theta \in \{0,1\}}$ of the estimator $y(t)$ of the signal $s(t)$, at the instant $t = t_0 + T$, in the presence of signal ($\theta = 1$) or in its absence ($\theta = 0$) in the observed process $x(t)$, is determined by the expression:

$$p_{\hat{s}}(y)\,|_{\theta \in \{0,1\}} = (2\pi D)^{-1/2} \exp\left\{\frac{(y - \theta E)^2}{2D}\right\}, \qquad (6.4.34)$$

where $D = EN_0$ is a noise variance in the output of detector; E is an energy of the signal $s(t)$; N_0 is power spectral density of interference (noise) in the input of processing unit.

Consider the estimator $y(t)$ of the signal $s(t)$ in the output of detector in its presence in the observed process $x(t) = s(t) + n(t)$ (6.4.30), and also the estimator $y(t)\,|_{n(t)=0}$ of the signal $s(t)$ in the output of detector in the absence of interference (noise) ($n(t) = 0$) in an observed process $x(t)$, $x(t) = s(t)$.

To characterize the quality of the estimator $y(t)$ of the signal $s(t)$ while solving the problem of its detection in signal space with the properties of the group $\mathbf{\Gamma}(+)$ of L-group $\mathbf{\Gamma}(+, \vee, \wedge)$ (6.4.30), we introduce the function $\mu(y_t, y_{t,0})$, which is analogous to metric (6.4.9):

$$\mu(y_t, y_{t,0}) = 2[\mathbf{P}(y_t \vee y_{t,0} > h) - \mathbf{P}(y_t \wedge y_{t,0} > h)], \qquad (6.4.35)$$

where y_t is a sample of the estimator $y(t)$ of the signal $s(t)$ in the output of detector at the instant $t = t_0 + T$ in the presence of the signal $s(t)$ and interference (noise) $n(t)$ in the observed process $x(t)$; $y_{t,0}$ is a sample of the estimator $y(t)\,|_{n(t)=0}$ of the signal $s(t)$ in the output of detector at the instant $t = t_0 + T$ in the absence of interference (noise) ($n(t) = 0$) in the observed process $x(t)$, $x(t) = s(t)$, which is equal to a signal energy E: $y_{t,0} = E$; h is some threshold level determined by an average of two mathematical expectations of the processes in the output of detector (6.4.32) and (6.4.33):

$$h = E/2. \qquad (6.4.36)$$

Joint fulfillment of the equality $y_{t,0} = E$ and the inequality $h < E$ implies that the probabilities appearing in the expression (6.4.35) are equal to:

$$\mathbf{P}(y_t \vee y_{t,0} > h) = \mathbf{P}(y_{t,0} > h) = 1; \qquad (6.4.37a)$$

$$\mathbf{P}(y_t \wedge y_{t,0} > h) = \mathbf{P}(y_t > h) = 1 - F_{\hat{s}}(h)\,|_{\theta=1}, \qquad (6.4.37b)$$

where $F_{\hat{s}}(y)\,|_{\theta=1}$ is the CDF of the estimator $y(t)$ in the output of detector at the instant $t = t_0 + T$ in the presence of the signal $s(t)$ ($\theta = 1$) in the observed process $x(t)$ (6.4.30), which, according to the PDF (6.4.34), is equal to:

$$F_{\hat{s}}(y)\,|_{\theta=1} = (2\pi D)^{-1/2} \int_{-\infty}^{y} \exp\left\{-\frac{(x - E)^2}{2D}\right\} dx. \qquad (6.4.38)$$

Substituting the formulas (6.4.37) into the expression (6.4.35), we obtain the resultant value for metric $\mu(y_t, y_{t,0})$ between the signal $s(t)$ and its estimator $y(t)$ while solving the problem of signal detection in the group $\mathbf{\Gamma}(+)$ of L-group $\mathbf{\Gamma}(+, \vee, \wedge)$:

$$\mu_{s\hat{s}} = \mu(y_t, y_{t,0}) = 2F_{\hat{s}}(h)\,|_{\theta=1} =$$

$$= 2[1 - \Phi\left(\frac{E-h}{\sqrt{D}}\right)] = 2[1 - \Phi(\sqrt{q^2/4})], \qquad (6.4.39)$$

where $F_{\hat{s}}(y)|_{\theta=1}$ is the CDF of the estimator $y(t)$ in the output of detector at the instant $t = t_0 + T$ in the presence of the signal $s(t)$ ($\theta = 1$) in the observed process $x(t)$ (6.4.30), which is determined by the formula (6.4.38); $\Phi(z) = (2\pi)^{-1/2} \int_{-\infty}^{z} \exp\left\{-\frac{x^2}{2}\right\} dx$ is the probability integral; $D = EN_0$ is a noise variance in the output of detector; E is an energy of the signal $s(t)$; h is a threshold level determined by the relationship (6.4.36); N_0 is a power spectral density of interference (noise) $n(t)$ in the input of detector; $q^2 = E/N_0$ is signal-to-noise ratio.

By the analogy with the quality index of signal classification (6.4.16), we define quality index of signal detection $\nu_{s\hat{s}}$.

Definition 6.4.2. By *quality index of estimator of the signal $\nu_{s\hat{s}}$ while solving the problem of its detection* we mean NMSI $\nu(y_t, y_{t,0})$ between the samples $y_{t,0}$ and y_t of stochastic processes $y(t)|_{n(t)=0}$, $y(t)$ in the output of detector, which is connected with metric $\nu(y_t, y_{t,0})$ (6.4.35) by the following relationship:

$$\nu_{s\hat{s}} = \nu(y_t, y_{t,0}) = 1 - \mu(y_t, y_{t,0}). \qquad (6.4.40)$$

Substituting the value of metric (6.4.39) into the coupling equation (6.4.40), we obtain the dependence of quality index of detection on signal-to-noise ratio $q^2 = E/N_0$:

$$\nu_{s\hat{s}}(q^2) = 2\Phi(\sqrt{q^2/4}) - 1. \qquad (6.4.41)$$

Based on the obtained formula, quality index of detection $\nu_{s\hat{s}}(q^2)$ (6.4.41) is identically equal to the quality index of classification of orthogonal signals $\nu_{s\hat{s}}(q^2)|_{\perp}$ (6.4.18a): $\nu_{s\hat{s}}(q^2) = \nu_{s\hat{s}}(q^2)|_{\perp}$, that, of course, is trivial, taking into account the known link between signal detection and signal classification.

Within the general model (6.4.29), consider the interaction of deterministic signal $s(t)$ and interference (noise) $n(t)$ in signal space with the properties of the lattice $\Gamma(\vee, \wedge)$ of L-group $\Gamma(+, \vee, \wedge)$, which is described by two binary operations of join and meet, respectively:

$$x(t) = \theta s(t) \vee n(t); \qquad (6.4.42a)$$

$$\tilde{x}(t) = \theta s(t) \wedge n(t), \qquad (6.4.42b)$$

where θ is an unknown nonrandom parameter that is able to take only two values $\theta \in \{0, 1\}$: $\theta = 0$ (signal is absent) or $\theta = 1$ (signal is present); $t \in T_s$, T_s is domain of definition of the detected signal $s(t)$, $T_s = [t_0, t_1]$; t_0 is known time of arrival of the signal $s(t)$; t_1 is time of signal ending; $T = t_1 - t_0$ is a duration of the signal $s(t)$.

Suppose that interference (noise) $n(t)$ is characterized by arbitrary probabilistic-statistical properties.

Consider the estimators $y(t)$, $\tilde{y}(t)$ of the signal $s(t)$ in the output of a detector

6.4 Quality Indices of Classification and Detection of Deterministic Signals

in the presence of the signal ($\theta = 1$) in the observed processes $x(t)$ (6.4.42a), $\tilde{x}(t)$ (6.4.42b):

$$y(t) = y(t)\big|_{x(t)=s(t)\vee n(t)} = s(t) \wedge x(t); \tag{6.4.43a}$$

$$\tilde{y}(t) = \tilde{y}(t)\big|_{\tilde{x}(t)=s(t)\wedge n(t)} = s(t) \vee \tilde{x}(t). \tag{6.4.43b}$$

Also consider the estimators $y(t)\big|_{n(t)=0}$, $\tilde{y}(t)\big|_{n(t)=0}$ of the signal $s(t)$ in the output of detector in the absence of interference (noise) ($n(t) = 0$) in the observed processes $x(t)$ (6.4.42a) and $\tilde{x}(t)$ (6.4.42b):

$$y(t)\big|_{n(t)=0} = s(t) \wedge x(t)\big|_{x(t)=s(t)\vee 0}; \tag{6.4.44a}$$

$$\tilde{y}(t)\big|_{n(t)=0} = s(t) \vee \tilde{x}(t)\big|_{\tilde{x}(t)=s(t)\wedge 0}. \tag{6.4.44b}$$

Determine the values of metric (6.4.35) to evaluate the quality index of the estimator of the signal (6.4.40) while solving the problem of detection in signal space with the properties of the lattice $\boldsymbol{\Gamma}(\vee, \wedge)$ of L-group $\boldsymbol{\Gamma}(+, \vee, \wedge)$:

$$\mu_{s\hat{s}} = \mu(y_t, y_{t,0}) = 2[\mathbf{P}(y_t \vee y_{t,0} > h) - \mathbf{P}(y_t \wedge y_{t,0} > h)]; \tag{6.4.45a}$$

$$\tilde{\mu}_{s\hat{s}} = \mu(\tilde{y}_t, \tilde{y}_{t,0}) = 2[\mathbf{P}(\tilde{y}_t \vee \tilde{y}_{t,0} > h) - \mathbf{P}(\tilde{y}_t \wedge \tilde{y}_{t,0} > h)], \tag{6.4.45b}$$

where y_t and $\tilde{y}_t$ are the samples of the estimators $y(t)$ (6.4.43a) and $\tilde{y}(t)$ (6.4.43b) of the signal $s(t)$ in the output of detector at the instant $t \in T_s$ in the presence of the signal ($\theta = 1$) in the observed processes $x(t)$ (6.4.42a) and $\tilde{x}(t)$ (6.4.42b); $y_{t,0}$ and $\tilde{y}_{t,0}$ are the samples of the estimators $y(t)\big|_{n(t)=0}$ (6.4.44a) and $\tilde{y}(t)\big|_{n(t)=0}$ (6.4.44b) of the signal $s(t)$ in the output of detector at the instant $t \in T_s$ in the absence of interference (noise) ($n(t) = 0$) in the observed processes $x(t)$ (6.4.42a) and $\tilde{x}(t)$ (6.4.42b); h is some threshold level.

The obtained relationships (6.4.43) and (6.4.44), and absorption axiom of a lattice imply that the samples y_t, $\tilde{y}_t$ of the estimators $y(t)$ (6.4.43a) and $\tilde{y}(t)$ (6.4.43b), and also the samples $y_{t,0}$, $\tilde{y}_{t,0}$ of the estimators $y(t)\big|_{n(t)=0}$ (6.4.44a) and $\tilde{y}(t)\big|_{n(t)=0}$ (6.4.44b) are identically equal to the received useful signal $s(t)$:

$$y_t = y(t) = s(t); \tag{6.4.46a}$$

$$\tilde{y}_t = \tilde{y}(t) = s(t); \tag{6.4.46b}$$

$$y_{t,0} = y(t)\big|_{n(t)=0} = s(t); \tag{6.4.47a}$$

$$\tilde{y}_{t,0} = \tilde{y}(t)\big|_{n(t)=0} = s(t). \tag{6.4.47b}$$

Substituting the obtained values of a pair of samples y_t (6.4.46a) and $y_{t,0}$ (6.4.47a) into the relationship (6.4.45a) and also substituting the values of a pair of samples $\tilde{y}_t$ (6.4.46b) and $\tilde{y}_{t,0}$ (6.4.47b) into the relationship (6.4.45b), we obtain that the values of metrics $\mu(y_t, y_{t,0})$ and $\mu(\tilde{y}_t, \tilde{y}_{t,0})$ between the signal $s(t)$ and its estimators $y(t)$, $\tilde{y}(t)$, while solving the problem of signal detection in signal space with the properties of the lattice $\boldsymbol{\Gamma}(\vee, \wedge)$ of L-group $\boldsymbol{\Gamma}(+, \vee, \wedge)$, are equal to zero:

$$\mu_{s\hat{s}} = \mu(y_t, y_{t,0}) = 0; \tag{6.4.48a}$$

$$\tilde{\mu}_{s\hat{s}} = \mu(\tilde{y}_t, \tilde{y}_{t,0}) = 0. \tag{6.4.48b}$$

The relationships (6.4.48) and the coupling equation (6.4.40) imply that quality indices of the estimators of the signals $\nu_{s\hat{s}}$ and $\tilde{\nu}_{s\hat{s}}$, while solving the problem of signal detection in signal space with the properties of the lattice $\mathbf{\Gamma}(\vee, \wedge)$ of L-group $\mathbf{\Gamma}(+, \vee, \wedge)$, take absolute values equal to 1:

$$\nu_{s\hat{s}} = 1 - \mu_{s\hat{s}} = 1; \tag{6.4.49a}$$

$$\tilde{\nu}_{s\hat{s}} = 1 - \tilde{\mu}_{s\hat{s}} = 1. \tag{6.4.49b}$$

The relationships (6.4.49) imply an important conclusion. While solving the problem of signal detection in signal space with the properties of the lattice $\mathbf{\Gamma}(\vee, \wedge)$ of L-group $\mathbf{\Gamma}(+, \vee, \wedge)$, there exists the possibility to provide absolute quality indices of the estimators of the signals $\nu_{s\hat{s}}$ and $\tilde{\nu}_{s\hat{s}}$. This creates the necessary conditions to detect deterministic signals in the presence of interference (noise) without information losses, that is uncharacteristic when solving the same problem within the signal space with the properties of the group $\mathbf{\Gamma}(+)$ of L-group $\mathbf{\Gamma}(+, \vee, \wedge)$ (particularly, in linear signal space). One more advantage for solving the problem of detection of deterministic signals in signal space with the properties of the lattice $\mathbf{\Gamma}(\vee, \wedge)$ of L-group $\mathbf{\Gamma}(+, \vee, \wedge)$, is the invariance property of both metrics (6.4.48) and quality indices of the estimators of the signals (6.4.49) with respect to parametric and nonparametric prior uncertainty conditions. The quality indices of signal detection in the signal space with lattice properties (6.4.49), as against the quality indices of signal detection in linear signal space (6.4.41), do not depend on signal-to-noise ratio and on interference (noise) distribution in the input of processing unit. The problem of synthesis of optimal signal detection algorithm in signal space with lattice properties demands an additional research that is the subject of consideration of the following chapter.

6.5 Capacity of Communication Channels Operating in Presence of Interference (Noise)

Based on formalized descriptions of the signals and their instantaneous values as the elements of metric space with special properties and measures of information quantity, and relying upon the results in Section 6.2, we consider the informational characteristics of communication channels operating in metric spaces with L-group properties to establish the most general regularities defining the processes of information transmitting and receiving in the presence of interference (noise).

In Section 5.2, we defined the relationships establishing upper bounds of the capacities of continuous and discrete noiseless channels. The goals of this section are: (1) obtain the relationships characterizing the capacities of continuous and discrete channels that operate in the presence of interference (noise) and (2) perform a brief comparative analysis of the obtained relationships determining channel capacity with known results of classical information theory.

6.5 Capacity of Communication Channels Operating in Presence of Interference (Noise)

Consider a generalized structural scheme of a communication channel functioning in the presence of interference (noise) as shown in Fig. 6.5.1. In the input of the receiving set of communication system, let the result $x(t)$ of interaction between useful signal $s(t)$ and interference (noise) $n(t)$ be determined by some binary operation of L-group $\boldsymbol{\Gamma}(+, \vee, \wedge)$:

$$x(t) = s(t) \oplus n(t); \ t \in T_s, \quad (6.5.1)$$

where $\oplus$ is some binary operation of L-group: $+$, $\vee$, $\wedge$; $T_s = [t_0, t_0 + T]$ is a domain of definition of useful signal $s(t)$; T is a duration of useful signal $s(t)$. The interference (noise) exerts a negative influence on the quantity of absolute information $I_{\hat{s}}$ contained in the estimator $\hat{s}(t) = f_{\hat{s}}[x(t)], \ t \in T_s$ of useful signal $s(t)$ formed by the proper processing of the interaction result $x(t)$ (6.5.1) known as filtering of useful signal (or its extraction) (see Fig. 6.5.1). The quantity of absolute information I_x contained in the observed process $x(t)$ can be arbitrarily large due to interference (noise) influence. The task of the receiving set is forming a processing function $f_{\hat{s}}[*]$, so that an obtained estimator $\hat{s}(t) = f_{\hat{s}}[x(t)], \ t \in T_s$ of useful signal $s(t)$ formed in the output of the receiving set would contain maximum information quantity concerning the signal $s(t)$. In other words, the main informational property of the estimator $\hat{s}(t)$ obtained as the result of extraction of useful signal $s(t)$ in the presence of interference (noise) $n(t)$ must provide the largest quantity of mutual information $I_{s\hat{s}}$ contained in the estimator $\hat{s}(t)$ concerning the useful signal $s(t)$:

$$\hat{s}(t) = f_{\hat{s}}[x(t)]\,|_{I_{s\hat{s}} \to \max}. \quad (6.5.2)$$

On the base of the relationship (6.5.2), we define the capacity C_n of communication channel functioning in the presence of interference (noise).

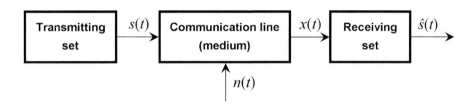

FIGURE 6.5.1 Generalized structural scheme of communication channel functioning in presence of interference (noise)

Definition 6.5.1. By *capacity C_n* of a *communication channel* operating in the presence of interference (noise) in physical signal space $\boldsymbol{\Gamma}$ with the properties of L-group $\boldsymbol{\Gamma}(+, \vee, \wedge)$, we mean the largest quantity of mutual information $I_{s\hat{s}}$ between useful signal $s(t)$ and its estimator $\hat{s}(t)$ that can be extracted from the interaction result $x(t)$ (6.5.1) in a signal duration T:

$$C_n = \max_{\hat{s} \in \boldsymbol{\Gamma}} \frac{I_{s\hat{s}}}{T} \ (\text{abit/s}). \quad (6.5.3)$$

According to Theorem 6.2.3, under interaction (6.5.1) between useful signal $s(t)$ and interference (noise) $n(t)$ in physical signal space Γ with the properties of L-group $\Gamma(+, \vee, \wedge)$, between the quantities of mutual information I_{sx} and $I_{s\hat{s}}$ contained in the pairs of the signals $s(t)$, $x(t)$ and $s(t)$, $\hat{s}(t)$, respectively, the inequality holds:

$$I_{s\hat{s}} \leq I_{sx}, \qquad (6.5.4)$$

so that the inequality (6.5.4) turns into identity if and only if the mapping $f_{\hat{s}}$ (6.5.2) belongs to a group $H = \{h_\alpha\}$ of continuous mappings $f_{\hat{s}} \in H$ preserving null (identity) element 0 of the group $\Gamma(+)$ of L-group $\Gamma(+, \vee, \wedge)$: $h_\alpha(0) = 0$, $h_\alpha \in H$.

In the first subsection, we investigate the informational characteristics of continuous communication channels and in the second one we consider the informational characteristics of discrete communication channels.

6.5.1 Capacity of Continuous Communication Channels Operating in Presence of Interference (Noise) in Metric Spaces with L-group Properties

According to the informational inequality of signal processing (6.5.4) of Theorem 6.2.3, the capacity C_n of a continuous communication channel functioning in the presence of interference (noise) in physical signal space Γ with the properties of L-group $\Gamma(+, \vee, \wedge)$ is determined by the quantity of mutual information I_{sx} contained in the observation result $x(t)$ (6.5.1) concerning the useful signal $s(t)$ on its domain of definition T_s with a duration T of this interval:

$$C_n = \frac{I_{sx}}{T} \text{ (abit/s).} \qquad (6.5.5)$$

According to the expression (6.2.4), determining the quantity of mutual information I_{sx}, the relationship (6.5.5) takes the form:

$$C_n = \frac{(\nu_{sx} I_s)}{T} = \nu_{sx} C, \text{ (abit/s)} \qquad (6.5.6)$$

where $\nu_{sx} = \nu(s_t, x_t)$ is normalized measure of statistical interrelationship (NMSI) of the samples s_t and x_t of stochastic signals $s(t)$ and $x(t)$; I_s is the quantity of absolute information contained in useful signal $s(t)$; $C = I_s/T$ is a continuous noiseless channel capacity.

Thus, the capacity C_n (abit/s) of a continuous channel functioning under interaction (6.5.1) between useful signal $s(t)$ and interference (noise) $n(t)$ in physical signal space Γ with the properties of L-group $\Gamma(+, \vee, \wedge)$, is determined by capacity C (abit/s) of noiseless continuous channel and by NMSI $\nu_{sx} = \nu(s_t, x_t)$ of the samples s_t and x_t of stochastic signals $s(t)$ and $x(t)$.

The capacity of the channel in the presence of interference (noise) C_n (bit/s) measured in bit/s, according to the expression (5.2.11), is determined by the relationship:

$$C_n \text{ (bit/s)} = C_n \text{ (abit/s)} \cdot \log_2[C_n \text{ (abit/s)} \cdot 1\,\text{s}]. \qquad (6.5.7)$$

6.5 Capacity of Communication Channels Operating in Presence of Interference (Noise)

On the base of the obtained general relationships (6.5.5) and (6.5.6), we consider the peculiarities of evaluating the capacities of continuous channels operating in the presence of interference (noise) for the cases of concrete kinds of interaction (6.5.1) between useful signal $s(t)$ and interference (noise) $n(t)$ in physical signal space Γ with the properties of L-group $\Gamma(+, \vee, \wedge)$, where $\oplus$ is some binary operation of L-group: $+, \vee, \wedge$.

For the case of additive interaction (6.5.1) between useful Gaussian signal $s(t)$ and interference (noise) $n(t)$ in the form of white Gaussian noise (WGN), we determine the capacity $C_{n,+}$ of continuous channel functioning in physical signal space Γ with the properties of additive commutative group $\Gamma(+)$ of L-group $\Gamma(+, \vee, \wedge)$:

$$x(t) = s(t) + n(t); \; t \in T_s. \tag{6.5.8}$$

Substituting the relationships (6.2.13) and (6.2.14) into the formula (6.5.5), and the relationships (6.2.12) and (6.2.14) into the formula (6.5.6), we obtain the expression for the capacity $C_{n,+}$ of Gaussian continuous channel with WGN:

$$C_{n,+} = I_{sx}/T = (\nu_{sx} I_s)/T = \nu_{sx} C =$$
$$= \frac{2C}{\pi} \arcsin[q/\sqrt{1+q^2}] \; \text{(abit/s)}, \tag{6.5.9}$$

where $\nu_{sx} = \frac{2}{\pi} \arcsin[\rho_{sx}]$ is NMSI between the samples s_t, x_t of Gaussian stochastic signals $s(t)$, $x(t)$; $\rho_{sx} = q/\sqrt{1+q^2}$ is the correlation coefficient between the samples s_t and x_t of Gaussian stochastic signals $s(t)$ and $x(t)$; $q^2 = S_0/N_0 = D_s/D_n$ is the ratio of the energy S_0 of Gaussian useful signal to power spectral density N_0 of the interference (noise) $n(t)$; $D_s = R(0) = \frac{1}{2\pi} \int\limits_{-\infty}^{\infty} S(\omega) d\omega$ is a variance of useful signal $s(t)$; $S_0 = S(0) = \int\limits_{-\infty}^{\infty} R(\tau) d\tau$; $D_n = N_0 (\int\limits_{-\infty}^{\infty} S(\omega) d\omega)/(2\pi S_0)$ is a variance of interference (noise) $n(t)$ brought to the energy of useful signal $s(t)$; $S(\omega)$ is power spectral density of useful signal $s(t)$; N_0 is the power spectral density of interference (noise) $n(t)$ in the form of WGN; C is the continuous noiseless channel capacity.

Thus, under additive interaction (6.5.8) between useful signal $s(t)$ and interference (noise) $n(t)$, the capacity $C_{n,+}$ (abit/s) of continuous channel functioning in physical signal space Γ with the properties of additive commutative group $\Gamma(+)$ of L-group $\Gamma(+, \vee, \wedge)$ is determined by the capacity C (abit/s) of the noiseless channel and the ratio $q^2 = S_0/N_0$ of the energy of the Gaussian useful signal $s(t)$ to the power spectral density of interference (noise) $n(t)$, but it does not exceed noiseless channel capacity C (abit/s).

The capacity $C_{n,+}$ (bit/s) of Gaussian continuous channel with additive WGN, according to the formulas (6.5.7) and (6.5.9), is determined by the relationship:

$$C_{n,+} \text{(bit/s)} = C_{n,+} \text{(abit/s)} \cdot \log_2[C_{n,+} \text{(abit/s)} \cdot 1 \text{s}], \tag{6.5.10}$$

where $C_{n,+} = \frac{2C}{\pi} \arcsin[q/\sqrt{1+q^2}]$ (abit/s); $q^2 = S_0/N_0 = D_s/D_n$ is the ratio of the energy of the Gaussian useful signal $s(t)$ to the power spectral density of interference (noise) $n(t)$.

We now determine the capacity $C_{n,\vee/\wedge}$ of continuous channels functioning in the presence of interference (noise) in the interaction (6.5.1) between useful signal $s(t)$ and interference (noise) $n(t)$ in physical signal space $\mathbf{\Gamma}$, where $\oplus$ is a binary operation of the lattice $\mathbf{\Gamma}(\vee,\wedge)$ of L-group $\mathbf{\Gamma}(+,\vee,\wedge)$ in the form of its join and meet, respectively:

$$x(t) = s(t) \vee n(t); \quad (6.5.11a)$$

$$\tilde{x}(t) = s(t) \wedge n(t), \quad (6.5.11b)$$

where $t \in T_s$; $T_s = [t_0, t_0 + T]$ is a domain of definition of useful signal $s(t)$; T is a duration of useful signal $s(t)$.

Substituting the identities (6.2.16a) and (6.2.16b) into the formula (6.5.5), we obtain the relationships for the capacity $C_{n,\vee}$ and $C_{n,\wedge}$ of continuous channel functioning under the interaction between useful signal $s(t)$ and interference (noise) $n(t)$ in the form of join and meet of the lattice $\mathbf{\Gamma}(\vee,\wedge)$:

$$C_{n,\vee} = I_{sx}/T = 0.5 I_s/T = 0.5C; \quad (6.5.12a)$$

$$C_{n,\wedge} = I_{s\tilde{x}}/T = 0.5 I_s/T = 0.5C, \quad (6.5.12b)$$

where C is continuous noiseless channel capacity; I_{sx} and $I_{s\tilde{x}}$ are the quantities of mutual information contained in the observed processes $x(t)$ (6.5.11a) and $\tilde{x}(t)$ (6.5.11b) concerning useful signal $s(t)$, respectively; T is a duration of useful signal $s(t)$.

Note that the sense of the relationship (6.2.16c) as the consequence of the identities (6.2.16a) and (6.2.16b), lies in the fact that while using the results of simultaneous processing of the observations in the form of join (6.5.11a) and meet (6.5.11b) of the lattice $\mathbf{\Gamma}(\vee,\wedge)$, one can avoid the losses of information on extracting a useful signal. Due to fulfillment of the relationship (3.2.39b) elucidating statistical independence of the processes $x(t)$ (6.5.11a), $\tilde{x}(t)$ (6.5.11b), each of them, according to the identities (6.2.16a) and (6.2.16b), contains half the information quantity contained in useful signal $s(t)$. On a qualitative level, this information is distinct with respect to its content, inasmuch as the processes $x(t)$ and $\tilde{x}(t)$ are statistically independent. However, when taken together (but not in the sum!), they contain complete information contained in the useful signal.

Substituting the identity (6.2.16c) into the formula (6.5.5), we obtain the relationship for the capacity $C_{n,\vee/\wedge}$ of continuous channel functioning in the presence of interference (noise), where we find the simultaneous processing of the observations in the form of join (6.5.11a) and meet (6.5.11b) of the lattice $\mathbf{\Gamma}(\vee,\wedge)$:

$$C_{n,\vee/\wedge} = (I_{sx} + I_{s\tilde{x}})/T = I_s/T = C, \quad (6.5.13)$$

where C is the continuous noiseless channel capacity.

The resultant capacity of continuous channel $C_{n,\vee/\wedge}$ (abit/s), functioning in the presence of interference (noise) under the interactions (6.5.11a,b) between useful signal $s(t)$ and interference (noise) $n(t)$ in physical signal space $\mathbf{\Gamma}$ with the properties of the lattice $\mathbf{\Gamma}(\vee,\wedge)$, is determined by the noiseless continuous channel capacity C (abit/s) and does not depend on energetic relationships between useful signal and interference (noise) and on their probabilistic-statistical properties. The last circumstance defines the invariance property of continuous channels with lattice properties with respect to parametric and nonparametric prior uncertainty conditions.

The capacity $C_{n,\vee/\wedge}$ (bit/s) of continuous channels, functioning in the presence of interference (noise), where we find the simultaneous processing of the observations in the form of join (6.5.11a) and meet (6.5.11b) of the lattice $\Gamma(\vee,\wedge)$, measured in bit/s, according to the formulas (6.5.7) and (6.5.13), is determined by the relationship:

$$C_{n,\vee/\wedge}(\text{bit/s}) = C_{n,\vee/\wedge}(\text{abit/s}) \cdot \log_2[C_{n,\vee/\wedge}(\text{abit/s}) \cdot 1\,\text{s}] =$$
$$= C(\text{abit/s}) \cdot \log_2[C(\text{abit/s}) \cdot 1\,\text{s}], \quad (6.5.14)$$

where C is continuous noiseless channel capacity.

Thus, the obtained results (6.5.9), (6.5.10) and (6.5.13), (6.5.14), determining the expressions for the capacities of continuous channels operating in linear signal spaces and the signal spaces with lattice properties respectively, imply that, independently of energetic relationships between useful signal and interference (noise), the capacity of a continuous channel is a finite quantity determined by the continuous noiseless channel capacity. These results, although contradictory to known results of classical information theory, confirm the remarkable phrase of Vadim Mityugov in 1976 [92]: "...on a given signal power, even in the absence of interference, information transmission rate is always a finite quantity".

6.5.2 Capacity of Discrete Communication Channels Operating in Presence of Interference (Noise) in Metric Spaces with L-group Properties

Consider now the peculiarities of evaluation of the capacity of discrete communication channels functioning in the presence of interference (noise).

In the most general case, the signal $s(t)$ transmitting a discrete message $u(t)$ in the form of a sequence of quasi-deterministic signals $\{s_j(t)\}$ can be represented in the following form:

$$s(t) = M[c(t), u(t)] = \sum_{j=0}^{n-1} s_j(t - j\tau_0), \ t \in T_s, \quad (6.5.15)$$

where $M[*,*]$ is a modulating function; $T_s = [t_0, t_0+T]$ is a domain of definition of useful signal $s(t)$; T is a duration time of useful signal $s(t)$; $s_j(t) = M[c(t), u_j(t)]$, $s_j(t) \in S = \{s_i(t)\}$, $i = 1, \ldots, m$; $c(t)$ is a carrier signal; $u(t)$ is a transmitted discrete message represented by a discrete random sequence of informational symbols $\{u_j(t)\}$:

$$u(t) = \sum_{j=0}^{n-1} u_j(t - j\tau_0), \ t \in T_s, \quad (6.5.16)$$

where τ_0 is a duration of a symbol $u_j(t)$; n is a number of informational symbols of random sequence $\{u_j(t)\}$.

Suppose that, at an arbitrary instant $t \in T_s$, each symbol $u_j(t)$ of discrete random sequence $u(t) = \{u_j(t)\}$ (16), $j = 0, 1, \ldots, n-1$ equiprobably takes the values from a set $\{u_i\}$, $i = 1, \ldots, m$.

We first define the discrete channel capacity measured in abit/s, and then define the capacity measured in units accepted by classical information theory, i.e., in bit/s.

Based on Definition 6.5.1, by *capacity* C_n of discrete channel, functioning in the presence of interference (noise) in signal space Γ with the properties of L-group $\Gamma(+, \vee, \wedge)$, we mean the largest quantity of mutual information $I_{s\hat{s}}$ between useful signal $s(t)$ and its estimator $\hat{s}(t)$ that can be extracted from the interaction result $x(t)$ (6.5.1) in a signal duration T:

$$C_n = \max_{\hat{s} \in \Gamma} \frac{I_{s\hat{s}}}{T} \text{ (abit/s)}. \tag{6.5.17}$$

According to the expression (6.2.4) determining the quantity of mutual information $I_{s\hat{s}}$, the relationship (6.5.17) takes the form:

$$C_n = \frac{(\nu_{sx} I_s)}{T} = \nu_{sx} C \text{ (abit/s)}, \tag{6.5.18}$$

where $\nu_{sx} = \nu(s_t, x_t)$ is the NMSI of the samples s_t and x_t of stochastic signals $s(t)$ and $x(t)$; I_s is a quantity of absolute information contained in the useful signal $s(t)$; $C = I_s/T$ is a discrete noiseless channel capacity.

We now determine the capacity $C_{n,+}$ of discrete channel functioning in the presence of interference (noise) in the additive interaction (6.5.8) between the useful Gaussian signal $s(t)$ and interference (noise) $n(t)$ in the form of white Gaussian noise (WGN) in physical signal space Γ with the properties of the additive commutative group $\Gamma(+)$ of L-group $\Gamma(+, \vee, \wedge)$.

Substituting the relationship (6.4.17) into the formula (6.5.18), we obtain the resultant expression for the capacity $C_{n,+}$ of a discrete channel with additive WGN:

$$C_{n,+} = [2\Phi\left(\sqrt{q^2(1-r_{ik})/2}\right) - 1]C \text{ (abit/s)}, \tag{6.5.19}$$

where $q^2 = E/N_0$ is signal-to-noise ratio; E is the energy of the signals $s_i(t)$ from a set $S = \{s_i(t)\}$, $i = 1, \ldots, m$; N_0 is a power spectral density of interference (noise) in the input of the processing unit (the receiving set); $\Phi(z) = (2\pi)^{-1/2} \int_{-\infty}^{z} \exp\left\{-\frac{x^2}{2}\right\} dx$ is the integral of probability; r_{ik} is cross-correlation coefficient between the signals $s_i(t)$, $s_k(t)$.

In a binary channel with additive WGN, where the transmission of opposite signals $s_1(t) = -s_2(t)$ ($r_{ik} = 1$) is realized, its capacity, according to the formula (6.5.19), is equal to:

$$C_{n,+} = [2\Phi\left(\sqrt{q^2}\right) - 1]C \text{ (abit/s)}, \tag{6.5.20}$$

6.5 Capacity of Communication Channels Operating in Presence of Interference (Noise)

where $C = I_s/T$ is binary noiseless channel capacity.

According to the relationship (5.1.13), information quantities transmitted by the signal (6.5.15) on the base of binary sequence (6.5.16), and measured in abits and in bits, are equivalent, so the capacity of a binary channel with additive WGN, measured in bit/s, exactly corresponds to the capacity measured in abit/s:

$$C_{n,+} \text{ (abit/s)} = C_{n,+} \text{ (bit/s)}.$$

Unfortunately, this correspondence does not hold when it concerns the capacity of a discrete channel transmitting discrete messages with cardinality $\text{Card}\{u_i\} = m$ of a set of values $\{u_i\}$ of discrete random sequence $u(t) = \{u_j(t)\}$ that is greater than 2: $\text{Card}\{u_i\} = m > 2$, i.e., while transmitting m-ary signals.

In this case, the capacity of discrete channels with additive noise, measured in bit/s, is connected with the capacity measured in abit/s by the relationship (5.2.10):

$$C_{n,+} \text{ (bit/s)} = \begin{cases} C_{n,+} \text{ (abit/s)} \cdot \log_2 m, & m < n; \\ C_{n,+} \text{ (abit/s)} \cdot \log_2 n, & m \geq n, \end{cases} \quad (6.5.21)$$

where m is the cardinality of a set of values $\{u_i\}$ of discrete random sequence $u(t) = \{u_j(t)\}$ (16): $m = \text{Card}\{u_i\}$; n is the length of discrete random sequence $u(t) = \{u_j(t)\}$ (6.5.16); $C_{n,+}$ (abit/s) is the capacity of a discrete channel with interference (noise) measured in abit/s determined by the formula (6.5.19).

According to formula (6.5.21), over the discrete channel transmitting the discrete messages with the length of n symbols, it is impossible to transmit more information per time unit, than the quantity $C_{n,+}$ (abit/s) $\cdot \log_2 n$ (bit/s).

We can determine the capacity $C_{n,\vee/\wedge}$ of discrete channel functioning in the presence of interference (noise), in the interaction between useful signal $s(t)$ and interference (noise) $n(t)$ in physical signal space $\mathbf{\Gamma}$ with the properties of the lattice $\mathbf{\Gamma}(\vee, \wedge)$ in the form of its join (6.5.11a) and meet (6.5.11b), respectively.

Substituting the relationships (6.4.28a) and (6.4.28b) into the formula (6.5.18), we obtain the expressions for the capacity $C_{n,\vee}$, $C_{n,\wedge}$ of discrete channel with interference (noise) for interactions between useful signal $s(t)$ and interference (noise) $n(t)$ in physical signal space $\mathbf{\Gamma}$ in the form of join (6.5.11a) and meet (6.5.11b) of the lattice $\mathbf{\Gamma}(\vee, \wedge)$, respectively:

$$C_{n,\vee} = \nu_{s\hat{s}} C = C \text{ (abit/s)}; \quad (6.5.22a)$$

$$C_{n,\wedge} = \tilde{\nu}_{s\hat{s}} C = C \text{ (abit/s)}, \quad (6.5.22b)$$

where $C = I_s/T$ (abit/s) is the discrete noiseless channel capacity measured in abit/s.

In the signal space $\mathbf{\Gamma}(\vee, \wedge)$ with lattice properties, the capacity $C_{n,\vee/\wedge}$ of a discrete channel with interference (noise), measured in bit/s, is connected with the discrete noiseless channel capacity C measured in abit/s by the relationship (5.2.10):

$$C_{n,\vee/\wedge} \text{ (bit/s)} = \begin{cases} C \text{ (abit/s)} \cdot \log_2 m, & m < n; \\ C \text{ (abit/s)} \cdot \log_2 n, & m \geq n, \end{cases} \quad (6.5.23)$$

where m is the cardinality of a set of values $\{u_i\}$ of discrete random sequence $u(t) = \{u_j(t)\}$ (6.5.16): $m = \mathrm{Card}\{u_i\}$; n is the length of discrete random sequence $u(t) = \{u_j(t)\}$ (6.5.16); $C = I_s/T$ (abit/s) is the discrete noiseless channel capacity measured in abit/s.

Thus, the obtained results (6.5.19) and (6.5.22a,b), determining the expressions of the capacities of discrete channels operating in the presence of interference (noise) in linear signal space and in signal space with lattice properties, imply that, regardless of energetic relationships between useful and interference signals, the capacity of discrete channel is a finite quantity determined by the discrete noiseless channel capacity.

The relationships (6.5.22a,b) and (6.5.23) imply an important conclusion. In signal space $\mathbf{\Gamma}(\vee, \wedge)$ with lattice properties, the capacity of discrete channel functioning in the presence of interference (noise) is equal to the discrete noiseless channel capacity and does not depend on signal-to-interference (signal-to-noise) ratio and interference (noise) distribution. The relationships (6.5.22) are the direct consequence from the absolute values of the quality indices of the signal estimators (6.4.28) while solving the problem of their classification from a set $S = \{s_i(t)\}$, $i = 1, \ldots, m$, that is stipulated by the possibility of extraction of deterministic signals $s_i(t)$ under interference (noise) background without information losses.

The results in this section allow us to draw the following conclusions.

1. The capacity of a channel functioning in the presence of interference (noise) in both linear signal space (6.5.9), (6.5.19) and a signal space with lattice properties (6.5.13), (6.5.22), is a finite quantity, whose upper bound is determined by the noiseless channel capacity.

2. The capacity of a channel functioning in the presence of interference (noise) in a signal space with lattice properties is equal to the noiseless channel capacity and does not depend on energetic relations between useful and interference signals and their probabilistic-statistical properties. The last circumstance defines the invariance property of a channel functioning in the signal space with lattice properties with respect to parametric and nonparametric prior uncertainty conditions.

6.6 Quality Indices of Resolution-Detection in Metric Spaces with L-group Properties

The general model of signal interaction related to the problem of their resolution is described by the formula (6.1.9). This problem is considered as a pared-down form in which the number of interacting useful signals does not exceed 2.

We consider a simple model of interaction between the signals $s_{i,k}(t)$ from a set of deterministic signals $S = \{s_i(t), s_k(t)\}$ and interference (noise) $n(t)$ in signal

6.6 Quality Indices of Resolution-Detection in Metric Spaces with L-group Properties

space with the properties of L-group $\mathbf{\Gamma}(+, \vee, \wedge)$:

$$x(t) = \theta_k s_k(t) \oplus \theta_i s_i(t) \oplus n(t), \ t \in T_s, \quad (6.6.1)$$

where $\oplus$ is some binary operation of L-group $\mathbf{\Gamma}(+, \vee, \wedge)$; $\theta_{i,k}$ is a random parameter that takes the values from the set $\{0,1\}$: $\theta_{i,k} \in \{0,1\}$; $T_s = [t_0, t_0+T]$ is the domain of definition of the signal $s_{i,k}(t)$; t_0 is the known time of arrival of the signal $s_{i,k}(t)$; T is a duration of the signal $s_{i,k}(t)$.

Let the signals from the set $S = \{s_i(t), s_k(t)\}$ be characterized by an energy $E_{i,k} = \int\limits_{t \in T_s} s_{i,k}^2(t) \mathrm{d}t$ and cross-correlation coefficient $r_{i,k}$:

$$r_{i,k} = \int\limits_{t \in T_s} s_i(t) s_k(t) \mathrm{d}t / \sqrt{E_i E_k}.$$

As noted in Section 6.1, the problem of resolution-classification (i.e., joint resolution and classification) of signals in the presence of interference (noise) lies in the plane, where one should decide based on some criterion, what the combination of useful signals from a set $S = \{s_i(t), s_k(t)\}$ is included in the observed process $x(t)$. In this section we consider the problem of resolution-detection (i.e., joint resolution and detection) of a useful signal $s_k(t)$ in the presence of interference (noise) $n(t)$, and also in the presence of interfering signal $s_i(t)$. Under this problem statement, the task of signal processing is establishing the presence of the k-th signal in the observed process $x(t)$ (6.6.1), i.e., its separate detection.

The first goal of this section is evaluation of quality indices of resolution-detection of deterministic signals on the base of metric (3.2.1). The second one is establishing general qualitative and quantitative regularities for resolution-detection of the signals in linear signal spaces and in signal spaces with lattice properties. As a model of signal space generalizing the properties of spaces of the mentioned types, we will use the signal space with the properties of L-group $\mathbf{\Gamma}(+, \vee, \wedge)$.

Further, on the base of metric (3.2.1), and considering the known interrelation between signal classification and signal resolution, based on material in Section 6.4, we will determine the quality indices of resolution-detection of the signals in signal space with the properties of both the group $\mathbf{\Gamma}(+)$ and the lattice $\mathbf{\Gamma}(\vee, \wedge)$ of L-group $\mathbf{\Gamma}(+, \vee, \wedge)$.

Within the framework of the general model (6.6.1), we consider the additive interaction between the signals $s_{i,k}(t)$ from a set $S = \{s_i(t), s_k(t)\}$ and interference (noise) $n(t)$ in signal space with the properties of the group $\mathbf{\Gamma}(+)$ of L-group $\mathbf{\Gamma}(+, \vee, \wedge)$:

$$x(t) = \theta_k s_k(t) + \theta_i s_i(t) + n(t), \ t \in T_s, \quad (6.6.2)$$

where $\theta_{i,k}$ is a random parameter taking the values from a set $\{0,1\}$: $\theta_{i,k} \in \{0,1\}$; $T_s = [t_0, t_0+T]$ is a domain of definition of the signal $s_{i,k}(t)$; t_0 is a known time of arrival of the signal $s_{i,k}(t)$; T is a duration of the signal $s_{i,k}(t)$.

Suppose that interference $n(t)$ is white Gaussian noise with power spectral density N_0.

Consider the estimator $y_k(t) = \hat{s}_k(t)$ of the signal $s_k(t)$ in the k-th processing channel in the presence of the signal $s_k(t)$ and in the absence of the signal $s_i(t)$ in the observed process $x(t) = s_k(t) + n(t)$ ($\theta_k = 1$, $\theta_i = 0$):

$$y_k(t) = \hat{s}_k(t) \big|_{x(t)=s_k(t)+n(t)}, \quad t = t_0 + T, \tag{6.6.3}$$

and also consider the estimator $y_k(t)\big|_{n(t)=0} = \hat{s}_k(t)\big|_{n(t)=0}$ of the signal $s_k(t)$ in the k-th processing channel in the absence of both interference (noise) ($n(t) = 0$) and the signal $s_i(t)$ in the observed process $x(t)=s_k(t)$:

$$y_k(t)\big|_{n(t)=0} = \hat{s}_k(t)\big|_{x(t)=s_k(t)}, \quad t = t_0 + T. \tag{6.6.4}$$

To evaluate the quality index of the estimator $y_k(t) = \hat{s}_k(t)$ of the signal $s_k(t)$, while solving the problem of resolution-detection of the signals in the group $\boldsymbol{\Gamma}(+)$ of L-group $\boldsymbol{\Gamma}(+, \vee, \wedge)$ in the presence of the signal $s_k(t)$ and in the absence of the signal $s_i(t)$ in the observed process $x(t) = s_k(t) + n(t)$ ($\theta_k = 1$, $\theta_i = 0$), we use the metric $\mu_{s_k \hat{s}_k} = \mu(y_t, y_{t,0})$ (6.4.9) introduced in Section 6.4, which characterizes the distinction between the estimator $y_k(t)$ (6.6.3) of the signal $s_k(t)$ in the k-th processing channel and the estimator $y_k(t)\big|_{n(t)=0}$ (6.6.4) of the signal $s_k(t)$ in the absence of interference (noise):

$$\mu_{s_k \hat{s}_k} = \mu(y_t, y_{t,0}) = 2[\mathbf{P}(y_t \vee y_{t,0} > h_k) - \mathbf{P}(y_t \wedge y_{t,0} > h_k)], \tag{6.6.5}$$

where y_t is the sample of the estimator $y_k(t)$ (6.6.3) of the signal $s_k(t)$ in the output of the k-th processing channel at the instant $t = t_0 + T$ in the presence of the signal $s_k(t)$ and in the absence of the signal $s_i(t)$ in the observed process $x(t)$; $y_{t,0}$ is the sample of the estimator $y_k(t)\big|_{n(t)=0}$ (6.6.4) of the signal $s_k(t)$ in the output of the k-th processing channel at the instant $t = t_0 + T$ in the absence of interference (noise) ($n(t) = 0$) and the signal $s_i(t)$ in the observed process $x(t)$, which is equal to a signal energy E_k: $y_{t,0} = E_k$; h_k is some threshold level $h_k < E_k$ determined by an average of two mathematical expectations of the processes in the k-th processing channel (6.4.32) and (6.4.33):

$$h_k = E_k/2. \tag{6.6.6}$$

The equality $y_{t,0} = E_k$ and the inequality $h_k < E_k$ imply that the probabilities appearing in the expression (6.6.5) are equal to:

$$\mathbf{P}(y_t \vee y_{t,0} > h_k) = \mathbf{P}(y_{t,0} > h_k) = 1; \tag{6.6.7a}$$

$$\mathbf{P}(y_t \wedge y_{t,0} > h_k) = \mathbf{P}(y_t > h_k) = 1 - F_{\hat{s}_k}(h_k)\big|_{\theta_k=1}, \tag{6.6.7b}$$

where $F_{\hat{s}_k}(y)\big|_{\theta_k=1}$ is the cumulative distribution function (CDF) of the estimator $y_k(t)$ in the output of detector at the instant $t = t_0 + T$ in the presence of the signal $s_k(t)$ ($\theta_k = 1$) in the observed process $x(t)$ (6.6.2), which, according to the PDF (6.4.34), is equal to:

$$F_{\hat{s}_k}(y)\big|_{\theta_k=1} = (2\pi D_k)^{-1/2} \int_{-\infty}^{y} \exp\left\{-\frac{(x - E_k)^2}{2D_k}\right\} dx, \tag{6.6.8}$$

6.6 Quality Indices of Resolution-Detection in Metric Spaces with L-group Properties

where $D_k = E_k N_0$ is a variance of a noise in the output of the detector; E_k is an energy of the signal $s_k(t)$; N_0 is the power spectral density of interference (noise) in the input of processing unit.

Substituting the formulas (6.6.7) into the expression (6.6.5), we obtain the resultant value for metric $\mu_{s_k \hat{s}_k} = \mu(y_t, y_{t,0})$ between the signal $s_k(t)$ and its estimator $y_k(t)$ while solving the problem of resolution-detection in the group $\Gamma(+)$ of L-group $\Gamma(+, \vee, \wedge)$:

$$\mu_{s_k \hat{s}_k} = \mu(y_t, y_{t,0}) = 2F_{\hat{s}_k}(h_k)|_{\theta_k=1} =$$
$$= 2[1 - \Phi\left(\frac{E_k - h_k}{\sqrt{D_k}}\right)] = 2[1 - \Phi\left(\sqrt{q_k^2/4}\right)], \quad (6.6.9)$$

where $F_{\hat{s}_k}(y)|_{\theta_k=1}$ is the CDF of the estimator $y_k(t)$ in the output of detector at the instant $t = t_0 + T$, in the presence of the signal $s_k(t)$ ($\theta_k = 1$) in the observed process $x(t)$ (6.6.2), which is determined by the formula (6.6.8); $\Phi(z) = (2\pi)^{-1/2} \int_{-\infty}^{z} \exp\left\{-\frac{x^2}{2}\right\} dx$ is the integral of probability; $D_k = E_k N_0$ is a variance of a noise in the k-th processing channel; E_k is an energy of the signal $s_k(t)$; $q_k^2 = E_k/N_0$ is a signal-to-noise ratio in the k-th processing channel; h_k is a threshold level determined by the relationship (6.6.6); N_0 is a power spectral density of interference (noise) in the input of processing unit.

By the analogy with the quality index of signal classification (6.4.16) and quality index of signal detection (6.4.40), we shall determine the quality index of resolution-detection of the signals $\nu_{s_k \hat{s}_k}$.

Definition 6.6.1. By *quality index of estimator of the signals* $\nu_{s_k \hat{s}_k}$, *while solving the problem of resolution-detection* in the presence of the signal $s_k(t)$ and in the absence of the signal $s_i(t)$ in the observed process $x(t)$ ($\theta_k = 1$, $\theta_i = 0$) (1), we mean the normalized measure of statistical interrelationship (NMSI) $\nu(y_t, y_{t,0})$ between the samples $y_{t,0}$ and y_t of stochastic processes $y_k(t)|_{n(t)=0}$ and $y_k(t)$ in the k-th processing channel of a unit of resolution-detection connected with metric $\mu_{s_k \hat{s}_k} = \mu(y_t, y_{t,0})$ (6.6.5) by the following equation:

$$\nu_{s_k \hat{s}_k} = \nu(y_t, y_{t,0}) = 1 - \mu_{s_k \hat{s}_k}. \quad (6.6.10)$$

Substituting the value of metric (6.6.9) into the coupling equation (6.6.10) between the NMSI and metric, we obtain the dependence of quality index of resolution-detection $\nu_{s_k \hat{s}_k}$ on the signal-to-noise ratio:

$$\nu_{s_k \hat{s}_k}(q_k^2) = 2\Phi\left(\sqrt{q_k^2/4}\right) - 1, \quad (6.6.11)$$

where $q_k^2 = E_k/N_0$ is the signal-to-noise ratio in the k-th processing channel.

Consider now the estimator $y_k(t) = \hat{s}_k(t)$ of the signal $s_k(t)$ in the k-th processing channel in the presence of signals $s_k(t)$ and $s_i(t)$ in the observed process $x(t) = s_k(t) + s_i(t) + n(t)$ ($\theta_k = 1$, $\theta_i = 1$):

$$y_k(t) = \hat{s}_k(t)|_{x(t)=s_k(t)+s_i(t)+n(t)}, \quad t = t_0 + T, \quad (6.6.12)$$

and also consider the estimator $y_k(t)\big|_{n(t)=0} = \hat{s}_k(t)\big|_{n(t)=0}$ of the signal $s_k(t)$ in the k-th processing channel in the absence of interference (noise) ($n(t) = 0$) in the observed process $x(t) = s_k(t) + s_i(t)$:

$$y_k(t)\big|_{n(t)=0} = \hat{s}_k(t)\big|_{x(t)=s_k(t)+s_i(t)}, \quad t = t_0 + T. \quad (6.6.13)$$

To determine the quality of the estimator $y_k(t) = \hat{s}_k(t)$ of the signal $s_k(t)$, while solving the problem of resolution-detection of the signals in the group $\Gamma(+)$ of L-group $\Gamma(+, \vee, \wedge)$ in the presence of signals $s_k(t)$ and $s_i(t)$ in the observed process $x(t) = s_k(t) + s_i(t) + n(t)$ ($\theta_k = 1$, $\theta_i = 1$), we use the metric $\mu_{s_k \hat{s}_k}\big|_{\theta_k=1,\theta_i=1} = \mu(y_t, y_{t,0})$ (6.6.5), which characterizes the distinction between the estimator $y_k(t)$ (6.6.12) of the signal $s_k(t)$ in the k-th processing channel and the estimator $y_k(t)\big|_{n(t)=0}$ (6.6.13) of the signal $s_k(t)$ in the absence of interference (noise):

$$\mu_{s_k \hat{s}_k}\big|_{\theta_k=1,\theta_i=1} = \mu(y_t, y_{t,0}) = 2[\mathbf{P}(y_t \vee y_{t,0} > h_k) - \mathbf{P}(y_t \wedge y_{t,0} > h_k)], \quad (6.6.14)$$

where y_t is the sample of the estimator $y_k(t)$ (6.6.12) of the signal $s_k(t)$ in the output of the k-th processing channel at the instant $t = t_0 + T$ in the presence of both the signals $s_k(t)$ and $s_i(t)$ in the observed process $x(t) = s_k(t) + s_i(t) + n(t)$ ($\theta_k = 1$, $\theta_i = 1$); $y_{t,0}$ is the sample of the estimator $y_k(t)\big|_{n(t)=0}$ (6.6.13) of the signal $s_k(t)$ in the output of the k-th processing signal at the instant $t = t_0 + T$ in the absence of interference (noise) ($n(t) = 0$) in the observed process $x(t) = s_k(t) + s_i(t)$, which is equal to the sum of the energy E_k of the signal $s_k(t)$ and mutual energy of both the signals $E_{ik} = r_{ik}\sqrt{E_i E_k}$: $y_{t,0} = E_k + E_{ik}$; h_k is some threshold level, $h_k < E_k$ determined by an average of two mathematical expectations of the processes in the k-th processing channel (6.4.32) and (6.4.33):

$$h_k = E_k/2. \quad (6.6.15)$$

The equality $y_{t,0} = E_k + E_{ik}$ and the inequality $h_k < E_k$ imply that the probabilities in the expression (6.6.14) are equal to:

$$\mathbf{P}(y_t \vee y_{t,0} > h_k) = \mathbf{P}(y_{t,0} > h_k) = 1; \quad (6.6.16a)$$

$$\mathbf{P}(y_t \wedge y_{t,0} > h_k) = \mathbf{P}(y_t > h_k) = 1 - F_{\hat{s}_k}(h_k)\big|_{\theta_k=1,\theta_i=1}, \quad (6.6.16b)$$

where $F_{\hat{s}_k}(y)\big|_{\theta_k=1,\theta_i=1}$ is the CDF of the estimator $y_k(t)$ in the k-th processing channel at the instant $t = t_0 + T$ in the presence of both the signals $s_k(t)$ and $s_i(t)$ in the observed process $x(t) = s_k(t) + s_i(t) + n(t)$ ($\theta_k = 1$, $\theta_i = 1$) (6.6.2), which, according to PDF (6.4.34), is equal to:

$$F_{\hat{s}_k}(y)\big|_{\theta_k=1,\theta_i=1} = (2\pi D_k)^{-1/2} \int_{-\infty}^{y} \exp\left\{-\frac{[x - (E_k + E_{ik})]^2}{2D_k}\right\} dx, \quad (6.6.17)$$

where $D_k = E_k N_0$ is a variance of a noise in the k-th processing channel; E_k is an energy of the signal $s_k(t)$; $E_{ik} = r_{ik}\sqrt{E_i E_k}$ is a mutual energy of the signals

6.6 Quality Indices of Resolution-Detection in Metric Spaces with L-group Properties

$s_k(t)$ and $s_i(t)$; N_0 is a power spectral density of interference (noise) in the input of processing unit.

Substituting the formulas (6.6.16) into the expression (6.6.14), we obtain the resultant value for metric $\mu_{s_k \hat{s}_k}|_{\theta_k=1,\theta_i=1} = \mu(y_t, y_{t,0})$ between the signal $s_k(t)$ and its estimator $y_k(t)$ while solving the problem of resolution-detection in the group $\Gamma(+)$ of L-group $\Gamma(+,\vee,\wedge)$:

$$\mu_{s_k \hat{s}_k}|_{\theta_k=1,\theta_i=1} = \mu(y_t, y_{t,0}) = 2F_{\hat{s}_k}(h_k)|_{\theta_k=1,\theta_i=1} =$$

$$= 2[1 - \Phi\left(\frac{(E_k + E_{ik}) - h_k}{\sqrt{D_k}}\right)] = 2[1 - \Phi(0.5q_k + r_{ik}q_i)], \qquad (6.6.18)$$

where $F_{\hat{s}_k}(y)|_{\theta_k=1,\theta_i=1}$ is the CDF of the estimator $y_k(t)$ in the k-th processing channel at the instant $t = t_0 + T$ in the presence of both the signals $s_k(t)$ and $s_i(t)$ in the observed process $x(t) = s_k(t) + s_i(t) + n(t)$ ($\theta_k = 1$, $\theta_i = 1$) (6.6.2), which is determined by the formula (6.6.17); $\Phi(z) = (2\pi)^{-1/2} \int\limits_{-\infty}^{z} \exp\left\{-\frac{x^2}{2}\right\} dx$ is the integral of probability; $D_k = E_k N_0$ is a variance of a noise in the k-th processing channel; E_k is an energy of the signal $s_k(t)$; $E_{ik} = r_{ik}\sqrt{E_i E_k}$ is mutual energy of the signals $s_k(t)$ and $s_i(t)$; r_{ik} is the cross-correlation coefficient of the signals $s_k(t)$ and $s_i(t)$; q_k^2 is signal-to-noise ratio in the k-th processing channel; q_i^2 is signal-to-noise ratio in the i-th processing channel; h_k is a threshold level determined by the relationship (6.6.15); N_0 is a power spectral density of interference (noise) in the input of processing unit.

By analogy with the quality index of resolution-detection of the signals (6.6.10), in the presence of the signal $s_k(t)$ and the absence of the signal $s_i(t)$ in the observed process $x(t) = s_k(t) + n(t)$ ($\theta_k = 1$, $\theta_i = 0$), we define the quality index of resolution-detection of the signals $\nu_{s_k \hat{s}_k}$ in the presence of both signals $s_k(t)$ and $s_i(t)$ in the observed process $x(t) = s_k(t) + s_i(t) + n(t)$ ($\theta_k = 1$, $\theta_i = 1$).

Definition 6.6.2. By *quality index of estimator of the signals* $\nu_{s_k \hat{s}_k}|_{\theta_k=1,\theta_i=1}$, *while solving the problem of resolution-detection* in the presence of signals $s_k(t)$ and $s_i(t)$ in the observed process $x(t) = s_k(t) + s_i(t) + n(t)$ ($\theta_k = 1$, $\theta_i = 1$) (6.6.1), we mean the NMSI $\nu(y_t, y_{t,0})$ between the samples $y_{t,0}$ and y_t of stochastic process $y_k(t)|_{n(t)=0}$, $y_k(t)$ in the k-th processing channel of the resolution-detection unit connected with metric $\mu_{s_k \hat{s}_k}|_{\theta_k=1,\theta_i=1} = \mu(y_t, y_{t,0})$ (6.6.14) by the following expression:

$$\nu_{s_k \hat{s}_k}|_{\theta_k=1,\theta_i=1} = \nu(y_t, y_{t,0}) = 1 - \mu_{s_k \hat{s}_k}|_{\theta_i=1,\theta_k=1}. \qquad (6.6.19)$$

Substituting the value of metric (6.6.18) into the coupling equation (6.6.19), we obtain the dependence of quality index of resolution-detection $\nu_{s_k \hat{s}_k}(q_k^2, q_i^2)|_{\theta_k=1,\theta_i=1}$ on the signal-to-noise ratio in the processing channel:

$$\nu_{s_k \hat{s}_k}(q_k^2, q_i^2)|_{\theta_k=1,\theta_i=1} = 2\Phi(0.5q_k + r_{ik}q_i) - 1, \qquad (6.6.20)$$

where r_{ik} is the cross-correlation coefficient of the signals $s_k(t)$ and $s_i(t)$; q_k^2, q_i^2 are signal-to-noise ratios in the k-th and the i-th processing channels, respectively.

From comparative analysis of quality indices of resolution-detection of the signals (6.6.11) and (6.6.20) in signal space with properties of the group $\Gamma(+)$ of L-group $\Gamma(+, \vee, \wedge)$, depending on the values of the cross-correlation coefficient r_{ik} between the signals $s_k(t)$ and $s_i(t)$, the following relationships hold:

$$\nu_{s_k \hat{s}_k}(q_k^2) \geq \nu_{s_k \hat{s}_k}(q_k^2, q_i^2)|_{\theta_k=1, \theta_i=1}, \text{ on } r_{ik} \leq 0;$$

$$\nu_{s_k \hat{s}_k}(q_k^2) \leq \nu_{s_k \hat{s}_k}(q_k^2, q_i^2)|_{\theta_k=1, \theta_i=1}, \text{ on } r_{ik} \geq 0.$$

Within the framework of the general model (6.6.1), consider the case of interaction between the signals $s_{i,k}(t)$ from a set $S = \{s_i(t), s_k(t)\}$ and interference (noise) $n(t)$ in the signal space with the properties of the lattice $\Gamma(\vee, \wedge)$ of L-group $\Gamma(+, \vee, \wedge)$:

$$x(t) = \theta_k s_k(t) \vee \theta_i s_i(t) \vee n(t), \; t \in T_s, \tag{6.6.21}$$

where $\vee$ and $\wedge$ are binary operations of join and meet of the lattice $\Gamma(\vee, \wedge)$ of L-group $\Gamma(+, \vee, \wedge)$, respectively; $\theta_{i,k}$ is a random parameter taking values from the set $\{0, 1\}$: $\theta_{i,k} \in \{0, 1\}$; $T_s = [t_0, t_0 + T]$ is a domain of definition of the signal $s_{i,k}(t)$; t_0 is known time of arrival of the signal $s_{i,k}(t)$; T is a duration of the signal $s_{i,k}(t)$.

Let the signals from a set $S = \{s_i(t), s_k(t)\}$ be characterized by the energies $E_{i,k} = \int\limits_{t \in T_s} s_{i,k}^2(t) dt$ and cross-correlation coefficient r_{ik}:

$$r_{ik} = \int\limits_{t \in T_s} s_i(t) s_k(t) dt / \sqrt{E_i E_k}.$$

Suppose also that interference (noise) $n(t)$ is characterized by arbitrary probabilistic-statistical properties.

Determine the quality indices of resolution-detection of the signals introduced by the Definitions 6.6.1, 6.6.2 and by the expressions (6.6.10), (6.6.19), respectively, while solving the problem of resolution-detection in signal space with the properties of the lattice $\Gamma(\vee, \wedge)$ of L-group $\Gamma(+, \vee, \wedge)$.

Consider the estimator $y_k(t)$ of the signal $s_k(t)$ in the k-th processing channel of a resolution-detection unit in the presence of the signal $s_k(t)$ and in the absence of the signal $s_i(t)$ in the observed process $x(t) = s_k(t) \vee 0 \vee n(t)$ ($\theta_k = 1, \theta_i = 0$) (6.6.21):

$$y_k(t) = y_k(t)\big|_{x(t)=s_k(t) \vee 0 \vee n(t)} = s_k(t) \wedge x(t), \tag{6.6.22}$$

and also consider the estimator $y_k(t)\big|_{n(t)=0}$ of the signal $s_k(t)$ in the k-th processing channel in the absence of interference (noise) ($n(t) = 0$) and the signal $s_i(t)$ in the observed process $x(t) = s_k(t) \vee 0$:

$$y_k(t)\big|_{n(t)=0} = s_k(t) \wedge x(t)\big|_{x(t)=s_k(t) \vee 0}. \tag{6.6.23}$$

Determine the value of function (6.6.5), which, if the signal space is lattice $\Gamma(\vee, \wedge)$, is also metric, and then determine the quality index of the estimator of the signals (6.6.10) while solving the problem of resolution-detection in signal space with the properties of the lattice $\Gamma(\vee, \wedge)$ of L-group $\Gamma(+, \vee, \wedge)$:

$$\mu_{s_k \hat{s}_k} = \mu(y_t, y_{t,0}) = 2[\mathbf{P}(y_t \vee y_{t,0} > h) - \mathbf{P}(y_t \wedge y_{t,0} > h)], \tag{6.6.24}$$

6.6 Quality Indices of Resolution-Detection in Metric Spaces with L-group Properties

where y_t is the sample of the estimator $y_k(t)$ (6.6.22) of the signal $s_k(t)$ in the output of the k-th processing channel at the instant $t \in T_s$ in the presence of the signal $s_k(t)$ and in the absence of the signal $s_i(t)$ in the observed process $x(t) = s_k(t) \vee 0 \vee n(t)$ ($\theta_k = 1$, $\theta_i = 0$) (6.6.21); $y_{t,0}$ is the sample of the estimator $y_k(t)|_{n(t)=0}$ (6.6.23) of the signal $s_k(t)$ in the output of the k-th processing channel at the instant $t \in T_s$ in the absence of interference (noise) $n(t) = 0$ and the signal $s_i(t)$ in the observed process $x(t) = s_k(t) \vee 0$; h is some threshold level determined by energetic and correlation relations between the signals $s_k(t)$ and $s_i(t)$.

The absorption axiom of the lattice $\Gamma(\vee, \wedge)$ of L-group $\Gamma(+, \vee, \wedge)$ contained in the third part of each of two multilink identities implies that the estimators $y_k(t)$ (6.6.22) and $y_k(t)|_{n(t)=0}$ (6.6.23) are identically equal to the received useful signal $s_k(t)$:

$$y_k(t) = s_k(t) \wedge x(t) = s_k(t) \wedge [s_k(t) \vee 0 \vee n(t)] = s_k(t); \qquad (6.6.25a)$$

$$y_k(t)|_{n(t)=0} = s_k(t) \wedge x(t) = s_k(t) \wedge [s_k(t) \vee 0] = s_k(t). \qquad (6.6.25b)$$

The obtained relationships imply that the sample y_t of the estimator $y_k(t)$ (6.6.22) and also the sample $y_{t,0}$ of the estimator $y_k(t)|_{n(t)=0}$ (6.6.23) are identically equal to the received signal $s_k(t)$:

$$y_t = y_k(t) = s_k(t); \qquad (6.6.26a)$$

$$y_{t,0} = y_k(t)|_{n(t)=0} = s_k(t). \qquad (6.6.26b)$$

Substituting the obtained values of the samples y_t (6.6.26a) and $y_{t,0}$ (6.6.26b) into the relationship (6.6.24), the value of metric $\mu_{s_k \hat{s}_k} = \mu(y_t, y_{t,0})$ between the signal $s_k(t)$ and its estimator $y_k(t)$ while solving the problem of resolution-detection of the signals in signal space with the properties of the lattice $\Gamma(\vee, \wedge)$ of L-group $\Gamma(+, \vee, \wedge)$ is equal to zero:

$$\mu_{s_k \hat{s}_k} = \mu(y_t, y_{t,0}) = 0. \qquad (6.6.27)$$

From (6.6.27) and the coupling equation (6.6.10), we obtain quality indices of the estimator of the signals $\nu_{s_k \hat{s}_k}$, while solving the problem of their resolution-detection in signal space with the properties of the lattice $\Gamma(\vee, \wedge)$ of L-group $\Gamma(+, \vee, \wedge)$ in the presence of the signal $s_k(t)$ and in the absence of the signal $s_i(t)$ in the observed process $x(t) = s_k(t) \vee 0 \vee n(t)$ ($\theta_k = 1$, $\theta_i = 0$), that take absolute values equal to 1:

$$\nu_{s_k \hat{s}_k} = 1 - \mu_{s_k \hat{s}_k} = 1. \qquad (6.6.28)$$

Consider now the estimator $y_k(t)$ of the signal $s_k(t)$ in the k-th processing channel of the resolution-detection unit in the presence of signals $s_k(t)$ and $s_i(t)$ in the observed process $x(t) = s_k(t) \vee s_i(t) \vee n(t)$ ($\theta_k = 1$, $\theta_i = 1$) (6.6.21):

$$y_k(t) = y_k(t)|_{x(t) = s_k(t) \vee s_i(t) \vee n(t)} = s_k(t) \wedge x(t), \qquad (6.6.29)$$

and also consider the estimator $y_k(t)|_{n(t)=0}$ of the signal $s_k(t)$ in the k-th processing channel in the absence of interference (noise) ($n(t) = 0$) in the observed process $x(t) = s_k(t) \vee s_i(t) \vee 0$:

$$y_k(t)\big|_{n(t)=0} = s_k(t) \wedge x(t)\big|_{x(t)=s_k(t)\vee s_i(t)\vee 0}. \qquad (6.6.30)$$

To determine the quality of the estimator $y_k(t) = \hat{s}_k(t)$ of the signal $s_k(t)$ while solving the problem of resolution-detection of the signals in signal space with the properties of the lattice $\mathbf{\Gamma}(\vee, \wedge)$ of L-group $\mathbf{\Gamma}(+, \vee, \wedge)$ in the presence of both the signals $s_k(t)$ and $s_i(t)$ in the observed process $x(t) = s_k(t) \vee s_i(t) \vee n(t)$ ($\theta_k = 1, \theta_i = 1$), we will use the metric $\mu_{s_k \hat{s}_k}\big|_{\theta_k=1, \theta_i=1} = \mu(y_t, y_{t,0})$ (6.6.24), which characterizes the distinctions between the estimator $y_k(t)$ (6.6.29) of the signal $s_k(t)$ in the k-th processing channel and the estimator $y_k(t)\big|_{n(t)=0}$ (6.6.30) of the signal $s_k(t)$ in the absence of interference (noise):

$$\mu_{s_k \hat{s}_k}\big|_{\theta_k=1, \theta_i=1} = \mu(y_t, y_{t,0}) = 2[\mathbf{P}(y_t \vee y_{t,0} > h) - \mathbf{P}(y_t \wedge y_{t,0} > h)], \qquad (6.6.31)$$

where y_t is the sample of the estimator $y_k(t)$ (6.6.29) of the signal $s_k(t)$ in the output of the k-th processing channel at the instant $t \in T_s$ in the presence of the signals $s_k(t)$ and $s_i(t)$ in the observed process $x(t) = s_k(t) \vee s_i(t) \vee n(t)$ ($\theta_k = 1, \theta_i = 1$) (6.6.21); $y_{t,0}$ is the sample of the estimator $y_k(t)\big|_{n(t)=0}$ (6.6.30) of the signal $s_k(t)$ in the output of the k-th processing channel at the instant $t \in T_s$ in the absence of interference (noise) ($n(t) = 0$) in the observed process $x(t) = s_k(t) \vee s_i(t) \vee 0$; h is some threshold level determined by energetic and correlation relations between the signals $s_k(t)$ and $s_i(t)$.

The absorption axiom of the lattice $\mathbf{\Gamma}(\vee, \wedge)$ contained in the third part of each of two multilink identities implies that the estimators $y_k(t)$ (6.6.29) and $y_k(t)\big|_{n(t)=0}$ (6.6.30) are identically equal to the received useful signal $s_k(t)$:

$$y_k(t) = s_k(t) \wedge x(t) = s_k(t) \wedge [s_k(t) \vee s_i(t) \vee n(t)] = s_k(t); \qquad (6.6.32a)$$

$$y_k(t)\big|_{n(t)=0} = s_k(t) \wedge x(t) = s_k(t) \wedge [s_k(t) \vee s_i(t) \vee 0] = s_k(t). \qquad (6.6.32b)$$

The obtained relationships imply that the sample y_t of the estimator $y_k(t)$ (6.6.29), and also the sample $y_{t,0}$ of the estimator $y_k(t)\big|_{n(t)=0}$ (6.6.30) are identically equal to the received useful signal $s_k(t)$:

$$y_t = y_k(t) = s_k(t); \qquad (6.6.33a)$$

$$y_{t,0} = y_k(t)\big|_{n(t)=0} = s_k(t). \qquad (6.6.33b)$$

Substituting the obtained values of the samples y_t (6.6.33a), $y_{t,0}$ (6.6.33b) into the relationship (6.6.31), the value of metric $\mu_{s_k \hat{s}_k}\big|_{\theta_k=1, \theta_i=1} = \mu(y_t, y_{t,0})$ between the signal $s_k(t)$ and its estimator $y_k(t)$ while solving the problem of resolution-detection of the signal in signal space with the properties of the lattice $\mathbf{\Gamma}(\vee, \wedge)$ of L-group $\mathbf{\Gamma}(+, \vee, \wedge)$ is equal to zero:

$$\mu_{s_k \hat{s}_k}\big|_{\theta_k=1, \theta_i=1} = \mu(y_t, y_{t,0}) = 0. \qquad (6.6.34)$$

The relationship (6.6.34) and the coupling equation (6.6.28) imply that quality indices of the estimator of the signals $\nu_{s_k \hat{s}_k}|_{\theta_k=1,\theta_i=1}$ while solving the problem of their resolution-detection in signal space with the properties of the lattice $\mathbf{\Gamma}(\vee,\wedge)$ of L-group $\mathbf{\Gamma}(+,\vee,\wedge)$ in the presence of both the signals $s_k(t)$ and $s_i(t)$ in the observed process $x(t) = s_k(t) \vee s_i(t) \vee n(t)$ ($\theta_k = 1$, $\theta_i = 1$), take absolute values that are equal to 1:

$$\nu_{s_k \hat{s}_k}|_{\theta_k=1,\theta_i=1} = 1 - \mu_{s_k \hat{s}_k}|_{\theta_k=1,\theta_i=1} = 1. \quad (6.6.35)$$

As follows from the comparative analysis of quality indices of signal resolution-detection (6.6.28), (6.6.35) in signal space with the properties of the lattice $\mathbf{\Gamma}(\vee,\wedge)$ of L-group $\mathbf{\Gamma}(+,\vee,\wedge)$, regardless of cross-correlation coefficient r_{ik} and energetic relations between the signals $s_k(t)$ and $s_i(t)$, the following identity holds:

$$\nu_{s_k \hat{s}_k}(q_k^2) = \nu_{s_k \hat{s}_k}(q_k^2, q_i^2)|_{\theta_k=1,\theta_i=1} = 1.$$

It fundamentally differs them from the quality indices of signal resolution-detection (6.6.11) and (6.6.20) in signal space with the properties of the group $\mathbf{\Gamma}(+)$ of L-group $\mathbf{\Gamma}(+,\vee,\wedge)$, which are essentially determined by both cross-correlation coefficient r_{ik} and energetic relations between the signals $s_k(t)$ and $s_i(t)$.

The relationships (6.6.28) and (6.6.35) imply an important conclusion. While analyzing the problem of signal resolution-detection in signal space with the properties of the lattice $\mathbf{\Gamma}(\vee,\wedge)$ of L-group $\mathbf{\Gamma}(+,\vee,\wedge)$ we face a possibility of providing absolute quality indices of signal resolution-detection $\nu_{s_k \hat{s}_k}$, $\nu_{s_k \hat{s}_k}|_{\theta_k=1,\theta_i=1}$. This creates the necessary conditions for resolution-detection of deterministic signals processed under interference (noise) background without losses of information. Note that this situation stipulated by the absence of information losses is not typical for solving a similar problem in signal space with group properties (particularly in linear signal space). One more advantage of signal resolution-detection in signal spaces with lattice properties is the invariance property of both metrics (6.6.27) and (6.6.34) and the quality indices of signal resolution-detection (6.6.28) and (6.6.35) with respect to parametric and nonparametric prior uncertainty conditions. The quality indices of signal resolution-detection in signal space with lattice properties (6.6.28) and (6.6.35), as against quality indices of signal resolution-detection in linear signal space (6.6.11) and (6.6.20), do not depend on signal-to-noise ratio (signal-to-interference ratio) and interference (noise) distribution in the input of the processing unit. The problem of synthesis of optimal signal resolution algorithm in signal space with lattice properties demands additional investigation that will be discussed in the next chapter.

6.7 Quality Indices of Resolution-Estimation in Metric Spaces with Lattice Properties

The problem of resolution-detection-estimation (i.e., joint resolution, detection, and estimation) requires us to establish: (1) is there the useful signal $s(t,\lambda)$ with an unknown nonrandom parameter λ in the input of a processing unit and (2) what is the value of parameter λ, so that both tasks are solved in the presence of interfering signal $s'(t, \lambda')$ and interference (noise) $n(t)$. Furthermore, the interfering signal $s'(t, \lambda')$ is a copy of a useful signal with an unknown parameter λ', which differs from λ: $\lambda' \neq \lambda$.

Consider the general model of interaction of the signals $s(t, \lambda)$ and $s'(t, \lambda')$ and interference (noise) $n(t)$ in signal space with the properties of the lattice $\mathbf{\Gamma}(\vee, \wedge)$:

$$x(t) = \theta s(t, \lambda) \oplus \theta' s'(t, \lambda') \oplus n(t), \ t \in T_{\text{obs}}, \tag{6.7.1}$$

where $\oplus$ is some binary operation of the lattice $\mathbf{\Gamma}(\vee, \wedge)$; θ, θ' is a random parameter that takes the values from the set $\{0, 1\}$: $\theta, \theta' \in \{0, 1\}$; T_{obs} is an observation interval, $T_{\text{obs}} = T_s \cup T'_s$; $T_s = [t'_0, t'_0 + T]$, $T'_s = [t'_0, t'_0 + T]$ are the domains of definitions of the signals $s(t, \lambda)$ and $s'(t, \lambda')$, respectively; t_0 and t'_0 are unknown arrival times of the signals $s(t, \lambda)$ and $s'(t, \lambda')$; T is a duration of the signals $s(t, \lambda)$ and $s'(t, \lambda')$.

Let the signals $s(t, \lambda)$ and $s'(t, \lambda')$ be periodic functions with a period T_0:

$$s(t, \lambda) = s(t + T_0, \lambda); \quad s'(t, \lambda') = s'(t + T_0, \lambda'), \tag{6.7.2}$$

and they are connected by the proportionality relation:

$$s'(t, \lambda') = as(t + \tau, \lambda + \Delta\lambda), \tag{6.7.3}$$

where a, τ, and $\Delta\lambda$ are some constants: a=const, τ=const, $\Delta\lambda$=const.

Consider also that interference $n(t)$ is quasi-white Gaussian noise with power spectral density N_0 and upper bound frequency F_n: $F_n \gg 1/T_0$.

The problem of joint resolution-detection-estimation can be formulated by testing statistical hypothesis $H_{\theta\theta'}$ concerning the presence (absence) of useful $s(t, \lambda)$ and interfering $s'(t, \lambda')$ signals observed under interference (noise) background. If either the hypothesis H_{11} or the hypothesis H_{10} holds ($\theta = 1$, i.e., in the observed process $x(t)$, the useful signal $s(t, \lambda)$ is present, whereas the interfering signal $s'(t, \lambda')$ can be present or absent: $\theta' \in \{0, 1\}$), then on the base of the relationship (6.1.7a), the estimator $\hat{\lambda}$ of a parameter λ is formed:

$$\hat{\lambda} = M^{-1}[c(t), \hat{s}(t)]; \tag{6.7.4a}$$

$$\hat{s}(t) = F_{\hat{s}}[x(t)], \tag{6.7.4b}$$

where $M[*, *]$ is a deterministic one-to-one function (modulating function); $c(t)$ is a carrier signal; $\hat{s}(t)$ is the estimator of useful signal $s(t, \lambda)$; $F_{\hat{s}}[*]$ is the estimator forming function.

The goal of this section is establishing the main potential characteristics of signal resolution-estimation in signal space with lattice properties on the basis of the estimators considered in Section 6.3.

The initial general model of signal interaction (6.6.1) in signal space with lattice properties can be written as two separate models, each of them is determined by the lattice operation of join and meet, respectively:

$$x(t) = \theta s(t, \lambda) \vee \theta' s'(t, \lambda') \vee n(t); \tag{6.7.5a}$$

6.7 Quality Indices of Resolution-Estimation in Metric Spaces with Lattice Properties

$$\tilde{x}(t) = \theta s(t, \lambda) \wedge \theta' s'(t, \lambda') \wedge n(t), \quad (6.7.5b)$$

where $t \in T_{\text{obs}}$; T_{obs} is an observation interval.

While formulating the problem of joint resolution-detection-estimation for interactions (6.7.5a), (6.7.5b), we consider two hypothesis $H_{11} = H_{\theta=1,\theta'=1}$, $H_{01} = H_{\theta=0,\theta'=1}$ concerning the presence of interfering signal $s'(t, \lambda')$ ($\theta' = 1$) and also the presence or absence of useful signal $s(t, \lambda)$ ($\theta \in \{0, 1\}$):

$$H_{11}: \quad x(t) = s(t, \lambda) \vee s'(t, \lambda') \vee n(t); \quad (6.7.6a)$$

$$H_{01}: \quad x(t) = 0 \vee s'(t, \lambda') \vee n(t); \quad (6.7.6b)$$

$$H_{11}: \quad \tilde{x}(t) = s(t, \lambda) \wedge s'(t, \lambda') \wedge n(t); \quad (6.7.7a)$$

$$H_{01}: \quad \tilde{x}(t) = 0 \wedge s'(t, \lambda') \wedge n(t), \quad (6.7.7b)$$

where $t \in T_{\text{obs}}$; T_{obs} is an observation interval; 0 is a null (neutral) element of the group $\mathbf{\Gamma}(+)$ of L-group $\mathbf{\Gamma}(+, \vee, \wedge)$.

As the estimator $\hat{s}_\wedge(t)$ of useful signal $s(t, \lambda)$ for the model (6.7.5a), we take the meet of lattice (6.3.3):

$$\hat{s}_\wedge(t) = \bigwedge_{j=1}^{N} x(t - jT_0), \quad (6.7.8)$$

where N is a number of periods of a carrier of useful signal $s(t, \lambda)$ in its domain of definition $T_s = [t_0, t_0 + T]$.

As the estimator $\hat{s}_\vee(t)$ of useful signal $s(t, \lambda)$ for the model (6.7.5b), we take the join of lattice (6.3.4):

$$\hat{s}_\vee(t) = \bigvee_{j=1}^{N} x(t - jT_0). \quad (6.7.9)$$

The questions dealing with obtaining the estimators $\hat{s}_\wedge(t)$ (6.7.8) and $\hat{s}_\vee(t)$ (6.7.9) on the basis of optimality criteria will be considered in Chapter 7.

Let some functions T_α and T_β of the observed process $x(t)$ be used as the estimators $\hat{s}_\alpha(t)$ and $\hat{s}_\beta(t)$ of the signals $s_\alpha(t)$ and $s_\beta(t)$ in signal space $\mathbf{\Gamma}(\oplus)$:

$$\hat{s}_\alpha(t) = T_\alpha[x(t)]; \quad (6.7.10a)$$

$$\hat{s}_\beta(t) = T_\beta[x(t)], \quad (6.7.10b)$$

where $x(t) = \theta_\alpha s_\alpha(t) \oplus \theta_\beta s_\beta(t) \oplus n(t)$; $\oplus$ is a binary operation in signal space $\mathbf{\Gamma}(\oplus)$; θ_α and θ_β are random parameters taking the values from a set $\{0, 1\}$: $\theta_\alpha, \theta_\beta \in \{0, 1\}$.

Let also the estimators $\hat{s}_\alpha(t)$ and $\hat{s}_\beta(t)$ of the signals $s_\alpha(t)$ and $s_\beta(t)$ be characterized by PDFs $p_{\hat{s}_\alpha}(z)$ and $p_{\hat{s}_\beta}(z)$, respectively.

Definition 6.7.1. A set $S_{\alpha\beta}$ of pairs of the estimators $\{\hat{s}_\alpha(t), \hat{s}_\beta(t)\}$ of the corresponding pairs of signals $\{s_\alpha(t), s_\beta(t)\}$ and $\{\hat{s}_\alpha(t), \hat{s}_\beta(t)\} \subset S_{\alpha\beta}$, $\alpha, \beta \in A$ is called an *estimator space* $(S_{\alpha\beta}, d_{\alpha\beta})$ with metric $d_{\alpha\beta}$:

$$d_{\alpha\beta} = d(\hat{s}_\alpha, \hat{s}_\beta) = \frac{1}{2} \int_{-\infty}^{\infty} |p_{\hat{s}_\alpha}(z) - p_{\hat{s}_\beta}(z)| dz, \quad (6.7.11)$$

where $d_{\alpha\beta}$ is a metric between PDFs $p_{\hat{s}_\alpha}(z)$ and $p_{\hat{s}_\beta}(z)$ of the estimators $\hat{s}_\alpha(t)$ and $\hat{s}_\beta(t)$ of the signals $s_\alpha(t)$ and $s_\beta(t)$ in signal space $\mathbf{\Gamma}(\oplus)$, respectively.

To characterize the quality of the estimators $\hat{s}_\alpha(t)$ and $\hat{s}_\beta(t)$ while solving the problem of joint resolution-estimation, we introduce the index based upon the metric (6.7.11).

Definition 6.7.2. By *quality index* $q_{\leftrightarrow}(\hat{s}_\alpha, \hat{s}_\beta)$ *of joint resolution-estimation of the signals* $s_\alpha(t)$ *and* $s_\beta(t)$, we mean the quantity equal to metric (6.7.11) between PDFs $p_{\hat{s}_\alpha}(z)$ and $p_{\hat{s}_\beta}(z)$ of the estimators $\hat{s}_\alpha(t)$ and $\hat{s}_\beta(t)$ of these signals:

$$q_{\leftrightarrow}(\hat{s}_\alpha, \hat{s}_\beta) = d(\hat{s}_\alpha, \hat{s}_\beta) = \frac{1}{2}\int_{-\infty}^{\infty} |p_{\hat{s}_\alpha}(z) - p_{\hat{s}_\beta}(z)| dz. \qquad (6.7.12)$$

Determine now the quality indices $q_{\leftrightarrow}(\hat{s}_\wedge|_{H_{11}}, \hat{s}_\wedge|_{H_{01}})$, $q_{\leftrightarrow}(\hat{s}_\vee|_{H_{11}}, \hat{s}_\vee|_{H_{01}})$ of joint resolution-estimation (6.7.12) for the estimators $\hat{s}_\wedge(t)$ (6.7.8) and $\hat{s}_\vee(t)$ (6.7.9) in the case of separate fulfillment of the hypotheses $H_{11} = H_{\theta=1,\theta'=1}$ and $H_{01} = H_{\theta=0,\theta'=1}$ (6.7.6) and (6.7.7), respectively.

By differentiating CDF (6.3.8c), under the condition that between the samples of instantaneous values s_t, s'_t of useful $s(t,\lambda)$ and interfering $s'(t,\lambda')$ signals, the two-sided inequality holds $0 < s'_t < s_t$, we obtain the PDF of the estimator $\hat{s}_\wedge(t)$:

$$p_{\hat{s}_\wedge}(z)|_{H_{11}} = \delta(z - s_t)[1 - (1 - F_n(s_t + 0))^N] + \\ + 1(z - s_t)N(1 - F_n(z))^{N-1}p_n(z); \qquad (6.7.13a)$$

$$p_{\hat{s}_\wedge}(z)|_{H_{01}} = \delta(z - s'_t)[1 - (1 - F_n(s'_t + 0))^N] + \\ + 1(z - s'_t)N(1 - F_n(z))^{N-1}p_n(z), \qquad (6.7.13b)$$

where s_t and s'_t are the samples of instantaneous values of the signals $s(t,\lambda)$ and $s'(t,\lambda')$, respectively: $s_t = s(t,\lambda)$, $s'_t = s'(t,\lambda')$, $t \in T_{\mathrm{obs}}$; $F_n(z)$, $p_n(z)$ are the univariate CDF and PDF of interference (noise), respectively; $\delta(t)$ is the Dirac delta function; $1(t)$ is the Heaviside step function; N is a number of periods of a carrier of useful signal $s(t,\lambda)$ on its domain of definition $T_s = [t_0, t_0 + T]$.

Similarly, differentiating CDF (6.3.8d), under the condition that between the samples of instantaneous values s_t, s'_t of useful $s(t,\lambda)$ and interfering $s'(t,\lambda')$ signals, the two-sided inequality holds $s_t < s'_t < 0$, we obtain the PDF of the estimator $\hat{s}_\vee(t)$:

$$p_{\hat{s}_\vee}(z)|_{H_{11}} = \delta(z - s_t)F_n^N(s_t - 0) + [1 - 1(z - s_t)]NF_n^{N-1}(z)p_n(z); \qquad (6.7.14a)$$

$$p_{\hat{s}_\vee}(z)|_{H_{01}} = \delta(z - s'_t)F_n^N(s'_t - 0) + [1 - 1(z - s'_t)]NF_n^{N-1}(z)p_n(z), \qquad (6.7.14b)$$

On the qualitative level, the forms of PDFs (6.7.13a,b) and (6.7.14a,b) of the estimators $\hat{s}_\wedge(t)|_{H_{11},H_{01}}$, $\hat{s}_\vee(t)|_{H_{11},H_{01}}$, in the case of separate fulfillment of the hypotheses H_{11}, H_{01} on $N = 3$, are shown in Fig. 6.7.1(a) and Fig. 6.7.1(b), respectively.

Substituting the values of PDFs of the estimators $\hat{s}_\wedge(t)|_{H_{11},H_{01}}$ (6.7.13a) and

6.7 Quality Indices of Resolution-Estimation in Metric Spaces with Lattice Properties

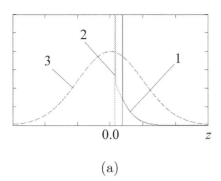
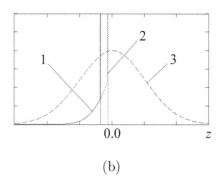

(a) (b)

FIGURE 6.7.1 PDFs of estimators: (a) $\hat{s}_\wedge(t)|_{H_{11},H_{01}}$; 1 = PDF $p_{\hat{s}_\wedge}(z)|_{H_{11}}$ (6.7.13a); 2 = PDF $p_{\hat{s}_\wedge}(z)|_{H_{01}}$ (6.7.13b); 3 = PDF of interference (noise) $p_n(z)$; (b) $\hat{s}_\vee(t)|_{H_{11},H_{01}}$. 1 = PDF $p_{\hat{s}_\vee}(z)|_{H_{11}}$ (6.7.14a); 2 = PDF $p_{\hat{s}_\vee}(z)|_{H_{01}}$ (6.7.14b); 3 = PDF of interference (noise) $p_n(z)$

(6.7.13b) into the formula (6.7.12), we obtain the quality index of joint resolution-estimation $q_\leftrightarrow(\hat{s}_\wedge|_{H_{11}}, \hat{s}_\wedge|_{H_{01}})$:

$$q_\leftrightarrow(\hat{s}_\wedge|_{H_{11}}, \hat{s}_\wedge|_{H_{01}}) = \frac{1}{2}\int_{-\infty}^{\infty} |p_{\hat{s}_\wedge}(z)|_{H_{11}} - p_{\hat{s}_\wedge}(z)|_{H_{01}}|dz =$$

$$= 1 - \int_{s_t}^{\infty} p_{\hat{s}_\wedge}(z)|_{H_{11}}\, dz = 1 - \int_{s_t}^{\infty} N(1-F_n(z))^{N-1}p_n(z)dz, \quad s_t \geq 0. \quad (6.7.15)$$

Similarly, substituting the values of PDFs of the estimators $\hat{s}_\vee(t)|_{H_{11},H_{01}}$ (6.7.14a,b) into the expression (6.7.12), we obtain the quality index of joint resolution-estimation $q_\leftrightarrow(\hat{s}_\vee|_{H_{11}}, \hat{s}_\vee|_{H_{01}})$:

$$q_\leftrightarrow(\hat{s}_\vee|_{H_{11}}, \hat{s}_\vee|_{H_{01}}) = \frac{1}{2}\int_{-\infty}^{\infty} |p_{\hat{s}_\vee}(z)|_{H_{11}} - p_{\hat{s}_\vee}(z)|_{H_{01}}|dz =$$

$$= 1 - \int_{-\infty}^{s_t} p_{\hat{s}_\wedge}(z)|_{H_{11}}\, dz = 1 - \int_{-\infty}^{s_t} NF_n^{N-1}(z)p_n(z)dz, \quad s_t < 0. \quad (6.7.16)$$

It is difficult to obtain the exact values of quality indices (6.7.15) and (6.7.16) even on the assumption of normalcy of interference (noise) PDF. Meanwhile, due to evenness of the function $p_n(z)$: $p_n(z) = p_n(-z)$, taking into account the positive (negative) definiteness of the samples s_t of instantaneous values of the signal $s(t)$ for the estimators $\hat{s}_\wedge(t)|_{H_{11},H_{01}}$ and $\hat{s}_\vee(t)|_{H_{11},H_{01}}$, it is easy to obtain the values of lower bounds of quality indices (6.7.15) and (6.7.16), which are determined by the following inequalities:

$$q_\leftrightarrow(\hat{s}_\wedge|_{H_{11}}, \hat{s}_\wedge|_{H_{01}}) \geq 1 - 2^{-N}; \quad (6.7.17a)$$

$$q_{\leftrightarrow}(\hat{s}_\vee|_{H_{11}}, \hat{s}_\vee|_{H_{01}}) \geq 1 - 2^{-N}. \qquad (6.7.17b)$$

The lower bounds of quality indices of joint resolution-estimation of the signals in signal space with lattice properties (6.7.17a) and (6.7.17b), determined by the metrics (6.7.15) and (6.7.16) between PDFs of the pairs of the estimators $\hat{s}_\wedge(t)|_{H_{11}}$, $\hat{s}_\wedge(t)|_{H_{01}}$ and $\hat{s}_\vee(t)|_{H_{11}}$, $\hat{s}_\vee(t)|_{H_{01}}$, respectively, do not depend on both signal-to-noise ratio and interference (noise) distribution. They are determined by the number N of independent samples of interference (noise) $n(t)$ used while forming the estimators (6.7.8) and (6.7.9) equal to the number of periods of carrier of the signal $s(t, \lambda)$. Thus, the quality indices of joint resolution-estimation of the signals (6.7.15) and (6.7.16) in signal space with lattice properties are invariant with respect to parametric and nonparametric prior uncertainty conditions.

The last circumstance, despite a fundamental distinction between statistical and deterministic approaches for evaluating the quality indices of signal resolution, indicates that the conclusions obtained upon their base coincide. The reason is that both statistical and deterministic characteristics of resolution depend strongly on the metric between the estimators $\hat{s}_\wedge(t)|_{H_{11}}$ and $\hat{s}_\wedge(t)|_{H_{01}}$ ($\hat{s}_\vee(t)|_{H_{11}}$ and $\hat{s}_\vee(t)|_{H_{01}}$) of the signals $s(t, \lambda) \vee s'(t, \lambda')$ and $s'(t, \lambda')$ ($s(t, \lambda) \wedge s'(t, \lambda')$ and $s'(t, \lambda')$) within the hypotheses (6.7.6a,b) and (6.7.7a,b), and also on the metric between the signals $s(t, \lambda) \vee s'(t, \lambda')$ and $s'(t, \lambda')$ ($s(t, \lambda) \wedge s'(t, \lambda')$ and $s'(t, \lambda')$).

Using the deterministic approach to evaluating the signal resolution, the last is characterized by a minimal difference $\Delta\lambda_{\min} = |\lambda - \lambda'|$ in the parameters of interacting signals $s(t, \lambda)$, $s'(t, \lambda')$, when two responses corresponding to these signals are separately observed in the output of the processing unit and two single maximums of these responses are distinguished.

Within the framework of the stated deterministic approach, we obtain the value of signal resolution in time delay for harmonic signals in signal space with lattice properties.

Consider, within the model (6.7.5a), the interaction of two harmonic signals $s_1(t)$ and $s_2(t)$ in signal space $\mathbf{\Gamma}(\vee, \wedge)$ with lattice properties in the absence of interference (noise) $n(t)$:

$$x(t) = s_1(t) \vee s_2(t) \vee 0. \qquad (6.7.18)$$

Let the model of the received signals $s_1(t)$ and $s_2(t)$ be determined by the expression:

$$\begin{cases} s_i(t) = A_i \cos(\omega_0 t + \varphi), & t \in T_i; \\ s_i(t) = 0, & t \notin T_i, \end{cases} \qquad (6.7.19)$$

where A_i is an unknown nonrandom amplitude of useful signal $s_i(t)$; $i = 1, 2$; $\omega_0 = 2\pi f_0$; f_0 is a known carrier frequency of the signal $s_i(t)$; φ is an unknown nonrandom initial phase of the signal $s_i(t)$, $\varphi \in [-\pi, \pi]$; T_i is a domain of definition of the signal $s_i(t)$, $T_i = [t_{0i}, t_{0i} + T]$; t_{0i} is an unknown time of arrival of the signal $s_i(t)$; T is a duration time of the signal; $T = NT_0$; T_0 is a period of a carrier; N is a number of periods of harmonic signal $s_i(t)$, $N = Tf_0$; $N \in \mathbf{N}$, $\mathbf{N}$ is the set of natural numbers.

Then the expression for the estimator $\hat{s}_\wedge(t)$ (6.7.8) of the signal $s_i(t)$, $i = 1, 2$,

6.7 Quality Indices of Resolution-Estimation in Metric Spaces with Lattice Properties

taking into account the observed process (6.7.18), is determined by the following relationship:

$$\hat{s}_\wedge(t) = \begin{cases} s_1(t) \vee s_2(t), & t \in T_{\hat{s}}; \\ 0, & t \notin T_{\hat{s}}, \end{cases} \quad (6.7.20)$$

$$T_{\hat{s}} = [\min[t_{01}, t_{02}] + (N-1)T_0, \ \max[t_{01}, t_{02}] + NT_0], \quad (6.7.20a)$$

where t_{0i} is an unknown time of arrival of the signal $s_i(t)$; N is a number of periods of the harmonic signal $s_i(t)$; T_0 is a period of a carrier; $i = 1, 2$.

The relationship (6.7.20) shows that the filter forming the estimator (6.7.8) provides the compression of a useful signal in N times. The result (6.7.20) can be interpreted as the potential signal resolution in time domain under the condition of extremely large signal-to-noise ratio in the input of the processing unit (filter). The expression (6.7.20) also implies that the resolution $\Delta\tau_{\min}$ of the filter in time parameter (in time delay) is a quantity of a quarter of a carrier period order T_0: $\Delta\tau_{\min} = T_0/4 + \varepsilon$, where ε is an arbitrarily small value.

Fig. 6.7.2 illustrates the signal $\hat{s}_\wedge(t)$ in the output of the filter forming the estimator (6.7.8) under the interaction of two harmonic signals $s_1(t)$ and $s_2(t)$ in the input of the processing unit (filter) in the absence of interference (noise) $n(t) = 0$, on the value of time delay that is equal to $t_{01} - t_{02} = T_0/3$.

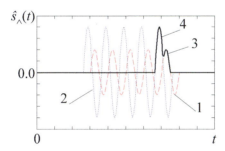

FIGURE 6.7.2 Signal $\hat{s}_\wedge(t)$ in the output of the filter forming the estimator (6.7.8)

The curves shown in the figure denote: 1 is the signal $s_1(t)$; 2 is the signal $s_2(t)$; 3 is the response $\hat{s}_1(t)$ of the signal $s_1(t)$; 4 is the response $\hat{s}_2(t)$ of the signal $s_2(t)$, the responses 3 and 4 of the $s_1(t)$, $s_2(t)$ are shown by the solid line.

The questions dealing with synthesis and analysis of signal resolution algorithm in signal space with lattice properties are investigated in Chapter 7.

7

Synthesis and Analysis of Signal Processing Algorithms

We demonstrated in Section 4.3, that under signal interaction in physical signal space, which is not isomorphic to a lattice, there inevitably exist the losses of some quantity of information contained in interacting signals ($a(t)$ or $b(t)$) with respect to the information quantity contained in the result of interaction $x(t) = a(t) \oplus b(t)$, where $\oplus$ is a commutative binary operation. At a qualitative level, these losses are described by informational inequalities (4.3.26). Thus, in linear space $\mathcal{LS}(+)$, signal interaction is always accompanied by losses of a part of information quantity contained in one of the interacting signals, as against the information quantity contained in the result of signal interaction.

As noted in Section 4.3, there are physical signal spaces, where the signal interaction is accompanied by the losses of information quantity between the signals that take place after their interaction, excluding the signal spaces with lattice properties. In such spaces, the signal interaction is characterized by the preservation of the quantity of mutual information contained in one of interacting signals ($a(t)$ or $b(t)$) and in the result of their interaction $x(t) = a(t) \oplus b(t)$, so that the informational identities (4.3.14) hold.

This chapter is a logical continuation of the main ideas stated in the fourth and the sixth chapters concerning signal processing theory. One can confirm or deny the results obtained in these chapters only on the basis of synthesis of optimal algorithms of signal processing in the spaces with lattice properties and further analysis of the quality indices of signal processing based on such algorithms. It is also of interest to carry out the comparative analysis of the synthesized algorithms and units of signal processing with the analogues obtained for linear signal spaces.

The known variants of the main signal processing problems have been formulated for the case of additive interaction $x(t) = s(t) + n(t)$ (in terms of linear space) between the signal $s(t)$ and interference (noise) $n(t)$. Nevertheless, some authors assume the statement of these problems on the basis of a more general model of interaction: $x(t) = \Phi[s(t), n(t)]$, where Φ is some deterministic function [155], [159], [163], [166]. The diversity of algebraic systems studied in modern mathematics allow investigating the properties and characteristics of both signal estimators and signal parameter estimators during interactions between useful signals and interference (noise) in signal spaces which differ from linear ones in their algebraic properties. The lattices have been described and investigated in algebraic literature [221], [223], [255].

The statement of a signal processing problem is often characterized by some level of prior uncertainty with respect to useful and/or interference signals. In his work [118], P.M. Woodward expressed an opinion that the question of prior distributions of informational and non-informational signal parameters will be a stumbling block on the way to the synthesis of optimal signal processing algorithms. This utterance is fully applicable to most signal processing problems, where prior knowledge of behavior and characteristics of useful and interference signals along with their parameters plays a significant role in synthesizing the optimal signal processing algorithms.

The methodology of the synthesis of signal processing algorithms in signal spaces with lattice properties has not been studied yet. Nevertheless, it is necessary to develop the such approaches to the synthesis to allow researchers to operate with minimum prior data concerning characteristics and properties of interacting useful and interference signals. First, no prior data concerning probabilistic distribution of useful signals and interference are supposed to be present. Second, a priori, the kind of useful signal (signals) is assumed to be known, i.e., it is either deterministic (quasi-deterministic) or stochastic. As to interference, a time interval τ_0 determining the independence of interference samples is known, and it is assumed that the relation $\tau_0 << T$ holds, where T is a signal duration.

Such constraints for prior data content concerning the processed signals impose their own peculiarities upon the approaches to the synthesis of signal processing algorithms in signal space with lattice properties. The choice of optimality criteria is not a mathematical problem.

Meanwhile, some considerations about the choice of criteria have to be stated. The first is that optimality criteria of solving a signal processing problem should not depend on possible distributions of useful and interference signals.

The other circumstance influencing the choice of appropriate optimality criteria is the presence of information regarding time interval τ_0 which, according to the Theorems 4.2.1 and/or 4.2.2, is the basis for the sampling (discretization) of the processed signals.

The third feature is the combination of the first and second ones implying that an optimality criterion has to be a function of a set of the samples of the observed process resulting from interactions of useful signal (signals) and interference (noise) in signal space with lattice properties. The additional consideration for a proper criterion of signal processing optimality in signal space with lattice properties is the need to consider the metric properties of nonlinear signal space.

The last two considerations take into account the fundamental feature of the considered approach to the synthesis of algorithms of optimal signal processing in signal space with lattice properties. Under optimization, one should take into account the metric relationships between the samples of received (processed) realization of the observed stochastic process. Thus, there is no need to take into account the possible properties and characteristics of unreceived realizations of signals that more completely reflect probabilistic-statistical characteristics and properties of the ensemble of signal realizations.

The last circumstance fundamentally distinguishes the considered approach to the synthesis from the classic one, where algorithms and units of signal processing are optimal on average with respect to entire statistical ensemble of the received (processed) signals. With the proposed approach, the algorithms and units of signal processing are optimal in the sense of the only realization of the observed stochastic process.

Before considering synthesis and analysis of signal processing algorithms in signal space with lattice properties, the algebraic properties of such spaces and their relations with linear signal spaces should be considered.

7.1 Signal Spaces with Lattice Properties

Within Section 4.1, we distinguished the notions of physical signal space and informational signal space established by Definitions 4.1.1 and 4.1.3, respectively. It should be useful to review the main features of physical signal spaces, inasmuch as the main subject contains physical signal spaces with lattice properties. Using two definitions that are similar to Definitions 4.1.1 and 4.1.2, we introduce the notions defining algebraic properties of physical signal space.

Definition 7.1.1. A set $\mathbf{\Gamma}$ of stochastic signals $\mathbf{\Gamma} = \{a(t), b(t), \ldots\}$ considered under their *physical interaction* with each other is a *physical signal space* $\mathbf{\Gamma}$, which forms a commutative *semigroup* $\mathbf{SG} = (\mathbf{\Gamma}, \oplus)$ with respect to a binary operation $\oplus$, introduced over it, with the following properties:

$$a(t) \oplus b(t) \in \mathbf{\Gamma} \qquad \text{(closure)};$$
$$a(t) \oplus b(t) = b(t) \oplus a(t) \qquad \text{(commutativity)};$$
$$a(t) \oplus (b(t) \oplus c(t)) = (a(t) \oplus b(t)) \oplus c(t) \qquad \text{(associativity)}.$$

Definition 7.1.2. A binary operation $\oplus$: $x(t) = a(t) \oplus b(t)$ between the elements $a(t)$ and $b(t)$ of semigroup $\mathbf{SG} = (\mathbf{\Gamma}, \oplus)$ is called a *physical interaction* between the signals $a(t)$ and $b(t)$ in physical signal space $\mathbf{\Gamma}$.

Any pair of stochastic signals $a(t)$ and $b(t)$ can be considered a partially ordered set $\mathbf{\Gamma}$ where, at each instant between two instantaneous values (samples) $a_{t_1} = a(t_1)$, $b_{t_2} = b(t_2)$ of the processes $a(t), b(t) \in \mathbf{\Gamma}$, there exists the relation of an order $a_t \leq b_t$ (or $a_t \geq b_t$). Then the partially ordered set $\mathbf{\Gamma}$ is a lattice with operations of join and meet, respectively: $a_t \vee b_t = \sup_{\mathbf{\Gamma}}\{a_t, b_t\}$, $a_t \wedge b_t = \inf_{\mathbf{\Gamma}}\{a_t, b_t\}$ and if $a(t) \leq b(t)$, then $a_t \wedge b_t = a_t$ and $a_t \vee b_t = b_t$ [221], [223]:

$$a_t \leq b_t \Leftrightarrow \begin{cases} a_t \wedge b_t = a_t; \\ a_t \vee b_t = b_t. \end{cases}$$

For a lattice $\mathbf{L} = (\mathbf{\Gamma}, \vee, \wedge)$, the following axioms hold [221], [223]:

$$a(t) \wedge a(t) = a(t), \qquad a(t) \vee a(t) = a(t) \qquad \text{(idempotency)};$$
$$a(t) \wedge b(t) = b(t) \wedge a(t), \quad a(t) \vee b(t) = b(t) \vee a(t) \quad \text{(commutativity)};$$
$$a(t) \wedge (b(t) \wedge c(t)) = (a(t) \wedge b(t)) \wedge c(t),$$
$$a(t) \vee (b(t) \vee c(t)) = (a(t) \vee b(t)) \vee c(t) \qquad \text{(associativity)};$$
$$a(t) \wedge (a(t) \vee b(t)) = a(t) \vee (a(t) \wedge b(t)) = a(t) \qquad \text{(absorption)}.$$

The axioms above defining a lattice are not independent, so, for instance, the property of idempotency follows from the axiom of absorption.

In this chapter, we deal mainly with physical signal spaces, which, according to Definition 7.1.1 (4.1.1), are both semigroups and lattices, where each group translation is isotonic. Such algebraic systems are called *lattice-ordered groups* **LG** or *L-groups* [221], [223].

One assumption concerning isotonic property of group translations has the following form [221]: for $\forall c(t), s(t) \in \Gamma$: $c(t) \leq s(t)$, the following relationship holds:

$$a(t) \oplus c(t) \oplus b(t) \leq a(t) \oplus s(t) \oplus b(t), \text{ for } \forall a(t), b(t) \in \Gamma. \tag{7.1.1}$$

Thus, any L-group $\mathbf{LG} = (\Gamma, \oplus, \vee, \wedge, 0)$ can be considered a universal algebra $(\Gamma, \oplus, \vee, \wedge, 0)$ with signature $(\oplus, \vee, \wedge, 0)$ determined by three binary operations $\oplus, \vee, \wedge$ and one 0-ary operation 0. Besides, $\mathbf{G} = (\Gamma, \oplus)$ is a commutative group, $\mathbf{L} = (\Gamma, \vee, \wedge)$ is a lattice, and $\mathbf{SG}_\vee = (\Gamma, \vee)$, $\mathbf{SG}_\wedge = (\Gamma, \wedge)$ are commutative semigroups, 0 is a null/neutral element of a group $\mathbf{G} = (\Gamma, \oplus)$: $\forall a(t) \in \Gamma$: $0 \oplus a(t) = a(t)$, $\exists a^{-1}(t)$: $a^{-1}(t) \oplus a(t) = 0$.

L-group is the only known algebraic structure in which along with lattice axiomatics, axioms of a group of a linear space $\mathcal{LS}$ hold. The last circumstance stipulates an extraordinary interest to L-groups from the direction of signal processing theory, inasmuch as such a diversity of algebraic (and also geometric) properties allows use of L-groups to describe, on the one hand, the signal spaces with group properties (for instance, linear signal spaces), and on the other hand, the signal spaces with lattice properties.

The most common model describing physical signal space is *linear space* $\mathcal{LS}$, by which we mean an additive commutative group $\mathbf{G}_{\mathcal{LS}} = (\Gamma, +)$ over a ring of scalars [233]. The physical signal space Γ can be constructed on the basis of L-group $\mathbf{LG} = (\Gamma, +, \vee, \wedge, 0)$, so that $\mathbf{G}_{\mathcal{LS}} = (\Gamma, +)$ is an additive commutative group, $\mathbf{L} = (\Gamma, \vee, \wedge)$ is a lattice, and $\mathbf{SG}_\vee = (\Gamma, \vee)$, $\mathbf{SG}_\wedge = (\Gamma, \wedge)$ are commutative semigroups, 0 is a null/neutral element of a group $\mathbf{G}_{\mathcal{LS}} = (\Gamma, +)$.

Further, to denote algebraic systems of L-group $\mathbf{LG} = (\Gamma, +, \vee, \wedge, 0)$, we use the following designations: $\mathcal{LS}(+)$ for a linear space; $\mathcal{L}(\vee, \wedge)$ for a lattice.

The main properties of L-group are stated, for instance, in [221, Section XIII.3].

Signal space with lattice properties $\mathcal{L}(\vee, \wedge)$ can be obtained by a transformation of the signals of a linear space $\mathcal{LS}(+)$ in such a way that in signal space $\mathcal{L}(\vee, \wedge)$ with lattice properties and operations of join and meet $\vee, \wedge$, the results of interaction between the signals $a(t)$ and $b(t)$ are realized according to the following relationships:

$$a(t) \vee b(t) = \{[a(t) + b(t)] + |a(t) - b(t)|\}/2; \tag{7.1.2a}$$

$$a(t) \wedge b(t) = \{[a(t) + b(t)] - |a(t) - b(t)|\}/2, \tag{7.1.2b}$$

which are the consequence of the equations cited in [221, Section XIII.3;(14)], [221, Section XIII.4;(22)].

The identities (7.1.2a,b) determine the mapping of a linear signal space $\mathcal{LS}(+)$ into a signal space with lattice properties $\mathcal{L}(\vee, \wedge)$: $T : \mathcal{LS}(+) \to \mathcal{L}(\vee, \wedge)$. The mapping T^{-1}, which is inverse to the initial one T: $T^{-1} : \mathcal{L}(\vee, \wedge) \to \mathcal{LS}(+)$, is determined by the known identity [221, Section XIII.3;(14)]:

$$a(t) + b(t) = a(t) \vee b(t) + a(t) \wedge b(t). \tag{7.1.3}$$

7.2 Estimation of Unknown Nonrandom Parameter in Sample Space with Lattice Properties

Estimation of the signals and their parameters is the most general problem of signal processing under interference (noise) background. Some other problems of signal processing relate to this one, for instance, signal detection, signal classification, and signal resolution. In the literature, the problems of signal processing in the presence of interference (noise) are formulated in terms of linear signal space, where the result of interaction x between the signal s and interference (noise) n is described by operation of addition of an additive commutative group: $x = s + n$. Quite similarly (i.e., by operation of addition), the literature describes the results of interaction between unknown nonrandom parameters of the signal with estimation errors (measurement errors) caused by the influence of interference (noise). However, in a number of cases, the interaction between useful signal and interference (noise), or the interaction between signal parameters and measurement errors can be nonlinear.

Characteristics and behavior of estimators under additive (in terms of linear sample space) interaction between the estimated parameter and the measurement errors from some arbitrary family of distributions, are covered in the corresponding literature [250], [231], [232], [251]. The examples of the estimators whose asymptotic variance never exceeds the Cramer-Rao lower bound, and on some values of estimated parameter, this variance is below it, were proposed by Hodges and Le Cam [250], [251], [256]. Such estimators are called superefficient. Here we consider an example of the estimators, which are close to the superefficient ones with respect to their properties, but the nature of their superefficiency is fundamentally another. It is stipulated by the differences of algebraic properties of sample spaces, where they take place, from the properties of linear sample space.

The subject for further consideration is comparative characteristics of the estimators of an unknown nonrandom location parameter in both linear sample space and sample space with lattice properties. Lattices are well investigated [221], [223]. *Sample space* $\mathcal{L}(\mathcal{Y}, \mathcal{B}_\mathcal{Y}; \vee, \wedge)$ *with lattice properties* is defined as a probabilistic space $(\mathcal{Y}, \mathcal{B}_\mathcal{Y})$ in which the axioms of distributive lattice $\mathcal{L}(\mathcal{Y}; \vee, \wedge)$ with operations of join and meet, respectively: $a \vee b = \sup_\mathcal{L}\{a, b\}$, $a \wedge b = \inf_\mathcal{L}\{a, b\}$; $a, b \in \mathcal{L}(\mathcal{Y}; \vee, \wedge)$ hold.

In most works on point estimation, the model of indirect measurement of an unknown nonrandom scalar location parameter λ is described by its additive interaction with statistically independent measurement errors in *linear sample space* $\mathcal{LS}(\mathcal{X}, \mathcal{B}_\mathcal{X}; +)$:

$$X_i = f(\lambda) + N_i,$$

where $f(\lambda)$ is some known one-to-one function of a measured parameter; $\{N_i\}$ are the independent measurement errors with a distribution from the distribution class with symmetric (even) probability density function $p_N(z) = p_N(-z)$ represented by the sample $N = (N_1, \ldots, N_n)$, $N_i \in N$, so that $N \in \mathcal{LS}(\mathcal{X}, \mathcal{B}_\mathcal{X}; +)$; $\{X_i\}$ are the measurement results represented by the sample $X = (X_1, \ldots, X_n)$, $X_i \in X$: $X \in \mathcal{LS}(\mathcal{X}, \mathcal{B}_\mathcal{X}; +)$; "+" is operation of addition of linear sample space $\mathcal{LS}(\mathcal{X}, \mathcal{B}_\mathcal{X}; +)$; $i = 1, \ldots, n$ is the index of the elements of statistical collections $\{N_i\}, \{X_i\}$; n is a size of the samples $N = (N_1, \ldots, N_n)$, $X = (X_1, \ldots, X_n)$.

The estimators, obtained on the basis of least squares method (LSM) and least modules method (LMM), according to the criteria of minimum of sums of squares and modules of measurement errors, respectively, are the first and simplest estimators [231], [257]:

$$\hat{\lambda}_{\text{LSM}} = \arg\min_\lambda \left\{ \sum_i (X_i - f(\lambda))^2 \right\}; \tag{7.2.1a}$$

$$\hat{\lambda}_{\text{LMM}} = \arg\min_\lambda \left\{ \sum_i |X_i - f(\lambda)| \right\}. \tag{7.2.1b}$$

Extrema of the functions $\sum_i (X_i - f(\lambda))^2$ and $\sum_i |X_i - f(\lambda)|$ determined by criteria (7.2.1a) and (7.2.1b) are found as the roots of the equations:

$$d\sum_i (X_i - f(\hat{\lambda}))^2 / d\hat{\lambda} = 0; \tag{7.2.2a}$$

$$d\sum_i |X_i - f(\hat{\lambda})| / d\hat{\lambda} = 0. \tag{7.2.2b}$$

The values of the estimators $\hat{\lambda}_{\text{LSM}}$ and $\hat{\lambda}_{\text{LMM}}$ are the solutions of the Equations (7.2.2a) and (7.2.2b) in the form of a function $f^{-1}[*]$ of the sample mean and the sample median med$\{*\}$ of the observations $\{X_i\}$, respectively:

$$\hat{\lambda}_{\text{LSM}} = f^{-1}\left(\frac{1}{n}\sum_{i=1}^n X_i\right); \tag{7.2.3a}$$

$$\hat{\lambda}_{\text{LMM}} = f^{-1}[\underset{i \in \mathbf{N} \cap [1,n]}{\text{med}} \{X_i\}], \tag{7.2.3b}$$

where $f^{-1}[*]$ is a function that is inverse with respect to a function $f(\lambda)$ of parameter λ; $\mathbf{N}$ is the set of natural numbers.

The estimators (7.2.3a) and (7.2.3b) are asymptotically effective in the case of Gaussian and Laplacian distributions of measurement errors, respectively.

Consider two models of indirect measurement of an unknown nonrandom scalar nonnegative location parameter $\lambda \in \mathbf{R}_+ = [0, \infty[$ in sample space with lattice properties $\mathcal{L}(\mathcal{Y}, \mathcal{B}_\mathcal{Y}; \vee, \wedge)$ respectively:

$$Y_i = f(\lambda) \vee N_i; \tag{7.2.4a}$$

$$\tilde{Y}_i = f(\lambda) \wedge N_i, \tag{7.2.4b}$$

7.2 Estimation of Unknown Nonrandom Parameter in Sample Space with Lattice Properties

where $f(\lambda)$ is some known one-to-one function of a measured parameter; $\{N_i\}$ are the independent measurement errors with a distribution from the distribution class with symmetric probability density function $p_N(z) = p_N(-z)$ represented by the sample $N = (N_1, \ldots, N_n)$, $N_i \in N$, and besides $N \in \mathcal{L}(\mathcal{Y}, \mathcal{B}_\mathcal{Y}; \vee, \wedge)$; $\{Y_i\}$, $\{\tilde{Y}_i\}$ are the measurement results represented by the samples $Y = (Y_1, \ldots, Y_n)$, $\tilde{Y} = (\tilde{Y}_1, \ldots, \tilde{Y}_n)$, $Y_i \in Y$, $\tilde{Y}_i \in \tilde{Y}$, respectively: $Y, \tilde{Y} \in \mathcal{L}(\mathcal{Y}, \mathcal{B}_\mathcal{Y}; \vee, \wedge)$; $\vee, \wedge$ are the operations of join and meet of sample space with lattice properties $\mathcal{L}(\mathcal{Y}, \mathcal{B}_\mathcal{Y}; \vee, \wedge)$, respectively; $i = 1, \ldots, n$ is the index of the elements of statistical collections $\{N_i\}$, $\{Y_i\}$, $\{\tilde{Y}_i\}$; n represents a size of the samples $N = (N_1, \ldots, N_n)$, $Y = (Y_1, \ldots, Y_n)$, $\tilde{Y} = (\tilde{Y}_1, \ldots, \tilde{Y}_n)$.

We can obtain the estimators $\hat{\lambda}_{n,\wedge}, \hat{\lambda}_{n,\vee}$ of parameter λ for the models of indirect measurement (7.2.4a) and (7.2.4b) according to criteria of minimum of meet/join of measurement errors, respectively:

$$\hat{\lambda}_{n,\wedge} = \arg\min_{\lambda \in \mathbf{R}_+} |\bigwedge_{i=1}^{n} (Y_i - f(\lambda))|; \tag{7.2.5a}$$

$$\hat{\lambda}_{n,\vee} = \arg\min_{\lambda \in \mathbf{R}_+} |\bigvee_{i=1}^{n} (\tilde{Y}_i - f(\lambda))|, \tag{7.2.5b}$$

where $\bigwedge_{i=1}^{n} Y_i = \inf_Y \{Y_i\}$ is the meet of a set $Y = (Y_1, \ldots, Y_n)$; $\bigvee_{i=1}^{n} \tilde{Y}_i = \sup_{\tilde{Y}} \{\tilde{Y}_i\}$ is the join of a set $\tilde{Y} = (\tilde{Y}_1, \ldots, \tilde{Y}_n)$.

We next find the extrema of the functions $|\bigwedge_{i=1}^{n}(Y_i - f(\lambda))|$ and $|\bigvee_{i=1}^{n}(\tilde{Y}_i - f(\lambda))|$ defined by the criteria (7.2.5a) and (7.2.5b), respectively, putting their derivatives at the estimator $\hat{\lambda}$ of a parameter λ to zero:

$$d|\bigwedge_{i=1}^{n}(Y_i - f(\hat{\lambda}))|/d\hat{\lambda} = -\operatorname{sign}[\bigwedge_{i=1}^{n}(Y_i - f(\hat{\lambda}))]f'(\hat{\lambda}) = 0; \tag{7.2.6a}$$

$$d|\bigvee_{i=1}^{n}(\tilde{Y}_i - f(\hat{\lambda}))|/d\hat{\lambda} = -\operatorname{sign}[\bigvee_{i=1}^{n}(\tilde{Y}_i - f(\hat{\lambda}))]f'(\hat{\lambda}) = 0. \tag{7.2.6b}$$

The values of the estimators $\hat{\lambda}_{n,\wedge}$ and $\hat{\lambda}_{n,\vee}$ are the solutions of the equations (7.2.6a) and (7.2.6b) in the form of the function $f^{-1}[*]$ of meet and join of the observation results $\{Y_i\}$ and $\{\tilde{Y}_i\}$, respectively:

$$\hat{\lambda}_{n,\wedge} = f^{-1}\left(\bigwedge_{i=1}^{n} Y_i\right); \tag{7.2.7a}$$

$$\hat{\lambda}_{n,\vee} = f^{-1}\left(\bigvee_{i=1}^{n} \tilde{Y}_i\right), \tag{7.2.7b}$$

where $f^{-1}[*]$ is an inverse function with respect to a function $f[*]$ of a parameter λ.

Derivatives of the functions $|\bigwedge_{i=1}^{n}(Y_i - f(\lambda))|$ and $|\bigvee_{i=1}^{n}(\tilde{Y}_i - f(\lambda))|$, according to the identities (7.2.6a) and (7.2.6b), change their signs from minus to plus at the points $\hat{\lambda}_{n,\wedge}, \hat{\lambda}_{n,\vee}$, so the extrema determined by the formulas (7.2.7a), (7.2.7b) are the minimum points of these functions and the solutions of Equations (7.2.5a), (7.2.5b) defining these estimation criteria.

We now carry out the comparative analysis of the quality characteristics of the estimation of an unknown nonrandom nonnegative location parameter $\lambda \in \mathbf{R}_+ = [0, \infty[$ for the model of the direct measurement in both linear sample space $\mathcal{LS}(\mathcal{X}, \mathcal{B}_\mathcal{X}; +)$ and sample space with lattice properties $\mathcal{L}(\mathcal{Y}, \mathcal{B}_\mathcal{Y}; \vee, \wedge)$, respectively:

$$X_i = \lambda + N_i; \tag{7.2.8}$$

$$Y_i = \lambda \vee N_i, \tag{7.2.9}$$

where $\{N_i\}$ are the independent measurement errors that are normally distributed with zero expectation and variance D, represented by the sample $N = (N_1, \ldots, N_n)$, $N_i \in N$, and $N \in \mathcal{LS}(\mathcal{X}, \mathcal{B}_\mathcal{X}; +)$ and $N \in \mathcal{L}(\mathcal{Y}, \mathcal{B}_\mathcal{Y}; \vee, \wedge)$; $\{X_i\}$, $\{Y_i\}$ are the measurement results represented by the samples $X = (X_1, \ldots, X_n)$, $Y = (Y_1, \ldots, Y_n)$; $X_i \in X$, $Y_i \in Y$, respectively: $X \in \mathcal{LS}(\mathcal{X}, \mathcal{B}_\mathcal{X}; +)$, $Y \in \mathcal{L}(\mathcal{Y}, \mathcal{B}_\mathcal{Y}; \vee, \wedge)$; "+" and "$\vee$" are operation of addition of linear sample space $\mathcal{LS}(\mathcal{X}, \mathcal{B}_\mathcal{X}; +)$ and operation of join of sample space with lattice properties $\mathcal{L}(\mathcal{Y}, \mathcal{B}_\mathcal{Y}; \vee, \wedge)$, respectively; $i = 1, \ldots, n$ is the index of the elements of statistical collections $\{N_i\}$, $\{X_i\}$, $\{Y_i\}$; n is a size of the samples $N = (N_1, \ldots, N_n)$, $X = (X_1, \ldots, X_n)$, $Y = (Y_1, \ldots, Y_n)$.

For the model (7.2.8), the estimator $\hat{\lambda}_{n,+}$ in the form of a sample mean is a uniformly minimum variance unbiased estimator [250], [251]:

$$\hat{\lambda}_{n,+} = \frac{1}{n} \sum_{i=1}^{n} X_i. \tag{7.2.10}$$

As the estimator $\hat{\lambda}_{n,\wedge}$ of parameter λ for the model (7.2.9), we take the meet (7.2.7a):

$$\hat{\lambda}_{n,\wedge} = \bigwedge_{i=1}^{n} Y_i, \tag{7.2.11}$$

where $\bigwedge_{i=1}^{n} Y_i = \inf_{Y} \{Y_i\}$ is the least value from the sample $Y = (Y_1, \ldots, Y_n)$.

Cumulative distribution function (CDF) $F_{\hat{\lambda}_{n,+}}(z)$ and probability density function (PDF) $p_{\hat{\lambda}_{n,+}}(z)$ of the estimator $\hat{\lambda}_{n,+}$ (7.2.10) for the model (7.2.8) are determined by the expressions [250], [251]:

$$F_{\hat{\lambda}_{n,+}}(z) = \int_{-\infty}^{z} p_{\hat{\lambda}_{n,+}}(x) dx; \tag{7.2.12}$$

$$p_{\hat{\lambda}_{n,+}}(z) = (2\pi D/n)^{-1/2} \exp\left(-(z-\lambda)^2/(2D/n)\right), \tag{7.2.13}$$

where D is a variance of a measurement error N_i; n is a size of the sample collection $\{X_i\}$.

It is also known that CDF $F_{\hat{\lambda}_{n,\wedge}}(z)$ of the estimator $\hat{\lambda}_{n,\wedge}$ (7.2.11) for the model (7.2.9) is determined by the expression [252], [253]:

$$F_{\hat{\lambda}_{n,\wedge}}(z) = 1 - [1 - F(z)]^n, \tag{7.2.14}$$

7.2 Estimation of Unknown Nonrandom Parameter in Sample Space with Lattice Properties

where $F(z)$ is CDF of random variable $Y_i = \lambda \vee N_i$:

$$F(z) = F_N(z) \cdot F_\lambda(z), \tag{7.2.15}$$

$F_N(z) = \int\limits_{-\infty}^{z} p_N(x)dx$ is CDF of measurement error N_i; $p_N(z) = \frac{1}{\sqrt{2\pi D}} \exp\left(\frac{-z^2}{2D}\right)$ is the PDF of measurement error N_i; $F_\lambda(z) = 1(z - \lambda)$ is CDF of an unknown nonrandom parameter $\lambda \geq 0$; $1(z)$ is Heaviside step function.

So, $F(z)$ (7.2.15) can be written in the form:

$$F(z) = \begin{cases} F_N(z), & z \geq \lambda; \\ 0, & z < \lambda. \end{cases} \tag{7.2.16}$$

Taking into account (7.2.16), formula (7.2.14) can be written in the form:

$$F_{\hat{\lambda}n,\wedge}(z) = \begin{cases} 1 - [1 - F_N(z)]^n, & z \geq \lambda; \\ 0, & z < \lambda, \end{cases}$$

or, more compactly:

$$F_{\hat{\lambda}n,\wedge}(z) = (1 - [1 - F_N(z)]^n) \cdot 1(z - \lambda). \tag{7.2.17}$$

According to its definition, the PDF $p_{\hat{\lambda}n,\wedge}(z)$ of the estimator $\hat{\lambda}_{n,\wedge}$ is a derivative of CDF $F_{\hat{\lambda}n,\wedge}(z)$:

$$p_{\hat{\lambda}n,\wedge}(z) = F'_{\hat{\lambda}n,\wedge}(z) =$$
$$= (1 - [1 - F_N(\lambda)]^n) \cdot \delta(z - \lambda) + n \cdot p_N(z)[1 - F_N(z)]^{n-1} \cdot 1(z - \lambda), \tag{7.2.18}$$

where $\delta(z)$ is the delta function.

The expression (7.2.18) can be written in the following form:

$$p_{\hat{\lambda}n,\wedge}(z) = \begin{cases} P_c \cdot \delta(z - \lambda), & z = \lambda; \\ n \cdot p_N(z)[1 - F_N(z)]^{n-1}, & z > \lambda, \end{cases} \tag{7.2.19}$$

where $P_c = 1 - P_e$ is the probability of correct formation of the estimator $\hat{\lambda}_{n,\wedge}$:

$$P_c = 1 - [1 - F_N(\lambda)]^n, \tag{7.2.20}$$

P_e is the probability of error formation of the estimator $\hat{\lambda}_{n,\wedge}$:

$$P_e = [1 - F_N(\lambda)]^n. \tag{7.2.21}$$

The PDFs $p_{\hat{\lambda}n,\wedge}(z)$ of the estimator $\hat{\lambda}_{n,\wedge}$ for $n = 1, 2$ are shown in Fig. 7.2.1. Each random variable N_i is determined by even PDF $p_N(z)$ with zero expectation, so the inequality holds:

$$F_N(\lambda) \geq 1/2. \tag{7.2.22}$$

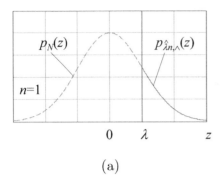 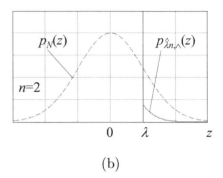

FIGURE 7.2.1 PDF $p_{\hat{\lambda}_{n,\wedge}}(z)$ of estimator $\hat{\lambda}_{n,\wedge}$: (a) $n = 1$; (b) $n = 2$

According to the formula (7.2.22), probability P_e of error formation of the estimator $\hat{\lambda}_{n,\wedge}$ is bounded above by the quantity:

$$P_e \leq 2^{-n}. \tag{7.2.23}$$

Correspondingly, the probability P_c of correct formation of the estimator $\hat{\lambda}_{n,\wedge}$ is bounded below by the quantity:

$$P_c \geq 1 - 2^n. \tag{7.2.24}$$

The relationship (7.2.19) implies that the estimator $\hat{\lambda}_{n,\wedge}$ is biased; nevertheless, it is both consistent and asymptotically unbiased, inasmuch as it converges in probability $\hat{\lambda}_{n,\wedge} \xrightarrow{P} \lambda$ and in distribution $\hat{\lambda}_{n,\wedge} \xrightarrow{p} \lambda$ to the estimated parameter λ:

$$\hat{\lambda}_{n,\wedge} \xrightarrow{P} \lambda : \lim_{n \to \infty} P\{|\hat{\lambda}_{n,\wedge} - \lambda| < \varepsilon\} = 1, \text{ for } \forall \varepsilon > 0;$$

$$\hat{\lambda}_{n,\wedge} \xrightarrow{p} \lambda : \lim_{n \to \infty} p_{\hat{\lambda}_{n,\wedge}}(z) = \delta(z - \lambda).$$

7.2.1 Efficiency of Estimator $\hat{\lambda}_{n,\wedge}$ in Sample Space with Lattice Properties with Respect to Estimator $\hat{\lambda}_{n,+}$ in Linear Sample Space

It is difficult to obtain an exact value of a variance $D\{\hat{\lambda}_{n,\wedge}\}$ of the estimator $\hat{\lambda}_{n,\wedge}$ directly on the basis of formula (7.2.19). The following theorem allows us to get an idea on a variance $D\{\hat{\lambda}_{n,\wedge}\}$ of the estimator $\hat{\lambda}_{n,\wedge}$.

Theorem 7.2.1. *For the model of the measurement (7.2.9), variance $D\{\hat{\lambda}_{n,\wedge}\}$ of the estimator $\hat{\lambda}_{n,\wedge}$ (7.2.11) of a parameter λ, is bounded above by the quantity:*

$$D\{\hat{\lambda}_{n,\wedge}\} \leq P_c P_e \lambda^2 + n \cdot P_e D \left[1 + \frac{\lambda \exp\left(-\lambda^2/2D\right)}{[1 - F_N(\lambda)]\sqrt{2\pi D}}\right],$$

where $P_c = 1 - [1 - F_N(\lambda)]^n$, $P_e = [1 - F_N(\lambda)]^n$.

7.2 Estimation of Unknown Nonrandom Parameter in Sample Space with Lattice Properties

Proof. The probability density function $p_{\hat{\lambda}n,\wedge}(z)$ (7.2.19) of the estimator $\hat{\lambda}_{n,\wedge}$ (7.2.11) can be represented in the following form:

$$p_{\hat{\lambda}n,\wedge}(z) = \begin{cases} P_c \cdot \delta(z-\lambda), & z = \lambda; \\ 2P_e \cdot p_N(z) \cdot K(z), & z > \lambda, \end{cases} \quad (7.2.25)$$

where $K(z)$ is a function determined by the formula:

$$K(z) = n[1 - F_N(z)]^{n-1}/(2 \cdot P_e). \quad (7.2.26)$$

It should be noted that for any $z \geq \lambda$, the inequality holds:

$$0 < [1 - F_N(z)] \leq [1 - F_N(\lambda)],$$

which implies the inequality:

$$0 < K(z) \leq n/(2 \cdot [1 - F_N(\lambda)]). \quad (7.2.27)$$

Taking into account the boundedness of the function $K(z)$ (7.2.26), appearing in PDF $p_{\hat{\lambda}n,\wedge}(z)$ (7.2.25) of the estimator $\hat{\lambda}_{n,\wedge}$, one can obtain the upper bound $\sup_{K(z)} D\{\hat{\lambda}_{n,\wedge}\}$ of its variance $D\{\hat{\lambda}_{n,\wedge}\}$:

$$\sup_{K(z)} D\{\hat{\lambda}_{n,\wedge}\} = \sup_{K(z)} m_2\{\hat{\lambda}_{n,\wedge}\} - \left(\inf_{K(z)} m_1\{\hat{\lambda}_{n,\wedge}\}\right)^2, \quad (7.2.28)$$

where $m_2\{\hat{\lambda}_{n,\wedge}\}$ is the second moment of the estimator $\hat{\lambda}_{n,\wedge}$:

$$m_2\{\hat{\lambda}_{n,\wedge}\} = \int_{-\infty}^{\infty} z^2 p_{\hat{\lambda}n,\wedge}(z) dz; \quad (7.2.29)$$

$m_1\{\hat{\lambda}_{n,\wedge}\}$ is an expectation of the estimator $\hat{\lambda}_{n,\wedge}$:

$$m_1\{\hat{\lambda}_{n,\wedge}\} = \int_{-\infty}^{\infty} z p_{\hat{\lambda}n,\wedge}(z) dz. \quad (7.2.30)$$

Substituting the formula (7.2.25) into the definition of the second moment (7.2.29) of the estimator $\hat{\lambda}_{n,\wedge}$, we obtain the following expression:

$$m_2\{\hat{\lambda}_{n,\wedge}\} = P_c \int_{-\infty}^{\infty} z^2 \delta(z-\lambda) dz + 2P_e \int_{\lambda}^{\infty} z^2 p_N(z) K(z) dz. \quad (7.2.31)$$

Applying the inequality (7.2.27) characterizing the boundedness of $K(z)$ above to the second component in the right part of the formula (7.2.31), we obtain the quantity of the upper bound $\sup_{K(z)} m_2\{\hat{\lambda}_{n,\wedge}\}$ of the second moment $m_2\{\hat{\lambda}_{n,\wedge}\}$:

$$\sup_{K(z)} m_2\{\hat{\lambda}_{n,\wedge}\} = P_c \lambda^2 + n \cdot P_e D \left[1 + \frac{\lambda \exp(-\lambda^2/2D)}{\sqrt{2\pi D}[1 - F_N(\lambda)]}\right]. \quad (7.2.32)$$

Similarly, substituting the formula (7.2.25) into the definition of expectation (7.2.30) of the estimator $\hat{\lambda}_{n,\wedge}$, we obtain the following expression:

$$m_1\{\hat{\lambda}_{n,\wedge}\} = P_c \int_{-\infty}^{\infty} z\delta(z-\lambda)dz + 2P_e \int_{\lambda}^{\infty} zp_N(z)K(z)dz. \qquad (7.2.33)$$

Applying the inequality (7.2.27) characterizing the boundedness below of $K(z)$ to the second component in the right part of the formula (7.2.33), we obtain the quantity of the lower bound $\inf_{K(z)} m_1\{\hat{\lambda}_{n,\wedge}\}$ of the expectation $m_1\{\hat{\lambda}_{n,\wedge}\}$:

$$\inf_{K(z)} m_1\{\hat{\lambda}_{n,\wedge}\} = P_c \lambda. \qquad (7.2.34)$$

Substituting the quantities determined by the formulas (7.2.32) and (7.2.34) into the formula (7.2.28), we obtain the following expression for the upper bound $\sup_{K(z)} D\{\hat{\lambda}_{n,\wedge}\}$ of a variance $D\{\hat{\lambda}_{n,\wedge}\}$ of the estimator $\hat{\lambda}_{n,\wedge}$:

$$\sup_{K(z)} D\{\hat{\lambda}_{n,\wedge}\} = P_c P_e \lambda^2 + n \cdot P_e D \left[1 + \frac{\lambda \exp(-\lambda^2/2D)}{[1 - F_N(\lambda)]\sqrt{2\pi D}}\right], \qquad (7.2.35)$$

where $P_c = 1 - [1 - F_N(\lambda)]^n$, $P_e = [1 - F_N(\lambda)]^n$. $\square$

Theorem 7.2.1 implies the corollary.

Corollary 7.2.1. *Variance $D\{\hat{\lambda}_{n,\wedge}\}$ of the estimator $\hat{\lambda}_{n,\wedge}$ (7.2.11) of a parameter λ for the model of measurement (7.2.9) is bounded above by the quantity:*

$$D\{\hat{\lambda}_{n,\wedge}\} \leq n \cdot 2^{-n} D. \qquad (7.2.36)$$

Proof. Investigate the behavior of the upper bound $\sup_{K(z)} D\{\hat{\lambda}_{n,\wedge}\}$ of variance $D\{\hat{\lambda}_{n,\wedge}\}$ of the estimator $\hat{\lambda}_{n,\wedge}$ on $\lambda \geq 0$. Show that $\sup_{K(z)} D\{\hat{\lambda}_{n,\wedge}\}$ is a monotone decreasing function of a parameter λ, so that this function takes a maximum value equal to $n \cdot 2^{-n} D$ in the point $\lambda = 0$:

$$\lim_{\lambda \to 0+0} \left(\sup_{K(z)} D\{\hat{\lambda}_{n,\wedge}\}\right) = n \cdot 2^{-n} D \geq \sup_{K(z)} D\{\hat{\lambda}_{n,\wedge}\} \geq D\{\hat{\lambda}_{n,\wedge}\}. \qquad (7.2.37)$$

We denote the function $\sup_{K(z)} D\{\hat{\lambda}_{n,\wedge}\}$ determined by the identity (7.2.35) by $y(\lambda)$:

$$y(\lambda) = (1 - Q^n(\lambda))Q^n(\lambda)\lambda^2 + nQ^n(\lambda)D(1 + \lambda f(\lambda)Q^{-1}(\lambda)), \qquad (7.2.38)$$

where $Q(x) = 1 - F_N(x)$; $f(x) = (2\pi D)^{-1/2} \exp\{-x^2/(2D)\}$.
Determine the derivative of the function (7.2.38), whose expression, omitting the intermediate computations, can be written in the form:

$$y'(\lambda) = -2n\lambda^2 f(\lambda) Q^{n-1}(\lambda)(1 - Q^n(\lambda)) - Df(\lambda)Q^{n-1}(\lambda)(n^2 - n) +$$

7.2 Estimation of Unknown Nonrandom Parameter in Sample Space with Lattice Properties

$$+\lambda Q^{n-2}(\lambda)[2Q^2(\lambda)(1-Q^n(\lambda)) - Df^2(\lambda)(n^2-n)]. \qquad (7.2.39)$$

It is obvious that the first two components in (7.2.39) are negative. Then, to prove the derivative $y'(\lambda)$ of the function $y(\lambda)$ (7.2.38) is negative, it is sufficient to show that the second multiplier in the third component is smaller than zero:

$$u(\lambda) = 2Q^2(\lambda)(1-Q^n(\lambda)) - Df^2(\lambda)(n^2-n) < 0. \qquad (7.2.40)$$

To prove the inequality (7.2.40) we use the following lemma.

Lemma 7.2.1. *For Gaussian random variable ξ with zero expectation and the variance D, the inequality holds:*

$$2(1-F(x)) \leq (2\pi D)^{1/2} f(x), \qquad (7.2.41)$$

where $f(x) = (2\pi D)^{-1/2} \exp\{-x^2/(2D)\}$ is a PDF of random variable ξ; $F(x) = \int_{-\infty}^{x} f(t)dt$ is CDF of random variable ξ.

Proof of lemma. Consider the ratio of Mills, $R(x)$ [258]:

$$R(x) = [1-F(x)]/f(x) \geq 0, \; R(0) = [\pi D/2]^{1/2}, \; x \geq 0.$$

Obviously, to prove the inequality (7.2.41), it is sufficient to prove that $R(x)$ is a monotone decreasing function on $x \geq 0$. Derivative $R'(x)$ of Mills' function is equal to:

$$R'(x) = xR(x)/D - 1. \qquad (7.2.42)$$

There exists the representation of Mills' function $R(x)$ in the form of a continued fraction [258]:

$$R(x) = \cfrac{1 \cdot D}{x + \cfrac{1 \cdot D}{x + \cfrac{2 \cdot D}{x + \cfrac{3 \cdot D}{x + \cfrac{4 \cdot D}{x + \ldots}}}}}. \qquad (7.2.43)$$

The relation (7.2.43) implies that $R(x) < D/x$, and the relation (7.2.42), in its turn, implies that $R'(x) < 0$. Thus, $R(x) \leq [\pi D/2]^{1/2}$ and $R'(x) < 0$, $x \geq 0$. Lemma 7.2.1 is proved.

We overwrite the statement (7.2.41) of Lemma 7.2.1 in a form convenient to prove the inequality (7.2.40):

$$2Q(\lambda) \leq (2\pi D)^{1/2} f(\lambda). \qquad (7.2.44)$$

Squaring both parts of the inequality (7.2.44), we obtain the intermediate inequality:

$$4Q^2(\lambda) \leq 2\pi D f(\lambda) \Rightarrow 2Q^2(\lambda) \leq \pi D f(\lambda),$$

which implies that the inequality (7.2.40) holds in the case of large samples $n \gg 1$ (it is sufficient that $n > 2$):

$$2Q^2(\lambda)(1 - Q^n(\lambda)) \leq 2Q^2(\lambda) \leq \pi Df(\lambda) < Df(\lambda)(n^2 - n). \tag{7.2.45}$$

Thus, the inequality (7.2.45) implies, that the function $y(\lambda)$ determined by the formula (38) is monotone decreasing function on $\lambda \geq 0$ and $n > 2$, so that $y(0) = n2^{-n}D$. Corollary 7.2.1 is proved. □

Compare the quality of the estimators $\hat{\lambda}_{n,+}$ (7.2.10) and $\hat{\lambda}_{n,\wedge}$ (7.2.11) of a parameter λ for the models of the measurement (7.2.8) and (7.2.9), respectively, using the relative efficiency $e\{\hat{\lambda}_{n,\wedge}, \hat{\lambda}_{n,+}\}$ equal to the ratio:

$$e\{\hat{\lambda}_{n,\wedge}, \hat{\lambda}_{n,+}\} = D\{\hat{\lambda}_{n,+}\}/D\{\hat{\lambda}_{n,\wedge}\}, \tag{7.2.46}$$

where $D\{\hat{\lambda}_{n,+}\} = D/n$ is a variance of the efficient estimator $\hat{\lambda}_{n,+}$ of a parameter λ in linear sample space; $D\{\hat{\lambda}_{n,\wedge}\}$ is a variance of the estimator $\hat{\lambda}_{n,\wedge}$ of a parameter λ in sample space with lattice properties; n is a size of the samples $X = (X_1, \ldots, X_n)$, $Y = (Y_1, \ldots, Y_n)$.

The comparative result of the quality of the estimators $\hat{\lambda}_{n,+}$ and $\hat{\lambda}_{n,\wedge}$ is determined by the following theorem.

Theorem 7.2.2. *The relative efficiency* $e\{\hat{\lambda}_{n,\wedge}, \hat{\lambda}_{n,+}\}$ *of the estimator* $\hat{\lambda}_{n,\wedge}$ *of a parameter* λ *in sample space with lattice properties* $\mathcal{L}(\mathcal{Y}, \mathcal{B}_\mathcal{Y}; \vee, \wedge)$ *with respect to the estimator* $\hat{\lambda}_{n,+}$ *of the same parameter* λ *in linear sample space* $\mathcal{LS}(\mathcal{X}, \mathcal{B}_\mathcal{X}; +)$ *is bounded below by the quantity:*

$$e\{\hat{\lambda}_{n,\wedge}, \hat{\lambda}_{n,+}\} \geq 2^n/n^2. \tag{7.2.47}$$

Proof. Using the statement (7.2.36) of Corollary 7.2.1 of Theorem 7.2.1 in the definition of relative efficiency $e\{\hat{\lambda}_{n,\wedge}, \hat{\lambda}_{n,+}\}$ (7.2.46), we obtain that:

$$D\{\hat{\lambda}_{n,\wedge}\} \leq n \cdot 2^{-n}D \Rightarrow e\{\hat{\lambda}_{n,\wedge}, \hat{\lambda}_{n,+}\} \geq (D/n)/(n \cdot 2^{-n}D) = 2^n/n^2.$$

□

Using the result (7.2.35) of Theorem 7.2.1, one can determine the lower bound $\inf_{K(z)} e\{\hat{\lambda}_{n,\wedge}, \hat{\lambda}_{n,+}\}$ of relative efficiency $e\{\hat{\lambda}_{n,\wedge}, \hat{\lambda}_{n,+}\}$ of the estimator $\hat{\lambda}_{n,\wedge}$ in the sample space with lattice properties (7.2.11) with respect to the estimator $\hat{\lambda}_{n,+}$ in linear sample space (7.2.10) for some limit cases:

$$\inf_{K(z)} e\{\hat{\lambda}_{n,\wedge}, \hat{\lambda}_{n,+}\} = D\{\hat{\lambda}_{n,+}\}/\sup_{K(z)} D\{\hat{\lambda}_{n,\wedge}\}, \tag{7.2.48}$$

where $D\{\hat{\lambda}_{n,+}\} = D/n$ is a variance of efficient estimator $\hat{\lambda}_{n,+}$ of a parameter λ in linear sample space; $\sup_{K(z)} D\{\hat{\lambda}_{n,\wedge}\}$ is the upper bound of a variance $D\{\hat{\lambda}_{n,\wedge}\}$ (7.2.35) of the estimator $\hat{\lambda}_{n,\wedge}$ of a parameter λ in sample space with lattice properties.

Assuming in the formula (7.2.35) that a parameter λ tends to zero and taking into account the equality $F_N(0) = 1/2$, we obtain the value of the limit of relative efficiency lower bound of the estimator $\lambda_{n,\wedge}$ with respect to the estimator $\hat{\lambda}_{n,+}$, which is identical to the statement (7.2.47) of Theorem 7.2.2:

$$\lim_{\lambda \to 0+0} \left[\inf_{K(z)} e\{\hat{\lambda}_{n,\wedge}, \hat{\lambda}_{n,+}\} \right] = 2^n/n^2. \qquad (7.2.49)$$

Analyzing Theorem 7.2.2, it is easy to be convinced that the value of asymptotic relative efficiency of the estimator $\lambda_{n,\wedge}$ with respect to the estimator $\hat{\lambda}_{n,+}$ on $n \to \infty$ is infinitely large on arbitrary values of an estimated parameter λ:

$$\lim_{n \to \infty} e\{\hat{\lambda}_{n,\wedge}, \hat{\lambda}_{n,+}\} \to \infty. \qquad (7.2.50)$$

Theorem 7.2.2 shows how large can be the worst value of relative efficiency of the estimator $\lambda_{n,\wedge}$ in sample space with lattice properties $\mathcal{L}(\mathcal{Y}, \mathcal{B}_\mathcal{Y}; \vee, \wedge)$ with respect to well known estimator $\hat{\lambda}_{n,+}$ in the form of a sample mean in linear sample space $\mathcal{LS}(\mathcal{X}, \mathcal{B}_\mathcal{X}; +)$.

Theorem 7.2.2 can be explained in the following way. While increasing the sample size n, the variance $D\{\hat{\lambda}_{n,+}\}$ of the estimator $\hat{\lambda}_{n,+}$ in linear sample space $\mathcal{LS}(\mathcal{X}, \mathcal{B}_\mathcal{X}; +)$ decreases with a rate proportional to n, whereas the variance of the estimator $D\{\hat{\lambda}_{n,\wedge}\}$ in sample space with lattice properties $\mathcal{L}(\mathcal{Y}, \mathcal{B}_\mathcal{Y}; \vee, \wedge)$ decreases with a rate proportional to $2^n/n$, i.e., almost exponentially.

Such striking distinctions concerning estimator behavior in linear sample space $\mathcal{LS}(\mathcal{X}, \mathcal{B}_\mathcal{X}; +)$ and estimator behavior in sample space with lattice properties $\mathcal{L}(\mathcal{Y}, \mathcal{B}_\mathcal{Y}; \vee, \wedge)$ can be elucidated by the fundamental difference between algebraic properties of these spaces is revealed by the best use of information contained in the statistical collection processed in one sample space as against another.

In summary, there exist estimators in nonlinear sample spaces characterized on Gaussian distribution of measurement errors by a variance that is noticeably smaller than an efficient estimator variance defined by the Cramer-Rao lower bound. This circumstance poses a question regarding the adequacy of estimator variance application to determine the efficiency of an unknown nonrandom parameter estimation on wide classes of sample spaces and on the families of symmetric distributions of measurement errors.

First, not all the distributions are characterized by a finite variance. Second, a variance does not contain all information concerning the properties of distribution. Third, the existence of superefficient estimators casts doubt on the correctness of a variance application to determine parameter estimation efficiency. Fourth, the analysis of the properties of the estimator sequences $\{\hat{\lambda}_n\}$ (n is a sample size) on the basis of their variances does not allow taking into account the topological properties of sample spaces and parameter estimators in these spaces.

On the ground of aforementioned considerations, another approach is proposed to determine the efficiency of an unknown nonrandom parameter estimation. Such an approach can determine the efficiency of unbiased and asymptotically unbiased estimators; it is based on metric properties of the estimators and is considered below.

7.2.2 Quality Indices of Estimators in Metric Sample Space

Let the estimation of an unknown nonrandom parameter $\lambda \geq 0$ be realized within the framework of the models of direct measurement in linear sample space $\mathcal{LS}(\mathcal{X}, \mathcal{B}_\mathcal{X}; +)$ and in sample space with the properties of a universal algebra $\mathcal{A}(\mathcal{Y}, \mathcal{B}_\mathcal{Y}; S)$ with a signature S, respectively [259], [260]:

$$X_i = \lambda + N_i; \qquad (7.2.51)$$

$$Y_i = \lambda \oplus N_i, \qquad (7.2.52)$$

where $\{N_i\}$ are independent measurement errors, each with a PDF $p_N^\alpha(z)$ from some indexed (over a parameter α) set $P = \{p_N^\alpha(z)\}$ represented by the sample $N = (N_1, \ldots, N_n)$, $N_i \in N$, so that $N \in \mathcal{LS}(\mathcal{X}, \mathcal{B}_\mathcal{X}; +)$ and $N \in \mathcal{A}(\mathcal{Y}, \mathcal{B}_\mathcal{Y}; S)$; $\{X_i\}$, $\{Y_i\}$ are the measurement results represented by the samples $X = (X_1, \ldots, X_n)$, $Y = (Y_1, \ldots, Y_n)$, $X_i \in X$, $Y_i \in Y$, respectively: $X \in \mathcal{LS}(\mathcal{X}, \mathcal{B}_\mathcal{X}; +)$, $Y \in \mathcal{A}(\mathcal{Y}, \mathcal{B}_\mathcal{Y}; S)$; "$+$" is a binary operation of additive commutative group $\mathcal{LS}(+)$ of linear sample space $\mathcal{LS}(\mathcal{X}, \mathcal{B}_\mathcal{X}; +)$; "$\oplus$" is a binary operation of additive commutative semigroup $\mathcal{A}(\oplus)$ of sample space with the properties of universal algebra $\mathcal{A}(\mathcal{Y}, \mathcal{B}_\mathcal{Y}; S)$ and a signature S; $i = 1, \ldots, n$ is an index of the elements of statistical collections $\{N_i\}$, $\{X_i\}$, $\{Y_i\}$; n represents a size of the samples $N = (N_1, \ldots, N_n)$, $X = (X_1, \ldots, X_n)$, $Y = (Y_1, \ldots, Y_n)$.

Let the estimators $\hat{\lambda}_{n,+}^\alpha$, $\tilde{\lambda}_{n,\oplus}^\alpha$ of an unknown nonrandom parameter λ, both in linear sample space $\mathcal{LS}(\mathcal{X}, \mathcal{B}_\mathcal{X}; +)$ and in sample space with the properties of universal algebra $\mathcal{A}(\mathcal{Y}, \mathcal{B}_\mathcal{Y}; S)$ and a signature S, be some functions $\hat{T}$ and $\tilde{T}$ of the samples $X = (X_1, \ldots, X_n)$, $Y = (Y_1, \ldots, Y_n)$, respectively, and, in general case, the functions $\hat{T}, \tilde{T}$ are different $\hat{T} \neq \tilde{T}$:

$$\hat{\lambda}_{n,+}^\alpha = \hat{T}[X]; \qquad (7.2.53)$$

$$\tilde{\lambda}_{n,\oplus}^\alpha = \tilde{T}[Y]. \qquad (7.2.54)$$

We also assume that the estimators $\hat{\lambda}_{n,+}^\alpha$, $\tilde{\lambda}_{n,\oplus}^\alpha$ of an unknown nonrandom parameter in sample spaces $\mathcal{LS}(\mathcal{X}, \mathcal{B}_\mathcal{X}; +)$ and $\mathcal{A}(\mathcal{Y}, \mathcal{B}_\mathcal{Y}; S)$ are characterized by PDFs $p_{\hat{\lambda}n,+}^\alpha(z)$, $p_{\tilde{\lambda}n,\oplus}^\alpha(z)$, respectively, under the condition that measurements errors $\{N_i\}$ are independent and are described by PDF $p_N^\alpha(z)$.

Definition 7.2.1. A set Λ^α of pairs of the estimators $\{\hat{\lambda}_{k,+}^\alpha; \tilde{\lambda}_{n,\oplus}^\alpha\}$, $k \leq n$, $\{\hat{\lambda}_{k,+}^\alpha; \tilde{\lambda}_{n,\oplus}^\alpha\} \subset \Lambda^\alpha$ is called *estimator space* $(\Lambda^\alpha, d^\alpha)$ with metric $d^\alpha(\hat{\lambda}_{k,+}^\alpha, \tilde{\lambda}_{n,\oplus}^\alpha)$:

$$d^\alpha(\hat{\lambda}_{k,+}^\alpha, \tilde{\lambda}_{n,\oplus}^\alpha) = \frac{1}{2}\int_{-\infty}^{\infty} |p_{\hat{\lambda}k,+}^\alpha(z) - p_{\tilde{\lambda}n,\oplus}^\alpha(z)|dz = d^\alpha(p_{\hat{\lambda}k,+}^\alpha(z), p_{\tilde{\lambda}n,\oplus}^\alpha(z)), \quad (7.2.55)$$

where $d^\alpha(p_{\hat{\lambda}k,+}^\alpha(z), p_{\tilde{\lambda}n,\oplus}^\alpha(z)) = \frac{1}{2}\int_{-\infty}^{\infty} |p_{\hat{\lambda}k,+}^\alpha(z) - p_{\tilde{\lambda}n,\oplus}^\alpha(z)|dz$ is a metric between PDFs $p_{\hat{\lambda}k,+}^\alpha(z)$, $p_{\tilde{\lambda}n,\oplus}^\alpha(z)$ of the estimators $\hat{\lambda}_{k,+}^\alpha$ and $\tilde{\lambda}_{n,\oplus}^\alpha$, respectively, under the condition that independent measurement errors $\{N_i\}$ are characterized by PDF $p_N^\alpha(z)$ from some indexed (over a parameter α) set $P = \{p_N^\alpha(z)\}$; k and n are the sizes of the samples $X = (X_1, \ldots, X_k)$, $Y = (Y_1, \ldots, Y_n)$; $X \in \mathcal{LS}(\mathcal{X}, \mathcal{B}_\mathcal{X}; +)$, $Y \in \mathcal{A}(\mathcal{Y}, \mathcal{B}_\mathcal{Y}; S)$, respectively, $k \leq n$.

To define the quality of the estimator $\tilde\lambda_{n,\oplus}^\alpha$ of an unknown nonrandom parameter λ in estimator space $(\Lambda^\alpha, d^\alpha)$, we introduce an index based on the metric (7.2.55).

Definition 7.2.2. By a *quality index of the estimator* $\tilde\lambda_{n,\oplus}^\alpha$ we mean a quantity $q\{\tilde\lambda_{n,\oplus}^\alpha\}$ equal to the metric (7.2.55) between the PDF $p_{\hat\lambda_{1,+}^\alpha}^\alpha(z)$ of the estimator $\hat\lambda_{1,+}^\alpha$ of a parameter λ obtained on the basis of the only measurement result X_1 within the model of additive interaction between the parameter λ and measurement errors (7.2.51), and PDF $p_{\tilde\lambda_{n,\oplus}^\alpha}^\alpha(z)$ of the estimator $\tilde\lambda_{n,\oplus}^\alpha$ of parameter λ obtained on the basis of all n measurement results of the observed sample collection $Y = (Y_1, \ldots, Y_n)$ within the model of investigated interaction between the parameter λ and measurement errors (7.2.52):

$$q\{\tilde\lambda_{n,\oplus}^\alpha\} = d^\alpha(p_{\hat\lambda_{1,+}^\alpha}^\alpha(z), p_{\tilde\lambda_{n,\oplus}^\alpha}^\alpha(z)). \qquad (7.2.56)$$

Now we compare how distinct from each other are the PDFs $p_{\hat\lambda_{1,+}^\alpha}^\alpha(z)$, $p_{\tilde\lambda_{n,\oplus}^\alpha}^\alpha(z)$ of the estimators $\hat\lambda_{1,+}^\alpha$ and $\tilde\lambda_{n,\oplus}^\alpha$, i.e., the estimators obtained on the basis of the only measurement result and n measurement results, respectively. By $\tilde\lambda_{n,\oplus}^\alpha$ we mean the estimator (7.2.54) obtained within the model (7.2.52), so that a linear sample space $\mathcal{LS}(\mathcal{X}, \mathcal{B}_\mathcal{X}; +)$ can be also used as a sample space with the properties of universal algebra $\mathcal{A}(\mathcal{Y}, \mathcal{B}_\mathcal{Y}; S)$.

The following theorem helps to determine a relation between the values of quality indices of the estimation $q\{\tilde\lambda_{n,+}\}$ and $q\{\tilde\lambda_{n,\wedge}\}$ (7.2.56) for the estimators $\tilde\lambda_{n,+}$ and $\tilde\lambda_{n,\wedge}$ defined by the expressions (7.2.10) and (7.2.11) within the models of direct measurement (7.2.8), (7.2.9) in both linear sample space $\mathcal{LS}(\mathcal{X}, \mathcal{B}_\mathcal{X}; +)$ and sample space with lattice properties $\mathcal{L}(\mathcal{Y}, \mathcal{B}_\mathcal{Y}; \vee, \wedge)$, respectively, on the assumption of normalcy of distribution of measurement errors $\{N_i\}$ with PDF $p_N(z)$ in the form: $p_N(z) = (2\pi D)^{-1/2} \exp\left(-z^2/2D\right)$.

Theorem 7.2.3. *The relation between quality indices $q\{\tilde\lambda_{n,+}\}$ and $q\{\tilde\lambda_{n,\wedge}\}$ of the estimators $\tilde\lambda_{n,+}$ and $\tilde\lambda_{n,\wedge}$ in linear sample space $\mathcal{LS}(\mathcal{X}, \mathcal{B}_\mathcal{X}; +)$ and in sample space with lattice properties $\mathcal{L}(\mathcal{Y}, \mathcal{B}_\mathcal{Y}; \vee, \wedge)$, respectively, on the assumption of normalcy of distribution of measurement errors, is determined by the following inequality:*

$$q\{\tilde\lambda_{n,\wedge}\} \geq 1 - 2^{-n} > 1 - [2\Phi(1) - 1]\sqrt{\frac{2}{n+1}} \geq q\{\tilde\lambda_{n,+}\}, \ n \geq 1, \qquad (7.2.57)$$

where $\Phi(x) = \frac{1}{\sqrt{2\pi}} \int\limits_{-\infty}^{x} \exp\left(-\frac{t^2}{2}\right) dt$.

Proof. As for the estimator $\tilde\lambda_{n,+}$, on the assumption of normalcy of distribution of measurement errors $\{N_i\}$, metric $d(p_{\hat{\lambda}1,+}(z), p_{\hat{\lambda}n,+}(z))$ is determined by the expression:

$$d(p_{\hat{\lambda}1,+}(z), p_{\hat{\lambda}n,+}(z)) = 2\left|\Phi\left(\sqrt{\frac{n}{n-1}\ln n}\right) - \Phi\left(\sqrt{\frac{1}{n-1}\ln n}\right)\right|, \qquad (7.2.58)$$

where $\Phi(x) = \frac{1}{\sqrt{2\pi}}\int_{-\infty}^{x}\exp\left(-\frac{t^2}{2}\right)dt$;

$$p_{\hat{\lambda}1,+}(z) = (2\pi D)^{-1/2}\exp\left(-(z-\lambda)^2/2D\right); \qquad (7.2.59)$$

$$p_{\hat{\lambda}n,+}(z) = (2\pi D/n)^{-1/2}\exp\left(-(z-\lambda)^2/(2D/n)\right). \qquad (7.2.60)$$

To make the following calculations convenient, we denote:

$$x' = \sqrt{\frac{n}{n-1}\ln n}, \quad x'' = \sqrt{\frac{1}{n-1}\ln n}.$$

Then the metric (7.2.58) is determined by the expression:

$$d(p_{\hat{\lambda}1,+}(z), p_{\hat{\lambda}n,+}(z)) = 2\left|\Phi(x') - \Phi(x'')\right|. \qquad (7.2.61)$$

On $x \geq 0$, the following inequality holds [258], [261]:

$$\Phi(x) \leq \Phi_1(x) = \frac{1}{2} + \frac{1}{2}\sqrt{1 - \exp(-2x^2/\pi)}. \qquad (7.2.62)$$

On $n \geq 1$, the relation $x'' = \sqrt{\frac{1}{n-1}\ln n} \leq 1$ is true. Thus, on $0 \leq x \leq 1$, the inequality holds:

$$\Phi(x) \geq \Phi_2(x) = \frac{1}{2} + \left[\Phi(1) - \frac{1}{2}\right]\cdot x. \qquad (7.2.63)$$

Using the inequalities (7.2.62) and (7.2.63), we obtain the following inequality:

$$d(p_{\hat{\lambda}1,+}(z), p_{\hat{\lambda}n,+}(z)) = 2\left|\Phi(x') - \Phi(x'')\right| \leq 2\left|\Phi_1(x') - \Phi_2(x'')\right|. \qquad (7.2.64)$$

The following series expansion of logarithmic function is known [262]:

$$\ln x = 2\sum_{k=1}^{\infty}\frac{(x-1)^{2k-1}}{(2k-1)(x+1)^{2k-1}}, \quad x > 0. \qquad (7.2.65)$$

Then the following inequality holds:

$$x'' = \sqrt{\frac{1}{n-1}\ln n} \geq \sqrt{\frac{2}{n+1}}, \quad n \geq 1. \qquad (7.2.66)$$

Applying the inequality (7.2.66) to the inequality (7.2.64) and taking into account that $\Phi_1(x') \leq 1$, we obtain the inequality:

$$d(p_{\hat{\lambda}1,+}(z), p_{\hat{\lambda}n,+}(z)) \leq 2\left|\Phi_1(x') - \Phi_2(x'')\right| \leq 1 - [2\Phi(1) - 1]\sqrt{\frac{2}{n+1}}. \qquad (7.2.67)$$

7.2 Estimation of Unknown Nonrandom Parameter in Sample Space with Lattice Properties

Quality index $q\{\hat{\lambda}_{n,+}\}$ of the estimator $\hat{\lambda}_{n,+}$, according to (7.2.56), is determined by the metric (7.2.58), which, by virtue of the inequality (7.2.67), is bounded above by the quantity:

$$q\{\hat{\lambda}_{n,+}\} = d(p_{\hat{\lambda}1,+}(z), p_{\hat{\lambda}n,+}(z)) \leq 1 - [2\Phi(1) - 1]\sqrt{\frac{2}{n+1}}. \qquad (7.2.68)$$

As for the estimator $\hat{\lambda}_{n,\wedge}$ of parameter λ in sample space with lattice properties $\mathcal{L}(\mathcal{Y}, \mathcal{B}_y; \vee, \wedge)$, the metric $d(p_{\hat{\lambda}1,+}(z), p_{\hat{\lambda}n,\wedge}(z))$ is determined by the expression:

$$d(p_{\hat{\lambda}1,+}(z), p_{\hat{\lambda}n,\wedge}(z)) = \frac{1}{2}\int_{-\infty}^{\infty} |p_{\hat{\lambda}1,+}(z) - p_{\hat{\lambda}n,\wedge}(z)|dz, \qquad (7.2.69)$$

where $p_{\hat{\lambda}n,\wedge}(z)$ is PDF of the estimator $\hat{\lambda}_{n,\wedge}$ determined by the formula (7.2.19) when the sample $Y = (Y_1, \ldots, Y_n)$ with a size n is used within the model (7.2.9); $p_{\hat{\lambda}1,+}(z)$ is the PDF of the estimator $\hat{\lambda}_{n,+}$ determined by the formula (7.2.59), when the only element X_1 from the sample $X = (X_1, \ldots, X_k)$ is used within the model (7.2.8).

Substituting the expressions (7.2.19) and (7.2.59) into the formula (7.2.69), we obtain the exact value of metric:

$$d(p_{\hat{\lambda}1,+}(z), p_{\hat{\lambda}n,\wedge}(z)) = P_c = 1 - [1 - F_N(\lambda)]^n, \qquad (7.2.70)$$

where P_c is a probability of a correct formation of the estimator $\hat{\lambda}_{n,\wedge}$ defined by the formula (7.2.20).

Since the estimated parameter λ takes nonnegative values ($\lambda \geq 0$), the quality index $q\{\hat{\lambda}_{n,\wedge}\}$ of the estimator $\hat{\lambda}_{n,\wedge}$, according to definition (7.2.56), the identity (7.2.70), and the inequality (7.2.24), is bounded below by the quantity $1 - 2^{-n}$:

$$q\{\hat{\lambda}_{n,\wedge}\} = d(p_{\hat{\lambda}1,+}(z), p_{\hat{\lambda}n,\wedge}(z)) \geq 1 - 2^{-n}. \qquad (7.2.71)$$

The joint fulfillment of the inequalities (7.2.68) and 7.2.71), and the inequality:

$$1 - 2^{-n} > 1 - [2\Phi(1) - 1]\sqrt{\frac{2}{n+1}},$$

implies fulfillment of the inequality (7.2.57). $\square$

Thus, Theorem 7.2.3 determines the upper $\Delta_{\wedge}(n)$ and lower $\Delta_{+}(n)$ bounds of the rates of a convergence of quality indices $q\{\hat{\lambda}_{n,\wedge}\}$ and $q\{\hat{\lambda}_{n,+}\}$ of the estimators $\hat{\lambda}_{n,\wedge}$ and $\hat{\lambda}_{n,+}$ to 1, described by the following functions, respectively:

$$\Delta_{\wedge}(n) = 2^{-n}; \quad \Delta_{+}(n) = [2\Phi(1) - 1]\sqrt{\frac{2}{n+1}},$$

whose plots are shown in Fig. 7.2.2.

The analysis of relative efficiency of the estimators $\hat{\lambda}_{n,\wedge}$, $\hat{\lambda}_{n,+}$ (7.2.47) and also the relation between their quality indices (7.2.57) allows us to draw the following conclusions.

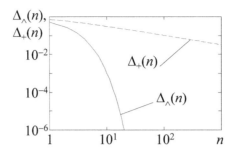

FIGURE 7.2.2 Upper $\Delta_\wedge(n)$ and lower $\Delta_+(n)$ bounds of quality indices $q\{\hat{\lambda}_{n,\wedge}\}$ and $q\{\hat{\lambda}_{n,+}\}$

1. The essential distinctions in characteristics and behavior of the estimators $\hat{\lambda}_{n,\wedge}$ and $\hat{\lambda}_{n,+}$ showing the advantage of the estimator $\hat{\lambda}_{n,\wedge}$ are elucidated by the features of the interaction between an unknown nonrandom parameter λ and measurement errors $\{N_i\}$ in sample space with lattice properties (model (7.2.9)). Under interactions between a parameter and measurement errors $\{N_i\}$ in sample space with lattice properties $\mathcal{L}(\mathcal{Y}, \mathcal{B}_\mathcal{Y}; \vee, \wedge)$, information contained in a sample is used with essentially lower losses than occur under their additive interaction in linear sample space $\mathcal{LS}(\mathcal{X}, \mathcal{B}_\mathcal{X}; +)$. The last circumstance stipulates higher efficiency of the estimation in sample space with lattice properties $\mathcal{L}(\mathcal{Y}, \mathcal{B}_\mathcal{Y}; \vee, \wedge)$ as against the estimation in linear sample space $\mathcal{LS}(\mathcal{X}, \mathcal{B}_\mathcal{X}; +)$.

2. Asymptotic properties of the estimator $\hat{\lambda}_{n,\wedge}$ in sample space with lattice properties $\mathcal{L}(\mathcal{Y}, \mathcal{B}_\mathcal{Y}; \vee, \wedge)$ are much better than these properties of the estimator $\hat{\lambda}_{n,+}$ in linear sample space $\mathcal{LS}(\mathcal{X}, \mathcal{B}_\mathcal{X}; +)$. The rate of a convergence of the quality index (7.2.56) to 1, as for the estimator $\hat{\lambda}_{n,\wedge}$, is determined by exponential dependence on a sample size n, and as for the estimator $\hat{\lambda}_{n,+}$, this rate does not exceed the square root of a sample size n.

3. The estimator $\hat{\lambda}_{n,\wedge}$ is more robust with respect to the kind of the distribution of measurement errors $\{N_i\}$ than the estimator $\hat{\lambda}_{n,+}$.

4. Consider the case of the estimation of an unknown nonrandom nonpositive parameter $\lambda \leq 0$ for the model of direct measurement in sample space with lattice properties $\mathcal{L}(\mathcal{Y}, \mathcal{B}_\mathcal{Y}; \vee, \wedge)$:

$$Z_i = \lambda \wedge N_i, \qquad (7.2.72)$$

where $\{N_i\}$ are independent measurement errors normally distributed with zero expectation and a variance D represented by the sample $N = (N_1, \ldots, N_n)$, $N_i \in \mathcal{N}$, so that $N \in \mathcal{L}(\mathcal{Y}, \mathcal{B}_\mathcal{Y}; \vee, \wedge)$; $\{Z_i\}$ are the measurements results represented by the sample $Z = (Z_1, \ldots, Z_n)$, $Z_i \in \mathcal{Z}$: $Z \in \mathcal{L}(\mathcal{Y}, \mathcal{B}_\mathcal{Y}; \vee, \wedge)$; $\wedge$ is operation of the meet of sample space with lattice properties $\mathcal{L}(\mathcal{Y}, \mathcal{B}_\mathcal{Y}; \vee, \wedge)$; $i = 1, \ldots, n$ is an index of the elements of statistical collections $\{N_i\}$, $\{Z_i\}$; n is a size of the samples $N = (N_1, \ldots, N_n)$, $Z = (Z_1, \ldots, Z_n)$.

If we take join of lattice (7.2.7b) as the estimator $\hat{\lambda}_{n,\vee}$ of a parameter λ within the model (7.2.72):

$$\hat{\lambda}_{n,\vee} = \bigvee_{i=1}^{n} Z_i,$$

where $\bigvee_{i=1}^{n} Z_i = \sup\{Z_i\}$ is the largest value from the sample $Z = (Z_1, \ldots, Z_n)$, then the estimator $\hat{\lambda}_{n,\vee}$ is characterized by quality index $q\{\hat{\lambda}_{n,\vee}\}$, which is determined according to the metric (7.2.56), and which is equal to quality index $q\{\hat{\lambda}_{n,\wedge}\}$ of the estimator $\hat{\lambda}_{n,\wedge}$: $q\{\hat{\lambda}_{n,\vee}\} = q\{\hat{\lambda}_{n,\wedge}\}$.

5. The proposed quality index (7.2.56) can be used successfully to determine the estimation quality on a wide class of sample spaces without constraints with respect to their algebraic and probabilistic-statistical properties.

7.3 Extraction of Stochastic Signal in Metric Space with Lattice Properties

The extraction (filtering) of a signal under interference (noise) background is the most general problem of signal processing theory. The known variants of an extraction problem statement are formulated for an additive interaction $x(t) = s(t) + n(t)$ (in terms of linear space) between useful signal $s(t)$ and interference (noise) $n(t)$ [254], [149], [163], [155]. Nevertheless, some authors assume the statement of this problem on the basis of a more general model of interaction: $x(t) = \Phi[s(t), n(t)]$, where Φ is some deterministic function [155, 163].

This section is intended to achieve a twofold goal. First, it is necessary to synthesize algorithm and unit of narrowband signal extraction in the presence of interference (noise) with independent samples in signal space with L-group properties. Second, it is necessary to describe characteristics and properties of the synthesized unit and compare them with known analogues solving the extraction (filtering) problem in linear signal space.

The synthesis and analysis of optimal algorithm of stochastic narrowband signal extraction (filtering) are realized under the following assumptions. While synthesizing, distributions of the signal $s(t)$ and interference (noise) $n(t)$ are considered arbitrary. Further, while analyzing a synthesized processing unit and its algorithm, both the signal $s(t)$ and interference (noise) $n(t)$ are considered Gaussian stochastic processes.

Let an interaction between stochastic signal $s(t)$ and interference (noise) $n(t)$ in signal space $\mathcal{L}(+, \vee, \wedge)$ with L-group properties be described by two binary operations $\vee$ and $\wedge$ in two receiving channels, respectively:

$$x(t) = s(t) \vee n(t); \qquad (7.3.1a)$$

$$\tilde{x}(t) = s(t) \wedge n(t). \qquad (7.3.1b)$$

Instantaneous values (time samples) of the signal $\{s(t_j)\}$ and interference (noise) $\{n(t_j)\}$ are the elements of signal space: $s(t_j), n(t_j) \in \mathcal{L}(+, \vee, \wedge)$. Time samples of interference (noise) $\{n(t_j)\}$ are considered independent. Let the samples $s(t_j)$ and $n(t_j)$ of the signal $s(t)$ and interference (noise) $n(t)$ be taken on the domain of definition T_s of the signal $s(t)$: $t_j \in T_s$ through the discretization (sampling) interval Δt, which provides the independence of the samples of interference (noise) $\{n(t_j)\}$, and $\Delta t << 1/f_0$, where f_0 is an unknown carrier frequency of the signal $s(t)$.

Taking into account the aforementioned constraints, the equations of observations in two receiving (processing) channels (7.3.1a) and (7.3.1b) look like:

$$x(t_j) = s(t_j) \vee n(t_j); \qquad (7.3.2a)$$

$$\tilde{x}(t_j) = s(t_j) \wedge n(t_j), \qquad (7.3.2b)$$

where $t_j = t - j\Delta t$, $j = 0, 1, \ldots, N-1$, $t_j \in T^*$, $T^* \subset T_s$; T^* is a processing interval: $T^* = [t - (N-1)\Delta t, t]$; $N \in \mathbf{N}$, $\mathbf{N}$ is the set of natural numbers.

7.3.1 Synthesis of Optimal Algorithm of Stochastic Signal Extraction in Metric Space with *L*-group Properties

While solving the problem of useful signal extraction (filtering), on the basis of processing of two random sequences $\{x(t_j)\}$ and $\{\tilde{x}(t_j)\}$, it is necessary to form such an estimator $\hat{s}(t)$ of the useful signal $s(t)$ in the output of processing unit, which in the best way (with respect to chosen criteria) corresponds to the receiving signal $s(t)$. The problem of extraction $Ext[s(t)]$ of stochastic signal $s(t)$ is formulated and solved through step-by-step processing of statistical collections $\{x(t_j)\}$ and $\{\tilde{x}(t_j)\}$ determined by the observation Equations (7.3.2a) and (7.3.2b):

$$Ext[s(t)] = \begin{cases} PF[s(t)]; & (a) \\ IP[s(t)]; & (b) \\ Sm[s(t)], & (c) \end{cases} \qquad (7.3.3)$$

where $PF[s(t)]$ is primary filtering; $IP[s(t)]$ is intermediate processing; $Sm[s(t)]$ is smoothing; they represent the successive stages of signal processing of the general algorithm of useful signal extraction (7.3.3).

The optimality criteria defining every processing stage $PF[s(t)]$, $IP[s(t)]$, and

7.3 Extraction of Stochastic Signal in Metric Space with Lattice Properties

$Sm[s(t)]$ that are interrelated and united into separate systems:

$$PF[s(t)] = \begin{cases} y(t) = \underset{\breve{y}(t) \in Y; t, t_j \in T^*}{\arg\min} \; |\overset{N-1}{\underset{j=0}{\wedge}} [x(t_j) - \breve{y}(t)]|; & (a) \\ \tilde{y}(t) = \underset{\widehat{y}(t) \in \tilde{Y}; t, t_j \in T^*}{\arg\min} \; |\overset{N-1}{\underset{j=0}{\vee}} [\tilde{x}(t_j) - \widehat{y}(t)]|; & (b) \\ w(t) = F[y(t), \tilde{y}(t)]; & (c) \\ \sum_{j=0}^{N-1} |w(t_j) - s(t_j)|\|_{n(t) \equiv 0, \Delta t \to 0} \to \underset{w(t_j) \in W}{\min}; & (d) \\ N = \underset{N' \in \mathbf{N} \cap]0, N^*]}{\arg\max} [\delta_d(N')]|_{N^*: \delta_d(N^*) = \delta_{d0}}, & (e) \end{cases} \quad (7.3.4)$$

where $y(t)$ and $\tilde{y}(t)$ functions are the solutions of minimization problems of metrics between the observed statistical collections $\{x(t_j)\}$ and $\{\tilde{x}(t_j)\}$ and optimization variables, i.e., the functions $\breve{y}(t)$ and $\widehat{y}(t)$, respectively; $w(t)$ is the function $F[*, *]$ uniting the results $y(t)$ and $\tilde{y}(t)$ of minimization of the functions of the observed collections $\{x(t_j)\}$ (7.3.2a) and $\{\tilde{x}(t_j)\}$ (7.3.2b); T^* is a processing interval; $N \in \mathbf{N}$, $\mathbf{N}$ is the set of natural numbers; N is a number of the samples of stochastic processes $x(t)$, $\tilde{x}(t)$ used in primary processing; $\delta_d(N)$ is a relative dynamic error of filtering as a function of sample number N; δ_{d0} is a given quantity of relative dynamic error of filtering:

$$IP[s(t)] = \begin{cases} \mathbf{M}\{u^2(t)\}|_{s(t) \equiv 0} \to \underset{L}{\min}; & (a) \\ \mathbf{M}\{[u(t) - s(t)]^2\}|_{n(t) \equiv 0, \Delta t \to 0} = \varepsilon; & (b) \\ u(t) = L[w(t)], & (c) \end{cases} \quad (7.3.5)$$

where $\mathbf{M}\{*\}$ is a symbol of mathematical expectation; $L[w(t)]$ is a functional transformation of the process $w(t)$ into the process $u(t)$; ε is a constant, which generally, is some function of signal power; for instance, in the case of Gaussian signal it is equal to: $\varepsilon = \varepsilon_0 D_s$, $\varepsilon_0 = $ const; $\varepsilon_0 << 1$, D_s is a variance of Gaussian signal $s(t)$;

$$Sm[s(t)] = \begin{cases} v(t) = \underset{v^\circ(t) \in V; t, t_k \in \tilde{T}}{\arg\min} \sum_{k=0}^{M-1} |u(t_k) - v^\circ(t)|; & (a) \\ \Delta \tilde{T}: \delta_d(\Delta \tilde{T}) = \delta_{d,\text{sm}}; & (b) \\ M = \underset{M' \in \mathbf{N} \cap [M^*, \infty[}{\arg\max} [\delta_f(M')]|_{M^*: \delta_f(M^*) = \delta_{f,\text{sm}}}, & (c) \end{cases} \quad (7.3.6)$$

where $v(t) = \hat{s}(t)$ is a result of filtering (the estimator $\hat{s}(t)$ of the signal $s(t)$) that is the solution of minimization of a metric between the instantaneous values of stochastic process $u(t)$ and optimization variable, i.e., the function $v^\circ(t)$; $t_j = t - j\Delta t$, $j = 0, 1, \ldots, N-1$, $t_j \in T^*$; $t_k = t - \frac{k}{M}\Delta \tilde{T}$, $k = 0, 1, \ldots, M-1$, $t_k \in \tilde{T}$, $\tilde{T} =]t - \Delta \tilde{T}, t]$; $\tilde{T}$ is an interval in which the smoothing of stochastic process $u(t)$ is realized; $\Delta \tilde{T}$ is a quantity of a smoothing interval $\tilde{T}$; $M \in \mathbf{N}$, $\mathbf{N}$ is the set of natural numbers; M is the number of samples of stochastic process $u(t)$ used during smoothing; $\delta_d(\Delta \tilde{T})$, $\delta_f(M)$ are relative dynamic and fluctuation errors of smoothing as the dependences on the quantity $\Delta \tilde{T}$ of the smoothing interval $\tilde{T}$ and the number of samples M, respectively; $\delta_{d,\text{sm}}$ and $\delta_{f,\text{sm}}$ are relative dynamic and fluctuation errors of smoothing, respectively.

The optimality criteria and single relations appearing in the systems (7.3.4), (7.3.5), and (7.3.6) define consecutive stages of processing $PF[s(t)]$, $IP[s(t)]$, $Sm[s(t)]$ of the general algorithm of useful signal extraction (7.3.3).

The equations (7.3.4a), (7.3.4b) define a criterion of minimum metric between the statistical sets of observations $\{x(t_j)\}$ and $\{\tilde{x}(t_j)\}$ and the results of primary filtering $y(t)$ and $\tilde{y}(t)$, respectively. The functions of metrics $|\bigwedge_{j=0}^{N-1}[x(t_j) - \breve{y}(t)]|$ and $|\bigvee_{j=0}^{N-1}[\tilde{x}(t_j) - \widehat{y}(t)]|$ are chosen taking into account the metric convergence and the convergence in probability of the sequences $y_{N-1} = \bigwedge_{j=0}^{N-1} x(t_j)$, $\tilde{y}_{N-1} = \bigvee_{j=0}^{N-1} \tilde{x}(t_j)$ to the estimated parameter for the interactions of the kind (7.2.2a) and (7.2.2b) (see Section 7.2). The equation (7.3.4d) defines the criterion of minimum metric $\sum_{j=0}^{N-1}|w(t_j) - s(t_j)|$ between the useful signal $s(t)$ and the process $w(t)$ in the processing interval $T^* = [t - (N-1)\Delta t, t]$. This criterion establishes the function $F[y(t), \tilde{y}(t)]$ (7.3.4c) uniting the results $y(t)$ and $\tilde{y}(t)$ of primary processing of the observed processes $x(t)$ and $\tilde{x}(t)$. Criterion (7.3.4d) is considered under two constraint conditions: (1) interference (noise) $n(t)$ is absent in the input of the signal processing unit; (2) the sample interval Δt tends to zero: $\Delta t \to 0$. The equation (7.3.4e) establishes the criterion of the choice of sample number N of stochastic processes $x(t)$ and $\tilde{x}(t)$ providing a given quantity of a relative dynamic error δ_{d0} of primary filtering.

The equations (7.3.5a) through (7.3.5c) establish the criterion of the choice of functional transformation $L[w(t)]$. The equation (7.3.5a) defines the criterion of minimum of the second moment of the process $u(t)$ in the absence of useful signal $s(t)$ in the input of the signal processing unit. The equation (7.3.5b) establishes the quantity of the second moment of the difference between the signals $u(t)$ and $s(t)$ under two constraint conditions: (1) interference (noise) $n(t)$ is absent in the input of the signal processing unit; (2) sample interval Δt tends to zero: $\Delta t \to 0$. The relation (7.3.5c) defines a coupling equation between the processes $u(t)$ and $w(t)$.

The equation (7.3.6a) defines the criterion of minimum metric $\sum_{k=0}^{M-1}|u(t_k) - v^\circ(t)|$ between instantaneous values of the process $u(t)$ and optimization variable $v^\circ(t)$ in the smoothing interval $\tilde{T} =]t - \Delta\tilde{T}, t]$, requiring the final processing of the signal $u(t)$ in the form of its smoothing. The equation (7.3.6b) establishes a criterion of the choice of the quantity $\Delta\tilde{T}$ of smoothing interval $\tilde{T}$ based on providing a given quantity of dynamic error of smoothing $\delta_{d,\text{sm}}$. The equation (7.3.6c) defines the criterion of the choice of sample number M of stochastic process $u(t)$ providing a given quantity of fluctuation error of smoothing $\delta_{f,\text{sm}}$.

We obtain the estimator $\hat{s}(t) = v(t)$ of the signal $s(t)$ in the output of the signal processing unit by consecutivly solving the optimization relationships of the system (7.3.4).

7.3 Extraction of Stochastic Signal in Metric Space with Lattice Properties

To solve the problem of minimization of the functions $|\bigwedge_{j=0}^{N-1}[x(t_j)-\breve{y}(t)]|$ (7.3.4a), $|\bigvee_{j=0}^{N-1}[\tilde{x}(t_j)-\widehat{y}(t)]|$ (7.3.4b), it is necessary to determine the extrema of these functions, setting their derivatives with respect to $\breve{y}(t)$ and $\widehat{y}(t)$ to zero, respectively:

$$d|\bigwedge_{j=0}^{N-1}[x(t_j)-\breve{y}(t)]|/d\breve{y}(t) = -\text{sign}(\bigwedge_{j=0}^{N-1}[x(t_j)-\breve{y}(t)]) = 0; \quad (7.3.7a)$$

$$d|\bigvee_{j=0}^{N-1}[\tilde{x}(t_j)-\widehat{y}(t)]|/d\widehat{y}(t) = -\text{sign}(\bigvee_{j=0}^{N-1}[\tilde{x}(t_j)-\widehat{y}(t)]) = 0. \quad (7.3.7b)$$

The solutions of Equations (7.3.7a) and (7.3.7b) are the values of the estimators $\breve{y}(t)$ and $\tilde{y}(t)$ in the form of meet and join of the observation results $\{x(t_j)\}$ and $\{\tilde{x}(t_j)\}$, respectively:

$$\breve{y}(t) = \bigwedge_{j=0}^{N-1} x(t_j) = \bigwedge_{j=0}^{N-1} x(t - j\Delta t); \quad (7.3.8a)$$

$$\tilde{y}(t) = \bigvee_{j=0}^{N-1} \tilde{x}(t_j) = \bigvee_{j=0}^{N-1} \tilde{x}(t - j\Delta t). \quad (7.3.8b)$$

The derivatives of the functions $|\bigwedge_{j=0}^{N-1}[x(t_j)-\breve{y}(t)]|$ and $|\bigvee_{j=0}^{N-1}[\tilde{x}(t_j)-\widehat{y}(t)]|$, according to the relationships (7.3.7a) and (7.3.7b), change their sign from minus to plus in the points $\breve{y}(t)$ and $\tilde{y}(t)$. Thus, the extrema determined by the formulas (7.3.8a), (7.3.8b) are the minimum points of these functions and, respectively, are the solutions of the equations (7.3.4a), (7.3.4b) defining these criteria of estimation (signal filtering).

The conditions of the criterion (7.3.4d) of the system (7.3.4): $n(t) \equiv 0$, $\Delta t \to 0$ imply the equations of observations (7.3.2a,b) of the following form: $x(t_j) = s(t_j) \vee 0$, $\tilde{x}(t_j) = s(t_j) \wedge 0$, and according to the relationships (7.3.8a), (7.3.8b), the identities hold:

$$\breve{y}(t)\big|_{n(t)\equiv 0, \Delta t \to 0} = \bigwedge_{j=0}^{N-1}[s(t_j) \vee 0] = s(t) \vee 0; \quad (7.3.9a)$$

$$\tilde{y}(t)\big|_{n(t)\equiv 0, \Delta t \to 0} = \bigvee_{j=0}^{N-1}[s(t_j) \wedge 0] = s(t) \wedge 0, \quad (7.3.9b)$$

where $t_j = t - j\Delta t$, $j = 0, 1, \ldots, N-1$.

To provide the criterion (7.3.4d) of the system (7.3.4) on joint fulfillment of the identities (7.3.9a), (7.3.9b), and (7.3.4c), it is necessary and sufficient that the coupling equation (7.3.4c) between the stochastic process $w(t)$ and a pair of the results of primary processing $\breve{y}(t)$, $\tilde{y}(t)$ has the form:

$$w(t)\big|_{n(t)\equiv 0, \Delta t \to 0} = \breve{y}(t)\big|_{n(t)\equiv 0, \Delta t \to 0} \vee 0 + \tilde{y}(t)\big|_{n(t)\equiv 0, \Delta t \to 0} \wedge 0 =$$
$$= s(t) \vee 0 \vee 0 + s(t) \wedge 0 \wedge 0 = s(t) \vee 0 + s(t) \wedge 0 = s(t). \quad (7.3.10)$$

Based on expression (7.3.10), the metric $\sum_{j=0}^{N-1}|w(t_j) - s(t_j)|$, that has to be minimized according to the criterion (7.3.4d) is minimal and equal to zero.

It is obvious that the coupling equation (7.3.4c) has to be invariant with respect to the presence (absence) of interference (noise) $n(t)$, so the final coupling equation can be written on the basis of the identity (7.3.10) in the form:

$$w(t) = y_+(t) + \tilde{y}_-(t); \tag{7.3.11}$$
$$y_+(t) = y(t) \vee 0; \tag{7.3.11a}$$
$$\tilde{y}_-(t) = \tilde{y}(t) \wedge 0. \tag{7.3.11b}$$

Thus, the identity (7.3.11) defines the kind of coupling equation (7.3.4c) obtained on the basis of joint fulfillment of criteria (7.3.4a), (7.3.4b), and (7.3.4d).

The solution $u(t)$ of the relationships (7.3.5a) through (7.3.5c) establishing the criterion of the choice of the functional transformation of the process $w(t)$ is the function $L[w(t)]$ defining the gain characteristic of the limiter:

$$u(t) = L[w(t)] = \begin{cases} a, & w(t) \geq a; \\ w(t), & -a < w(t) < a; \\ -a, & w(t) \leq -a, \end{cases} \tag{7.3.12}$$

and its linear part provides the condition (7.3.5b), and its clipping part (above and below) provides the minimization of the second moment of the process $u(t)$ according to the criterion (7.3.5a).

The relationship (7.3.12) can be written in terms of L-group $\mathcal{L}(+, \vee, \wedge)$ in the form:

$$u(t) = L[w(t)] = [(w(t) \wedge a) \vee 0] + [(w(t) \vee (-a)) \wedge 0], \tag{7.3.12a}$$

where, in the case of a Gaussian signal $s(t)$ with a variance D_s that is smaller than D: $D_s < D$, the limiter parameter a can be chosen proportional to $\sqrt{D}$: $a^2 \sim D$ providing the equation (7.3.5b) holds.

We can finally obtain the estimator $\hat{s}(t) = v(t)$ of the signal $s(t)$ in the output of the filtering unit by solving the minimization equation on the basis of criterion (7.3.6a). We find the extremum of the function $\sum_{k=0}^{M-1} |u(t_k) - v°(t)|$, setting its derivative with respect to $v°(t)$ to zero:

$$d\left\{\sum_{k=0}^{M-1} |u(t_k) - v°(t)|\right\}/dv°(t) = -\sum_{k=0}^{M-1} \text{sign}[u(t_k) - v°(t)] = 0.$$

The solution of the last equation is the value of the estimator $v(t)$ in the form of the sample median $\text{med}\{*\}$ of a collection of the samples $\{u(t_k)\}$ of stochastic process $u(t)$:

$$v(t) = \underset{t_k \in \tilde{T}}{\text{med}}\{u(t_k)\}, \tag{7.3.13}$$

where $t_k = t - \frac{k}{M}\Delta\tilde{T}$, $k = 0, 1, \ldots, M-1$; $t_k \in \tilde{T} =]t - \Delta\tilde{T}, t]$; $\tilde{T}$ is an interval in which smoothing of the stochastic process $u(t)$ is realized.

The derivative of the function $\sum_{k=0}^{M-1} |u(t_k) - v°(t)|$ in the point $v(t)$ changes its

7.3 Extraction of Stochastic Signal in Metric Space with Lattice Properties

sign from minus to plus. Thus, the extremum determined by the function (7.3.13) is a minimum of this function, and correspondingly, is the solution of Equation (7.3.6a) determining this estimation criterion.

Thus, summarizing the relationships (7.3.13), (7.3.11), (7.3.11a,b), (7.3.8a,b), one can draw a conclusion that the estimator $\hat{s}(t) = v(t)$ of the signal $s(t)$, extracted in the presence of interference (noise) $n(t)$, is the function of smoothing of the stochastic process $u(t)$ obtained by limiting the process $w(t)$ that is the sum of the results $y(t)$ and $\tilde{y}(t)$ of the corresponding primary processing of the observed stochastic processes $x(t)$ and $\tilde{x}(t)$ in the interval $T^* = [t - (N-1)\Delta t, t]$.

A block diagram of the processing unit, according to the general algorithm $Ext[s(t)]$, its stages $PF[s(t)]$, $IP[s(t)]$, $Sm[s(t)]$, and the relationships (7.3.8a,b), (7.3.11a,b), (7.3.11), (7.3.13), includes two processing channels, each containing transversal filters; units of evaluation of positive $y_+(t)$ and negative $\tilde{y}_-(t)$ parts of the processes $y(t)$ and $\tilde{y}(t)$, respectively; an adder uniting the results of signal processing in both channels; a limiter $L[w(t)]$, and median filter (MF) (see Fig. 7.3.1).

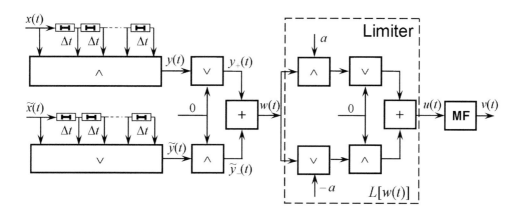

FIGURE 7.3.1 Block diagram of processing unit realizing general algorithm $Ext[s(t)]$

Transversal filters in both processing channels perform primary filtration $PF[s(t)]$ $y(t) = \bigwedge_{j=0}^{N-1} x(t - j\Delta t)$, $\tilde{y}(t) = \bigvee_{j=0}^{N-1} \tilde{x}(t - j\Delta t)$ of the observed stochastic processes $x(t)$ and $\tilde{x}(t)$ according to Equations (7.3.8a,b) providing fulfillment of the criteria (7.3.4a), (7.3.4b). The units of evaluation of positive $y_+(t)$ and negative $\tilde{y}_-(t)$ parts of the processes $y(t)$ and $\tilde{y}(t)$ in two processing channels compute the values of these functions according to the identities (7.3.11a) and (7.3.11b), respectively. The adder unites the results of signal processing in both channels according to the equality (7.3.11), providing the criteria (7.3.4c) and (7.3.4d) to hold. The limiter $L[w(t)]$ realizes intermediate processing $IP[s(t)]$ in the form of clipping of the signal $w(t)$ in the output of the adder, according to the criteria (7.3.5a), (7.3.5b), to exclude from further processing the interference (noise) overshoots, whose instantaneous values exceed a given value a. A median filter (MF) performs smoothing $Sm[s(t)]$ of stochastic process $u(t)$ according to the formula (7.3.13), providing fulfillment of criterion (7.3.6a).

7.3.2 Analysis of Optimal Algorithm of Signal Extraction in Metric Space with L-group Properties

The further analysis of the synthesized signal processing algorithm and unit will be realized so that the probabilistic-statistical characteristics of stochastic processes at various processing stages will be obtained, and then the signal processing quality indices of this unit will be determined.

To simplify the analysis of the synthesized unit, we assume that interference $n(t)$ is a quasi-white Gaussian noise with a power spectral density $N(\omega) = N_0 =$ const, $0 \leq \omega \leq \omega_{n,\max}$, $\omega_{n,\max} = 2\pi f_{n,\max}$ and the corresponding normalized correlation function $r_n(\tau)$:

$$r_n(\tau) = \sin(2\pi f_{n,\max}\tau)/(2\pi f_{n,\max}\tau), \qquad (7.3.14)$$

where $f_{n,\max}$ is an upper bound frequency of the power spectral density of interference (noise).

We also consider that the received signal $s(t)$ is a narrowband stochastic process with unknown dependences on amplitude $A(t)$ and phase $\varphi(t)$ modulation:

$$s(t) = A(t)\cos(\omega_0 t + \varphi(t)), \ t \in T_s, \qquad (7.3.15)$$

where $\omega_0 = 2\pi f_0$, f_0 is an unknown frequency of signal carrier; $\omega_{n,\max} \gg \omega_0$; T_s is a domain of definition of the signal $s(t)$.

Determine, respectively, bivariate conditional probability distribution function (PDF) and cumulative distribution function (CDF) of the instantaneous values $y(t_{1,2})$ of statistics $y(t)$ (7.3.8a) under receiving the realization $s^*(t)$ of the signal $s(t)$:

$$p_y(z_1, z_2; t_1, t_2/s^*) \equiv p_y(z_1, z_2/s^*);$$
$$F_y(z_1, z_2; t_1, t_2/s^*) \equiv F_y(z_1, z_2/s^*).$$

It is known that the CDF $F_y(z_1, z_2/s^*)$ of the meet $y(t_{1,2})$ of N independent random variables $\{x(t_{1,2} - l\Delta t)\}$ (7.3.8a) is determined by the expression [243]:

$$F_y(z_1, z_2/s^*) = 1 - [1 - F(z_1, z_2/s^*)]^N, \qquad (7.3.16)$$

where $F(z_1, z_2/s^*)$ is a conditional CDF of bivariate random variable $x(t_{1,2}) = s^*(t_{1,2}) \vee n(t_{1,2})$:

$$F(z_1, z_2/s^*) = F_n(z_1, z_2) \cdot F_s(z_1, z_2),$$

where $F_n(z_1, z_2) = \int_{-\infty}^{z_2} \int_{-\infty}^{z_1} p_n(x_1, x_2) dx_1 dx_2$ is the bivariate CDF of instantaneous values of interference (noise) $n(t_{1,2})$; $p_n(z_1, z_2)$ is the bivariate PDF of instantaneous values of interference (noise) $n(t_{1,2})$; $F_s(z_1, z_2) = 1(z_1 - s^*(t_1)) \cdot 1(z_2 - s^*(t_2))$ is the bivariate CDF of instantaneous values of realization $s^*(t_{1,2})$ of the signal $s(t_{1,2})$; $1(t)$ is Heaviside step function.

Taking into account the above, the formula (7.3.16) can be represented in the form:

$$F_y(z_1, z_2/s^*) = \left(1 - [1 - F_n(z_1, z_2)]^N\right) 1(z_1 - s^*(t_1)) \cdot 1(z_2 - s^*(t_2)). \qquad (7.3.17)$$

7.3 Extraction of Stochastic Signal in Metric Space with Lattice Properties

According to its definition, the PDF $p_y(z_1, z_2/s^*)$ of instantaneous values $y(t_{1,2})$ of statistics (7.3.8a) is a mixed derivative of CDF $F_y(z_1, z_2/s^*)$, so the expression evaluating it can be written in the following form:

$$p_y(z_1, z_2/s^*) = \frac{\partial^2 F_y(z_1, z_2/s^*)}{\partial z_1 \partial z_2} =$$
$$= \left(1 - [1 - F_n(z_1, z_2)]^N\right) \delta(z_1 - s^*(t_1)) \cdot \delta(z_2 - s^*(t_2)) +$$
$$+ N[1 - F_n(z_1, z_2)]^{N-1} \{F_n(z_2/z_1) p_n(z_1) 1(z_1 - s^*(t_1)) \cdot \delta(z_2 - s^*(t_2)) +$$
$$+ F_n(z_1/z_2) p_n(z_2) \delta(z_1 - s^*(t_1)) \cdot 1(z_2 - s^*(t_2))\} +$$
$$+ N[1 - F_n(z_1, z_2)]^{N-1} \{\frac{N-1}{1 - F_n(z_1, z_2)} F_n(z_2/z_1) F_n(z_1/z_2) p_n(z_1) p_n(z_2) +$$
$$+ p_n(z_1, z_2)\} \cdot 1(z_1 - s^*(t_1)) \cdot 1(z_2 - s^*(t_2)), \quad (7.3.18)$$

where $\delta(z)$ is Dirac delta function; $F_n(z_1/z_2)$ and $F_n(z_2/z_1)$ are conditional CDFs of instantaneous values of interference (noise) $n(t_{1,2})$, $F_n(z_i/z_2) = \int\limits_{-\infty}^{z_i} p_n(x_i, z_j) dx_i / p_n(z_j)$; $p_n(z_{1,2})$ is a univariate PDF of instantaneous values of interference (noise) $n(t_{1,2})$.

With help from analogous reasoning, one can obtain the univariate conditional CDF $F_y(z/s^*)$:

$$F_y(z/s^*) = \left(1 - [1 - F_n(z)]^N\right) \cdot 1(z - s^*(t)),$$

and also the univariate conditional PDF $p_y(z/s^*)$ of the instantaneous value $y(t)$ of statistics (7.3.8a):

$$p_y(z/s^*) = \frac{dF_y(z/s^*)}{dz} = \begin{cases} P(C_c) \cdot \delta(z - s^*(t)), & z \leq s^*(t); \\ N \cdot p_n(z)[1 - F_n(z)]^{N-1}, & z > s^*(t), \end{cases} \quad (7.3.19)$$

where $P(C_c)$ and $P(C_e)$ are the probabilities of correct and error formation of the signal estimator $\hat{s}(t)$, respectively:

$$P(C_c) = 1 - [1 - F_n(s^*(t))]^N; \quad (7.3.19a)$$

$$P(C_e) = [1 - F_n(s^*(t))]^N. \quad (7.3.19b)$$

Qualitative analysis of the expression (7.3.19) for the PDF $p_y(z/s^*)$ allows us to draw the following conclusion. On nonnegative values of the signal $s(t) \geq 0$, the process $y(t)$ rather accurately ($P(C_c) \geq 1 - 2^{-n}$) reproduces the signal $s(t)$. On the other hand, from (7.3.19), the initial distribution of the noise component of stochastic process $y(t)$ in the output of transversal filter is essentially changed. On negative values of the signal $s(t) < 0$, the event $y(t) < s(t) < 0$ is impossible. Generally, on $s(t) \leq y(t) < 0$, the process $y(t)$ reproduces the useful signal $s(t)$ with distortions. However, for sufficiently small signal-interference (signal-noise) ratio, regardless of the sign of $s(t)$, the process $y(t)$ accurately ($P(C_c) \approx 1 - 2^{-n}$) reproduces the signal $s(t)$. If this condition does not hold, then, while processing negative instantaneous values of the signal $s(t) < 0$, the probability $P(C_c)$ of a correct formation of the estimator $\hat{s}(t)$, according to (7.3.19), becomes smaller than the quantity $1 - 2^{-n}$.

To overcome this disadvantage (when $s(t) < 0$), within the filtering unit shown in Fig. 7.3.1, we foresee the further processing of exclusively nonnegative values $y_+(t)$ of the process in the output of the filter $y(t)$ (7.3.11a): $y_+(t) = y(t) \vee 0$.

Applying similar reasoning, one can elucidate the necessity of formation of the statistics $\tilde{y}_-(t)$ determined by the expression (7.3.11b): $\tilde{y}_-(t) = \tilde{y}(t) \wedge 0$.

The realization $s^*(t)$ of useful signal $s(t)$ acting in the input of this filtering unit, and also possible realization $w^*(t)$ of the process $w(t)$ in the output of the adder, obtained via statistical modeling, are shown in Fig. 7.3.2.

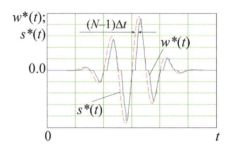

FIGURE 7.3.2 Realization $s^*(t)$ of useful signal $s(t)$ acting in input of filtering unit and possible realization $w^*(t)$ of process $w(t)$ in output of adder

The example corresponds to the following conditions. The signal is a narrowband pulse with a bell-shaped envelope; interference is a quasi-white Gaussian noise with the ratio of maximum frequency of power spectral density to signal carrier frequency $f_{n,\max}/f_0 = 64$; the signal-interference (signal-noise) ratio E_s/N_0 is equal to $E_s/N_0 = 10^{-6}$ (E_s is a signal energy, N_0 is power spectral density of interference (noise)). Transversal filters include time-delay lines with 16 taps (the number of the samples N of the signals $x(t)$ and $\tilde{x}(t)$ used while processing is equal to 16). The delay of the leading edge of semi-periods of the process $w(t)$ with respect to the signal $s(t)$ stipulated by the dynamic error of formation of the signal estimator $\hat{s}(t)$, is equal to $(N-1)\Delta t$, $\Delta t = 1/(2f_{n,\max}) \ll 1/f_0$.

The probability of error formation $P(C_e)$ of the signal estimator $\hat{s}(t)$ (probability of noise overshoot formation) is too small, $P(C_e) \approx 2^{-16}$, so there are no noise overshoots in the interval of modeling ($m = 2^{10} = 1024$ samples). While varying the signal-interference (signal-noise) ratio E_s/N_0 within wide bounds ($10 \ldots 10^{-10}$), the cross-correlation coefficient of the signals $s(t)$ and $w(t)$ takes the values $0.97 \ldots 0.98$ elucidated by invariance property of the investigated algorithm (unit) with respect to the conditions of parametric prior uncertainty.

It should be noted that probabilistic-statistical properties of the processes $y_+(t)$ (7.3.11a) and $\tilde{y}_-(t)$ (7.3.11b) are identical; in particular, their univariate and bivariate PDFs are reflective symmetrical. Thus, the expression for the bivariate conditional PDF $p_w(z_1, z_2/s^*)$ of the process $w(t)$ in the output of the filter can be written on the basis of the PDF (7.3.18) of the process $y(t)$ passed through the signal limiter in the following form:

$$p_w(z_1, z_2/s^*) = \left(1 - [1 - F_n(|z_1|, |z_2|)]^N\right) \delta(z_1 - s^*(t_1)) \cdot \delta(z_2 - s^*(t_2)) +$$
$$+ N[1 - F_n(|z_1|, |z_2|)]^{N-1} \{F_n(|z_2|/z_1) p_n(z_1) h_s(z_1) \delta(z_2 - s^*(t_2)) +$$
$$+ F_n(|z_1|/z_2) p_n(z_2) h_s(z_2) \delta(z_1 - s^*(t_1))\} + N[1 - F_n(|z_1|, |z_2|)]^{N-1} \times$$
$$\times \{\frac{N-1}{1 - F_n(|z_1|, |z_2|)} F_n(|z_2|/z_1) F_n(|z_1|/z_2) p_n(z_1) p_n(z_2) +$$
$$+ p_n(z_1, z_2)\} \cdot h_s(z_1) h_s(z_2), \quad (7.3.20)$$

where $F_n(|z_2|/z_1)$, $F_n(|z_1|/z_2)$ are the conditional CDFs of instantaneous values of interference (noise) $n(t_{1,2})$: $F_n(|z_i|/z_j) = \int\limits_{-\infty}^{|z_i|} p_n(x_i, z_j) dx_i / p_n(z_j)$; $p_n(z_{1,2})$ is the univariate PDF of instantaneous values of interference (noise) $n(t_{1,2})$; $h_s(z_{1,2})$ is the function taking into account the sign of instantaneous values of realization $s^*(t_{1,2})$ equal to:

$$h_s(z_{1,2}) = [1 - \mathrm{sign}(s^*(t_{1,2}))]/2 + \mathrm{sign}(s^*(t_{1,2})) \cdot 1(z_{1,2} - s^*(t_{1,2})). \quad (7.3.21)$$

Univariate conditional PDF $p_w(z/s^*)$ of instantaneous values of the resulting process $w(t)$ in the output of the processing unit determined by the expression (7.3.11), can be written on the basis of the formula (7.3.19) determining PDF $p_y(z/s^*)$ of the process $y(t)$ (7.3.8a):

$$p_w(z/s^*) = P^*(C_c) \cdot \delta(z - s^*(t)) + P^*(C_e) p_0(z); \quad (7.3.22)$$
$$P^*(C_c) = 1 - [1 - F_n(|s^*(t)|)]^N \geq 1 - 2^{-N}; \quad (7.3.22a)$$
$$P^*(C_e) = [1 - F_n(|s^*(t)|)]^N \leq 2^{-N}, \quad (7.3.22b)$$

where $p_0(z)$ is PDF of noise component in the output of the processing unit equal to:

$$p_0(z) = p^*(z) \cdot h_s(z); \quad (7.3.22c)$$
$$p^*(z) = N \cdot p_n(z)[1 - F_n(|z|)]^{N-1}/P^*(C_e),$$

$h_s(z)$ is the function determined by the formula (7.3.21) that takes into account the sign of $s^*(t)$.

The analysis of PDFs (7.3.20) and (7.3.22) implies that the process $w(t)$ in the output of the adder in the presence of the signal $(s(t) \neq 0)$ is non-stationary; whereas if the signal is absent $(s(t) = 0)$, then this process is stationary. In the absence of a signal, the stochastic process $w(t)$ possesses an ergodic property with respect to the univariate conditional PDF $p_w(z/s^*)$ (7.3.22). The random variables $w(t)$ and $w(t + \tau)$ are independent under condition $\tau \to \infty$, since the samples $n(t)$ and $n(t + \tau)$ of quasi-white Gaussian noise with normalized correlation function (7.3.14), acting in the input of the processing unit, are asymptotically independent; the condition holds [115]:

$$\lim_{\tau \to \infty} p_w(z, z; \tau/s^*) = p_w^2(z/s^*).$$

On nonnegative instantaneous values of the process $w(t) \geq 0$, the PDF $p_w(z/s^*)$ corresponds to the PDF $p_w^+(z/s^*)$ of its positive part, i.e., to the PDF of the process $y_+(t)$:

$$p_w^+(z/s^*) = p_w(z/s^*), \ z \geq 0. \tag{7.3.23}$$

Based on the expression (7.3.22), on negative instantaneous values of the process $w(t) < 0$, PDF $p_w^-(z/s^*)$ of its negative part $\tilde{y}_-(t)$ is reflectively symmetrical with respect to PDF $p_w^+(z/s^*)$ determined by the formula (7.3.23). We now analyze the properties of distribution $p_w^+(z/s^*)$ (20) of positive part $y_+(t)$ of stochastic process $w(t)$ in the output of the adder. The PDFs $p_w(z/s^*)$, $p_w^+(z/s^*)$ and $p_w^-(z/s^*)$ belong to a class of so-called ε-contaminated distributions that are widely used in probability theory and mathematical statistics [251], [263]. Characteristic function (CF) $\Phi_w^+(u)$ corresponding to PDF $p_w^+(z/s^*)$ (7.3.23) is represented in the form:

$$\Phi_w^+(u) = \Phi_{w_n}^+(u) e^{jus^*(t)}, \tag{7.3.24}$$

where $\Phi_{w_n}^+(u) = P^*(C_c) + P^*(C_e)\Phi_0(u); \ \Phi_0(u) = \int\limits_{-\infty}^{\infty} p_0(z + s^*(t))e^{juz} dz$.

The relationship (7.3.24) implies that CF $\Phi_w^+(u)$ of stochastic process $w_+(t) = w(t) \vee 0 = y_+(t)$ is the product of two CFs: $e^{jus^*(t)}$ and $\Phi_{w_n}^+(u)$, so the process $w_+(t)$ in the output of the adder can be represented as the sum of two independent processes, i.e., the signal $w_{s+}(t)$ and noise $w_{n+}(t)$ components:

$$w_+(t) = w_{s+}(t) + w_{n+}(t).$$

The signal component $w_{s+}(t)$ is a stochastic process whose instantaneous values of its realization $w_{s+}^*(t)$ take values equal to (1) the least positive instantaneous value of the signal from the set $\{s(t_j)\}$, $t_j = t - j\Delta t$, $j = 0, 1, \ldots, N-1$ with the probability (7.3.22a) on $s(t) \geq 0$ or (2) zero with the probability (7.3.22b) on $s(t) < 0$. The noise component $w_{n+}(t)$ is a stochastic process whose instantaneous values of its realization take values equal to (1) the least positive instantaneous value of interference (noise) from the set $\{n(t_j)\}$, $t_j = t - j\Delta t$, $j = 0, 1, \ldots, N-1$ with the probability 2^{-N} or (2) zero with the probability $1 - 2^{-N}$.

Similar reasoning could be extended onto negative part $w_-(t) = w(t) \wedge 0 = \tilde{y}_-(t)$ of the stochastic process $w(t)$ in the output of the adder, that also can be represented as the sum of two independent processes, i.e., the signal $w_{s-}(t)$ and the noise $w_{n-}(t)$ components: $w_-(t) = w_{s-}(t) + w_{n-}(t)$. Thus, the analysis of the distribution $p_w(z/s^*)$ (7.3.22) of the process $w(t)$ in the output of the adder implies that this process can be represented in the form of the signal and noise components:

$$w(t) = w_s(t) + w_n(t) = w_+(t) + w_-(t); \tag{7.3.25}$$

$$w_s(t) = w_{s+}(t) + w_{s-}(t); \tag{7.3.25a}$$

$$w_n(t) = w_{n+}(t) + w_{n-}(t). \tag{7.3.25b}$$

We now evaluate the correlation function $R_{w_n}(\tau)$ of the noise component $w_n(t)$ of the signal $w(t)$ in the output of the adder. Based on the evenness of the PDF

7.3 Extraction of Stochastic Signal in Metric Space with Lattice Properties

$p_w(z/s^*)$ on $s^*(t) = 0$, the expectation of the process $w_n(t)$ in the output of the processing unit is equal to zero. Then $R_{w_n}(\tau)$ is determined on the basis of PDF $p_w(z_1, z_2/s^*)$ (7.3.20) by the expression:

$$R_{w_n}(\tau) = \int_{-\infty}^{\infty} \int_{-\infty}^{\infty} z_1 z_2 p_w(z_1, z_2/s^*) dz_1 dz_2 \Big|_{s^*(t)=0}. \quad (7.3.26)$$

Computation of exact value of the function (7.3.26) is accompanied by considerable difficulties, so we simply use an upper bound of $R_{w_n}(\tau)$ determined by the inequality:

$$R_{w_n}(\tau) \leq N \cdot 2^{-(N-1)} \cdot D_n [r_n(\tau)]^{2(1+(N \ln 2)/2)}, \quad (7.3.27)$$

where D_n is a variance of interference (noise) in the input of processing unit equal to $D_n = N_0 f_{n,\max}$; $r_n(\tau)$ is a normalized correlation function of interference (noise) $n(t)$ in the input of the processing unit determined by the identity (7.3.14); N is a number of the samples of interference (noise) $n(t)$ which simultaneously take part in processing.

The analysis of the relationship (7.3.27) permits us to draw the following conclusions. The variance D_{w_n} of the noise component $w_n(t)$ is bounded above by the quantity:

$$D_{w_n} = R_{w_n}(0) \leq N \cdot 2^{-(N-1)} \cdot D_n. \quad (7.3.28)$$

The normalized correlation function $r_{w_n}(\tau)$ of the noise component $w_n(t)$ is bounded above by the quantity:

$$r_{w_n}(\tau) \leq [r_n(\tau)]^{2(1+(N \ln 2)/2)}. \quad (7.3.29)$$

When $N = 10$, the correlative relations of instantaneous values of the noise component $w_n(t)$ decrease to 10^{-8} over the interval $\tau = 5/4 f_{n,\max}$. This suggests that the rate of their destroying is considerable, and the filter possesses asymptotically whitening properties.

Next we determine the mean $N^+(H = 0)$ of the positive overshoots of stochastic process $w(t)$ in the output of the adder per time unit at the level $H = 0$ in the absence of the signal ($s(t) = 0$). Generally, for the stationary stochastic process $w(t)$, the quantity $N^+(H)$ is determined by the formula [116], [264]:

$$N^+(H) = \int_0^{\infty} w' p(H, w') dw' \Big|_{s(t)=0}, \quad (7.3.30)$$

where w' is a variable corresponding to derivative $w'(t)$ of the stochastic process $w(t)$; $p(w, w')$ is a joint PDF of the process $w(t)$ and its derivative $w'(t)$.

Method of evaluation of PDF $p(w, w')$ is considered in [115], [116]. It is based upon the computation of the limit: $p(w, w') = \lim_{\Delta \to 0} \Delta \cdot p_w(z - \frac{\Delta}{2} w', z + \frac{\Delta}{2} w'/s^*)$, where $p_w(z_1, z_2/s^*)$ is a bivariate PDF of the process $w(t)$ (7.3.20). As a result, mean $N^+(H = 0)$ of the positive overshoots of stochastic process $w(t)$ in the output of the adder per time unit at the level $H = 0$ in the absence of the signal ($s(t) = 0$) is determined by the expression:

$$N^+(H = 0) = N \cdot 2^{-(N-1)} f_{n,\max}/\sqrt{3}, \qquad (7.3.31)$$

where N is a number of the samples of interference (noise) $n(t)$, which simultaneously take part in signal processing; $f_{n,\max}$ is maximum frequency of power spectral density of interference (noise) $n(t)$.

It should be noted that the result (7.3.31) can be obtained as a mean of coincidences of stochastic process overshoots from N, according to the method considered, for instance, in [116], [264].

The average duration $\bar{\tau}(H)$ of positive overshoots of stationary ergodic stochastic process $w(t)$ with respect to the level H is determined over the probability of its exceeding $P(w(t) > H)$ by the formula [116], [264]:

$$\bar{\tau}(H) \cdot N^+(H) = P(w(t) > H), \qquad (7.3.32)$$

where $P(w(t) > 0)$, according to the expression (7.3.22b), is equal to $P^*(C_e) = 2^{-N}$.

Then the average duration $\bar{\tau}(0)$ of positive overshoots of the process $w(t)$ with respect to the level $H = 0$, taking into account the expression (7.3.31), is determined by the formula:

$$\bar{\tau}(0) = \sqrt{3}/(2N f_{n,\max}). \qquad (7.3.33)$$

The noise component $w_n(t)$ of the signal $w(t)$ in the output of the adder, according to (7.3.25b), is equal to $w_n(t) = w_{n+}(t) + w_{n-}(t)$. The component $w_{n+}(t)$ ($w_{n-}(t)$) is a pulse stochastic process with average duration of noise overshoots $\bar{\tau}(0)$ that is small compared to the correlation interval of interference (noise) $\Delta t = 1/2 f_{n,\max}$: $\bar{\tau}(0) = \sqrt{3}\Delta t/N$. Distribution of instantaneous values of noise overshoots of noise component $w_n(t)$ is characterized by PDF $p_0(z)$ (7.3.22c). The component $w_{n+}(t)$ ($w_{n-}(t)$) can be described by homogeneous Poisson process with a constant overshoot flow density $\lambda = N^+(H = 0)$ determined by the relationship (7.3.31).

While applying the known smoothing procedures [263], [265], [266] built upon the basis of median estimators, a variance D_{v_n} of noise component $v_n(t)$ of the process $v(t)$ in the output of the filter (7.3.13) (under condition that the signal $s(t)$ is a Gaussian narrowband process, and interference (noise) $n(t)$ is very strong) can be reduced to the quantity:

$$D_{v_n} \leq D_{w_n} \int_{\Delta \tilde{T}}^{\infty} p_{w_n}(\tau) d\tau = a^2 \exp\left\{-\frac{\Delta \tilde{T}^2 N^2 f_{n,\max}^2}{8\sqrt{\pi}}\right\}, \qquad (7.3.34)$$

where a is a parameter of the limiter $L[w(t)]$, which, for instance, on $\varepsilon \approx 10^{-5}$ (see (7.3.5b) of the system (7.3.5) can be equal to: $a^2 = 16 D_s$; D_s is a variance of Gaussian signal $s(t)$; $\Delta \tilde{T}$ is a quantity of the smoothing interval $\tilde{T}$; N is a number of the samples of the interference (noise) $n(t)$, which simultaneously take part in signal processing; $f_{n,\max}$ is maximum frequency of power spectral density of interference

7.3 Extraction of Stochastic Signal in Metric Space with Lattice Properties

(noise) $n(t)$; $p_{w_n}(\tau)$ is PDF of duration of overshoots of the noise component $w_n(t)$ of the process $w(t)$, which is approximated by experimentally obtained dependence:

$$p_{w_n}(\tau) = \frac{\tau}{D_{w_n,\tau}} \exp\left\{-\frac{\tau^2}{2D_{w_n,\tau}}\right\},$$

where $D_{w_n,\tau} = \sqrt{\pi}/(N^2 f_{n,\max}^2)$.

The formula (7.3.34) implies that the signal-to-noise ratio q_{out}^2 in the output of the filter is determined by the quantity:

$$q_{\text{out}}^2 = D_s/D_{v_n} \geq \frac{1}{16} \exp\left\{\frac{\Delta \tilde{T}^2 N^2 f_{n,\max}^2}{8\sqrt{\pi}}\right\}. \tag{7.3.35}$$

The formula (7.3.34) also implies that *fluctuation component of signal estimator error* (*relative filtering error*) δ_f is bounded by the quantity:

$$\delta_f = \mathbf{M}\{(s(t)-v(t))^2\}/2D_s\big|_{s(t)\equiv 0} \leq 8\exp\left\{-\frac{\Delta \tilde{T}^2 N^2 f_{n,\max}^2}{8\sqrt{\pi}}\right\}, \tag{7.3.36}$$

where $\mathbf{M}(*)$ is a symbol of mathematical expectation.

In the interval of primary processing $T^* = [t - (N-1)\Delta t, t]$, the instantaneous values of the signal $s(t)$ are slightly varied that cause dynamic error of signal primary filtering δ_{d0}. This error can be evaluated. Assume there is no interference (noise) in the input of the processing unit ($n(t) = 0$). Then the domain of definition of the signal $s(t)$ can be represented by the partition of subsets $T_{s+}^{s'+}, T_{s+}^{s'-}, T_{s-}^{s'-}, T_{s-}^{s'+}$:

$$T_s = T_{s+}^{s'+} \cup T_{s+}^{s'-} \cup T_{s-}^{s'-} \cup T_{s-}^{s'+}, \quad T_{s\pm}^{s'\pm} \cap T_{s\mp}^{s'\mp} = \emptyset, \quad T_{s\pm}^{s'\mp} \cap T_{s\mp}^{s'\pm} = \emptyset;$$

$$T_{s+}^{s'+} = \{t : s(t) > 0 \,\&\, s'(t) > 0\}, \quad T_{s+}^{s'-} = \{t : s(t) > 0 \,\&\, s'(t) \leq 0\};$$

$$T_{s-}^{s'-} = \{t : s(t) \leq 0 \,\&\, s'(t) \leq 0\}, \quad T_{s-}^{s'+} = \{t : s(t) \leq 0 \,\&\, s'(t) > 0\};$$

$$m(T_{s+}^{s'+}) = m(T_{s+}^{s'-}) = m(T_{s-}^{s'-}) = m(T_{s-}^{s'+}), \tag{7.3.37}$$

where $m(X)$ is a measure of a set X; $s'(t)$ is a derivative of the narrowband signal $s(t)$.

Then the process $v(t)$ in the output of the processing unit is determined by the relationship:

$$v(t) = \begin{cases} s(t-(N-1)\Delta t), & t \in T_{s+}^{s'+} \cup T_{s-}^{s'-}; \\ s(t), & t \in T_{s+}^{s'-} \cup T_{s-}^{s'+}, \end{cases} \tag{7.3.38}$$

i.e., in the absence of interference (noise) ($n(t) = 0$), the stochastic signal $s(t)$ is exactly reproduced within drooping parts ($s'(t) \leq 0$) on $s(t) > 0$ and also within ascending parts ($s'(t) > 0$) on $s(t) \leq 0$, and reproduced with a delay equal to $(N-1)\Delta t$, within ascending parts ($s'(t) > 0$) on $s(t) > 0$ and also within drooping parts ($s'(t) \leq 0$) on $s(t) \leq 0$ (see Fig. 7.3.2). Thus, *dynamic error of signal filtering* δ_{d0}, taking into account the identity (7.3.37), is equal to the quantity:

$$\delta_{d0} = \mathbf{M}\{(s(t)-v(t))^2\}/2D_s\big|_{n(t)=0} = 1 - r_s[(N-1)\Delta t], \tag{7.3.39}$$

where $r_s(\tau)$ is normalized correlation function of the signal $s(t)$; $\Delta t = 1/(2f_{n,\max}) \ll 1/f_0$.

Smoothing of the process $u(t)$ in a median filter is accompanied by *dynamic error of signal smoothing* $\delta_{d,\mathrm{sm}}$ stipulated by partial smoothing of local extrema of the signal $s(t)$. Then, taking into account the relationship (7.3.38), dynamic error of signal smoothing $\delta_{d,\mathrm{sm}}$ is equal to:

$$\delta_{d,\mathrm{sm}} = \mathbf{M}\{(s(t) - u(t))^2\}/2D_s|_{n(t)=0} = 1 - r_s[\Delta \tilde{T}], \qquad (7.3.40)$$

where $\Delta \tilde{T}$ is a quantity of the smoothing interval $\tilde{T}$ of the useful signal $s(t)$.

Thus, the relative error δ of the estimator of Gaussian narrowband signal (7.3.15) in the presence of interference (noise) under their interaction in the signal space with lattice properties is bounded by the errors determined by the relationships (7.3.36), (7.3.39), and (7.3.40):

$$\delta \leq \delta_f + \delta_{d0} + \delta_{d,\mathrm{sm}}. \qquad (7.3.41)$$

7.3.3 Possibilities of Further Processing of Narrowband Stochastic Signal Extracted on Basis of Optimal Filtering Algorithm in Metric Space with *L*-group Properties

According to the formula (6.1.2), the useful signal $s(t)$, as a rule, is a deterministic one-to-one function $M[*,*]$ of carrier signal $c(t)$ and vector of informational parameters $\boldsymbol{\lambda}(t)$ varying on a domain of definition T_s of the signal $s(t)$:

$$s(t) = M[c(t), \boldsymbol{\lambda}(t)], \qquad (7.3.42)$$

where $\boldsymbol{\lambda}(t) = [\lambda_i(t)] = [\lambda_1(t), \ldots, \lambda_m(t)]$, $i = 1, \ldots, m$, $\boldsymbol{\lambda}(t) \in \boldsymbol{\Lambda}$ is vector of informational parameters of the signal $s(t)$.

If the modulating function $M[*,*]$ is known, the estimator $\hat{s}(t) = v(t)$ of the signal $s(t)$ extracted in the presence of interference (noise) $n(t)$ determined by the formula (7.3.13) allows us to obtain the estimator $\hat{\boldsymbol{\lambda}}(t)$ of informational parameters $\boldsymbol{\lambda}(t)$ of the signal which, according to the formula (7.3.42), is determined by an inverse function with respect to $M[*,*]$ (see Fig. 7.3.3):

$$\hat{\boldsymbol{\lambda}}(t) = [\hat{\lambda}_i(t)] = M^{-1}[c(t), \hat{s}(t)], \ \hat{\boldsymbol{\lambda}}(t) \in \boldsymbol{\Lambda}. \qquad (7.3.43)$$

If all the parameters $\boldsymbol{\lambda}(t) = [\lambda_i(t)]$, $i = 1, \ldots, m$ are unknown nonrandom and constant on a domain of definition T_s of the signal $s(t)$ ($\lambda_i(t) = \mathrm{const} = \lambda_i$, $t \in T_s$), then application of the relationship (7.3.43), together with the formula (7.3.13), defines an algorithm of signal parameter estimation. If the dependence $\lambda_i(t)$ cannot be neglected and it is necessary to trace the instantaneous values of varying informational parameters $\lambda_i(t)$ that are stochastic processes, the application of the relationship (7.3.43) together with the formula (7.3.13) defines an algorithm of signal parameter filtering.

Inasmuch as the estimator $\hat{s}(t) = v(t)$ of the useful signal and the estimator $\hat{\boldsymbol{\lambda}}(t)$ of informational parameters $[\lambda_i(t)]$ of useful signal $s(t)$ are connected by a one-to-one relation (7.3.43), the information quantity contained in the estimator $\hat{\boldsymbol{\lambda}}(t)$ and

7.3 Extraction of Stochastic Signal in Metric Space with Lattice Properties

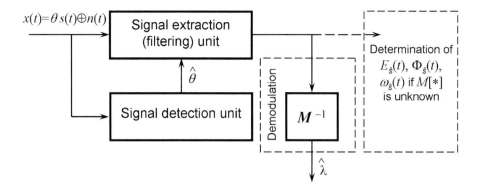

FIGURE 7.3.3 Generalized block diagram of signal processing unit

information quantity contained in the estimator $\hat{s}(t)$ are the same. Quality indices of these estimators are the same too.

If, while extracting the stochastic signal $s(t)$, the addressee does not know the modulating function $M[*,*]$, then, on the basis of the estimator $\hat{s}(t)$ of the signal $s(t)$, one can determine its envelope $E_{\hat{s}}(t)$, phase $\Phi_{\hat{s}}(t)$, and instantaneous frequency $\omega_{\hat{s}}(t)$ (see Fig. 7.3.3):

$$E_{\hat{s}}(t) = \sqrt{\hat{s}^2(t) + \tilde{s}^2(t)}, \qquad (7.3.44)$$

where $\tilde{s}(t) = \mathcal{H}[\hat{s}(t)] = \dfrac{1}{\pi}\displaystyle\int_{-\infty}^{\infty}\dfrac{\hat{s}(\tau)\mathrm{d}\tau}{t-\tau}$ is Hilbert transform of the estimator $\hat{s}(t)$;

$$\Phi_{\hat{s}}(t) = \operatorname{arctg}[\tilde{s}(t)/\hat{s}(t)]; \qquad (7.3.45)$$

$$\omega_{\hat{s}}(t) = \Phi'_{\hat{s}}(t) = [\tilde{s}'(t)\hat{s}(t) - \tilde{s}(t)\hat{s}'(t)]/E_{\hat{s}}^2(t). \qquad (7.3.46)$$

On the basis of the obtained relationships (7.3.44) through (7.3.46), a modulating function $M[*,*]$ can be determined.

The investigation of optimal filtering algorithm in signal space with L-group properties led to the following conclusions.

1. The problem formulation of the synthesis of stochastic signal extraction algorithm on the basis of successive minimization of the functions of the observed statistical collections (7.3.2a,b) with simultaneous rejection of using prior data concerning possible behavior and characteristics of the signal and interference (noise) distribution provides the invariance property of signal processing with respect to the conditions of both parametric and nonparametric prior uncertainty, confirmed by the obtained quality indices of signal processing (7.3.22a,b), (7.3.35), (7.3.36), (7.3.39), (7.3.40), and (7.3.41).

2. The feature of signal filtering in space with L-group properties is that both useful and interference signals can be extracted from the received mixture

with the same or almost the same quality regardless of their energetic relationships. The relation of the prior information quantity contained in these signals does not matter.

3. The obtained signal extraction unit operating in the space with L-group properties almost does not distort the structure of the received useful signal regardless of the conditions of parametric prior uncertainty. High quality of the obtained signal estimator $\hat{s}(t)$ allows solving the main signal processing problems.

4. The essential distinctions of signal extracting quality between linear space $\mathcal{LS}(+)$ and signal space $\mathcal{L}(+, \vee, \wedge)$ with L-group properties may be explained by the fundamental differences in the types of interactions between the useful signal $s(t)$ and interference (noise) $n(t)$ within these spaces: additive $s(t) + n(t)$, and interactions in the form of join $s(t) \vee n(t)$ and meet $s(t) \wedge n(t)$, respectively.

7.4 Signal Detection in Metric Space with Lattice Properties

The known variants of signal detection problem statement are mostly formulated for the case of an additive interaction $x(t) = s(t) + n(t)$ (in terms of linear space) between the signal $s(t)$ and interference (noise) $n(t)$ [254], [163], [155], [153]. Nevertheless, this problem can be formulated on the basis of a more general model of signal interaction: $x(t) = \Phi[s(t), n(t)]$, where Φ is some deterministic function [159], [166]. As a rule, signal detection is not realized without simultaneous solving of other signal processing problems. Meanwhile, signal detection is an independent signal processing problem that has to be solved while constructing of signal processing fundamentals in signal spaces with new properties that fundamentally differ from the properties of linear signal space.

This section has a twofold goal. First, it is necessary to synthesize algorithm of signal detection in the presence of interference (noise) in signal space with lattice properties. Second, the synthesized algorithm of signal detection should be analyzed; besides, characteristics and properties of this algorithm should be compared with the known analogues solving the signal detection problem in linear signal space.

In linear signal space, signal detection problem is solved by signal energy estimation on the basis of comparison of signal energy estimator with some threshold [254], [163]. This problem formulated in signal space with lattice properties is solved in a different way. As shown below, the estimator of the received useful signal is assumed to be formed.

7.4.1 Deterministic Signal Detection in Presence of Interference (Noise) in Metric Space with Lattice Properties

Deterministic signal application is bounded by the framework of theoretical analysis, whose results, in the form of potential signal processing quality indices, may be used to compare them with other signal processing algorithms. The known literature on signal detection usually considers the probabilistic character of the observed stochastic process causes any signal detector operating under interference (noise) background is characterized by nonzero probabilities of error decisions. However, as shown below, that is not quite so, for instance, when we consider the case of deterministic signal detection in signal space with lattice properties.

For signal detection, it is necessary to find out whether the signal $s(t)$ is present or absent in the received realization of stochastic process $x(t)$ within the observation interval T_{obs}. Consider the model of interaction between the signal $s(t)$ with completely known parameters and interference (noise) $n(t)$ in signal space in the form of distributive lattice $\mathcal{L}(\vee, \wedge)$ with binary operations of join $a \vee b$ and meet $a \wedge b$, respectively: $a \vee b = \sup_{\mathcal{L}}(a,b)$, $a \wedge b = \inf_{\mathcal{L}}(a,b)$; $a, b \in \mathcal{L}(\vee, \wedge)$:

$$x(t) = \theta s(t) \vee n(t); \quad t \in T_s \equiv T_{\text{obs}}; \quad \theta \in \{0, 1\}, \quad (7.4.1)$$

where $T_s = [t_0, t_0 + T]$ is the domain of definition of the signal $s(t)$; t_0 is the known time of arrival of the signal $s(t)$; T is a duration of the signal $s(t)$; θ is an unknown nonrandom parameter taking value $\theta = 0$ (the signal is absent) or $\theta = 1$ (the signal is present).

Thus, deterministic signal detection is equivalent to the estimation of an unknown nonrandom parameter θ. For a solution, one has to obtain both algorithm and block diagram of the optimal signal detector and its quality indices of detection (conditional probabilities of correct detection and false alarm) have to be determined.

Assume that interference (noise) $n(t)$ is characterized by an arbitrary distribution with an even univariate probability density function $p_n(x) = p_n(-x)$.

To synthesize the detectors in linear signal space $\mathcal{LS}(+)$, i.e., when the interaction equality $x(t) = \theta s(t) + n(t)$ holds, $\theta \in \{0, 1\}$, the part of the theory of statistical inference called statistical hypotheses testing is used. The strategies of decision making (within signal detection problem solving) considered in the literature suppose a likelihood ratio computation deemed necessary to determine a likelihood function. Such a function is determined by multivariate probability density function of interference (noise) $n(t)$. During signal detection problem solving in linear signal space $\mathcal{LS}(+)$, i.e., when the interaction equality $x(t) = s(t) + n(t)$ holds, the trick of the variable change is used to obtain likelihood function: $n(t) = x(t) - s(t)$ [254], [163]. However, it is impossible to use this subterfuge to determine likelihood ratio under interaction (7.4.1) between the signal and interference (noise), inasmuch as the equation is unsolvable with respect to the variable $n(t)$ because the lattice $\mathcal{L}(\vee, \wedge)$ does not possess the group properties; another approach is necessary.

As applied to the case (7.4.1), solving the signal detection problem in the presence of interference (noise) $n(t)$ with an arbitrary distribution lies in formation of an estimator $\hat{s}(t)$ of the received signal $s(t)$ which (on the basis of the chosen criteria) would allow the observer to distinguish two possible situations of signal receiving determined by the parameter θ. We formulate the problem of detection $Det[s(t)]$ of the signal $s(t)$ on the basis of minimization of squared metric $\int_{T_s} |y(t) - s(t)|^2 dt$ between the function $y(t) = F[x(t)]$ of the observed process $x(t)$ and the signal $s(t)$ under the condition that the observed process $x(t)$ includes the signal $\theta s(t) = s(t)$ ($\theta = 1$): $x(t) = s(t) \vee n(t)$:

$$Det[s(t)] = \begin{cases} y(t) = F[x(t)] = \hat{s}(t); & (a) \\ \int_{T_s} |y(t) - s(t)|^2 dt \,|_{\theta=1} \to \min_{y(t) \in Y}; & (b) \\ \hat{\theta} = 1[\max_{t \in T_s}[\int_{T_s} y(t)s(t)dt]\,|_{\theta \in \{0,1\}} - l_0]; & (c) \\ \int_{T_s} y(t)s(t)dt\,|_{\theta=1} \neq \int_{T_s} y(t)s(t)dt\,|_{\theta=0}, & (d) \end{cases} \quad (7.4.2)$$

where $y(t) = \hat{s}(t)$ is the signal estimator $\theta s(t)$, $\theta \in \{0, 1\}$ in the observed process $x(t)$: $x(t) = \theta s(t) \vee n(t)$; $F[*]$ is some deterministic function; $\int_{T_s} |y(t) - s(t)|^2 dt = \|y(t) - s(t)\|^2$ is a squared metric between the signals $y(t), s(t)$ in Hilbert space $\mathcal{HS}$; $\hat{\theta}$ is an estimator of an unknown nonrandom parameter θ, $\theta \in \{0, 1\}$; $\int_{T_s} y(t)s(t)dt = (y(t), s(t))$ is scalar product of the signals $y(t), s(t)$ in Hilbert space $\mathcal{HS}$; $1[x]$ is Heaviside step function; l_0 is some threshold value.

The relation (7.4.2a) of the system (7.4.2) specifies the rule of formation of the estimator $\hat{s}(t)$ of the received signal in the form of some deterministic function $F[x(t)]$ of the observed process $x(t)$. The relation (7.4.2b) defines the criterion of minimum of squared metric $\int_{T_s} |y(t) - s(t)|^2 dt\,|_{\theta=1}$ in Hilbert space $\mathcal{HS}$ between the signals $y(t)$ and $s(t)$; this criterion is applied when the signal is present in the observed process $\theta s(t) = s(t)$ ($\theta = 1$): $x(t) = s(t) \vee n(t)$. The relation (7.4.2c) of the system (7.4.2) determines the rule of forming the estimator $\hat{\theta}$ of parameter θ, which lies in computation of maximum value of correlation integral between the estimator $y(t) = \hat{s}(t)$ and useful signal $s(t)$ and in further comparison with a threshold value l_0. The inequality (7.4.2d) determines a constraint, which must hold unconditionally to successfully solve the detection problem (to form the estimator $\hat{\theta}$) according to the rule (7.4.2c).

The solution of the problem of minimization of squared metric

$$\int_{T_s} |y(t) - s(t)|^2 dt\,|_{\theta=1} \to \min$$

between the function $y(t) = F[x(t)]$ and the signal $s(t)$ in its presence in the observed process $x(t)$: $x(t) = s(t) \vee n(t)$ follows directly from the absorption axiom of the lattice $\mathcal{L}(\vee, \wedge)$ (see page 269) contained in the third part of the multilink identity:

$$y(t) = s(t) \wedge x(t) = s(t) \wedge [s(t) \vee n(t)] = s(t). \quad (7.4.3)$$

7.4 Signal Detection in Metric Space with Lattice Properties

The identity (7.4.3) directly implies, first, the kind of function $F[x(t)]$ from the relation (7.4.2a) of the system (7.4.2):

$$y(t) = F[x(t)] = s(t) \wedge x(t). \qquad (7.4.4)$$

Second, the relationship (7.4.3) implies that squared metric $\|y(t) - s(t)\|^2$ is identically equal to zero:

$$\int_{T_s} |y(t) - s(t)|^2 dt \,|_{\theta=1} = 0.$$

The identity (7.4.4) implies that in the presence of the signal $s(t)$ in the observed process $x(t) = s(t) \vee n(t)$, at the instant $t = t_0 + T$, correlation integral $\int_{T_s} y(t)s(t)dt\,|_{\theta=1}$ takes a maximum value equal to the energy E of the signal $s(t)$:

$$\max_{t=t_0+T}[\int_{T_s} y(t)s(t)dt\,|_{\theta=1}] = \int_{t_0}^{t_0+T} s(t)s(t)dt = E. \qquad (7.4.5)$$

The set of values of the estimator $y(t)\,|_{\theta=0}$, according to the identity (7.4.4), is determined by the expression:

$$y(t)\,|_{\theta=0} = s(t) \wedge [0 \vee n(t)]. \qquad (7.4.6)$$

The identity (7.4.6) implies that under joint fulfillment of the inequalities $s(t) > 0$ and $n(t) \leq 0$, the estimator $y(t)\,|_{\theta=0}$ takes values equal to zero:

$$y(t)\,|_{\theta=0} = 0,$$

while under joint fulfillment of the inequalities $s(t) > 0$ and $n(t) > 0$, when the inequality $s(t) \leq 0$ holds, the estimator $y(t)\,|_{\theta=0}$ takes values equal to the instantaneous values of useful signals $s(t)$:

$$y(t)\,|_{\theta=0} = s(t).$$

Based on a prior assumption concerning evenness of univariate probability density function of interference (noise) $p_n(x) = p_n(-x)$, the estimator $y(t)\,|_{\theta=0}$ determined by the identity (7.4.6) takes values equal to zero $y(t)\,|_{\theta=0} = 0$ with probability $\mathbf{P} = 1/4$, and values equal to useful signal $y(t)\,|_{\theta=0} = s(t)$ with probability $\mathbf{P} = 3/4$. However, despite rather high cross-correlation between the estimator $y(t)\,|_{\theta=0}$ and useful signal $s(t)$, the maximum value of correlation integral $\int_{T_s} y(t)s(t)dt\,|_{\theta=0}$ on $\theta = 0$ at the instant $t = t_0 + T$ is equal to $3/4$ of the energy E of the signal $s(t)$:

$$\max_{t=t_0+T}[\int_{T_s} y(t)s(t)dt\,|_{\theta=0}] = 3\int_{t_0}^{t_0+T} s(t)s(t)dt/4 = 3E/4. \qquad (7.4.7)$$

The identities (7.4.5) and (7.4.7) confirm fulfillment of the constraint (7.4.2d) of the system (7.4.2), and specify the upper and lower bounds for the threshold level l_0, respectively:

$$3E/4 < l_0 < E. \qquad (7.4.8)$$

Thus, summarizing the relationships (7.4.2a) through (7.4.2c) of the system (7.4.2), one can draw the conclusion that the deterministic signal detector has to form the estimator $y(t) = \hat{s}(t)$ that, according to (7.4.4), is equal to $\hat{s}(t) = s(t) \wedge x(t)$; also it has to compute correlation integral $\int_{T_s} y(t)s(t)dt$ in the interval $T_s = [t_0, t_0 + T]$ and to determine the presence or absence of the useful signal $s(t)$. According to the equation (7.4.2a) of the system (7.4.2), the decision $\hat{\theta} = 1$ concerning the presence of the signal $s(t)$ ($\theta = 1$) in the observed process $x(t)$ is made if, at the instant $t = t_0 + T$, the maximum value of correlation integral $\int_{T_s} y(t)s(t)dt \,|_{\theta=1} = E$ exceeds the threshold level l_0. The decision $\hat{\theta} = 0$ concerning the absence of the useful signal $s(t)$ ($\theta = 0$) in the observed process $x(t)$ is made if the maximum value of correlation integral $\int_{T_s} y(t)s(t)dt \,|_{\theta=0} = 3E/4$ observed in the instant $t = t_0 + T$ does not exceed the threshold level l_0.

The block diagram of a deterministic signal detection unit synthesized in signal space with lattice properties includes the signal estimator $\hat{s}(t) = y(t)$ formation unit; correlation integral $\int_{T_s} y(t)s(t)dt$ computing unit; the strobing circuit (SC) and decision gate (DG) (see Fig 7.4.1). The correlation integral computing unit consists of multiplier and integrator.

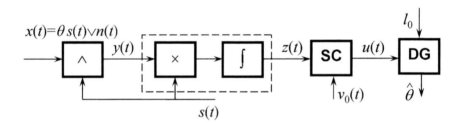

FIGURE 7.4.1 Block diagram of deterministic signal detection unit

We now analyze metric relations between the signals $\theta s(t) = s(t)$ ($\theta = 1$), $\theta s(t) = 0$ ($\theta = 0$) and their estimators $\hat{s}(t)|_{\theta=1} = y(t)|_{\theta=1}$, $\hat{s}(t)|_{\theta=0} = y(t)|_{\theta=0}$. For the signals $\theta s(t) = s(t)$, $\theta s(t) = 0$ and their estimators $y(t)|_{\theta=1}$, $y(t)|_{\theta=0}$, the following metric relationships hold:

$$\|0 - y(t)|_{\theta=1}\|^2 + \|s(t) - y(t)|_{\theta=1}\|^2 = \|0 - s(t)\|^2 = \|s(t)\|^2; \qquad (7.4.9a)$$

$$\|0 - y(t)|_{\theta=0}\|^2 + \|s(t) - y(t)|_{\theta=0}\|^2 = \|0 - s(t)\|^2 = \|s(t)\|^2, \qquad (7.4.9b)$$

where $\|a(t) - b(t)\|^2 = \|a(t)\|^2 + \|b(t)\|^2 - 2(a(t), b(t))$ is a squared metric between the functions $a(t)$, $b(t)$ in Hilbert space $\mathcal{HS}$; $\|a(t)\|^2$ is a squared norm of a function

7.4 Signal Detection in Metric Space with Lattice Properties

$a(t)$ in Hilbert space $\mathcal{HS}$; $(a(t), b(t)) = \int_{T^*} a(t)b(t)dt$ is a scalar product of the functions $a(t)$ and $b(t)$ in Hilbert space $\mathcal{HS}$; T^* is a domain of definition of the functions $a(t)$ and $b(t)$.

The relationships (7.4.4) and (7.4.9a) imply that on an arbitrary signal-to-interference (signal-to-noise) ratio in the input of a detection unit, the cross-correlation coefficient $\rho[s(t), y(t)|_{\theta=1}]$ between the signal $s(t)$ and the estimator $y(t)|_{\theta=1}$ of the signal $s(t)$, according to their identity $(y(t)|_{\theta=1} = s(t))$ is equal to 1:

$$\rho[s(t), y(t)|_{\theta=1}] = 1,$$

and the squared metrics taken from (7.4.9a) are determined by the following relationships:

$$\|s(t) - y(t)|_{\theta=1}\|^2 = 0; \quad \|0 - y(t)|_{\theta=1}\|^2 = \|0 - s(t)\|^2 = \|s(t)\|^2 = E,$$

where E is the energy of the signal $s(t)$.

The relationships (7.4.7) and (7.4.9b) imply that in the absence of the signal in the observed process $x(t) = 0 \vee n(t)$ in the input of a detector unit, the correlation coefficient $\rho[s(t), y(t)|_{\theta=0}]$ between the signal $s(t)$ and the estimator $y(t)|_{\theta=0}$ is equal to:

$$\rho[s(t), y(t)|_{\theta=0}] = \frac{(y(t)|_{\theta=0}, s(t))}{\sqrt{\|y(t)|_{\theta=0}\|^2 \|s(t)\|^2}} = \sqrt{3/4},$$

and the squared metrics taken from (7.4.9b) are determined by the following relationships:

$$\|0 - y(t)|_{\theta=0}\|^2 = 3E/4; \quad (7.4.10a)$$

$$\|s(t) - y(t)|_{\theta=0}\|^2 = E/4; \quad (7.4.10b)$$

$$\|0 - s(t)\|^2 = E. \quad (7.4.10c)$$

The relationships (7.4.10) indicate that the difference $s(t) - y(t)|_{\theta=0}$ between the signal $s(t)$ and the estimator $y(t)|_{\theta=0}$ and the estimator $y(t)|_{\theta=0}$ alone are orthogonal in Hilbert space:

$$(s(t) - y(t))|_{\theta=0} \perp y(t)|_{\theta=0}.$$

Figure 7.4.2 illustrates the signals $z(t)$ and $u(t)$ in the outputs of the correlation integral computing unit (dash line) and strobing circuit (solid line), respectively, and also the strobing pulses $v_0(t)$ (dot line), obtained by statistical modeling under the condition that the mixture $x(t) = \theta s(t) \vee n(t)$ contains signals $\theta s(t) = s(t)$, $\theta s(t) = 0$, $\theta s(t) = s(t)$, $\theta s(t) = 0$ in the input of a detection unit distributed in series over the intervals $[0, T]$, $[T, 2T]$, $[2T, 3T]$, $[3T, 4T]$, respectively.

The signal-to-noise ratio E/N_0 is equal to $E/N_0 = 10^{-8}$, where E is the energy of the signal $s(t)$; N_0 is a power spectral density of noise $n(t)$. The interference $n(t)$ is quasi-white Gaussian noise with independent samples. In the output of the correlation integral computing unit, the function $z(t)$ is linear.

As seen in Fig. 7.4.2, despite an extremely small signal-to-noise ratio, in the input of the decision gate (in the output of strobing circuit), the signals $u(t)$ are distinguished by their amplitude without any error. The identity (7.4.5) implies that in the presence of the signal $s(t)$ in $x(t) = s(t) \vee n(t)$ ($\theta = 1$) in the output of the integrator at the instant $t = t_0 + jT$, $j = 1, 3$, the maximum value of correlation integral (7.4.5) is equal to $E\rho[s(t), y(t)|_{\theta=1}] = E$. In the absence of the signal in the observed process $x(t) = 0 \vee n(t)$ ($\theta = 0$) in the output of the integrator (see Fig. 7.4.1) at the instant $t = t_0 + jT$, $j = 2, 4$, the value of correlation integral is equal to $E\rho[s(t), y(t)|_{\theta=0}] = 3E/4$.

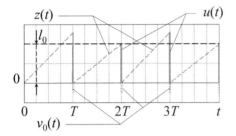 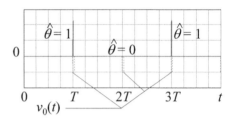

FIGURE 7.4.2 Signals $z(t)$ and $u(t)$ in outputs of correlation integral computing unit (dash line) and strobing circuit (solid line), respectively; strobing pulses $v_0(t)$ (dot line)

FIGURE 7.4.3 Estimator $\hat{\theta}$ of unknown nonrandom parameter θ characterizing presence ($\theta = 1$) or absence ($\theta = 0$) of useful signal $s(t)$ in output of decision gate

Figure 7.4.3 shows the result of formation of the estimator $\hat{\theta}$ of unknown nonrandom parameter θ, which characterizes the presence ($\theta = 1$) or absence ($\theta = 0$) of the useful signal $s(t)$ in the observed process $x(t)$. The decision gate (DG) (see Fig. 7.4.1) forms the estimator $\hat{\theta}$ according to the rule (7.4.2c) of the system (7.4.2) by means of the comparison of the signal $u(t)$ at the instants $t = t_0 + jT$, $j = 1, 2, 3, \ldots$ with the threshold value l_0 chosen according to two-sided inequality (7.4.8).

Thus, regardless of the conditions of parametric and nonparametric prior uncertainty and, respectively, independently of probabilistic-statistical properties of interference (noise), the optimal deterministic signal detector in signal space with lattice properties accurately detects the signals with the conditional probabilities of the correct detection $D = 1$ and false alarm $F = 0$. Absolute values of quality indices of signal detection $D = 1$, $F = 0$ are stipulated by the fact that the estimator $y(t) = \hat{s}(t)$ of the received signal $\theta s(t)$, $\theta \in \{0, 1\}$ formed in the input of the correlation integral computing unit (see Fig. 7.4.1), regardless of instantaneous values of interference (noise) $n(t)$, can take only two values from a set $\{0, s(t)\}$.

The results of the investigation of algorithm and unit of deterministic signal detection in signal space with lattice properties allow us to draw the following conclusions.

1. Formulation of the synthesis problem of deterministic signal detection algorithm, on the basis of minimization of squared metric in Hilbert space $\mathcal{HS}$ (7.4.2b), with simultaneous rejection of the use of interference (noise)

7.4 Signal Detection in Metric Space with Lattice Properties

distribution prior data, provides the invariance property of the algorithm and unit of signal detection with respect to the conditions of parametric and nonparametric prior uncertainty confirmed by absolute values of the conditional probabilities of correct detection D and false alarm F.

2. The existing distinctions between deterministic signal detection quality indices for detection in linear signal space $\mathcal{LS}(+)$ and signal space with lattice properties $\mathcal{L}(\vee, \wedge)$, are elucidated by fundamental differences in the kinds of interactions between the signal $s(t)$ and interference (noise) $n(t)$ in these spaces: additive $s(t) + n(t)$ and the interaction in the form of operations of join $s(t) \vee n(t)$ and meet $s(t) \wedge n(t)$, respectively, and as a consequence, by the distinctions of the properties of these signal spaces.

7.4.2 Detection of Harmonic Signal with Unknown Nonrandom Amplitude and Initial Phase with Joint Estimation of Time of Signal Arrival (Ending) in Presence of Interference (Noise) in Metric Space with L-group Properties

As a rule, if time of signal arrival (ending) is known, the detection problem makes no sense from the standpoint of obtaining information concerning the presence (or absence) of the signal. Within this subsection, the signal detection problem is considered along with the problem of estimation of time of signal arrival (ending).

Synthesis and analysis of optimal algorithm of detection of harmonic signal with unknown nonrandom amplitude and initial phase with joint estimation of time of signal arrival (ending) in the presence of interference (noise) in signal space with lattice properties are fulfilled under the following assumptions. During synthesis, the distribution of interference (noise) $n(t)$ is considered arbitrary. During the further analysis of the synthesized algorithm and unit of signal processing, interference (noise) $n(t)$ is considered Gaussian.

Let the interaction between the harmonic signal $s(t)$ and interference (noise) $n(t)$ in signal space $\mathcal{L}(+, \vee, \wedge)$ with L-group properties be described by two binary operations $\vee$ and $\wedge$ in two receiving channels, respectively:

$$x(t) = \theta s(t) \vee n(t); \tag{7.4.11a}$$

$$\tilde{x}(t) = \theta s(t) \wedge n(t), \tag{7.4.11b}$$

where θ is an unknown nonrandom parameter that takes only two values $\theta \in \{0, 1\}$: $\theta = 0$ (the signal is absent) and $\theta = 1$ (the signal is present); $t \in T_s$, T_s is a domain of definition of the signal $s(t)$, $T_s = [t_0, t_1]$; t_0 is an unknown time of arrival of the signal $s(t)$; t_1 is an unknown time of signal ending; $T = t_1 - t_0$ is known signal duration; $T_s \subset T_{\text{obs}}$; T_{obs} is an observation interval of the signal: $T_{\text{obs}} = [t'_0, t'_1]$; $t'_0 < t_0$, $t_1 < t'_1$.

Let the model of the received harmonic signal $s(t)$ be determined by the expression:

$$s(t) = \begin{cases} A\cos(\omega_0 t + \varphi), & t \in T_s; \\ 0, & t \notin T_s, \end{cases} \tag{7.4.12}$$

where A is an unknown nonrandom amplitude of the useful signal $s(t)$; $\omega_0 = 2\pi f_0$; f_0 is known carrier frequency of the signal $s(t)$; φ is an unknown nonrandom initial phase of the useful signal, $\varphi \in [-\pi, \pi]$; T_s is a domain of definition of the signal $s(t)$, $T_s = [t_0, t_1]$; t_0 is an unknown time of arrival of the signal $s(t)$; t_1 is an unknown time of signal ending; $T = t_1 - t_0$ is known signal duration, $T = N_s T_0$; N_s is a number of periods of the harmonic signal $s(t)$, $N_s \in \mathbf{N}$, $\mathbf{N}$ is the set of natural numbers; T_0 is a period of signal carrier.

We assume that interference (noise) $n(t)$ is characterized by such statistical properties that two neighbor samples of interference (noise) $n(t_j)$ and $n(t_{j\pm 1})$, which are distant from each other over the interval $|t_{j\pm 1} - t_j| = 1/f_0 = T_0$, are independent. Instantaneous values (the samples) of the signal $\{s(t_j)\}$ and interference (noise) $\{n(t_j)\}$ are the elements of signal space: $s(t_j), n(t_j) \in \mathcal{L}(+, \vee, \wedge)$. The samples $s(t_j)$ and $n(t_j)$ of the signal $s(t)$ and interference (noise) $n(t)$ are taken on a domain of definition T_s of the signal $s(t)$: $t_j \in T_s$ through the period of signal carrier $T_0 = 1/f_0$ providing the independence of interference (noise) samples $\{n(t_j)\}$.

Taking into account the aforementioned considerations, the equations of the observations in both processing channels (7.4.11a) and (7.4.11b) take the form:

$$x(t_j) = \theta s(t_j) \vee n(t_j); \quad (7.4.13a)$$

$$\tilde{x}(t_j) = \theta s(t_j) \wedge n(t_j), \quad (7.4.13b)$$

where $t_j = t - jT_0$, $j = 0, 1, \ldots, J-1$, $t_j \in T_s \subset T_{\text{obs}}$; $T_s = [t_0, t_1]$, $T_{\text{obs}} = [t'_0, t'_1]$; $t'_0 < t_0$, $t_1 < t'_1$; $J \in \mathbf{N}$, $\mathbf{N}$ is the set of natural numbers.

In this subsection, the problem of joint detection of the signal $s(t)$ and estimation of its time of ending t_1 is formulated and solved on the basis of matched filtering of the signal $s(t)$, after which the joint detection of the signal $s(t)$ and estimation of its time of ending t_1 are realized.

The content of matched filtering problem in the signal space with L-group (or lattice) properties essentially differs from the similar problem solved in linear signal space $\mathcal{LS}$. As is well known, the base of the formulation and solving the matched filtering problem in linear signal space $\mathcal{LS}$ is the criterion of maximum of signal-to-noise ratio in the filter output [254], [163], or the criterion of minimum of average risk, which determines the proper structure of a matched filter. In this subsection the signal is not assumed to be known, and signal processing is not considered linear; that defines certain features of processing.

The problem of *matched filtering* $MF[s(t)]$ in signal space $\mathcal{L}(+, \vee, \wedge)$ with L-group properties is solved through step-by-step processing of statistical collections $\{x(t_j)\}$ and $\{\tilde{x}(t_j)\}$ determined by the observation equations (7.4.13a) and (7.4.13b):

$$MF[s(t)] = \begin{cases} PF[s(t)]; & (a) \\ IP[s(t)]; & (b) \\ Sm[s(t)], & (c) \end{cases} \quad (7.4.14)$$

where $PF[s(t)]$ is primary filtering; $IP[s(t)]$ is intermediate processing; $Sm[s(t)]$

7.4 Signal Detection in Metric Space with Lattice Properties

is smoothing; they all form the processing stages of a general algorithm of useful signal matched filtering (7.4.14).

The criteria of optimality determining every processing stage $PF[s(t)]$, $IP[s(t)]$, $Sm[s(t)]$ are united into the single systems:

$$PF[s(t)] = \begin{cases} y(t) = \underset{\breve{y}(t)\in Y; t, t_j \in T_{obs}}{\arg\min} \; |\bigwedge_{j=0}^{J-1}[x(t_j) - \breve{y}(t)]|; & (a) \\ \tilde{y}(t) = \underset{\widehat{y}(t)\in \widehat{Y}; t, t_j \in T_{obs}}{\arg\min} \; |\bigvee_{j=0}^{J-1}[\tilde{x}(t_j) - \widehat{y}(t)]|; & (b) \\ J = \underset{y(t)\in Y; \; T_{obs}}{\arg\min}[\int |y(t)|dt]\big|_{n(t)\equiv 0}, \; \int_{T_{obs}} |y(t)|dt \neq 0; & (c) \\ w(t) = F[y(t), \tilde{y}(t)]; & (d) \\ \int_{[t_1-T_0, t_1]} |w(t) - s(t)|dt]\big|_{n(t)\equiv 0} \to \underset{w(t)\in W}{\min}, & (e) \end{cases} \quad (7.4.15)$$

where $y(t)$, $\tilde{y}(t)$ are the solution functions for minimization of the metric between the observed statistical collections $\{x(t_j)\}$, $\{\tilde{x}(t_j)\}$ and optimization variables, i.e., the functions $\breve{y}(t)$, $\widehat{y}(t)$, respectively; $w(t)$ is the function $F[*,*]$ of uniting the results $y(t)$, $\tilde{y}(t)$ of minimization of the functions of the observed collections $\{x(t_j)\}$ and $\{\tilde{x}(t_j)\}$; T_{obs} is an observation interval of the signal; J is a number of samples of stochastic processes $x(t)$, $\tilde{x}(t)$ used under processing $J \in \mathbf{N}$; $\mathbf{N}$ is the set of natural numbers;

$$IP[s(t)] = \begin{cases} \mathbf{M}\{u^2(t)\}\big|_{s(t)\equiv 0} \underset{L}{\to} \min; & (a) \\ \mathbf{M}\{[u(t) - s(t)]^2\}\big|_{n(t)\equiv 0} = \varepsilon; & (b) \\ u(t) = L[w(t)], & (c) \end{cases} \quad (7.4.16)$$

where $\mathbf{M}\{*\}$ is the symbol of mathematical expectation; $L[w(t)]$ is a functional transformation of the process $w(t)$ into the process $u(t)$; ε is a constant that is some function of a power of the signal $s(t)$;

$$Sm[s(t)] = \begin{cases} v(t) = \underset{v^\circ(t)\in V; t, t_k \in \tilde{T}}{\arg\min} \sum_{k=0}^{M-1} |u(t_k) - v^\circ(t)|; & (a) \\ \Delta\tilde{T} : \delta_d(\Delta\tilde{T}) = \delta_{d,sm}; & (b) \\ M = \underset{M'\in \mathbf{N}\cap[M^*,\infty[}{\arg\max}[\delta_f(M')]\big|_{M^*:\delta_f(M^*)=\delta_{f,sm}}, & (c) \end{cases} \quad (7.4.17)$$

where $v(t) = \hat{s}(t)$ is the result of filtering (the estimator $\hat{s}(t)$ of the signal $s(t)$) that is the solution of the problem of minimizing the metric $\sum_{k=0}^{M-1} |u(t_k) - v^\circ(t)|$ between the instantaneous values of stochastic process $u(t)$ and optimization variable, i.e., the function $v^\circ(t)$; $t_k = t - \frac{k}{M}\Delta\tilde{T}$, $k = 0, 1, \ldots, M-1$, $t_k \in \tilde{T} =]t - \Delta\tilde{T}, t]$; $\tilde{T}$ is an interval, in which smoothing of stochastic process $u(t)$ is realized; $M \in \mathbf{N}$, $\mathbf{N}$ is the set of natural numbers; M is a number of the samples of stochastic process $u(t)$ used under smoothing on the interval $\tilde{T}$; $\delta_d(\Delta\tilde{T})$ and $\delta_f(M)$ are relative dynamic and fluctuation errors of smoothing as the dependence on the quantity $\Delta\tilde{T}$ of a smoothing interval $\tilde{T}$ and a number of the samples M, respectively; $\delta_{d,\text{sm}}$, $\delta_{f,\text{sm}}$ are the quantities of the relative dynamic and fluctuation errors of smoothing, respectively.

The problems of joint detection $Det[s(t)]$ of the signal $s(t)$ and estimation $Est[t_1]$ of time of its ending t_1 are based on the detection and estimation criteria united into one system, which is a logical continuation of the system (7.4.14):

$$Det[s(t)]/Est[t_1] = \begin{cases} E_v(\hat{t}_1 - \frac{T_0}{2}) \overset{d_1}{\underset{d_0}{\gtrless}} l_0(F); & (a) \\ \hat{t}_1 = \underset{\varphi \in \Phi_1 \vee \Phi_2; t^\circ \in T_{\text{obs}}}{\arg\min} \mathbf{M}_\varphi\{(t_1 - t^\circ)^2\}\big|_{n(t) \equiv 0}, & (b) \end{cases} \quad (7.4.18)$$

where $E_v(\hat{t}_1 - \frac{T_0}{2})$ is an instantaneous value of the envelope $E_v(t)$ of the estimator $v(t) = \hat{s}(t)$ of useful signal $s(t)$ at the instant $t = \hat{t}_1 - \frac{T_0}{2}$: $E_v(t) = \sqrt{v^2(t) + v_{\mathcal{H}}^2(t)}$; $v_{\mathcal{H}}(t) = \mathcal{H}[v(t)]$ is Hilbert transform; d_1, d_0 are the decisions made concerning true value of an unknown nonrandom parameter θ, $\theta \in \{0, 1\}$; $l_0(F)$ is some threshold level as a function of a given conditional probability of false alarm F; $\hat{t}_1$ is the estimator of time of signal ending t_1; $\mathbf{M}_\varphi\{(t_1 - t^\circ)^2\}$ is a mean squared difference between true value of time of signal ending t_1 and optimization variable t°; $\mathbf{M}_\varphi\{*\}$ is a symbol of mathematical expectation with averaging over the initial phase φ of the signal; Φ_1 and Φ_2 are possible domains of definition of the initial phase φ of the signal: $\Phi_1 = [-\pi/2, \pi/2]$ and $\Phi_2 = [\pi/2, 3\pi/2]$.

We now explain the optimality criteria and single relationships appearing in the systems (7.4.15), (7.4.16), (7.4.17) determining the successive stages $PF[s(t)]$, $IP[s(t)]$, and $Sm[s(t)]$ of the general algorithm of useful signal processing (7.4.14).

Equations (7.4.15a) and (7.4.15b) of the system (7.4.15) define the criteria of minimum of metrics between the statistical sets of the observations $\{x(t_j)\}$ and $\{\tilde{x}(t_j)\}$ and the results of primary processing $y(t)$ and $\tilde{y}(t)$, respectively. The functions of metrics $|\bigwedge_{j=0}^{J-1}[x(t_j) - \breve{y}(t)]|$, $|\bigvee_{j=0}^{J-1}[\tilde{x}(t_j) - \widehat{y}(t)]|$ are chosen to provide the metric convergence and the convergence in probability to the useful signal $s(t)$ of the sequences $y(t) = \bigwedge_{j=0}^{J-1} x(t_j)$ and $\tilde{y}(t) = \bigvee_{j=0}^{J-1} \tilde{x}(t_j)$ for the interactions of both kinds (7.4.13a) and (7.4.13b).

The relationship (7.4.15c) determines the criterion of the choice of a number J of the samples of stochastic processes $x(t)$ and $\tilde{x}(t)$ used during signal processing based on the minimization of the norm $\int_{T_{\text{obs}}} |y(t)| dt$. The criterion (7.4.15c) is considered under two constraints: (1) interference (noise) is identically equal to zero: $n(t) \equiv 0$; (2) the norm $\int_{T_{\text{obs}}} |y(t)| dt$ of the function $y(t)$ is not equal to zero: $\int_{T_{\text{obs}}} |y(t)| dt \neq 0$.

Equation (7.4.15e) defines the criterion of minimum of the metric between the useful signal $s(t)$ and the function $w(t)$, i.e., $\int_{[t_1 - T_0, t_1]} |w(t) - s(t)| dt \big|_{n(t) \equiv 0}$ in the interval $[t_1 - T_0, t_1]$. This criterion establishes the kind of function $F[y(t), \tilde{y}(t)]$ (7.4.15d) uniting the results $y(t)$ and $\tilde{y}(t)$ of primary processing of the observed

7.4 Signal Detection in Metric Space with Lattice Properties

processes $x(t)$ and $\tilde{x}(t)$. The criterion (7.4.15e) is considered when interference (noise) is identically equal to zero: $n(t) \equiv 0$.

The equations (7.4.16a), (7.4.16b), (7.4.16c) of the system (7.4.16) define the criterion of the choice of functional transformation $L[w(t)]$. The equation (7.4.16a) defines the criterion of minimum of the second moment of the process $u(t)$ in the absence of the useful signal $s(t)$ in the input of signal processing unit. The equation (7.4.16b) determines the quantity of the second moment of the difference between the signals $u(t)$ and $s(t)$ in the absence of interference (noise) $n(t)$ in the input of a signal processing unit.

The equation (7.4.17a) of the system (7.4.17) defines the criterion of minimum of metric $\sum_{k=0}^{M-1} |u(t_k) - v°(t)|$ between the instantaneous values of the process $u(t)$ and optimization variable $v°(t)$ within the smoothing interval $\tilde{T} =]t - \Delta\tilde{T}, t]$, requiring the final processing of the signal $u(t)$ in the form of smoothing under the condition that the useful signal is identically equal to zero: $s(t) \equiv 0$. The relationship (7.4.17b) establishes the rule of the choice of the quantity $\Delta\tilde{T}$ of smoothing interval $\tilde{T}$ based on a relative dynamic error $\delta_{d,\text{sm}}$ of smoothing. The equation (7.4.17c) defines the criterion of the choice of a number M of the samples of stochastic process $u(t)$ based on a relative fluctuation error $\delta_{f,\text{sm}}$ of smoothing.

The equation (7.4.18a) of the system (7.4.18) defines the criterion of the decision d_1 concerning the signal presence (if $E_v(\hat{t}_1 - \frac{T_0}{2}) > l_0(F)$) or the decision d_0 regarding its absence (if $E_v(\hat{t}_1 - \frac{T_0}{2}) < l_0(F)$). The relationship (7.4.18b) defines the criterion of forming the estimator $\hat{t}_1$ of time of signal ending t_1 based on minimization of the mean squared difference $\mathbf{M}_\varphi\{(t_1 - t°)^2\}$ between true value of time of signal ending t_1 and optimization variable $t°$ under the conditions that averaging is realized over the initial phase φ of the signal taken in one of two intervals: $\Phi_1 = [-\pi/2, \pi/2]$ or $\Phi_2 = [\pi/2, 3\pi/2]$, and interference (noise) is absent: $n(t) \equiv 0$.

To solve the problem of minimizing the functions $|\bigwedge_{j=0}^{J-1}[x(t_j) - \breve{y}(t)]|$ (7.4.15a), $|\bigvee_{j=0}^{J-1}[\tilde{x}(t_j) - \widehat{y}(t)]|$ (7.4.15b), we find the extrema of these functions, setting their derivatives with respect to $\breve{y}(t)$ and $\widehat{y}(t)$ to zero, respectively:

$$d|\bigwedge_{j=0}^{J-1}[x(t_j) - \breve{y}(t)]|/d\breve{y}(t) = -\text{sign}(\bigwedge_{j=0}^{J-1}[x(t_j) - \breve{y}(t)]) = 0; \quad (7.4.19a)$$

$$d|\bigvee_{j=0}^{J-1}[\tilde{x}(t_j) - \widehat{y}(t)]|/d\widehat{y}(t) = -\text{sign}(\bigvee_{j=0}^{J-1}[\tilde{x}(t_j) - \widehat{y}(t)]) = 0. \quad (7.4.19b)$$

The solutions of Equations (7.4.19a) and (7.4.19b) are the values of the estimators $y(t)$ and $\tilde{y}(t)$ in the form of meet and join of the observation results $\{x(t_j)\}$ and $\{\tilde{x}(t_j)\}$, respectively:

$$y(t) = \bigwedge_{j=0}^{J-1} x(t_j) = \bigwedge_{j=0}^{J-1} x(t - jT_0); \quad (7.4.20a)$$

$$\tilde{y}(t) = \bigvee_{j=0}^{J-1} \tilde{x}(t_j) = \bigvee_{j=0}^{J-1} \tilde{x}(t - jT_0). \quad (7.4.20b)$$

The derivatives of the functions $|\bigwedge_{j=0}^{J-1}[x(t_j)-\breve{y}(t)]|$ and $|\bigvee_{j=0}^{J-1}[\tilde{x}(t_j)-\widehat{y}(t)]|$, according to the relationships (7.4.19a) and (7.4.19b) at the points $y(t)$ and $\tilde{y}(t)$, change their sign from minus to plus. Thus, the extrema determined by the formulas (7.4.20a) and (7.4.20b) are minimum points of these functions, and the solutions of the equations (7.4.15a) and (7.4.15b) determining these estimation criteria.

The condition $n(t) \equiv 0$ of the criterion (7.4.15c) of the system (7.4.15) implies the corresponding changes in the observation equations (7.4.13a,b): $x(t_j) = s(t_j) \vee 0$, $\tilde{x}(t_j) = s(t_j) \wedge 0$. Thus, according to the relationships (7.4.20a) and (7.4.20b), the identities hold:

$$y(t)\big|_{n(t)\equiv 0} = \bigwedge_{j=0}^{J-1}[s(t_j) \vee 0] =$$
$$= [s(t) \vee 0] \wedge [s(t - (J-1)T_0) \vee 0]; \quad (7.4.21a)$$

$$\tilde{y}(t)\big|_{n(t)\equiv 0} = \bigvee_{j=0}^{J-1}[s(t_j) \wedge 0] =$$
$$= [s(t) \wedge 0] \vee [s(t - (J-1)T_0) \wedge 0], \quad (7.4.21b)$$

where $t_j = t - jT_0$, $j = 0, 1, \ldots, J-1$. Based on (7.4.21a), we obtain the value of the norm $\int_{T_{obs}} |y(t)|dt$ from the criterion (7.4.15c):

$$\int_{T_{obs}} |y(t)|dt = \begin{cases} 4(N_s - J + 1)A/\pi, & J \leq N_s; \\ 0, & J > N_s, \end{cases} \quad (7.4.22)$$

where N_s is a number of periods of harmonic signal $s(t)$; A is an unknown nonrandom amplitude of the useful signal $s(t)$.

From Equation (7.4.22), according to the criterion (7.4.15c), we obtain optimal value of J samples of stochastic processes $x(t)$ and $\tilde{x}(t)$ used during primary processing (7.4.20a,b) equal to the number of signal periods N_s:

$$J = N_s. \quad (7.4.23)$$

When the last identity holds, the processes determined by the relationships (7.4.21a) and (7.4.21b) in the interval $[t_1 - T_0, t_1]$ are, respectively, equal to:

$$y(t)\big|_{n(t)\equiv 0} = s(t) \vee 0; \quad (7.4.24a)$$

$$\tilde{y}(t)\big|_{n(t)\equiv 0} = s(t) \wedge 0. \quad (7.4.24b)$$

To realize the criterion (7.4.15e) of the system (7.4.15) under joint fulfillment of the identities (7.4.24a), (7.4.24b) and (7.4.15d), it is necessary and sufficient that the coupling equation (7.4.15d) between stochastic process $w(t)$ and results of primary processing $y(t)$ and $\tilde{y}(t)$ has the form:

$$w(t)\big|_{n(t)\equiv 0} = y(t)\big|_{n(t)\equiv 0} \vee 0 + \tilde{y}(t)\big|_{n(t)\equiv 0} \wedge 0 =$$
$$= s(t) \vee 0 \vee 0 + s(t) \wedge 0 \wedge 0 = s(t), \ t \in [t_1 - T_0, t_1]. \quad (7.4.25)$$

As follows from the expression (7.4.25), the metric $\int_{[t_1-T_0,t_1]} |w(t) - s(t)| dt \big|_{n(t)\equiv 0}$, that must be minimized according to the criterion (7.4.15e), is minimal and equal to zero.

Obviously, the coupling equation (7.4.15d) has to be invariant with respect to the presence (the absence) of interference (noise) $n(t)$, so the final variant of the coupling equation can be written on the base of (7.4.25) in the form:

$$w(t) = y_+(t) + \tilde{y}_-(t); \tag{7.4.26}$$
$$y_+(t) = y(t) \vee 0; \tag{7.4.26a}$$
$$\tilde{y}_-(t) = \tilde{y}(t) \wedge 0. \tag{7.4.26b}$$

Thus, the identity (7.4.26) determines the kind of the coupling equation (7.4.15d) obtained from the joint fulfillment of the criteria (7.4.15a), (7.4.15b), and (7.4.15e).

The solution $u(t)$ of the relationships (7.4.16a), (7.4.16b), (7.4.16c) of the system (7.4.16), defining the criterion of the choice of functional transformation of the process $w(t)$, is the function $L[w(t)]$ determining the gain characteristic of the limiter:

$$u(t) = L[w(t)] = \begin{cases} a, & w(t) \geq a; \\ w(t), & -a < w(t) < a; \\ -a, & w(t) \leq -a, \end{cases} \tag{7.4.27}$$

its linear part provides the condition (7.4.16b); its clipping part (above and below) provides minimization of the second moment of the process $u(t)$, according to the criterion (7.4.16a).

The relationship (7.4.27) can be written in terms of L-group $\mathcal{L}(+, \vee, \wedge)$ in the form:

$$u(t) = L[w(t)] = [(w(t) \wedge a) \vee 0] + [(w(t) \vee (-a)) \wedge 0], \tag{7.4.27a}$$

where the limiter parameter a is chosen to be equal to: $a = \sup_A \Delta_A = A_{\max}$, $\Delta_A =]0, A_{\max}]$; $A_{\max}$ is maximum possible value of useful signal amplitude.

We obtain the estimator $v(t) = \hat{s}(t)$ of the signal $s(t)$ by solving the equation of minimization of the function on the base of criterion (7.4.17a) of the system (7.4.17). We find the extremum of the function $\sum_{k=0}^{M-1} |u(t_k) - v^\circ(t)|$, setting its derivative with respect to $v^\circ(t)$ to zero:

$$d\{\sum_{k=0}^{M-1} |u(t_k) - v^\circ(t)|\}/dv^\circ(t) = -\sum_{k=0}^{M-1} \text{sign}[u(t_k) - v^\circ(t)] = 0.$$

The solution of the last equation is the value of the estimator $v(t)$ in the form of sample median med$\{*\}$ of the collection $\{u(t_k)\}$ of stochastic process $u(t)$:

$$v(t) = \underset{t_k \in \tilde{T}}{\text{med}} \{u(t_k)\}, \tag{7.4.28}$$

where $t_k = t - \frac{k}{M}\Delta\tilde{T}$, $k = 0, 1, \ldots, M-1$; $t_k \in \tilde{T} =]t - \Delta\tilde{T}, t]$; $\tilde{T}$ is a smoothing interval of stochastic process $u(t)$; $\Delta\tilde{T}$ is a quantity of the smoothing interval $\tilde{T}$.

A derivative of the function $\sum_{k=0}^{M-1} |u(t_k) - v°(t)|$ at the point $v(t)$ changes its sign from minus to plus. Thus, the extremum determined by the formula (7.4.28) is minimum of this function, and the solution of Equation (7.4.17a) defining this criterion of signal processing.

The rule of making the decision d_1 concerning the presence of the signal (if $E_v(\hat{t}_1 - \frac{T_0}{2}) > l_0(F)$) or the decision d_0 concerning the absence of the signal (if $E_v(\hat{t}_1 - \frac{T_0}{2}) < l_0(F)$), stated by the equation (7.4.18a), suggests (1) formation of the envelope $E_v(t)$ of the estimator $v(t) = \hat{s}(t)$ of useful signal $s(t)$, and (2) comparison of the value of the envelope $E_v(\hat{t}_1 - \frac{T_0}{2})$ with a threshold value $l_0(F)$ at the instant $t = \hat{t}_1 - \frac{T_0}{2}$ determined by the estimator $\hat{t}_1$; as the result, the decision making is realized:

$$E_v(\hat{t}_1 - \frac{T_0}{2}) \overset{d_1}{\underset{d_0}{\gtrless}} l_0(F). \quad (7.4.29)$$

The relationship (7.4.18b) defines the criterion of forming the estimator $\hat{t}_1$ of time of signal ending t_1 based on minimizing the mean squared difference $\mathbf{M}_\varphi\{(t_1 - t°)^2\}$ between true value of time of signal ending t_1 and optimization variable $t°$ under the conditions that averaging is realized over the initial phase φ of the signal taken in one of two intervals: $\Phi_1 = [-\pi/2, \pi/2]$ or $\Phi_2 = [\pi/2, 3\pi/2]$, and interference (noise) is identically equal to zero: $n(t) \equiv 0$.

The solution of optimization equation (7.4.18b) of the system (7.4.18) is determined by the identity:

$$\hat{t}_1 = \begin{cases} \hat{t}_- + (T_0/2) + (T_0\hat{\varphi}/2\pi), & \varphi \in \Phi_1 = [-\pi/2, \pi/2]; \\ \hat{t}_+ + (T_0/2) - (T_0\hat{\varphi}/2\pi), & \varphi \in \Phi_2 = [\pi/2, 3\pi/2], \end{cases} \quad (7.4.30)$$

where $\hat{t}_\pm = (\int_{T_{obs}} tv_\pm(t)dt)/(\int_{T_{obs}} v_\pm(t)dt)$ is the estimator of barycentric coordinate of the positive $v_+(t)$ or the negative $v_-(t)$ parts of the smoothed stochastic process $v(t)$, respectively; $v_+(t) = v(t) \vee 0$, $v_-(t) = v(t) \wedge 0$; T_{obs} is an observation interval of the signal: $T_{obs} = [t'_0, t'_1]$; $t'_0 < t_0$, $t_1 < t'_1$; $\hat{\varphi} = \arcsin[2(\hat{t}_+ - \hat{t}_-)/T_0]$ is the estimator of an unknown nonrandom initial phase φ of the useful signal $s(t)$.

The value of mean squared difference $\mathbf{M}_\varphi\{(t_1 - \hat{t}_1)^2\}$ between the true value of time of signal ending t_1 and its estimator $\hat{t}_1$ determined by Equation (7.4.30), under the given conditions, is equal to zero:

$$\mathbf{M}_\varphi\{(t_1 - \hat{t}_1)^2\}\big|_{n(t)=0} = 0, \text{ on } \varphi \in \Phi_1 = [-\pi/2, \pi/2] \text{ or } \varphi \in \Phi_2 = [\pi/2, 3\pi/2].$$

If the initial phase φ of the signal can change within the interval from $-\pi$ to π, i.e., it is not known beforehand to which interval ($\Phi_1 = [-\pi/2, \pi/2]$ or $\Phi_2 = [\pi/2, 3\pi/2]$ from the relationships (7.4.18b) and (7.4.30)) the phase φ belongs, then the estimators $\hat{t}_1$ of time of signal ending t_1 can be satisfactory and determined by the identities:

$$\hat{t}_1 = \max_{\hat{t}_\pm \in T_{obs}} [\hat{t}_-, \hat{t}_+] + (T_0/4), \quad \varphi \in [-\pi, \pi]; \quad (7.4.31a)$$

$$\text{or}: \quad \hat{t}_1 = \frac{1}{2}(\hat{t}_- + \hat{t}_+) + \frac{T_0}{2}, \quad \varphi \in [-\pi, \pi]. \quad (7.4.31b)$$

7.4 Signal Detection in Metric Space with Lattice Properties

Summarizing the relationships (7.4.30), (7.4.29), (7.4.28), (7.4.27), (7.4.26), (7.4.25), (7.4.20a,b), one can conclude that the estimator $\hat{t}_1$ of time of signal ending and its envelope $E_v(t)$ are formed from further processing of the estimator $v(t) = \hat{s}(t)$ of harmonic signal $s(t)$ detected in the presence of interference (noise) $n(t)$, which is the function of smoothing stochastic process $u(t)$ obtained by limitation of the process $w(t)$ that combines the results $y(t)$ and $\tilde{y}(t)$ corresponding to primary processing of the stochastic processes $x(t)$ and $\tilde{x}(t)$ in the observation interval T_{obs}.

The block diagram of the processing unit, according to the relationships (7.4.20a,b), (7.4.23), (7.4.26), (7.4.27), (7.4.28), (7.4.29), and (7.4.30), includes two processing channels, each containing transversal filters; the units of formation of positive $y_+(t)$ and negative $\tilde{y}_-(t)$ parts of the processes $y(t)$ and $\tilde{y}(t)$, respectively; an adder uniting the results of signal processing in both channels; a limiter $L[w(t)]$; a median filter (MF), an estimator formation unit (EFU), an envelope computation unit (ECU), and a decision gate (DG) (see Fig. 7.4.4).

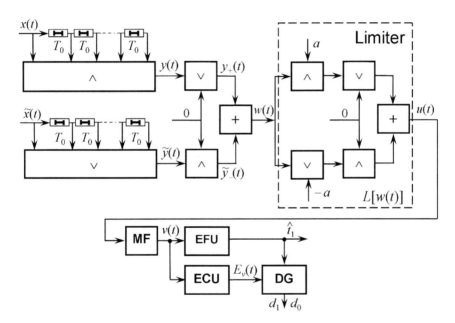

FIGURE 7.4.4 Block diagram of processing unit that realizes harmonic signal detection with joint estimation of time of signal arrival (ending)

Transversal filters in two processing channels realize primary filtering $PF[s(t)]$: $y(t) = \bigwedge_{j=0}^{N-1} x(t - jT_0)$, $\tilde{y}(t) = \bigvee_{j=0}^{N-1} \tilde{x}(t - jT_0)$ of the observed stochastic processes $x(t)$, $\tilde{x}(t)$, according to the equations (7.4.20a,b) and (7.4.23), providing fulfillment of the criteria (7.4.15a), (7.4.15b), and (7.4.15c) of the system (7.4.15). In two processing channels, the units of formation of positive $y_+(t)$ and negative $\tilde{y}_-(t)$ parts of the processes $y(t)$ and $\tilde{y}(t)$ form the values of this functions, according to the identities (7.4.26a) and (7.4.26b), respectively. The adder unites the results of signal processing in two channels, according to the equality (7.4.26), providing fulfillment of the criteria (7.4.15d) and (7.4.15e) of the system (7.4.15).

The limiter $L[w(t)]$ realizes intermediate processing $IP[s(t)]$ in the form of limitation of the signal $w(t)$ in the output of the adder, according to the criteria (7.4.16a), (7.4.16b) of the system (7.4.16), to exclude from further processing noise overshoots whose instantaneous values exceed a given value a. The median filter (MF) realizes smoothing $Sm[s(t)]$ of the process $u(t)$, according to formula (7.4.28), fulfilling criterion (7.4.17a) of the system (7.4.17).

Envelope computation unit (ECU) forms the envelope $E_v(t)$ of the signal $v(t)$ in the output of the median filter (MF). The estimator formation unit (EFU) forms the estimator $\hat{t}_1$ of time of signal ending, according to the equation (7.4.30), providing fulfillment of the criterion (7.4.18b) of the system (7.4.18).

The decision gate (DG) compares, at the instant $t = \hat{t}_1 - \frac{T_0}{2}$, the instantaneous value of the envelope $E_v(t)$ with threshold value $l_0(F)$, and makes the decision d_1 concerning the presence of the signal (if $E_v(\hat{t}_1 - \frac{T_0}{2}) > l_0(F)$) or the decision d_0 concerning the absence of the signal (if $E_v(\hat{t}_1 - \frac{T_0}{2}) < l_0(F)$), according to the rule (7.4.29) of the criterion (7.4.18a) of the system (7.4.18).

Figures 7.4.5 through 7.4.8 illustrate the results of statistical modeling of signal processing by a synthesized unit. The useful signal $s(t)$ is harmonic with a number of periods $N_s = 16$, with the initial phase $\varphi = \pi/3$. Signal-to-noise ratio E/N_0 is equal to $E/N_0 = 10^{-10}$, where E is signal energy; N_0 is power spectral density. Product $T_0 f_{n,\max} = 32$, where T_0 is a period of a carrier of the signal $s(t)$; $f_{n,\max}$ is a maximum frequency of power spectral density of noise $n(t)$ in the form of quasi-white Gaussian noise.

Fig. 7.4.5 illustrates the useful signal $s(t)$ and realizations $w^*(t)$, $v^*(t)$ of the signals $w(t)$, $v(t)$ in the outputs of the adder and median filter. As shown in the figure, the matched filter provides the compression of the useful harmonic signal $s(t)$ in such a way that the duration of the signal $v(t)$ in the output of matched filter is equal to a period T_0 of the harmonic signal $s(t)$; thus, compressing the useful signal N_s times, where N_s is a number of periods of harmonic signal $s(t)$.

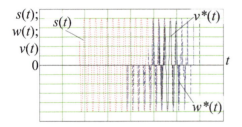

FIGURE 7.4.5 Useful signal $s(t)$ (dot line); realization $w^*(t)$ of signal $w(t)$ in output of adder (dash line); realization $v^*(t)$ of signal $v(t)$ in output of median filter (solid line)

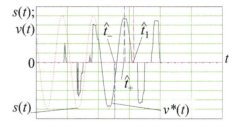

FIGURE 7.4.6 Useful signal $s(t)$ (dot line); realization $v^*(t)$ of signal in output of median filter $v(t)$ (solid line); δ-pulse determining time position of estimator $\hat{t}_1$ of time of signal ending t_1

Figure 7.4.6 illustrates: the useful signal $s(t)$; realization $v^*(t)$ of the signal in the output of median filter $v(t)$; δ-pulses determining time position of the estimators $\hat{t}_\pm$ of barycentric coordinates of positive $v_+(t)$ or negative $v_-(t)$ parts of the smoothed stochastic process $v(t)$; and δ-pulse determining time position of the estimator $\hat{t}_1$ of time of signal ending t_1, according to the formula (7.4.30).

Figure 7.4.7 illustrates: the useful signal $s(t)$; realization $v^*(t)$ of the signal in the output of median filter $v(t)$; δ-pulses determining time position of the estimators $\hat{t}_\pm$ of barycentric coordinates of positive $v_+(t)$ or negative $v_-(t)$ parts of the smoothed stochastic process $v(t)$; and δ-pulse determining time position of the estimator $\hat{t}_1$ of time of signal ending t_1, according to the formulas (7.4.30), (7.4.31a), and (7.4.31b).

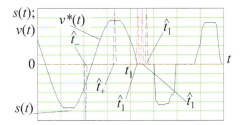
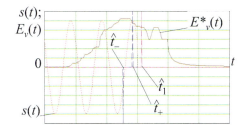

FIGURE 7.4.7 Useful signal $s(t)$ (dot line); realization $v^*(t)$ of signal in output of median filter $v(t)$ (solid line); δ-pulses determining time position of estimators $\hat{t}_\pm$; δ-pulse determining time position of estimator $\hat{t}_1$ of time of signal ending t_1

FIGURE 7.4.8 Useful signal $s(t)$ (dot line); realization $E_v^*(t)$ of envelope $E_v(t)$ of signal $v(t)$ (solid line); δ-pulses determining time position of estimators $\hat{t}_\pm$; δ-pulse determining time position of estimator $\hat{t}_1$ of time of signal ending t_1

Figure 7.4.8 illustrates: the useful signal $s(t)$; realization $E_v^*(t)$ of the envelope $E_v(t)$ of the signal $v(t)$ in the output of median filter; δ-pulses determining time position of the estimators $\hat{t}_\pm$ of barycentric coordinates of positive $v_+(t)$ or negative $v_-(t)$ parts of the smoothed stochastic process $v(t)$; and δ-pulse determining time position of the estimator $\hat{t}_1$ of time of signal ending t_1 according to the formula (7.4.30).

We can determine the quality indices of harmonic signal detection by a synthesized processing unit (Fig. 7.4.4). We use the theorem from [251] indicating that the median estimator $v(t)$ formed by median filter (MF) converges in distribution to a Gaussian random variable with zero mean.

Along with the signal $w(t)$ in the output of the adder (see formula (7.3.25)), the process $v(t)$ in the output of a median filter can be represented as the sum of signal $v_s(t)$ and noise $v_n(t)$ components:

$$v(t) = v_s(t) + v_n(t). \qquad (7.4.32)$$

As mentioned in Section 7.3, variance D_{v_n} (7.3.34) of the noise component $v_n(t)$ of the process $v(t)$ in the output of the filter (7.4.28) (under the condition that the signal $s(t)$ is a harmonic oscillation in the form (7.4.12)) can be decreased to the

quantity:
$$D_{v_n} \leq D_{v,\max} = a^2 \exp\left\{-\frac{\Delta \tilde{T}^2 N_s^2 f_{n,\max}^2}{8\sqrt{\pi}}\right\}, \qquad (7.4.33)$$

where a is a parameter of the limiter $L[w(t)]$; $\Delta \tilde{T}$ is a value of smoothing interval $\tilde{T}$; N_s is a number of periods of harmonic signal $s(t)$; $f_{n,\max}$ is a maximum frequency of power spectral density of interference (noise) $n(t)$.

Under additive interaction between signal $v_s(t)$ and noise $v_n(t)$ components in the output of median filter (7.4.32), assuming the normalcy of distribution of noise component $v_n(t)$ with zero mean, the conditional PDF $p_v(z/\theta = 1)$ of the envelope $E_v(t)$ of the process $v(t)$ in the output of median filter can be represented in the form of Rice distribution:

$$p_v(z/\theta = 1) = \frac{z}{D_{v,\max}} \exp\left[-\frac{(z^2 + E^2)}{2D_{v,\max}}\right] I_0\left(\frac{E \cdot z}{D_{v,\max}}\right), \quad z \geq 0, \qquad (7.4.34)$$

where E is a value of the envelope $E_v(t)$; $D_{v,\max}$ is a maximum value of a variance of noise component $v_n(t)$ (7.4.33); θ is an unknown nonrandom parameter, $\theta \in \{0, 1\}$, determining the presence or absence of a useful signal in the observed stochastic process (7.4.11a,b).

Formula (7.4.343), on $E = 0$ (the signal is absent), represents Rayleigh distribution:

$$p_v(z/\theta = 0) = \frac{z}{D_{v,\max}} \exp\left[-\frac{z^2}{2D_{v,\max}}\right], \quad z \geq 0. \qquad (7.4.35)$$

The conditional probability of false alarm F, according to (7.4.35), is equal to:

$$F = \int_{l_0}^{\infty} p_v(z/\theta = 0) dz = \exp\left(-\frac{l_0^2}{2D_{v,\max}}\right), \qquad (7.4.36)$$

where l_0 is a threshold value determined by the inverse dependence:

$$l_0 = \sqrt{-2D_{v,\max} \ln(F)}, \qquad (7.4.37)$$

$D_{v,\max}$ is maximum value of a variance of noise component $v_n(t)$ determined by the formula (7.4.33).

The conditional probability of correct detection D, according to (7.4.34), is equal to:

$$D = \int_{l_0}^{\infty} p_v(z/\theta = 1) dz, \qquad (7.4.38)$$

where the value E of the envelope $E_v(t)$ is equal to $E = A$; A is an amplitude of useful signal.

It is too difficult to obtain an exact quantity $D[\delta \hat{t}_1]$ of a variance of relative error $\delta \hat{t}_1 = (\hat{t}_1 - t_1)/T_0$ of the estimator $\hat{t}_1$ of time of signal ending t_1. For three estimators determining by the formulas (7.4.30), (7.4.31a), and (7.4.31b), we point

7.4 Signal Detection in Metric Space with Lattice Properties

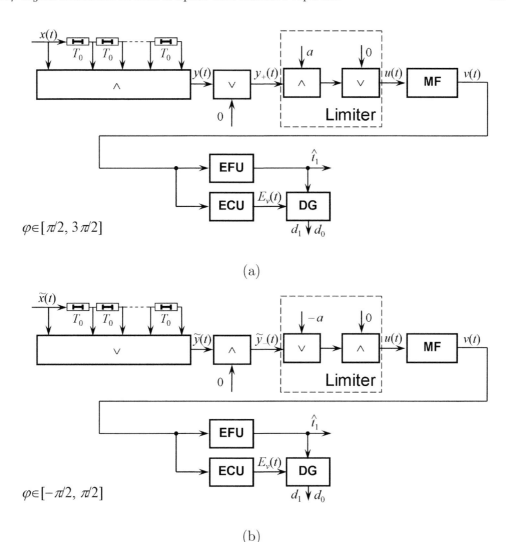

FIGURE 7.4.9 Block diagram of unit processing observations defined by Equations: (a) (7.4.11a); (b) (7.4.11b)

to the upper bound $|\delta \hat{t}_1|_{\max}$ of absolute quantity of relative error $|\delta \hat{t}_1| = |\hat{t}_1 - t_1|/T_0$ of the estimator $\hat{t}_1$ of time of signal ending t_1, obtained by the statistical modeling method:

$$|\delta \hat{t}_1| = |\hat{t}_1 - t_1|/T_0 \leq |\delta \hat{t}_1|_{\max} = 0.5. \qquad (7.4.39)$$

The quantity of the upper bound $|\delta \hat{t}_1|_{\max}$ of absolute quantity of relative error $|\delta \hat{t}_1|$ of the estimator $\hat{t}_1$ of time of signal ending t_1 along with detection quality indices (7.4.36) and (7.4.38) do not depend on energetic relationships between the useful signal and interference (noise).

When we have prior information concerning belonging of the initial phase of the signal to intervals $\Phi_1 = [-\pi/2, \pi/2]$ or $\Phi_2 = [\pi/2, 3\pi/2]$, which determine the methods of formation of the estimators $\hat{t}_1$ of time of signal ending t_1, according to (7.4.30), to construct the processing unit, enough information is contained only in one of the observation equations (7.4.11a) or (7.4.11b), respectively. The block diagrams of the processing unit are shown in Fig. 7.4.9a and Fig.7.4.9b, respectively.

7.4.3 Detection of Harmonic Signal with Unknown Nonrandom Amplitude and Initial Phase with Joint Estimation of Amplitude, Initial Phase, and Time of Signal Arrival (Ending) in Presence of Interference (Noise) in Metric Space with *L*-group Properties

Synthesis and analysis of optimal algorithm of detection of harmonic signal with unknown nonrandom amplitude and initial phase with joint estimation of amplitude, initial phase, and time of signal arrival (ending) in the presence of interference (noise) in metric space with L-group properties are fulfilled by the following assumptions. Under the synthesis, the interference distribution is considered to be an arbitrary one. While carrying out the further analysis, the interference (noise) $n(t)$ is considered to be quasi-white Gaussian noise.

The interaction between harmonic signal $s(t)$ and interference (noise) $n(t)$ in metric space $\mathcal{L}(+, \vee, \wedge)$ with L-group properties is described by two binary operations $\vee$ and $\wedge$ in two receiving channels, respectively:

$$x(t) = \theta s(t) \vee n(t); \tag{7.4.40a}$$

$$\tilde{x}(t) = \theta s(t) \wedge n(t), \tag{7.4.40b}$$

where θ is an unknown nonrandom parameter that take only two values $\theta \in \{0, 1\}$: $\theta = 0$ (the signal is absent) and $\theta = 1$ (the signal is present); $t \in T_s$, T_s is a domain of definition of the signal $s(t)$, $T_s = [t_0, t_1]$; t_0 is an unknown time of arrival of the signal $s(t)$; t_1 is an unknown time of signal ending; $T = t_1 - t_0$ is a known signal duration; $T_s \subset T_{\text{obs}}$; T_{obs} is an observation interval of the signal $T_{\text{obs}} = [t'_0, t'_1]$; $t'_0 < t_0$, $t_1 < t'_1$.

Let the model of the received harmonic signal $s(t)$ be determined by the expression:

$$s(t) = \begin{cases} A\cos(\omega_0 t + \varphi), & t \in T_s; \\ 0, & t \notin T_s, \end{cases} \tag{7.4.41}$$

where A is an unknown nonrandom amplitude of the useful signal $s(t)$; $\omega_0 = 2\pi f_0$; f_0 is a known carrier frequency of the signal $s(t)$; φ is an unknown nonrandom initial phase of the useful signal, $\varphi \in [-\pi, \pi]$; T_s is a domain of definition of the signal $s(t)$, $T_s = [t_0, t_1]$; t_0 is an unknown time of arrival of the signal $s(t)$; t_1 is an unknown time of signal ending; $T = t_1 - t_0$ is known signal duration, $T = N_s T_0$; N_s is a number of periods of harmonic signal $s(t)$, $N_s \in \mathbf{N}$, $\mathbf{N}$ is the set of natural numbers; T_0 is a carrier period.

The instantaneous values (the samples) of the signal $\{s(t_j)\}$ and interference (noise) $\{n(t_j)\}$ are the elements of the signal space $s(t_j), n(t_j) \in \mathcal{L}(+, \vee, \wedge)$. The samples of interference (noise) $n(t_j)$ are considered independent. The samples $s(t_j)$ and $n(t_j)$ of the signal $s(t)$ and interference (noise) $n(t)$ are taken on the domain of definition T_s of the signal $s(t)$: $t_j \in T_s$ over the sample interval Δt that provides

7.4 Signal Detection in Metric Space with Lattice Properties

the independence of the samples of interference (noise) $\{n(t_j)\}$; $\Delta t \ll 1/f_0$, where f_0 is known carrier frequency of the signal $s(t)$.

Taking into account these considerations, the equations of observations in two processing channels (7.4.40a) and (7.4.40b) take the form:

$$x(t_j) = \theta s(t_j) \vee n(t_j); \qquad (7.4.42a)$$

$$\tilde{x}(t_j) = \theta s(t_j) \wedge n(t_j), \qquad (7.4.42b)$$

where $t_j = t - jT_0$, $j = 0, 1, \ldots, N-1$, $t_j \in T^* \subset T_s \subset T_{\text{obs}}$; T^* is the interval of primary processing $T^* = [t - (N-1)\Delta t, t]$; $T_s = [t_0, t_1]$, $T_{\text{obs}} = [t'_0, t'_1]$; $t'_0 < t_0$, $t_1 < t'_1$; $N \in \mathbf{N}$, $\mathbf{N}$ is the set of natural numbers.

In this subsection, as in Subsection 7.4.2, the problems of joint detection of the signal $s(t)$ and estimation of its time of ending t_1, and also estimation of both the amplitude A and the initial phase φ are formulated and solved on the basis of the extraction $Ext[s(t)]$ of useful signal $s(t)$ in the presence of interference (noise) $n(t)$, with further matched filtering $MF[s(t)]$. After that joint detection of the signal $s(t)$ and estimation of its time of ending t_1 are realized, and estimation of the amplitude A and the initial phase φ is fulfilled.

The problem of extraction $Ext[s(t)]$ of the useful signal $s(t)$ in the presence of interference (noise) $n(t)$ is formulated and solved similarly as shown in Section 7.3 on the basis of step-by-step processing of the statistical collections $\{x(t_j)\}$ and $\{\tilde{x}(t_j)\}$ determined by the observation equations (7.4.42a) and (7.4.42b):

$$Ext[s(t)] = \begin{cases} PF[s(t)]; & (a) \\ IP[s(t)]; & (b) \\ ISm[s(t)], & (c) \end{cases} \qquad (7.4.43)$$

where $PF[s(t)]$ is primary filtering; $IP[s(t)]$ is intermediate processing; $ISm[s(t)]$ is intermediate smoothing that form the successive stages of processing of the general signal extraction algorithm (7.4.43).

The used optimality criteria determining every stage of processing $PF[s(t)]$, $IP[s(t)]$, $ISm[s(t)]$ are interrelated and involved into the single systems:

$$PF[s(t)] = \begin{cases} y(t) = \underset{\breve{y}(t) \in Y; t, t_j \in T^*}{\arg\min} |\bigwedge_{j=0}^{N-1} [x(t_j) - \breve{y}(t)]|; & (a) \\ \tilde{y}(t) = \underset{\widehat{y}(t) \in \tilde{Y}; t, t_j \in T^*}{\arg\min} |\bigvee_{j=0}^{N-1} [\tilde{x}(t_j) - \widehat{y}(t)]|; & (b) \\ w(t) = F[y(t), \tilde{y}(t)]; & (c) \\ \sum_{j=0}^{N-1} |w(t_j) - s(t_j)|\big|_{n(t) \equiv 0, \Delta t \to 0} \to \underset{w(t_j) \in W}{\min}; & (d) \\ N = \underset{N' \in \mathbf{N} \cap]0, N^*]}{\arg\max} [\delta_d(N')]\big|_{N^*: \delta_d(N^*) = \delta_{d0}}, & (e) \end{cases} \qquad (7.4.44)$$

where $y(t)$, $\tilde{y}(t)$ are the solution functions of the problem of minimization of a metric between the observed statistical collections $\{x(t_j)\}$, $\{\tilde{x}(t_j)\}$ and the optimization variables, i.e., the functions $\breve{y}(t)$ and $\widehat{y}(t)$, respectively; $w(t)$ is the function $F[*,*]$

of uniting the results $y(t)$ and $\tilde{y}(t)$ of minimization of the functions of the observed collections $\{x(t_j)\}$ (7.4.42a) and $\{\tilde{x}(t_j)\}$ (7.4.42b); T^* is the processing interval; $N \in \mathbf{N}$, $\mathbf{N}$ is the set of natural numbers; N is a number of the samples of the processes $x(t)$ and $\tilde{x}(t)$ used under their primary processing; $\delta_d(N)$ is the relative dynamic error of filtering as the dependence on the number of the samples N; δ_{d0} is the relative dynamic error of filtering;

$$IP[s(t)] = \begin{cases} \mathbf{M}\{u^2(t)\}\big|_{s(t)\equiv 0} \to \min_L; & (a) \\ \mathbf{M}\{[u(t) - s(t)]^2\}\big|_{n(t)\equiv 0,\, \Delta t \to 0} = \varepsilon; & (b) \\ u(t) = L[w(t)], & (c) \end{cases} \quad (7.4.45)$$

where $\mathbf{M}\{*\}$ is the symbol of mathematical expectation; $L[w(t)]$ is a functional transformation of stochastic process $w(t)$ into the process $u(t)$; ε is a constant, which, in general case, is some function of a power of the signal $s(t)$;

$$ISm[s(t)] =$$
$$= \begin{cases} v(t) = \underset{v^\circ(t)\in V;\, t, t_k \in \tilde{T}}{\arg\min} \sum_{k=0}^{M-1} |u(t_k) - v^\circ(t)|; & (a) \\ \Delta\tilde{T} : \delta_d(\Delta\tilde{T}) = \delta_{d,\mathrm{sm}}; & (b) \\ M = \underset{M' \in \mathbf{N}\cap[M^*,\infty[}{\arg\max} [\delta_f(M')]\big|_{M^*:\delta_f(M^*)=\delta_{f,\mathrm{sm}}}, & (c) \end{cases} \quad (7.4.46)$$

where $v(t) = \hat{s}(t)$ is the result of filtering (the estimator $\hat{s}(t)$ of the signal $s(t)$), that is the solution of the problem of minimization of metric between the instantaneous values of stochastic process $u(t)$ and optimization variable, i.e., the function $v^\circ(t)$; $t_k = t - \frac{k}{M}\Delta\tilde{T}$, $k = 0, 1, \ldots, M-1$, $t_k \in \tilde{T}$, $\tilde{T} =]t - \Delta\tilde{T}, t]$; $\tilde{T}$ is an interval on which smoothing of stochastic process $u(t)$ is realized; $\Delta\tilde{T}$ is a quantity of a smoothing interval $\tilde{T}$; $M \in \mathbf{N}$, $\mathbf{N}$ is the set of natural numbers; M is a number of the samples of stochastic process $u(t)$ used under smoothing; $\delta_d(\Delta\tilde{T})$ and $\delta_f(M)$ are the relative dynamic and fluctuation errors of smoothing as the dependences on the quantity $\Delta\tilde{T}$ of the smoothing interval $\tilde{T}$ and the number of the samples M respectively; $\delta_{d,\mathrm{sm}}$, $\delta_{f,\mathrm{sm}}$ are relative dynamic and fluctuation errors of smoothing, respectively.

We now explain the optimality criteria and the single relationships involved into the systems (7.4.44), (7.4.45), (7.4.46) determining the successive stages of processing $PF[s(t)]$, $IP[s(t)]$, $ISm[s(t)]$ of the general algorithm of signal extraction (7.4.43).

Equations (7.4.44a) and (7.4.44b) of the system (7.4.44) define the criteria of minimum of metrics between the statistical sets of the observations $\{x(t_j)\}$ and $\{\tilde{x}(t_j)\}$ and the results of primary filtering $y(t)$ and $\tilde{y}(t)$, respectively. The functions of metrics $|\bigwedge_{j=0}^{N-1}[x(t_j) - \breve{y}(t)]|$ and $|\bigvee_{j=0}^{N-1}[\tilde{x}(t_j) - \widehat{y}(t)]|$ are chosen to provide the metric convergence and the convergence in probability to the estimated parameter of the sequences $y_{N-1} = \bigwedge_{j=0}^{N-1} x(t_j)$, $\tilde{y}_{N-1} = \bigvee_{j=0}^{N-1} \tilde{x}(t_j)$ for the interactions in the form (7.2.2a) and (7.2.2b) (see Section 7.2).

The equation (7.4.44d) of the system (7.4.44) defines the criterion of minimum

7.4 Signal Detection in Metric Space with Lattice Properties

of the metric $\sum_{j=0}^{N-1} |w(t_j) - s(t_j)|$ between the useful signal $s(t)$ and the function $w(t)$ in the processing interval $T^* = [t - (N-1)\Delta t, t]$. This criterion establishes the kind of the function $F[y(t), \tilde{y}(t)]$ (7.4.44c) uniting the primary processing results $y(t)$ and $\tilde{y}(t)$ obtained from the observed processes $x(t)$ and $\tilde{x}(t)$. The criterion (7.4.44d) is considered under two constraint conditions: (1) interference (noise) is identically equal to zero: $n(t) \equiv 0$; (2) the sampling interval Δt tends to zero: $\Delta t \to 0$. The equation (7.4.44e) of the system (7.4.44) determines the criterion of the choice of the number of the samples N of stochastic processes $x(t)$, $\tilde{x}(t)$, based on the quantity δ_{d0} of relative dynamic error of primary filtering.

Equations (7.4.45a), (7.4.45b), (7.4.45c) of the system (7.4.45) define the choice of the functional transformation $L[w(t)]$. Equation (7.4.45a) determines the criterion of minimum of the second moment of the process $u(t)$ in the absence of useful signal $s(t)$ in the input of the signal processing unit. Equation (7.4.45b) determines the second moment of the difference between the signals $u(t)$ and $s(t)$ under two constraint conditions: (1) interference (noise) $n(t)$ in the input of signal processing unit is absent; (2) the sampling interval Δt tends to zero: $\Delta t \to 0$.

Equation (7.4.46a) of the system (7.4.46) defines the criterion of minimum of metric $\sum_{k=0}^{M-1} |u(t_k) - v^\circ(t)|$ between the instantaneous values of the process $u(t)$ and optimization variable $v^\circ(t)$ in the smoothing interval $\tilde{T} =]t - \Delta \tilde{T}, t]$, requiring intermediate smoothing of the signal $u(t)$. Equation (7.4.46b) determines the criterion of the choice of the quantity $\Delta \tilde{T}$ of the smoothing interval $\tilde{T}$, based on a given quantity of dynamic error of intermediate smoothing $\delta_{d,\mathrm{sm}}$. Equation (7.4.46c) determines the criterion of the choice of a number of the samples M of stochastic process $u(t)$, based on a given quantity of fluctuation error of primary smoothing $\delta_{f,\mathrm{sm}}$.

Thus, the criteria (7.4.44) through (7.4.46) define the signal extraction algorithm in the presence of interference in the form of quasi-white noise with independent samples (see Equation (7.3.3) from Section 7.3).

The further problem of matched filtering $MF[s(t)]$ in signal space $\mathcal{L}(+, \vee, \wedge)$ with L-group properties is solved similarly, as shown in Subsection 7.4.2, with the only difference that the statistical collections $\{v_+(t_j)\}$ and $\{v_-(t_j)\}$ of the positive $v_+(t) = v(t) \vee 0$ and negative $v_-(t) = v(t) \wedge 0$ parts of the process $v(t) = \hat{s}(t)$, obtained on the basis of the criterion (7.4.46a) of the system (7.4.46), take part in signal processing. The intermediate processing in the limiter is excluded from the algorithm $MF[s(t)]$, inasmuch as it is foreseen in the signal extraction algorithm $Ext[s(t)]$. Thus, the problem of matched filtering $MF[s(t)]$ of useful signal $s(t)$ in the presence of interference (noise) $n(t)$ is formulated and solved on the basis of step-by-step processing of statistical collections $\{v_+(t_j)\}$ and $\{v_-(t_j)\}$ determined by the equations $v_+(t) = v(t) \vee 0$ and $v_-(t) = v(t) \wedge 0$:

$$MF[s(t)] = \begin{cases} IF[s(t)]; & (a) \\ Sm[s(t)], & (b) \end{cases} \quad (7.4.47)$$

where $IF[s(t)]$ is intermediate filtering; $Sm[s(t)]$ is smoothing that together form the successive processing stages of the general algorithm of the useful signal matched filtering (7.4.47).

The optimality criteria determining every stage of processing $IF[s(t)]$ and $Sm[s(t)]$ are interrelated and involved in the separate systems:

$$IF[s(t)] = \begin{cases} Y(t) = \underset{\breve{Y}(t)\in Y; t, t_j \in T_{obs}}{\arg\min} \quad |\bigwedge_{j=0}^{J-1}[v_+(t_j) - \breve{Y}(t)]|; & (a) \\ \tilde{Y}(t) = \underset{\widehat{Y}(t)\in \tilde{Y}; t, t_j \in T_{obs}}{\arg\min} \quad |\bigvee_{j=0}^{J-1}[\tilde{v}_-(t_j) - \widehat{Y}(t)]|; & (b) \\ J = \underset{Y(t)\in Y; T_{obs}}{\arg\min}[\int |Y(t)|dt]\Big|_{n(t)\equiv 0}, \int_{T_{obs}} |Y(t)|dt \neq 0; & (c) \\ W(t) = F[Y(t), \tilde{Y}(t)]; & (d) \\ \int_{[t_1-T_0, t_1]} |W(t) - s(t)|dt \Big|_{n(t)\equiv 0} \to \underset{W(t)\in W}{\min}, & (e) \end{cases} \quad (7.4.48)$$

where $\{v_+(t_j)\}$ and $\{v_-(t_j)\}$ are the processed statistical collections of the positive $v_+(t) = v(t) \vee 0$ and the negative $v_-(t) = v(t) \wedge 0$ parts of the process $v(t) = \hat{s}(t)$, obtained on the basis of the criterion (7.4.46a) of the system (7.4.46); $Y(t)$ and $\tilde{Y}(t)$ are the solution functions of the problem of minimization of metric between the observed statistical collections $\{v_+(t_j)\}$, $\{v_-(t_j)\}$ and optimization variables, i.e., the functions $\breve{Y}(t)$ and $\widehat{Y}(t)$, respectively; $W(t)$ is the function $F[*, *]$ of uniting the results $Y(t)$ and $\tilde{Y}(t)$ of minimizing the functions of the observed collections $\{v_+(t_j)\}$ and $\{v_-(t_j)\}$;

$$Sm[s(t)] = \begin{cases} V(t) = \underset{V^\circ(t)\in V; t, t_k \in \tilde{T}}{\arg\min} \sum_{k=0}^{M-1} |W(t_k) - V^\circ(t)|\Big|_{s(t)\equiv 0}; & (a) \\ \Delta \tilde{T}_{MF} : \delta_d(\Delta \tilde{T}_{MF}) = \delta_{d,sm}^{MF}; & (b) \\ M_{MF} = \underset{M'\in \mathbf{N}\cap[M^*,\infty[}{\arg\max}[\delta_f(M')]\Big|_{M^*:\delta_f(M^*)=\delta_{f,sm}^{MF}}, & (c) \end{cases} \quad (7.4.49)$$

where $V(t)$ is the result of matched filtering that is the solution of the problem of minimizing the metric between the instantaneous values of stochastic process $W(t)$ and optimization variable, i.e., the function $V^\circ(t)$; $t_k = t - \frac{k}{M_{MF}}\Delta \tilde{T}_{MF}$, $k = 0, 1, \ldots, M_{MF} - 1$, $t_k \in \tilde{T}_{MF} =]t - \Delta \tilde{T}_{MF}, t]$; $\tilde{T}_{MF}$ is the interval in which smoothing of the stochastic process $W(t)$ is realized; $M \in \mathbf{N}$, $\mathbf{N}$ is the set of natural numbers; M_{MF} is a number of the samples of stochastic process $W(t)$ used during smoothing in the interval $\tilde{T}_{MF}$; $\delta_d(\Delta \tilde{T}_{MF})$ and $\delta_f(M_{MF})$ are relative dynamic and fluctuation errors of smoothing, as the dependences on the quantity $\Delta \tilde{T}_{MF}$ of the smoothing interval $\tilde{T}_{MF}$ and a number of the samples M_{MF}, respectively; $\delta_{d,sm}^{MF}$ and $\delta_{f,sm}^{MF}$ are the given quantities of relative dynamic and fluctuation errors of smoothing, respectively.

We now explain the optimality criteria and the single relationships involved

7.4 Signal Detection in Metric Space with Lattice Properties

in the systems (7.4.48) and (7.4.49) determining the successive processing stages $IF[s(t)]$, $Sm[s(t)]$ of the general matched filtering algorithm $MF[s(t)]$ of the useful signal (7.4.47).

Equations (7.4.48a) and (7.4.48b) of the system (7.4.48) define the criteria of minimum of the metric between statistical sets of the observations $\{v_+(t_j)\}$ and $\{v_-(t_j)\}$ and the results of primary processing $Y(t)$ and $\tilde{Y}(t)$, respectively. The functions of metrics $|\bigwedge_{j=0}^{J-1}[v_+(t_j) - \breve{Y}(t)]|$ and $|\bigvee_{j=0}^{J-1}[v_-(t_j) - \widehat{Y}(t)]|$ are chosen to provide the metric convergence and the convergence in probability to the useful signal $s(t)$ of the sequences $Y(t) = \bigwedge_{j=0}^{J-1} v_+(t_j)$ and $\tilde{Y}(t) = \bigvee_{j=0}^{J-1} v_-(t_j)$ based on the interactions in the form (7.4.42a) and (7.4.42b). The relationship (7.4.48c) determines the criterion of the choice of a number of the samples J of stochastic process $v_+(t)$, $v_-(t)$ used during signal processing on the basis of minimizing the norm $\int_{T_{obs}} |Y(t)|dt$. The criterion (7.4.48c) is considered under two constraint conditions: (1) interference (noise) is identically equal to zero: $n(t) \equiv 0$; (2) the norm $\int_{T_{obs}} |Y(t)|dt$ of the function $Y(t)$ is not equal to zero: $\int_{T_{obs}} |\dot{Y}(t)|dt \neq 0$. Equation (7.4.48e) defines the criterion of minimum of the metric $\int_{[t_1-T_0,t_1]} |W(t) - s(t)|dt \big|_{n(t) \equiv 0}$ between the useful signal $s(t)$ and the function $W(t)$ in the interval $[t_1 - T_0, t_1]$. This criterion establishes the kind of the function $F[Y(t), \tilde{Y}(t)]$ (7.4.48d) uniting the results $Y(t)$ and $\tilde{Y}(t)$ of primary processing of the observed processes $v_+(t)$ and $v_-(t)$. The criterion (7.4.48e) is considered under the condition that interference (noise) is identically equal to zero: $n(t) \equiv 0$.

Equation (7.4.49a) of the system (7.4.49) determines the criterion of minimum of the metric $\sum_{k=0}^{M-1} |W(t_k) - V^\circ(t)|$ between the instantaneous values of the process $W(t)$ and optimization variable $V^\circ(t)$ in the smoothing interval $\tilde{T}_{MF} =]t - \Delta\tilde{T}_{MF}, t]$, requiring final processing of the signal $W(t)$ by smoothing under the condition that the useful signal $s(t)$ is identically equal to zero: $s(t) \equiv 0$. The relationship (7.4.49b) determines the rule of the choice of the quantity $\Delta\tilde{T}_{MF}$ of the smoothing interval $\tilde{T}_{MF}$, based on a given quantity of relative dynamic error $\delta_{d,sm}^{MF}$ of smoothing. Equation (7.4.49c) determines the criterion of the choice of a number of the samples M_{MF} of the stochastic process $W(t)$, based on a given quantity of relative fluctuation error $\delta_{f,sm}^{MF}$ of its smoothing.

The problem of joint detection $Det[s(t)]$ of the signal $s(t)$ and estimation $Est[t_1]$ of time of its ending t_1 is formulated on the detection and estimation criteria involved in one system, which is a logical continuation of (7.4.43):

$$Det[s(t)]/Est[t_1] = \begin{cases} E_V(\hat{t}_1 - \frac{T_0}{2}) \overset{d_1}{\underset{d_0}{\gtrless}} l_0(F); & (a) \\ \hat{t}_1 = \arg\min_{\varphi \in \Phi_1; t^\circ \in T_{obs}} \mathbf{M}_\varphi\{(t_1 - t^\circ)^2\}\big|_{n(t) \equiv 0}, & (b) \end{cases} \quad (7.4.50)$$

where $E_V(\hat{t}_1 - \frac{T_0}{2})$ is an instantaneous value of the envelope $E_V(t)$ of the result $V(t)$ of matched filtering of the useful signal $s(t)$ at the instant $t = \hat{t}_1 - \frac{T_0}{2}$: $E_V(t) = \sqrt{V^2(t) + V_\mathcal{H}^2(t)}$; $V_\mathcal{H}(t) = \mathcal{H}[V(t)]$ is Hilbert transform; d_1 and d_0 are the decisions concerning true values of unknown nonrandom parameter θ, $\theta \in \{0,1\}$; $l_0(F)$ is some threshold level as the dependence on a given conditional probability of false alarm F; $\hat{t}_1$ is the estimator of the time of signal ending t_1; $\mathbf{M}_\varphi\{(t_1 - t^\circ)^2\}$ is a mean squared difference between true value of time of signal ending t_1 and optimization variable t°; $\mathbf{M}_\varphi\{*\}$ is the symbol of mathematical expectation with averaging over the initial phase φ of the signal; Φ_1 is a domain of definition of the initial phase φ of the signal: $\Phi_1 = [-\pi/2, \pi/2]$.

Equation (7.4.50a) of the system (7.4.50) determines the rule for making the decision d_1 concerning the presence of the signal (if $E_V(\hat{t}_1 - \frac{T_0}{2}) > l_0(F)$) or the decision d_0 concerning the absence of the signal (if $E_V(\hat{t}_1 - \frac{T_0}{2}) < l_0(F)$). The relationship (7.4.50b) determines the criterion of formation of the estimator $\hat{t}_1$ of time of signal ending t_1 on the basis of minimization of mean squared difference $\mathbf{M}_\varphi\{(t_1 - t^\circ)^2\}$ between true value of the time of signal ending t_1 and optimization variable t°, when averaging is realized over the initial phase φ of the signal taken in the interval $\Phi_1 = [-\pi/2, \pi/2]$, and interference (noise) is identically equal to zero: $n(t) \equiv 0$.

The problem of estimation of the amplitude A and the initial phase φ of useful signal $s(t)$ is formulated and solved on the basis of two estimation criteria within one system that is a logical continuation of (7.4.44) through (7.4.46), (7.4.48) through (7.4.50):

$$Est[A, \varphi] = \begin{cases} \hat{\varphi} = \underset{\varphi \in \Phi_1; t \in T_{obs}}{\arg\max} \int_{T_s} v(t)\cos(\omega_0 t + \varphi)dt; & (a) \\ \hat{A} = \underset{A \in \Delta_A; t \in T_{obs}}{\arg\min} \int_{T_s} [v(t) - A\cos(\omega_0 t + \varphi)]^2 dt, & (b) \end{cases} \quad (7.4.51)$$

where $\hat{\varphi}$, $\hat{A}$ are the estimators of the amplitude A and the initial phase φ of the useful signal $s(t)$, respectively; $v(t) = \hat{s}(t)$ is the result of useful signal extraction (the estimator $\hat{s}(t)$ of the signal $s(t)$) obtained on the basis of the criterion (7.4.46a) of the system (7.4.46); T_s is a domain of definition of the signal $s(t)$, $T_s = [t_0, t_1]$; t_0 is an unknown time of arrival of the signal $s(t)$; t_1 is an unknown time of signal ending; T_{obs} is the interval of the observation of the signal $T_{obs} = [t'_0, t'_1]$; $t'_0 < t_0$, $t_1 < t'_1$; Φ_1 is a domain of definition of the initial phase of the signal: $\Phi_1 = [-\pi/2, \pi/2]$; Δ_A is a domain of definition of the signal amplitude: $\Delta_A =]0, A_{max}]$.

Solving of the signal processing problems determined by the equation systems (7.4.43), (7.4.47), (7.4.50), (7.4.51) is realized in the following way.

Solving the problem of the extraction $Ext[s(t)]$ of the useful signal $s(t)$ in the presence of interference (noise) $n(t)$ is described in detail in Section 7.3. Here we consider the intermediate results that determine the structure-forming elements of the general processing algorithm.

The solutions of optimization equations (7.4.44a) and (7.4.44b) of the system

7.4 Signal Detection in Metric Space with Lattice Properties

(7.4.44) are the values of the estimators $y(t)$, $\tilde{y}(t)$ in the form of meet and join of the observation results $\{x(t_j)\}$ and $\{\tilde{x}(t_j)\}$, respectively:

$$y(t) = \bigwedge_{j=0}^{N-1} x(t_j) = \bigwedge_{j=0}^{N-1} x(t - j\Delta t); \tag{7.4.52a}$$

$$\tilde{y}(t) = \bigvee_{j=0}^{N-1} \tilde{x}(t_j) = \bigvee_{j=0}^{N-1} \tilde{x}(t - j\Delta t). \tag{7.4.52b}$$

The coupling equation (7.4.44c) is obtained from criteria (7.4.44a), (7.4.44b), (7.4.44d), and takes the form:

$$w(t) = y_+(t) + \tilde{y}_-(t); \tag{7.4.53}$$
$$y_+(t) = y(t) \vee 0; \tag{7.4.53a}$$
$$\tilde{y}_-(t) = \tilde{y}(t) \wedge 0. \tag{7.4.53b}$$

The solution $u(t)$ of the relationships (7.4.45a), (7.4.45b), (7.4.45c) of the system (7.4.45), that define the criterion of the choice of transformation of the process $w(t)$, is the function $L[w(t)]$ that determines the gain characteristic of the limiter:

$$u(t) = L[w(t)] = [(w(t) \wedge a) \vee 0] + [(w(t) \vee (-a)) \wedge 0], \tag{7.4.54}$$

where the parameter of the limiter a is chosen to be equal to $a = \sup_A \Delta_A = A_{\max}$ and $\Delta_A =]0, A_{\max}]$; $A_{\max}$ is a maximum possible value of the useful signal amplitude.

The solution of optimization equation (7.4.46a) of the system (7.4.46) is a value of the estimator $v(t)$ in the form of the sample median med$\{*\}$ of the sample collection $\{u(t_k)\}$ of stochastic process $u(t)$:

$$v(t) = \underset{t_k \in \tilde{T}}{\operatorname{med}}\{u(t_k)\}, \tag{7.4.55}$$

where $t_k = t - \frac{k}{M}\Delta\tilde{T}$, $k = 0, 1, \ldots, M - 1$; $t_k \in \tilde{T} =]t - \Delta\tilde{T}, t]$; $\tilde{T}$ is the interval, in which smoothing of stochastic process $u(t)$ is realized; $\Delta\tilde{T}$ is a quantity of the smoothing interval $\tilde{T}$.

Summarizing the relationships (7.4.52) through (7.4.55), one can draw the conclusion that the estimator $v(t) = \hat{s}(t)$ of the signal $s(t)$ extracted in the presence of interference (noise) $n(t)$ is the function of smoothing of stochastic process $u(t)$ obtained by limiting the process $w(t)$ that combines the results $y(t)$ and $\tilde{y}(t)$ of a proper primary processing of the observed stochastic processes $x(t)$, $\tilde{x}(t)$ in the interval $T^* = [t - (N-1)\Delta t, t]$.

Solving the problem of matched filtering $MF[s(t)]$ of useful signal $s(t)$ in the signal space $\mathcal{L}(+, \vee, \wedge)$ with L-group properties in the presence of interference (noise) $n(t)$ is described in detail in Subsection 7.4.2, so similarly we consider the intermediate results, which determine structure-forming elements of the general processing algorithm.

The solutions of the optimization equations (7.4.48a) and (7.4.48b) of the system

(7.4.48) are the values of the estimators $Y(t)$ and $\tilde{Y}(t)$ in the form of meet and join of the observation results $\{v_+(t_j)\}$ and $\{v_-(t_j)\}$, respectively:

$$Y(t) = \bigwedge_{j=0}^{J-1} v_+(t_j) = \bigwedge_{j=0}^{J-1} v_+(t - jT_0); \quad (7.4.56a)$$

$$\tilde{Y}(t) = \bigvee_{j=0}^{J-1} \tilde{v}_-(t_j) = \bigvee_{j=0}^{J-1} \tilde{v}_-(t - jT_0). \quad (7.4.56b)$$

According to the criterion (7.4.48c), the optimal value of a number of the samples J of stochastic process $v_+(t)$, $v_-(t)$ used during primary processing (7.4.56a,b) is equal to the number of periods N_s of the signal:

$$J = N_s. \quad (7.4.57)$$

The coupling equation (7.4.48d) satisfying the criterion (7.4.48e) takes the form:

$$W(t) = Y(t) + \tilde{Y}(t); \quad Y(t) \geq 0, \ \tilde{Y}(t) \leq 0. \quad (7.4.58)$$

The solution of optimization equation (7.4.49a) of the system (7.4.49) is the value of the estimator $V(t)$ in the form of the sample median $\text{med}\{*\}$ of the collection of the samples $\{W(t_k)\}$ of stochastic process $W(t)$:

$$V(t) = \underset{t_k \in \tilde{T}}{\text{med}}\{W(t_k)\}, \quad (7.4.59)$$

where $t_k = t - \frac{k}{M_{MF}}\Delta \tilde{T}_{MF}$, $k = 0, 1, \ldots, M_{MF} - 1$; $t_k \in \tilde{T}_{MF} =]t - \Delta \tilde{T}_{MF}, t]$; $\tilde{T}_{MF}$ is the interval in which smoothing of stochastic process $W(t)$ is realized; $\Delta \tilde{T}_{MF}$ is a quantity of the smoothing interval $\tilde{T}_{MF}$.

The sense of the obtained relationships (7.4.56) through (7.4.59) lies in the fact that the result of matched filtering $V(t)$ of useful signal $s(t)$ extracted and detected in the presence of interference (noise) $n(t)$ is the function of smoothing of stochastic process $W(t)$ that is a combination of the results $Y(t)$ and $\tilde{Y}(t)$ of signal processing of the positive $v_+(t)$ and the negative $v_-(t)$ parts of the observed stochastic process $v(t)$ in the interval $T^* = [t - (N_s - 1)\Delta t, t]$.

Solving the problem of joint detection $Det[s(t)]$ of the signal $s(t)$ and estimation $Est[t_1]$ of its time of ending t_1 is described in detail in Subsection 7.4.2, so here we note only the intermediate results which determine the structure-forming elements of the general processing algorithm.

The rule of making the decision d_1 concerning the presence of the signal (if $E_V(\hat{t}_1 - \frac{T_0}{2}) > l_0(F)$) or the decision d_0 concerning the absence of the signal (if $E_V(\hat{t}_1 - \frac{T_0}{2}) < l_0(F)$), determined by Equation (7.4.50a) of the system (7.4.50), supposes formation of the envelope $E_V(t)$ of the estimator $V(t) = \hat{s}(t)$ of the useful signal $s(t)$ and the comparison of the value of the envelope $E_V(\hat{t}_1 - \frac{T_0}{2})$ with the threshold value $l_0(F)$ at the instant $t = \hat{t}_1 - \frac{T_0}{2}$ determined by the estimator $\hat{t}_1$, and as the result, the decision is made:

$$E_V(\hat{t}_1 - \frac{T_0}{2}) \underset{d_0}{\overset{d_1}{\gtrless}} l_0(F). \quad (7.4.60)$$

7.4 Signal Detection in Metric Space with Lattice Properties

The relationship (7.4.50b) of the system (7.4.50) determines the criterion of forming the estimator $\hat{t}_1$ of time of signal ending t_1 on the basis of minimization of mean squared difference $\mathbf{M}_\varphi\{(t_1-t^\circ)^2\}$ between true value of time of signal ending t_1 and optimization variable t° when averaging is realized over the initial phase φ of the signal taken in the interval $\Phi_1 = [-\pi/2, \pi/2]$, and interference (noise) is identically equal to zero: $n(t) \equiv 0$.

Generally, as it is shown in Subsection 7.4.2, the solution of optimization equality (7.4.50b) of the system (7.4.50) is determined by the identity:

$$\hat{t}_1 = \begin{cases} \hat{t}_- + (T_0/2) + (T_0\hat{\varphi}/2\pi), & \varphi \in \Phi_1 = [-\pi/2, \pi/2]; \\ \hat{t}_+ + (T_0/2) - (T_0\hat{\varphi}/2\pi), & \varphi \in \Phi_2 = [\pi/2, 3\pi/2], \end{cases} \quad (7.4.61)$$

where $\hat{t}_\pm = (\int_{T_{\text{obs}}} tV_\pm(t)dt)/(\int_{T_{\text{obs}}} V_\pm(t)dt)$ is the estimator of the barycentric coordinate of the positive $V_+(t)$ or the negative $V_-(t)$ parts of the smoothed stochastic process $V(t)$, respectively; $V_+(t) = V(t) \vee 0$, $V_-(t) = V(t) \wedge 0$; T_{obs} is an interval of the observation of the signal $T_{\text{obs}} = [t'_0, t'_1]$; $t'_0 < t_0$, $t_1 < t'_1$; $\hat{\varphi} = \arcsin[2(\hat{t}_+ - \hat{t}_-)/T_0]$ is the estimator of an unknown nonrandom initial phase φ of the useful signal $s(t)$.

However, as mentioned above, the initial phase is determined in the interval $\Phi_1 = [-\pi/2, \pi/2]$, that, according to (7.4.61), supposes the estimator $\hat{t}_1$ takes the form:

$$\hat{t}_1 = \hat{t}_- + (T_0/2) + (T_0\hat{\varphi}/2\pi), \quad \varphi \in \Phi_1 = [-\pi/2, \pi/2]. \quad (7.4.61a)$$

The sense of the obtained relationships (7.4.61a) and (7.4.60) lies in the fact that the estimator $\hat{t}_1$ of time of signal ending and the envelope $E_V(t)$ are formed on the base of the proper processing of the result of matched filtering $V(t)$ of the useful signal $s(t)$ extracted and detected in the presence of interference (noise) $n(t)$. Signal detection is fixed on the base of comparing the instantaneous value of the envelope $E_V(t)$ with a threshold value at the instant $t = \hat{t}_1 - \frac{T_0}{2}$ determined by the estimator $\hat{t}_1$.

Consider, finally, the problem of the estimation of unknown nonrandom amplitude and initial phase of the signal stated, for instance, in [149], [155].

Equation (7.4.51a) of the system (7.4.51) implies that the estimator $\hat{\varphi}$ of initial phase φ is found by maximizing the expression (with respect to φ):

$$Q(\varphi) = \int_{T_s} v(t)\cos(\omega_0 t + \varphi)dt \to \max_{\varphi \in \Phi_1}, \quad (7.4.62)$$

where $v(t)$ is the result of extraction $Ext[s(t)]$ of useful signal $s(t)$ in the presence of interference (noise), whose algorithm is described by the system (7.4.43); T_s is a domain of definition of the signal $s(t)$.

Factoring the cosine of the sum, we can compute the derivative of the function $Q(\varphi)$ with respect to φ, which we put to zero to determine the extremum:

$$\frac{dQ(\hat{\varphi})}{d\hat{\varphi}} = -\sin\hat{\varphi}\int_{T_s} v(t)\cos(\omega_0 t + \varphi)dt - \cos\hat{\varphi}\int_{T_s} v(t)\sin(\omega_0 t + \varphi)dt = 0. \quad (7.4.63)$$

Equation (7.4.63) has a unique solution that is the maximum of the function $Q(\varphi)$:

$$\hat{\varphi} = -\text{arctg}(v_s/v_c); \qquad (7.4.64)$$

$$v_s = \int_{T_s} v(t)\sin(\omega_0 t + \varphi)dt; \qquad (7.4.64\text{a})$$

$$v_c = \int_{T_s} v(t)\cos(\omega_0 t + \varphi)dt. \qquad (7.4.64\text{b})$$

Equation (7.4.51b) of the system (7.4.51) implies that the estimator $\hat{A}$ of the amplitude A can be found on the basis of obtaining a minimum with respect to variable A of the expression:

$$Q(A) = \int_{T_s} [v(t) - A\cos(\omega_0 t + \varphi)]^2 dt \to \max_{A \in \Delta_A}. \qquad (7.4.65)$$

We find the derivative of the function $Q(A)$ with respect to A, setting it to zero to find the extremum:

$$\frac{dQ(\hat{A})}{d\hat{A}} = -2\int_{T_s} v(t)\cos(\omega_0 t + \varphi)dt + 2\hat{A}\int_{T_s}\cos^2(\omega_0 t + \varphi)dt = 0. \qquad (7.4.66)$$

Equation (7.4.66) has a unique solution which determines the minimum of the function $Q(A)$:

$$\hat{A} = \frac{\int_{T_s} v(t)\cos(\omega_0 t + \varphi)dt}{\int_{T_s}\cos^2(\omega_0 t + \varphi)dt} = 2Q(\varphi)/T, \qquad (7.4.67)$$

where $Q(\varphi)$ is the function determined by the relationship (7.4.62); T is the known duration of the signal.

It is easy to make sure that the function $Q(\varphi)$ can be represented in the form:

$$Q(\varphi) = \sqrt{v_c^2 + v_s^2}, \qquad (7.4.68)$$

where v_s and v_c are the quantities determined by the integrals (7.4.64a) and (7.4.64b), respectively.

Taking into account (7.4.68), we write the final expression for the estimator $\hat{A}$ of the amplitude A of the signal:

$$\hat{A} = 2\sqrt{v_c^2 + v_s^2}/T. \qquad (7.4.69)$$

Thus, the solution of the problem of estimation of unknown nonrandom amplitude and initial phase of harmonic signal is described by the relationships (7.4.64) and (7.4.69). These estimators coincide with the estimators obtained by the maximum likelihood method; see, for instance, [155, (3.3.32)], [149, (6.10)].

7.4 Signal Detection in Metric Space with Lattice Properties

The block diagram of the processing unit, according to the results of solving the optimization equations involved into the systems (7.4.43), (7.4.47), (7.4.50), (7.4.51), is described by the relationships (7.4.52) through (7.4.60); (7.4.61a); (7.4.64) and (7.4.69) and includes: signal extraction unit (SEU), matched filtering unit (MFU), signal detection unit (SDU), and also, amplitude and initial phase estimator formation unit (EFU) (see Fig. 7.4.10).

The SEU realizes signal processing, according to the relationships (7.4.52) through (7.4.55) and includes two processing channels, each containing transversal filter realizing primary filtering of the observed stochastic processes $x(t)$ and $\tilde{x}(t)$; the units of formation of the positive $y_+(t)$ and the negative $\tilde{y}_-(t)$ parts of the processes $y(t)$ and $\tilde{y}(t)$, respectively; an adder summing the results of signal processing in two processing channels; a limiter; a median filter (MF) realizing intermediate smoothing of the process $u(t)$ (see Fig. 7.4.10).

Matched filtering unit realizes signal processing, according to the relationships (7.4.56) through (7.4.59), and contains: two processing channels, each including the units of formation of the positive $v_+(t)$ and the negative $v_-(t)$ parts of the process $v(t)$; transversal filters realizing primary filtering of the observed stochastic process $v_+(t)$ and $v_-(t)$; an adder summing the results of signal processing $Y(t)$, $\tilde{Y}(t)$ in two channels; a median filter (MF) that smooths the process $W(t) = Y(t) + \tilde{Y}(t)$ (see Fig. 7.4.10).

The SDU realizes signal processing, according to the relationships (7.4.60) and (7.4.61a), and includes time of ending estimator formation unit (EFU), envelope computation unit (ECU), and decision gate (DG).

The amplitude and initial phase estimator formation unit realizes signal processing according to the relationships (7.4.64) and (7.4.69).

Transversal filters of signal extraction unit in two processing channels realize primary filtering $PF[s(t)]$: $y(t) = \bigwedge_{j=0}^{N-1} x(t - j\Delta t)$, $\tilde{y}(t) = \bigvee_{j=0}^{N-1} \tilde{x}(t - j\Delta t)$ of the stochastic processes $x(t)$ and $\tilde{x}(t)$, according to the equations (7.4.52a,b), fulfilling criteria (7.4.44a) and (7.4.44b) of the system (7.4.44). The units of formation of the positive $y_+(t)$ and the negative $\tilde{y}_-(t)$ parts of the processes $y(t)$ and $\tilde{y}(t)$ in two processing channels form the values of these functions according to the identities (7.4.53a) and (7.4.53b), respectively. The adder sums the results of signal processing in two processing channels, according to the equality (7.4.53), providing fulfillment of the criteria (7.4.44c) and (7.4.44d) of the system (7.4.44).

The limiter $L[w(t)]$ realizes intermediate processing $IP[s(t)]$ by clipping the signal $w(t)$ in the output of the adder, according to the criteria (7.4.45a), (7.4.45b) of the system (7.4.45) to exclude noise overshoots whose instantaneous values exceed value a from further processing.

The median filter (MF) realizes intermediate smoothing $ISm[s(t)]$ of $w(t) = y_+(t) + \tilde{y}_-(t)$, according to the formula (7.4.55), providing fulfillment of the criterion (7.4.46a) of the system (7.4.46).

In a matched filtering unit (MFU), the units of formation of the positive $v_+(t)$ and the negative $v_-(t)$ parts of the process $v(t)$ form the values of these functions, according to the identities $v_+(t) = v(t) \vee 0$ and $v_-(t) = v(t) \wedge 0$, respectively.

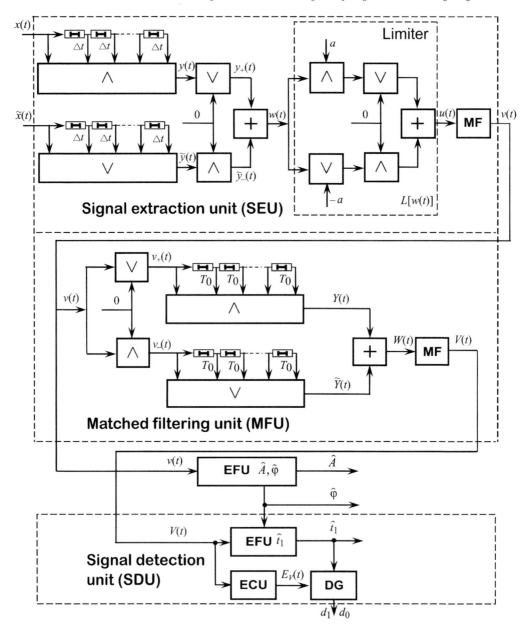

FIGURE 7.4.10 Block diagram of processing unit that realizes harmonic signal detection with joint estimation of amplitude, initial phase, and time of signal arrival (ending)

Transversal filters of matched filtering unit in two processing channels realize intermediate filtering $IF[s(t)]$: $Y(t) = \bigwedge_{j=0}^{N_s-1} v_+(t-jT_0)$ and $\tilde{Y}(t) = \bigvee_{j=0}^{N_s-1} v_-(t-jT_0)$ of the observed stochastic processes $v_+(t)$ and $v_-(t)$, according to Equations (7.4.56a,b), providing fulfillment of the criteria (7.4.48a) and (7.4.48b) of the system (7.4.48). The adder sums the results of signal processing in two processing channels, ac-

7.4 Signal Detection in Metric Space with Lattice Properties

cording to the equality (7.4.58), providing fulfillment of the criteria (7.4.48d) and (7.4.48e) of the system (7.4.48).

The median filter (MF) realizes smoothing $Sm[s(t)]$ of the process $W(t) = Y(t) + \tilde{Y}(t)$ according to the formula (7.4.59), providing fulfillment of the criterion (7.4.49a) of the system (7.4.49).

In a signal detection unit (SDU), the envelope computation unit (ECU) forms the envelope $E_V(t)$ of the signal $V(t)$ in the output of the median filter (MF) of the matched filtering unit. The time of the signal ending estimator formation unit (EFU $\hat{t}_1$) forms the estimator $\hat{t}_1$, according to Equation (7.4.61a), providing fulfillment of the criterion (7.4.50b) of the system (7.4.50). At the instant $t = \hat{t}_1 - \frac{T_0}{2}$, the decision gate (DG) compares the instantaneous value of the envelope $E_V(t)$ with the threshold value $l_0(F)$, and as the result, it makes the decision d_1 concerning the presence of the signal (if $E_V(\hat{t}_1 - \frac{T_0}{2}) > l_0(F)$) or the decision d_0 concerning the absence of the signal (if $E_V(\hat{t}_1 - \frac{T_0}{2}) < l_0(F)$), according to the rule (7.4.60) of the criterion (7.4.50a) of the system (7.4.50).

Amplitude and initial phase estimator formation units compute the estimators $\hat{A}$, $\hat{\varphi}$ according to the formulas (7.4.69) and (7.4.64), respectively, so that the estimator $\hat{\varphi}$ is used to form the time of signal ending estimator $\hat{t}_1$.

Figures 7.4.11 through 7.4.14 illustrate the results of statistical modeling of signal processing by a synthesized unit under the following conditions: the useful signal $s(t)$ is harmonic with number of periods $N_s = 8$, and also with initial phase $\varphi = \pi/3$. Signal-to-noise ratio E/N_0 is equal to $E/N_0 = 10^{-10}$, where E is an energy of the signal; N_0 is the power spectral density of noise. The product $T_0 f_{n,\max} = 64$, where T_0 is the period of a carrier of the signal $s(t)$; $f_{n,\max}$ is maximum frequency of power spectral density of noise $n(t)$ in the form of quasi-white Gaussian noise.

Figure 7.4.11 illustrates the useful signal $s(t)$ and realization $w^*(t)$ of the signal $w(t)$ in the output of the adder of the signal extraction unit (SEU). The noise overshoots appear in the form of short pulses of considerable amplitude.

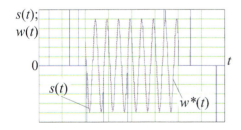

FIGURE 7.4.11 Useful signal $s(t)$ (dot line) and realization $w^*(t)$ of signal $w(t)$ in output of adder of signal extraction unit (SEU) (solid line)

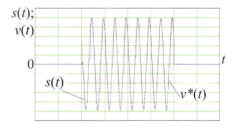

FIGURE 7.4.12 Useful signal $s(t)$ (dot line) and realization $v^*(t)$ of signal $v(t)$ in output of median filter of signal extraction unit (SEU) (solid line)

Figure 7.4.12 illustrates the useful signal $s(t)$ and realization $v^*(t)$ of the signal $v(t)$ in the output of median filter of the signal extraction unit (SEU). Noise overshoots in comparison with the previous figure are removed by median filter.

Figure 7.4.13 illustrates the useful signal $s(t)$ and realization $W^*(t)$ of the signal $W(t)$ in the output of the adder of matched filtering unit (MFU). Comparing the signal $W^*(t)$ with the signal $v^*(t)$, one can conclude that the remnants of noise overshoots observed in the output of the median filter of signal extraction unit (SEU), were removed during processing of the signal $v(t)$ in the transversal filters of matched filtering unit (MFU). The MFU compresses the harmonic signal $s(t)$ in such a way that duration of the signal $W(t)$ in the input of median filter of the MFU is equal to the period T_0 of harmonic signal $s(t)$, thus, compressing the useful signal N_s times where N_s is a number of periods of harmonic signal $s(t)$.

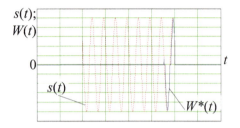 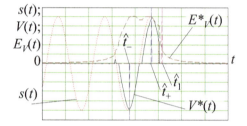

FIGURE 7.4.13 Useful signal $s(t)$ (dot line) and realization $W^*(t)$ of signal $W(t)$ in output of adder of matched filtering unit (MFU) (solid line)

FIGURE 7.4.14 Useful signal $s(t)$ (dot line); realization $V^*(t)$ of signal $V(t)$ (solid line); realization $E_V^*(t)$ of its envelope (dash line)

Figure 7.4.14 illustrates the useful signal $s(t)$; realization $V^*(t)$ of the signal $V(t)$ in the output of the median filter of the MFU; realization $E_V^*(t)$ of the envelope $E_V(t)$ of the signal $V(t)$; δ-pulses determining the time position of the estimators $\hat{t}_\pm$ of barycentric coordinates of the positive $v_+(t)$ or the negative $v_-(t)$ parts of the smoothed stochastic process $v(t)$; δ-pulse determining the time position of the estimator $\hat{t}_1$ of time of signal ending t_1, according to the formula (7.4.61a). As can be seen from the figure, the leading edge of realization $V^*(t)$ of the signal $V(t)$ retards with respect to the useful signal $s(t)$ for time $(N-1)/(2f_{n,\max})$, where N is the number of the samples of stochastic processes $x(t)$ and $\tilde{x}(t)$ used during primary processing in transversal filters of the SEU; $f_{n,\max}$ is maximum frequency of power spectral density of noise $n(t)$ in the form of quasi-white Gaussian noise.

We can determine the quality indices of estimation of unknown nonrandom amplitude A and initial phase φ of harmonic signal $s(t)$. The errors of the estimators of these parameters are determined by dynamic and fluctuation components. Dynamic errors of the estimators $\hat{A}$ and $\hat{\varphi}$ of the amplitude A and the initial phase φ are caused by the method of their obtaining, while assuming the interference (noise) absence in the input of signal processing unit. Fluctuation errors of these estimators are caused by remains of noise in the input of the estimator formation unit.

We first find the dynamic error $\Delta_d \hat{\varphi}$ of the estimator of initial phase $\hat{\varphi}$ of

7.4 Signal Detection in Metric Space with Lattice Properties

harmonic signal $s(t)$ determined by the relationship (7.4.64):

$$\Delta_d \hat{\varphi} = |\hat{\varphi} - \varphi|. \tag{7.4.70}$$

The signal $v(t)$ in the output of median filter of signal extraction unit in the absence of interference (noise) in the input of signal processing unit can be represented in the form:

$$v(t) = \begin{cases} s(t), & t \in T_{s+}^- \cup T_{s-}^+; \\ s(t - \Delta T), & t \in T_{s+}^+ \cup T_{s-}^-, \end{cases} \tag{7.4.71}$$

where $T_s = T_{s+}^+ \cup T_{s+}^- \cup T_{s-}^- \cup T_{s-}^+$ is the domain of definition of the signals $v(t)$ and $s(t)$; ΔT is the quantity of the interval of primary processing T^*: $T^* = [t - (N-1)\Delta t, t]$, $\Delta T = (N-1)\Delta t$; Δt is the sample interval providing independence of the interference (noise) samples $\{n(t_j)\}$; T_{s+}^+, T_{s+}^-, T_{s-}^-, T_{s-}^+ are the domains with the following properties:

$$\begin{cases} T_{s+}^+ = \{t: (0 \leq v(t) < s(t)) \& (s'(t - \Delta T) > 0)\}; \\ T_{s+}^- = \{t: (0 \leq v(t) = s(t)) \& (s'(t) < 0)\}; \\ T_{s-}^- = \{t: (s(t) < v(t) \leq 0) \& (s'(t - \Delta T) < 0)\}; \\ T_{s-}^+ = \{t: (s(t) = v(t) \leq 0) \& (s'(t) > 0)\}. \end{cases} \tag{7.4.72}$$

The smallness of the quantity ΔT of the interval of primary processing T^* implies that the value of the signal $v(t)$ in the output of the median filter of the signal extraction unit is equal to: $v(t) = s(t - \Delta T)$ on the domain $t \in T_{s+}^+ \cup T_{s-}^-$, can be approximately represented by two terms of expansion in Taylor series in the neighborhood of the point t:

$$v(t) = s(t - \Delta T) \approx s(t) - \Delta T \cdot s'(t), \ t \in T_{s+}^+ \cup T_{s-}^-. \tag{7.4.73}$$

The quantity v_s determined by the integral (7.4.64a) can be approximately represented in the form of the following sum:

$$v_s = \int_{T_s} v(t) \sin(\omega_0 t + \varphi) dt \approx \int_{T_s} s(t) \sin(\omega_0 t + \varphi) dt + Q_+ + Q_-, \tag{7.4.74}$$

where $Q_+ = \Delta T \int_{T_{s+}^+} s'(t) \sin(\omega_0 t + \varphi) dt$, $Q_- = \Delta T \int_{T_{s-}^-} s'(t) \sin(\omega_0 t + \varphi) dt$.

Taking into account that the derivation $s'(t)$ of the useful signal $s(t)$ is equal to: $s'(t) = -A\omega_0 \sin(\omega_0 t + \varphi)$, the sum of the last two terms of the expansion (7.4.74) is equal to:

$$Q_+ + Q_- = \frac{2\pi A \Delta T}{T_0}(0.25T_0 - \Delta T)\cos\varphi. \tag{7.4.75}$$

Substituting the value of sum $Q_+ + Q_-$ (7.4.75) into the relationship (7.4.74), we obtain the final approximating expression for the quantity v_s:

$$v_s \approx \int_{T_s} s(t)\sin(\omega_0 t + \varphi)dt + \frac{2\pi A \Delta T}{T_0}(0.25T_0 - \Delta T)\cos\varphi. \tag{7.4.76}$$

Similarly, the quantity v_c determined by the integral (7.4.64b) can be approximately represented in the form of the following sum:

$$v_c = \int_{T_s} v(t)\cos(\omega_0 t + \varphi)dt \approx \int_{T_s} s(t)\cos(\omega_0 t + \varphi)dt + R_+ + R_-, \quad (7.4.77)$$

where $R_+ = \Delta T \int_{T_{s+}^+} s'(t)\cos(\omega_0 t + \varphi)dt$, $R_- = \Delta T \int_{T_{s-}^-} s'(t)\cos(\omega_0 t + \varphi)dt$.

The sum of the last two terms of expansion (7.4.77) is equal to:

$$R_+ + R_- = \frac{2\pi A \Delta T}{T_0}(0.25T_0 - \Delta T)\sin\varphi. \quad (7.4.78)$$

Substituting the value of the sum $R_+ + R_-$ (7.4.78) into the relationship (7.4.77), we obtain the final approximating expression for the quantity v_c:

$$v_c \approx \int_{T_s} s(t)\cos(\omega_0 t + \varphi)dt + \frac{2\pi A \Delta T}{T_0}(0.25T_0 - \Delta T)\sin\varphi. \quad (7.4.79)$$

Taking into account the obtained approximations for the quantities v_s and v_c (7.4.76) and (7.4.79), the approximating expression for the estimator $\hat\varphi$ of initial phase φ (7.4.64) takes the form:

$$\hat\varphi \approx -\text{arctg}\left(\frac{-Q\sin\varphi + q\cos\varphi}{Q\cos\varphi + q\sin\varphi}\right) = -\text{arctg}\left(\frac{r\sin(\alpha - \varphi)}{r\cos(\alpha - \varphi)}\right) = \varphi - \alpha, \quad (7.4.80)$$

where $Q = AT/2$; $q = \frac{2\pi A\Delta T}{T_0}(0.25T_0 - \Delta T)$; $r = \sqrt{Q^2 + q^2}$, $\alpha = \text{tg}(q/Q)$.

Substituting the approximate value of the estimator $\hat\varphi$ of initial phase φ (7.4.80) into the initial formula (7.4.70), we obtain approximate value of the dynamic error $\Delta_d\hat\varphi$ of the estimator of initial phase $\hat\varphi$ of the harmonic signal $s(t)$:

$$\Delta_d\hat\varphi = \text{tg}\left(\frac{\pi\Delta T}{T}\left(1 - \frac{4\Delta T}{T_0}\right)\right), \quad \Delta T << T_0 < T, \quad (7.4.81)$$

where T is known signal duration; $T = N_s T_0$; N_s is a number of periods of harmonic signal $s(t)$, $Ns = Tf_0$; f_0 is a carrier frequency of the signal $s(t)$; T_0 is a period of a carrier: $T_0 = 1/f_0$.

We now find the relative dynamic error $\delta_d\hat A$ of the estimator $\hat A$ of the amplitude A of the useful signal $s(t)$:

$$\delta_d\hat A = \frac{|A - \hat A|}{A}. \quad (7.4.82)$$

Taking into account the obtained approximations for the quantities v_s (7.4.76) and v_c (7.4.79), the approximating expression for the estimator $\hat A$ of the amplitude A (7.4.69) takes the form:

$$\hat A \approx 2\sqrt{(-Q\sin\varphi + q\cos\varphi)^2 + (Q\cos\varphi + q\sin\varphi)^2}/T = $$
$$= 2\sqrt{Q^2 + q^2}/T, \quad (7.4.83)$$

7.4 Signal Detection in Metric Space with Lattice Properties

where $Q = AT/2$; $q = \frac{2\pi A \Delta T}{T_0}(0.25T_0 - \Delta T)$.

Substituting the approximate value of the estimator $\hat{A}$ of the amplitude A (7.4.83) into the initial formula (7.4.82), we obtain the approximate value of the relative dynamic error $\delta_d \hat{A}$ of the amplitude A of the harmonic signal $s(t)$:

$$\delta_d \hat{A} \approx \left| \sqrt{1 + \left(\frac{2\pi \Delta T}{T}\right)^2 \left(1 - \frac{4\Delta T}{T_0}\right)} - 1 \right|, \quad \Delta T \ll T_0 < T, \quad (7.4.84)$$

Find the fluctuation errors of the estimators $\hat{\varphi}$ and $\hat{A}$ of the initial phase φ and the amplitude A of the harmonic signal $s(t)$, which characterize the synthesized signal processing unit (Fig. 7.4.10). We use the theorem from [251] concerning the median estimator $v(t)$ obtained by the median filter (MF) converges in distribution to the Gaussian random variable with zero mean.

As noted in Section 7.3, the variance D_{v_n} (see Formula (7.3.34)) of the noise component $v_n(t)$ of the process $v(t)$ in the output of median filter can be decreased to the quantity:

$$D_{v_n} \leq D_{v,\max} = a^2 \exp\left\{-\frac{\Delta \tilde{T}^2 N^2 f_{n,\max}^2}{8\sqrt{\pi}}\right\}, \quad (7.4.85)$$

where a is a parameter of the limiter $L[w(t)]$; $\Delta \tilde{T}$ is a quantity of the smoothing interval $\tilde{T}$; N is a number of the samples of interference (noise) $n(t)$ that simultaneously take part in signal processing; $f_{n,\max}$ is a maximum frequency of power spectral density of interference (noise) $n(t)$.

The variances of fluctuation errors of the estimators $\hat{\varphi}$ and $\hat{A}$ of the initial phase φ and the amplitude A of the harmonic signal $s(t)$ are determined on the basis of the variances of the envelope and the phase of a vector with independent Gaussian components v_s and v_c.

To determine expectations and variances of Gaussian components v_s and v_c, we use the relationships for the first moments of distribution of the process in the output of correlation device [115, (3.6.65)], from which, while substituting the initial values, we obtain the quantities of the variances D_s and D_c and the expectations m_s and m_c of the components v_s and v_c, respectively:

$$D_s = 0.5 A^2 \cos^2 \varphi + D_{v,\max}, \quad D_c = 0.5 A^2 \sin^2 \varphi + D_{v,\max}; \quad (7.4.86a)$$

$$m_s = A \sin \varphi / \sqrt{2}, \quad m_c = A \cos \varphi / \sqrt{2}. \quad (7.4.86b)$$

Then the variance $D_{\hat{\varphi}}$ of the estimator $\hat{\varphi}$ of the initial phase φ is determined by the formula [113, (3.93)], and the variance $D_{\hat{A}}$ of the estimator of the amplitude $\hat{A}$ is determined on the base of the formula [113, (3.73)] for the moments of the envelope, and in the case of strong interference (noise) they are, respectively, equal to:

$$D_{\hat{\varphi}} \approx D_{v,\max}/A^2; \quad (7.4.87a)$$

$$D_{\hat{A}} \approx 4 D_{v,\max}[1 - D_{v,\max}/2A^2]/T^2, \quad (7.4.87b)$$

where $D_{v,\max}$ is the maximum variance of the noise component $v_n(t)$ of the process $v(t)$ in the output of the median filter of the signal extraction unit.

7.4.4 Features of Detection of Linear Frequency Modulated Signal with Unknown Nonrandom Amplitude and Initial Phase with Joint Estimation of Time of Signal Arrival (Ending) in Presence of Interference (Noise) in Metric Space with L-group Properties

In this subsection we consider features of optimal algorithm synthesis of a detection of linear frequency modulated (LFM) signal with unknown nonrandom amplitude and initial phase with joint estimation of time of signal arrival (ending) in the presence of interference (noise) in metric space with L-group properties. Synthesis of the optimal detection algorithm assumes interference (noise) distribution is arbitrary.

Let the interaction between LFM signal $s(t)$ and interference (noise) $n(t)$ in metric space $\mathcal{L}(+, \vee, \wedge)$ with L-group properties be described by two binary operations of join $\vee$ and meet $\wedge$ in two receiving channels, respectively:

$$x(t) = \theta s(t) \vee n(t); \tag{7.4.88a}$$

$$\tilde{x}(t) = \theta s(t) \wedge n(t), \tag{7.4.88b}$$

where θ is an unknown nonrandom parameter that can take only two values $\theta \in \{0, 1\}$: $\theta = 0$ (the signal is absent) or $\theta = 1$ (the signal is present); $t \in T_s$, T_s is a domain of definition of the signal $s(t)$, $T_s = [t_0, t_1]$; t_0 is an unknown time of arrival of the signal $s(t)$; t_1 is an unknown time of signal ending; $T = t_1 - t_0$ is known signal duration; $T_s \subset T_{\text{obs}}$; T_{obs} is the observation interval of the signal: $T_{\text{obs}} = [t'_0, t'_1]$; $t'_0 < t_0$, $t_1 < t'_1$.

Let also the model of the received LFM signal $s(t)$ be determined by the expression:

$$s(t) = \begin{cases} A\cos([(\Delta\omega/T)t + \omega_0]t + \varphi), & t \in T_s; \\ 0, & t \notin T_s, \end{cases} \tag{7.4.89}$$

where A is an unknown nonrandom amplitude of the useful signal $s(t)$; $\omega_0 = 2\pi f_0$; f_0 is the known carrier frequency of the signal $s(t)$; $\Delta\omega$ is the known deviation of LFM signal; φ is an unknown nonrandom initial phase of the useful signal, $\varphi \in [-\pi, \pi]$; T_s is a domain of definition of the signal $s(t)$, $T_s = [t_0, t_1]$; t_0 is an unknown time of arrival of the signal $s(t)$; t_1 is an unknown time of signal ending; $T = t_1 - t_0$ is known signal duration, $T = N_s T_0$; N_s is integer number of periods of LFM signal $s(t)$, $N_s \in \mathbf{N}$, $\mathbf{N}$ is the set of natural numbers; T_0 is a period of signal carrier.

We suppose that two neighbor samples $n(t_j)$ and $n(t_{j\pm1})$ of interference (noise) $n(t)$ are statistically independent if a time distance between them is greater than or equal to the interval:

$$|t_{j\pm1} - t_j| \geq T_{0,\min}, \tag{7.4.90}$$

where $T_{0,\min}$ is the minimal period of the oscillation of LFM signal $s(t)$.

The instantaneous values (the samples) of the signal $\{s(t_j)\}$ and interference (noise) $\{n(t_j)\}$ are the elements of signal space: $s(t_j), n(t_j) \in \mathcal{L}(+, \vee, \wedge)$. The samples $s(t_j)$ and $n(t_j)$ of the signal $s(t)$ and interference (noise) $n(t)$ are taken on the domain of definition T_s of the signal $s(t)$: $t_j \in T_s$ over the variable interval $T_j^{\pm} \geq T_{0,\min}$ that provides the independence of interference (noise) samples $\{n(t_j)\}$.

7.4 Signal Detection in Metric Space with Lattice Properties

Taking into account the aforementioned considerations, the observation equations in two processing channels (7.4.88a) and (7.4.88b) take the form:

$$x(t_j) = \theta s(t_j) \vee n(t_j); \qquad (7.4.91a)$$

$$\tilde{x}(t_j) = \theta s(t_j) \wedge n(t_j), \qquad (7.4.91b)$$

where $t_j = t - T_j^{\pm}$, $j = 0, 1, \ldots, N-1$, $t_j \in T_s \subset T_{\text{obs}}$; $T_s = [t_0, t_1]$, $T_{\text{obs}} = [t_0', t_1']$; $t_0' < t_0$, $t_1 < t_1'$; $J \in \mathbf{N}$, $\mathbf{N}$ is the set of natural numbers; the interval $T_j^{\pm}$ may change over duration depending on t_j.

In this subsection and Subsection 7.4.2, the problem of joint detection of useful LFM signal $s(t)$ and estimation of its time of ending t_1 is formulated and solved on the basis of matched filtering of the signal $s(t)$ and then the joint detection of the signal $s(t)$ and estimation of time of its ending t_1 are performed.

The matched filtering problem $MF[s(t)]$ in the signal space $\mathcal{L}(+, \vee, \wedge)$ with L-group properties is solved via step-by-step processing of statistical collections $\{x(t_j)\}$ and $\{\tilde{x}(t_j)\}$ determined by the observation equations (7.4.91a) and (7.4.91b), as shown in Subsection 7.4.2 for harmonic signal, with the only distinction that the features of time structure of LFM signal are taken into account:

$$MF[s(t)] = \begin{cases} PF[s(t)]; & (a) \\ IP[s(t)]; & (b) \\ Sm[s(t)], & (c) \end{cases} \qquad (7.4.92)$$

where $PF[s(t)]$ is primary filtering; $IP[s(t)]$ is intermediate processing; $Sm[s(t)]$ is smoothing; together they form successive processing stages of the general algorithm of matched filtering $MF[s(t)]$ of the useful LFM signal (7.4.92).

The optimality criteria defining every single processing stage $PF[s(t)]$, $IP[s(t)]$, $Sm[s(t)]$ are involved in single systems:

$$PF[s(t)] = \begin{cases} y(t) = \underset{\breve{y}(t) \in Y; t, t_j \in T_{\text{obs}}}{\arg\min} |\bigwedge_{j=0}^{J-1}[x(t_j) - \breve{y}(t)]|; & (a) \\ \tilde{y}(t) = \underset{\widehat{y}(t) \in \tilde{Y}; t, t_j \in T_{\text{obs}}}{\arg\min} |\bigvee_{j=0}^{J-1}[\tilde{x}(t_j) - \widehat{y}(t)]|; & (b) \\ J = \underset{y(t) \in Y;\, T_{\text{obs}}}{\arg\min}[\int |y(t)|dt]\big|_{n(t)\equiv 0}, \int_{T_{\text{obs}}} |y(t)|dt \neq 0; & (c) \\ w(t) = F[y(t), \tilde{y}(t)]; & (d) \\ \int_{[t_1 - T_{0,\min}, t_1]} |w(t) - s(t)|dt\big|_{n(t)\equiv 0} \rightarrow \underset{w(t) \in W}{\min}, & (e) \end{cases} \qquad (7.4.93)$$

where $y(t)$ and $\tilde{y}(t)$ are the solution functions of the problem of minimization of metrics between the observed statistical collections $\{x(t_j)\}$, $\{\tilde{x}(t_j)\}$ and optimization variables, i.e., the functions $\breve{y}(t)$ and $\widehat{y}(t)$, respectively; $w(t)$ is a function $F[*,*]$ of uniting the results $y(t)$ and $\tilde{y}(t)$ of minimization of the functions of the observed collections $\{x(t_j)\}$ and $\{\tilde{x}(t_j)\}$; T_{obs} is an observation interval of the signal; J is a number of the samples of stochastic processes $x(t)$ and $\tilde{x}(t)$ used under signal processing, $J \in \mathbf{N}$; $\mathbf{N}$ is the set of natural numbers;

$$IP[s(t)] = \begin{cases} \mathbf{M}\{u^2(t)\}\big|_{s(t)\equiv 0} \to \min_L; & (a) \\ \mathbf{M}\{[u(t)-s(t)]^2\}\big|_{n(t)\equiv 0} = \varepsilon; & (b) \\ u(t) = L[w(t)], & (c) \end{cases} \quad (7.4.94)$$

where $\mathbf{M}\{*\}$ is the symbol of mathematical expectation; $L[w(t)]$ is a functional transformation of the process $w(t)$ into the process $u(t)$; ε is a constant which is some function of a power of the signal $s(t)$;

$$Sm[s(t)] = \begin{cases} v(t) = \underset{v^\circ(t)\in V;\, t, t_k \in \tilde{T}}{\arg\min} \sum_{k=0}^{M-1} |u(t_k) - v^\circ(t)|; & (a) \\ \Delta\tilde{T} : \delta_d(\Delta\tilde{T}) = \delta_{d,\mathrm{sm}}; & (b) \\ M = \underset{M'\in \mathbf{N}\cap [M^*,\infty[}{\arg\max} [\delta_f(M')]\big|_{M^*:\delta_f(M^*)=\delta_{f,\mathrm{sm}}}, & (c) \end{cases} \quad (7.4.95)$$

where $v(t) = \hat{s}(t)$ is the result of filtering (the estimator $\hat{s}(t)$ of the signal $s(t)$) that is the solution of minimization of the metric between the instantaneous values of stochastic process $u(t)$ and optimization variable, i.e., the function $v^\circ(t)$; $t_k = t - \frac{k}{M}\Delta\tilde{T}$, $k = 0,1,\ldots,M-1$, $t_k \in \tilde{T} =]t - \Delta\tilde{T}, t]$; $\tilde{T}$ is an interval, in which smoothing of stochastic process $u(t)$ is realized; $M \in \mathbf{N}$, $\mathbf{N}$ is the set of natural numbers; M is a number of the samples of stochastic process $u(t)$ used during smoothing in the interval $\tilde{T}$; $\delta_d(\Delta\tilde{T})$, $\delta_f(M)$ are relative dynamic and fluctuation errors of smoothing, as the dependences on the quantity $\Delta\tilde{T}$ of the smoothing interval $\tilde{T}$ and the number of the samples M respectively; $\delta_{d,\mathrm{sm}}$, $\delta_{f,\mathrm{sm}}$ are given quantities of the relative dynamic and fluctuation errors of smoothing, respectively.

The problem of joint detection $Det[s(t)]$ of LFM signal $s(t)$ and estimation $Est[t_1]$ of time of its ending t_1 is formulated on the basis of the detection and estimation criteria involved in the same system, which is a logical continuation of the system (7.4.92):

$$Det[s(t)]/Est[t_1] =$$
$$= \begin{cases} E_v(\hat{t}_1 - \frac{T_{0,\min}}{2}) \gtrless_{d_0}^{d_1} l_0(F); & (a) \\ \hat{t}_1 = \underset{\varphi\in\Phi_1\vee\Phi_2;\, t^\circ \in T_{\mathrm{obs}}}{\arg\min} \mathbf{M}_\varphi\{(t_1 - t^\circ)^2\}\big|_{n(t)\equiv 0}, & (b) \end{cases} \quad (7.4.96)$$

where $E_v(\hat{t}_1 - \frac{T_{0,\min}}{2})$ is the instantaneous value of the envelope $E_v(t)$ of the estimator $v(t) = \hat{s}(t)$ of the useful signal $s(t)$ at the instant $t = \hat{t}_1 - \frac{T_{0,\min}}{2}$: $E_v(t) = \sqrt{v^2(t) + v_\mathcal{H}^2(t)}$, $v_\mathcal{H}(t) = \mathcal{H}[v(t)]$ is a Hilbert transform; d_1 and d_0 are the decisions made about the true values of an unknown nonrandom parameter θ, $\theta \in \{0,1\}$; $l_0(F)$ is some threshold level dependent on a given conditional probability of false alarm F; $\hat{t}_1$ is the estimator of time of signal ending t_1; $T_{0,\min}$ is the minimal period of the oscillation of the LFM signal $s(t)$; $\mathbf{M}_\varphi\{(t_1 - t^\circ)^2\}$ is the mean squared difference between true value of time of signal ending t_1 and optimization variable $t°$; $\mathbf{M}_\varphi\{*\}$ is the symbol of mathematical expectation with the averaging with respect to the initial phase φ of the signal; Φ_1 and Φ_2 are possible domains of definition of the initial phase φ of the signal: $\Phi_1 = [-\pi/2, \pi/2]$ and $\Phi_2 = [\pi/2, 3\pi/2]$.

We now explain the optimality criteria and some relationships appearing in the systems (7.4.93), (7.4.94), (7.4.95) that define the successive stages of signal processing $PF[s(t)]$, $IP[s(t)]$, $Sm[s(t)]$ of the general algorithm of matched filtering $MF[s(t)]$ of the useful signal $s(t)$ (7.4.92).

Equations (7.4.93a) and (7.4.93b) of the system (7.4.93) determine the criteria of minimum of metrics between statistical sets of the observations $\{x(t_j)\}$, $\{\tilde{x}(t_j)\}$ and the results of primary processing $y(t)$, $\tilde{y}(t)$, respectively.

The functions of metrics $|\bigwedge_{j=0}^{J-1}[x(t_j)-\breve{y}(t)]|$, $|\bigvee_{j=0}^{J-1}[\tilde{x}(t_j)-\widehat{y}(t)]|$ are chosen to provide the metric convergence and the convergence in probability to the useful signal $s(t)$ of the sequences $y(t) = \bigwedge_{j=0}^{J-1} x(t_j)$, $\tilde{y}(t) = \bigvee_{j=0}^{J-1} \tilde{x}(t_j)$ based on the interactions of the kind (7.4.91a), (7.4.91b).

The relationship (7.4.93c) determines the criterion of the choice of a number of the samples J of stochastic processes $x(t)$, $\tilde{x}(t)$ used during signal processing on the basis of minimization of the norm $\int_{T_{obs}} |y(t)|dt$. The criterion (7.4.93c) is considered under two constraint conditions: (1) interference (noise) is identically equal to zero: $n(t) \equiv 0$; (2) the norm $\int_{T_{obs}} |y(t)|dt$ of the function $y(t)$ is not equal to zero: $\int_{T_{obs}} |y(t)|dt \neq 0$.

Equation (7.4.93e) determines the criterion of minimum of metric $\int_{[t_1-T_0, t_1]} |w(t) - s(t)|dt \big|_{n(t) \equiv 0}$ between the useful signal $s(t)$ and the function $w(t)$ in the interval $[t_1 - T_{0,\min}, t_1]$. This criterion establishes the kind of the function $F[y(t), \tilde{y}(t)]$ (7.4.93d) uniting the results $y(t)$ and $\tilde{y}(t)$ of primary processing of the observed stochastic processes $x(t)$ and $\tilde{x}(t)$. The criterion (7.4.93e) is considered under the condition that interference (noise) is identically equal to zero: $n(t) \equiv 0$.

Equations (7.4.94a), (7.4.94b), (7.4.94c) of the system (7.4.94) determine the criterion of the choice of functional transformation $L[w(t)]$. The equation (7.4.94a) determines the criterion of minimum of the second moment of the process $u(t)$ in the absence of the LFM signal $s(t)$ in the input of the signal processing unit. The equation (7.4.94b) determines the quantity of the second moment of the difference between the signals $u(t)$ and $s(t)$ in the absence of interference (noise) $n(t)$ in the input of the signal processing unit.

Equation (7.4.95a) of the system (7.4.95) determines the criterion of minimum of the metric $\sum_{k=0}^{M-1} |u(t_k) - v°(t)|$ between the instantaneous values of the process $u(t)$ and optimization variable $v°(t)$ in the smoothing interval $\tilde{T} =]t - \Delta\tilde{T}, t]$, requiring the final processing of the signal $u(t)$ in the form of smoothing under the condition that the useful signal is identically equal to zero: $s(t) \equiv 0$. The relationship (7.4.95b) determines the rule of the choice of the quantity $\Delta \tilde{T}$ of the smoothing interval $\tilde{T}$, based on providing a given quantity of relative dynamic error $\delta_{d,\mathrm{sm}}$ of smoothing.

Equation (7.4.95c) determines the criterion of the choice of a number of the samples M of stochastic process $u(t)$, based on providing a given quantity of relative fluctuation error $\delta_{f,\mathrm{sm}}$ of its smoothing.

Equation (7.4.96a) of the system (7.4.96) determines the rule of making the decision d_1 concerning the presence of the signal (if $E_v(\hat{t}_1 - \frac{T_{0,\min}}{2}) > l_0(F)$) or the decision d_0 concerning the absence of the signal (if $E_v(\hat{t}_1 - \frac{T_{0,\min}}{2}) < l_0(F)$).

The relationship (7.4.96b) determines the criterion of formation of the estimator $\hat{t}_1$ of time of signal ending t_1 on the basis of minimization of the mean squared difference $\mathbf{M}_\varphi\{(t_1 - t^\circ)^2\}$ between true value of time of signal ending t_1 and optimization variable t° under the conditions that the averaging is realized over the initial phase φ of the signal, taken in one of two intervals: $\Phi_1 = [-\pi/2, \pi/2]$ or $\Phi_2 = [\pi/2, 3\pi/2]$, and interference (noise) is identically equal to zero: $n(t) \equiv 0$.

Solving the problem of matched filtering $MF[s(t)]$ of the useful signal $s(t)$ in signal space $\mathcal{L}(+, \vee, \wedge)$ with L-group properties in the presence of interference (noise) $n(t)$ is described in detail in Subsection 7.4.2, so we consider intermediate results, which determine the structure-forming elements of the general signal processing algorithm, paying the attention only to the features of LFM signal processing.

The solutions of optimization equations (7.4.93a), (7.4.93b) of the system (7.4.93) are the values of the estimators $y(t)$, $\tilde{y}(t)$ in the form of meet and join of the observation results $\{x(t_j)\}$, $\{\tilde{x}(t_j)\}$, respectively:

$$y(t) = \bigwedge_{j=0}^{J-1} x(t_j) = \bigwedge_{j=0}^{J-1} x(t - T_j^+); \tag{7.4.97a}$$

$$\tilde{y}(t) = \bigvee_{j=0}^{J-1} \tilde{x}(t_j) = \bigvee_{j=0}^{J-1} \tilde{x}(t - T_j^-), \tag{7.4.97b}$$

where the variable intervals $T_j^\pm$ are determined by the following relationships:

$$T_j^+ = t_{m,J}^+ - t_{m,j}^+; \; \{t_{m,j}^+\} : (s'(t_{m,j}^+) = 0) \& (s''(t_{m,j}^+) < 0);$$
$$t_{m,J}^+ = \max_j\{t_{m,j}^+\}, \; T_{j=0}^+ \equiv 0; \tag{7.4.98a}$$

$$T_j^- = t_{m,J}^- - t_{m,j}^-; \; \{t_{m,j}^-\} : (s'(t_{m,j}^-) = 0) \& (s''(t_{m,j}^-) > 0);$$
$$t_{m,J}^- = \max_j\{t_{m,j}^-\}, \; T_{j=0}^- \equiv 0, \tag{7.4.98b}$$

where $\{t_{m,j}^+\}$, $\{t_{m,j}^-\}$ are the positions of local maximums and minimums of the LFM signal $s(t)$ on the time axis, respectively; $s'(t)$ and $s''(t)$ are the first and the second derivatives of the useful signal $s(t)$ with respect to time.

The condition $s(t) \equiv 0$ of the criterion (7.4.93c) of the system (7.4.93) implies the corresponding specifications of the observation equations (7.4.91a,b): $x(t_j) = s(t_j) \vee 0$ and $\tilde{x}(t_j) = s(t_j) \wedge 0$. So, according to the relationships (7.4.97a), (7.4.97b),

7.4 Signal Detection in Metric Space with Lattice Properties

the identities hold:

$$y(t)\big|_{n(t)\equiv 0} = \bigwedge_{j=0}^{J-1}[s(t_j) \vee 0] =$$
$$= [s(t) \vee 0] \wedge [s(t-T^+_{J-1}) \vee 0]; \quad (7.4.99a)$$

$$\tilde{y}(t)\big|_{n(t)\equiv 0} = \bigvee_{j=0}^{J-1}[s(t_j) \wedge 0] =$$
$$= [s(t) \wedge 0] \vee [s(t-T^-_{J-1}) \wedge 0], \quad (7.4.99b)$$

where $t_j = t - T^\pm_j$, $j = 0, 1, \ldots, J-1$; $T^\pm_{j=0} \equiv 0$.

According to the criterion (7.4.93c) of the system (7.4.93), the optimal value of a number of the samples J of stochastic processes $x(t)$, $\tilde{x}(t)$ used under primary processing (7.4.97a,b) is equal to an integer number of periods of LFM signal N_s:

$$J = N_s. \quad (7.4.100)$$

The final coupling equation can be written in the form:

$$w(t) = y_+(t) + \tilde{y}_-(t); \quad (7.4.101)$$
$$y_+(t) = y(t) \vee 0; \quad (7.4.101a)$$
$$\tilde{y}_-(t) = \tilde{y}(t) \wedge 0. \quad (7.4.101b)$$

Thus, the identity (7.4.101) determines the kind of the coupling equation (7.4.93d) obtained from joint fulfillment of the criteria (7.4.93a), (7.4.93b), and (7.4.93e) of the system (7.4.93).

The solution $u(t)$ of the relationships (7.4.94a), (7.4.94b), (7.4.94c) of the system (7.4.94), determining the criterion of the choice of functional transformation of the process $w(t)$, is the function $L[w(t)]$ that determines the gain characteristic of the limiter:

$$u(t) = L[w(t)] = [(w(t) \wedge a) \vee 0] + [(w(t) \vee (-a)) \wedge 0], \quad (7.4.102)$$

where a is a parameter of the limiter chosen to be equal to: $a = \sup_A \Delta_A = A_{\max}$, $\Delta_A =]0, A_{\max}]$; $A_{\max}$ is a maximum possible value of amplitude of the useful signal.

The solution of optimization equation (7.4.95a) of the system (7.4.95) is the value of the estimator $v(t)$ in the form of a sample median $\text{med}\{*\}$ of the collection of the samples $\{u(t_k)\}$ of stochastic process $u(t)$:

$$v(t) = \underset{t_k \in \tilde{T}}{\text{med}}\{u(t_k)\}, \quad (7.4.103)$$

where $t_k = t - \frac{k}{M}\Delta\tilde{T}$, $k = 0, 1, \ldots, M-1$; $t_k \in \tilde{T} =]t - \Delta\tilde{T}, t]$; $\tilde{T}$ is an interval in which smoothing of stochastic process $u(t)$ is realized; $\Delta\tilde{T}$ is a quantity of the smoothing interval $\tilde{T}$.

Solving the problem of joint detection $Det[s(t)]$ of the signal $s(t)$ and estimation $Est[t_1]$ of time of its ending t_1 is described in Subsection 7.4.2. We consider only the intermediate results determining the structure-forming elements of the general signal processing algorithm.

The rule of making the decision d_1 concerning the presence of the signal (if $E_v(\hat{t}_1 - \frac{T_{0,\min}}{2}) > l_0(F)$) or the decision d_0 concerning the absence of the signal (if $E_v(\hat{t}_1 - \frac{T_{0,\min}}{2}) < l_0(F)$), determined by the equation (7.4.96a) of the system (7.4.96), includes (1) forming the envelope $E_v(t)$ of the estimator $v(t) = \hat{s}(t)$ of the useful signal $s(t)$, and (2) comparison of the value of the envelope $E_v(t)$ with a threshold value $l_0(F)$ at the instant $t = \hat{t}_1 - \frac{T_{0,\min}}{2}$ determined by the estimator $\hat{t}_1$. Thus, decision making is realized in the following form:

$$E_v(\hat{t}_1 - \frac{T_{0,\min}}{2}) \underset{d_0}{\overset{d_1}{\gtrless}} l_0(F). \tag{7.4.104}$$

The solution of optimization equation (7.4.96b) of the system (7.4.96) is determined by the identity:

$$\hat{t}_1 = \begin{cases} \hat{t}_- + (T_{0,\min}/2) + (T_{0,\min}\hat{\varphi}/2\pi), & \varphi \in \Phi_1 = [-\pi/2, \pi/2]; \\ \hat{t}_+ + (T_{0,\min}/2) - (T_{0,\min}\hat{\varphi}/2\pi), & \varphi \in \Phi_2 = [\pi/2, 3\pi/2], \end{cases} \tag{7.4.105}$$

where $\hat{t}_\pm = (\int_{T_{\text{obs}}} t v_\pm(t) dt)/(\int_{T_{\text{obs}}} v_\pm(t) dt)$ is the estimator of barycentric coordinate of the positive $v_+(t)$ or the negative $v_-(t)$ parts of the smoothed stochastic process $v(t)$, respectively; $v_+(t) = v(t) \vee 0$, $v_-(t) = v(t) \wedge 0$; T_{obs} is an observation interval of the signal: $T_{\text{obs}} = [t'_0, t'_1]$; $t'_0 < t_0$, $t_1 < t'_1$; $\hat{\varphi} = \arcsin[2(\hat{t}_+ - \hat{t}_-)/T_{0,\min}]$ is the estimator of an unknown nonrandom initial phase φ of the useful signal $s(t)$; $T_{0,\min}$ is a minimal period of the oscillation of the LFM signal $s(t)$.

If the initial phase φ of the signal can change from $-\pi$ to π, i.e., it is unknown beforehand, to which interval ($\Phi_1 = [-\pi/2, \pi/2]$ or $\Phi_2 = [\pi/2, 3\pi/2]$) from the relationships (7.4.96b) and (7.4.105) φ belongs, we use the estimators $\hat{t}_1$ of time of signal ending t_1 determined by the identities:

$$\hat{t}_1 = \max_{\hat{t}_\pm \in T_{\text{obs}}} [\hat{t}_-, \hat{t}_+] + (T_{0,\min}/4), \quad \varphi \in [-\pi, \pi]; \tag{7.4.106a}$$

$$\text{or}: \quad \hat{t}_1 = \frac{1}{2}(\hat{t}_- + \hat{t}_+) + \frac{T_{0,\min}}{2}, \quad \varphi \in [-\pi, \pi]. \tag{7.4.106b}$$

Thus, summarizing the relationships (7.4.97), (7.4.100) through (7.4.105), one can conclude that the estimator $\hat{t}_1$ of time of signal ending and the envelope $E_v(t)$ are formed on the basis of the further processing of the estimator $v(t) = \hat{s}(t)$ of the LFM signal $s(t)$ detected in the presence of interference (noise) $n(t)$. The estimator $\hat{s}(t)$ is the smoothing function of the stochastic process $u(t)$ obtained by limitation of the process $w(t)$ that combines the results $y(t)$ and $\tilde{y}(t)$ of the corresponding primary processing of the observed stochastic process $x(t)$ and $\tilde{x}(t)$ in the interval of observation T_{obs}.

The block diagram of signal processing unit, according to the relationships (7.4.97), (7.4.100) through (7.4.105), includes: two processing channels, each containing transversal filter; the units of formation of the positive $y_+(t)$ and the negative $\tilde{y}_-(t)$ parts of the processes $y(t)$ and $\tilde{y}(t)$, respectively; the adder that sums the results of signal processing in two channels; the limiter; median filter (MF); estimator formation unit (EFU), envelope computation unit (ECU), and decision gate (DG) (see Fig. 7.4.15).

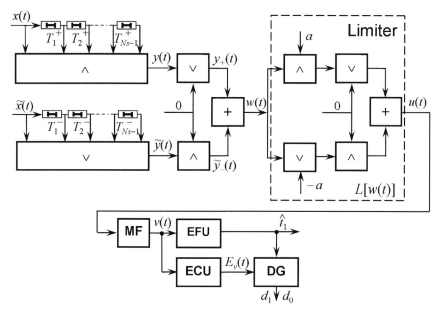

FIGURE 7.4.15 Block diagram of processing unit that realizes LFM signal detection with joint estimation of time of signal arrival (ending)

Transversal filters in two processing channels realize primary filtering $PF[s(t)]$: $y(t) = \bigwedge_{j=0}^{N-1} x(t - jT_0)$ and $\tilde{y}(t) = \bigvee_{j=0}^{N-1} \tilde{x}(t - jT_0)$ of the observed stochastic processes $x(t)$ and $\tilde{x}(t)$, according to Equations (7.4.97a,b) and (7.4.100), providing fulfillment of the criteria (7.4.93a), (7.4.93b), and (7.4.93c) of the system (7.4.93). The units of formation of the positive $y_+(t)$ and the negative $\tilde{y}_-(t)$ parts of the processes $y(t)$ and $\tilde{y}(t)$ in two processing channels form the values of these functions, according to the identities (7.4.101a) and (7.4.101b), respectively. The adder unites the results of signal processing in two channels, according to the equality (7.4.101), providing fulfillment of the criteria (7.4.93d) and (7.4.93e) of the system (7.4.93).

The limiter $L[w(t)]$ realizes intermediate processing $IP[s(t)]$ by means of limiting the signal $w(t)$ in the output of the adder, according to the criteria (7.4.94a), (7.4.94b) of the system (7.4.94), to exclude the noise overshoots whose instantaneous values exceed value a from further processing.

The median filter (MF) realizes smoothing $Sm[s(t)]$ of the process $w(t) = y_+(t) + \tilde{y}_-(t)$, according to the formula (7.4.103), providing fulfillment of the criterion (7.4.95a) of the system (7.4.95). The envelope computation unit (ECU) forms the envelope $E_v(t)$ of the signal $v(t)$ in the output of median filter (MF).

The estimator formation unit (EFU) forms the estimator $\hat{t}_1$ of the time of signal ending, according to Equation (7.4.105), providing fulfillment of the criterion (7.4.96b) of the system (7.4.96). At the instant $t = \hat{t}_1 - \frac{T_{0,\min}}{2}$, the decision gate (DG) compares the instantaneous value of the envelope $E_v(t)$ with the threshold value $l_0(F)$. The decision d_1 concerning the presence of the signal (if $E_v(\hat{t}_1 - \frac{T_{0,\min}}{2}) > l_0(F)$) or the decision d_0 concerning the absence of the signal (if $E_v(\hat{t}_1 - \frac{T_{0,\min}}{2}) < l_0(F)$) is made according to the rule (7.4.104) of the criterion (7.4.96a) of the system (7.4.96).

Figures 7.4.16 through 7.4.19 illustrate the results of statistical modeling of signal processing realized by the synthesized unit under the following conditions: the useful signal $s(t)$ is LFM with an integer number of periods $N_s = 10$ and deviation $\Delta\omega = \omega_0/2$, where ω_0 is known carrier frequency of the signal $s(t)$; and also with initial phase $\varphi = -\pi/6$. The signal-to-noise ratio E/N_0 is equal to $E/N_0 = 10^{-10}$, where E is a signal energy; N_0 is the power spectral density of noise. The product $T_{0,\min} f_{n,\max} = 125$, where $T_{0,\min}$ is a minimal period of the oscillation of the carrier of the LFM signal $s(t)$; $f_{n,\max}$ is the maximum frequency of power spectral density of noise $n(t)$ in the form of quasi-white Gaussian noise.

Figure 7.4.16 illustrates the useful signal $s(t)$; realization $w^*(t)$ of the signal $w(t)$ in the output of the adder; realization $v^*(t)$ of the signal in the output of median filter $v(t)$.

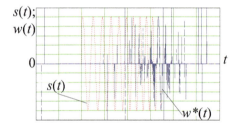 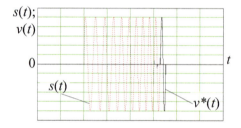

FIGURE 7.4.16 Useful signal $s(t)$ (dot line) and realization $w^*(t)$ of signal $w(t)$ in output of adder (dash line)

FIGURE 7.4.17 Useful signal $s(t)$ (dot line) and realization $v^*(t)$ of signal $v(t)$ in output of median filter (solid line)

Figure 7.4.17 illustrates the useful signal $s(t)$ and realization $v^*(t)$ of the signal $v(t)$ in the output of median filter. The matched filter provides the compression of the LFM signal $s(t)$ in such a way that a duration of the signal $v(t)$ in the output of matched filter is equal to minimal period of the oscillation $T_{0,\min}$ of the LFM signal $s(t)$.

Figure 7.4.18 illustrates: the useful signal $s(t)$; realization $v^*(t)$ of the signal $v(t)$ in the output of median filter; δ-pulses determining time positions of the estimators $\hat{t}_\pm$ of barycentric coordinates of the positive $v_+(t)$ and negative $v_-(t)$ parts of the smoothed stochastic process; δ-pulse determining time position of the estimator $\hat{t}_1$ of time of signal ending t_1, according to the formula (7.4.105).

Figure 7.4.19 illustrates the useful signal $s(t)$; realization $E_v^*(t)$ of the envelope $E_v(t)$ of the signal $v(t)$ in the output of median filter; δ-pulses determining time position of the estimators $\hat{t}_\pm$ of barycentric coordinates of the positive $v_+(t)$ or the negative $v_-(t)$ parts of the smoothed stochastic process $v(t)$; δ-pulse determining

7.5 Signal Detection in Metric Space with Lattice Properties

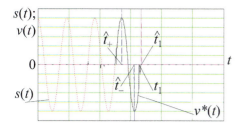

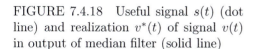

FIGURE 7.4.18 Useful signal $s(t)$ (dot line) and realization $v^*(t)$ of signal $v(t)$ in output of median filter (solid line)

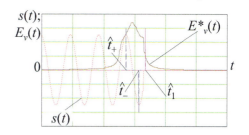

FIGURE 7.4.19 Useful signal $s(t)$ (dot line) and realization $E_v^*(t)$ of envelope $E_v(t)$ of signal $v(t)$ (solid line)

time position of the estimator $\hat{t}_1$ of time of signal ending t_1, according to the formula (7.4.105).

The results of the investigations of algorithms and units of signal detection in signal space with lattice properties allow us to draw the following conclusions.

1. The proposed approach to the synthesis problem of deterministic and quasi-deterministic signal detection algorithms, with rejection of the use of prior data concerning interference (noise) distribution, provides the invariance properties of algorithms and units of signal detection with respect to the conditions of parametric and nonparametric prior uncertainty confirmed by the obtained signal processing quality indices.

2. While solving the problem of deterministic signal detection in the presence of interference (noise) in signal space with lattice properties, no constraints are imposed on a lower bound of the signal-to-interference (signal-to-noise) ratio E/N_0 providing absolute values of signal detection quality indices, i.e., the conditional probabilities of correct detection D and false alarm F equal to $D = 1$, $F = 0$ on arbitrarily small ratio $E/N_0 \neq 0$.

3. The existing distinctions between signal detection quality indices describing the detection in linear signal space $\mathcal{LS}(+)$ and signal space with lattice properties $\mathcal{L}(\vee, \wedge)$ are elucidated by fundamental differences in the kinds of interactions between the signal $s(t)$ and interference (noise) $n(t)$ in these spaces: the additive $s(t) + n(t)$ and the interaction in the form of join $s(t) \vee n(t)$ and meet $s(t) \wedge n(t)$, respectively, and, therefore, by the distinctions of the properties of these signal spaces.

7.5 Classification of Deterministic Signals in Metric Space with Lattice Properties

The known variants of the statement of signal classification problem are formulated mainly for the case of additive interaction (in terms of linear space): $x(t) = s_i(t) + n(t)$ between the signal $s_i(t)$, $s_i(t) \in S$ and interference (noise) $n(t)$ [155], [159], [166], [127], [158]. Nevertheless, the statement of this problem is admissible on the basis of more general model of interaction: $x(t) = \Phi[s_i(t), n(t)]$, where Φ is some deterministic function [159], [166]. This section has a twofold goal. First, it is necessary to synthesize the algorithm and unit of classification of the deterministic signals in the presence of interference (noise) in a signal space with lattice properties. Second, it is necessary to describe characteristics and properties of a synthesized unit and to compare them with known analogues that solve the problem of signal classification in linear signal space.

The synthesis of optimal algorithm of deterministic signal classification is fulfilled on the assumption of an arbitrary distribution of interference (noise) $n(t)$. Signal processing theory usually assumes that probabilistic character of the observed stochastic process causes any algorithm of signal classification in the presence of interference (noise) to be characterized by nonzero probabilities of error decisions. However, as shown below, this is not always true.

During signal classification, it is necessary to determine which one from m useful signals in the set $S = \{s_i(t)\}$, $i = 1, \ldots, m$ is received in the input of a signal classification unit at a given instant. Consider the model of the interaction between the signal $s_i(t)$ from the set of deterministic signals $S = \{s_i(t)\}$, $i = 1, \ldots, m$ and interference (noise) $n(t)$ in signal space with the properties of distributive lattice $\mathcal{L}(\vee, \wedge)$ with binary operations of join $a(t) \vee b(t)$ and meet $a(t) \wedge b(t)$, respectively: $a(t) \vee b(t) = \sup_L(a(t), b(t))$, $a(t) \wedge b(t) = \inf_L(a(t), b(t))$; $a(t), b(t) \in \mathcal{L}(\vee, \wedge)$:

$$x(t) = s_i(t) \vee n(t), \ t \in T_s, \ i = 1, \ldots, m, \qquad (7.5.1)$$

where $T_s = [t_0, t_0 + T]$ is a domain of definition of the signal $s_i(t)$; t_0 is known time of arrival of the signal $s_i(t)$; T is a duration of the signal $s_i(t)$; $m \in \mathbf{N}$, $\mathbf{N}$ is the set of natural numbers.

Let the signals from the set $S = \{s_i(t)\}$, $i = 1, \ldots, m$ be characterized by the same energy $E_i = \int_{t \in T_s} s_i^2(t) \mathrm{d}t = E$ and cross-correlation coefficients r_{ik} equal to $r_{ik} = \int_{t \in T_s} s_i(t) s_k(t) \mathrm{d}t / E$. Assume that interference (noise) $n(t)$ is characterized by arbitrary probabilistic-statistical properties.

To synthesize classification algorithms in linear signal space, i.e., when the relation $x(t) = s_i(t) + n(t)$ holds, signal processing theory uses the part of statistical inference theory called statistical hypothesis testing. Within the signal classification problem, any strategy of decision making in the literature supposes the likelihood ratio to be computed and the likelihood function to be determined. Likelihood function is determined by the multivariate probability density function of interference

7.5 Classification of Deterministic Signals in Metric Space with Lattice Properties

(noise) $n(t)$. Thus, while solving the problem of signal classification in linear space, i.e., when the interaction equation $x(t) = s_i(t) + n(t)$ holds, $i = 1, \ldots, m$, $m \in \mathbf{N}$, to determine the likelihood function, the methodical trick is used that supposes a change of variable: $n(t) = x(t) - s_i(t)$. However, it is not possible to use the same subterfuge to determine likelihood ratio when the interaction between the signal and interference (noise) takes the form (7.5.1), inasmuch as the equation is unsolvable with respect to the variable $n(t)$ since the lattice $\mathcal{L}(\vee, \wedge)$ has no group properties; thus, another approach is necessary here.

Based on (7.5.1), the solution of the problem of classification of the signal $s_i(t)$ from the set of deterministic signals $S = \{s_i(t)\}$, $i = 1, \ldots, m$ in the presence of interference (noise) $n(t)$ lies in formation of an estimator $\hat{s}_i(t)$ of the received signal, which best allows (from the standpoint of the chosen criteria) an observer to classify these signals. In this section, the problem of classification of the signals from the set $S = \{s_i(t)\}$, $i = 1, \ldots, m$ is based on minimization of the squared metric $\int_{t \in T_s} |y_i(t) - s_i(t)|^2 \mathrm{d}t \big|_{i=k}$ between the function $y_i(t) = F_i[x(t)]$ of the observed process $x(t)$ and the signal $s_i(t)$ in the presence of the signal $s_k(t)$ in $x(t)$: $x(t) = s_k(t) \vee n(t)$:

$$\begin{cases} y_i(t) = F_i[x(t)] = \hat{s}_i(t); & (a) \\ \int_{t \in T_s} |y_k(t) - s_k(t)|^2 \mathrm{d}t \big|_{x(t)=s_k(t) \vee n(t)} \to \min_{y_k(t) \in Y}; & (b) \\ \hat{k} = \arg\max_{i \in I; s_i(t) \in S} \left[\int_{t \in T_s} y_i(t) s_i(t) \mathrm{d}t \right] \big|_{x(t)=s_k(t) \vee n(t)}; & (c) \\ i \in I,\ I = \mathbf{N} \cap [0, m],\ m \in \mathbf{N}, & (d) \end{cases} \quad (7.5.2)$$

where $y_i(t) = \hat{s}_i(t)$ is the estimator of the signal $s_i(t)$ in the presence of the signal $s_k(t)$ in the process $x(t)$: $x(t) = s_k(t) \vee n(t)$, $1 \leq k \leq m$; $F_i[*]$ is some deterministic function; $\int_{t \in T_s} |y_k(t) - s_k(t)|^2 \mathrm{d}t = \|y_k(t) - s_k(t)\|^2$ is the squared metric between the signals $y_k(t)$ and $s_k(t)$ in Hilbert space $\mathcal{HS}$; $\hat{k}$ is the decision concerning the number of processing channel, which corresponds to the received signal $s_k(t)$ from the set of deterministic signals $S = \{s_i(t)\}$, $i = 1, \ldots, m$; $1 \leq k \leq m$, $k \in I$, $I = \mathbf{N} \cap [0, m]$, $m \in \mathbf{N}$, $\mathbf{N}$ is the set of natural numbers; $\int_{t \in T_s} y_i(t) s_i(t) \mathrm{d}t = (y_i(t), s_i(t))$ is a scalar product of the signals $y_i(t)$ and $s_i(t)$ in Hilbert space $\mathcal{HS}$.

The relationship (7.5.2a) of the system (7.5.2) defines the rule of formation of the estimator $\hat{s}_i(t)$ of the received signal in the i-th processing channel in the form of some deterministic function $F_i[x(t)]$ of the process $x(t)$. The relationship (7.5.2b) determines the criterion of minimum squared metric $\int_{t \in T_s} |y_k(t) - s_k(t)|^2 \mathrm{d}t$ in Hilbert space $\mathcal{HS}$ between the signals $y_k(t)$ and $s_k(t)$ in the k-th processing channel. This criterion is considered under the condition when reception of the signal $s_k(t)$: $x(t) = s_k(t) \vee n(t)$ is realized. The relationship (7.5.2c) of the system (7.5.2) determines the criterion of maximum value of the correlation integral between the estimator $y_i(t) = \hat{s}_i(t)$ of the received signal $s_k(t)$ in the i-th processing channel and the signal $s_i(t)$. According to this criterion, the choice of a channel number $\hat{k}$

corresponds to maximum value of correlation integral $\int_{t \in T_s} y_i(t)s_i(t)\mathrm{d}t$, $i \in I$. The relationship (7.5.2d) determines a domain of definition I of processing channel i.

The solution of the problem of minimization of the squared metric (7.5.2b) between the function $y_k(t) = F_k[x(t)]$ and the signal $s_k(t)$ in its presence in the process $x(t)$: $x(t) = s_k(t) \vee n(t)$, follows directly from the absorption axiom of the lattice $\mathcal{L}(\vee, \wedge)$ (see page 269) contained in the third part of multilink identity:

$$y_k(t) = s_k(t) \wedge x(t) = s_k(t) \wedge [s_k(t) \vee n(t)] = s_k(t). \tag{7.5.3}$$

The identity (7.5.3) directly implies the type of the function $F_i[x(t)]$ from the relationship (7.5.2a) of the system (7.5.2):

$$y_i(t) = F_i[x(t)] = s_i(t) \wedge x(t) = \hat{s}_i(t). \tag{7.5.4}$$

Also, the identity (7.5.3) directly implies that a squared metric is identically equal to zero:

$$\int_{t \in T_s} |y_k(t) - s_k(t)|^2 \mathrm{d}t \big|_{x(t)=s_k(t)\vee n(t)} = 0. \tag{7.5.5}$$

The identity (7.5.4) implies that in the presence of the signal $s_k(t)$ in the process $x(t) = s_k(t) \vee n(t)$, the solution of optimization equation (7.5.2c) of the system (7.5.2) is equal to:

$$\arg\max_{i \in I;\, s_i(t) \in S} [\int_{t \in T_s} y_i(t)s_i(t)\mathrm{d}t] \big|_{x(t)=s_k(t)\vee n(t)} = \hat{k}, \tag{7.5.6}$$

and at the instant $t = t_0 + T$, correlation integral $\int_{t \in T_s} y_i(t)s_i(t)\mathrm{d}t$ takes maximum value on $i = k$ equal to energy E of the signal $s_i(t)$:

$$\int_{t \in T_s} y_i(t)s_i(t)\mathrm{d}t \big|_{i=k} = \int_{t \in T_s} s_k(t)s_k(t)\mathrm{d}t = E. \tag{7.5.7}$$

To summarize the relationships (7.5.4) through (7.5.7), one can conclude that the signal processing unit has to form the estimator $y_i(t) = \hat{s}_i(t)$ of the signal $s_i(t)$ in each of m processing channels, that is equal, according to (7.5.4), to $\hat{s}_i(t) = s_i(t) \wedge x(t)$; compute the correlation integral $\int_{t \in T_s} y_i(t)s_i(t)\mathrm{d}t$ in the interval $T_s = [t_0, t_0 + T]$ in each processing channel; and, according to Equation (7.5.2c), make the decision that in the process $x(t) = s_k(t) \vee n(t)$, there exists the signal $s_k(t)$, which corresponds to the channel where the maximum value of correlation integral $\int_{t \in T_s} y_k(t)s_k(t)\mathrm{d}t$ is formed at the instant $t = t_0 + T$ equal to the signal energy E.

The block diagram of the unit of classification of deterministic signals in a signal space with lattice properties includes the decision gate (DG) and m parallel processing channels, each containing the circuit of formation of the estimator $\hat{s}_i(t)$ of the signal $s_i(t)$; the correlation integral computation circuit $\int_{t \in T_s} \hat{s}_i(t)s_i(t)\mathrm{d}t$; and

7.5 Classification of Deterministic Signals in Metric Space with Lattice Properties

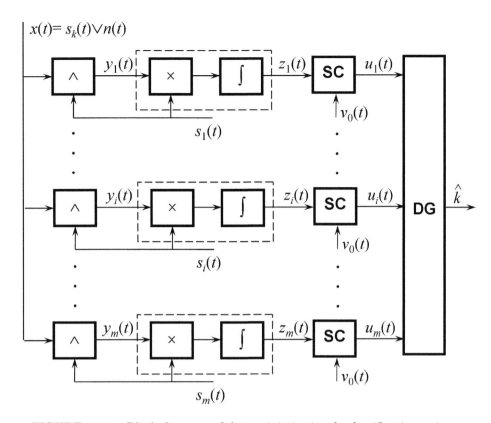

FIGURE 7.5.1 Block diagram of deterministic signals classification unit

strobing circuit (SC) (see Fig. 7.5.1). The correlation integral computation circuit consists of a multiplier and an integrator.

We can analyze the relations between the signals $s_i(t)$ and $s_k(t)$ and their estimators $\hat{s}_i(t)$ and $\hat{s}_k(t)$ in the corresponding processing channels in the presence of the signal $s_k(t)$: $x(t) = s_k(t) \vee n(t)$ in the process $x(t)$. Let the signals from the set $S = \{s_i(t)\}$, $i = 1, \ldots, m$ be characterized by the same energy $E_i = E$. For the signals $s_i(t)$ and $s_k(t)$ and their estimators $\hat{s}_i(t)$ and $\hat{s}_k(t)$ in the corresponding processing channels, on an arbitrary signal-to-interference (signal-to-noise) ratio in the process $x(t) = s_k(t) \vee n(t)$ observed in the input of classification unit, the following metric relationships hold:

$$\|s_i(t) - \hat{s}_k(t)\|^2 + \|s_k(t) - \hat{s}_k(t)\|^2 = \|s_i(t) - s_k(t)\|^2; \qquad (7.5.8a)$$

$$\|s_i(t) - \hat{s}_i(t)\|^2 + \|s_k(t) - \hat{s}_i(t)\|^2 = \|s_i(t) - s_k(t)\|^2, \qquad (7.5.8b)$$

where $\|a(t) - b(t)\|^2 = \|a(t)\|^2 + \|b(t)\|^2 - 2(a(t), b(t))$ is a squared metric between the functions $a(t)$ and $b(t)$ in Hilbert space $\mathcal{HS}$; $\|a(t)\|^2$ is a squared norm of the function $a(t)$ in Hilbert space $\mathcal{HS}$; $(a(t), b(t)) = \int\limits_{t \in T^*} a(t)b(t)dt$ is scalar product of the functions $a(t)$ and $b(t)$ in Hilbert space $\mathcal{HS}$; T^* is a domain of definition of the functions $a(t)$ and $b(t)$.

The relationships (7.5.4) and (7.5.8a) directly imply that on an arbitrary signal-to-interference (signal-to-noise) ratio in $x(t) = s_k(t) \vee n(t)$, the correlation coefficients $\rho[s_k(t), \hat{s}_k(t)]$ and $\rho[s_i(t), \hat{s}_k(t)]$ between the signals $s_i(t)$ and $s_k(t)$ and the estimator $\hat{s}_k(t)$ of the signal $s_k(t)$ in the k-th processing channel are, respectively, equal to:

$$\rho[s_k(t), \hat{s}_k(t)] = 1; \qquad (7.5.9a)$$

$$\rho[s_i(t), \hat{s}_k(t)] = r_{ik}, \qquad (7.5.9b)$$

and the squared metrics, according to (7.5.8a), are determined by the following relationships:

$$\|s_k(t) - \hat{s}_k(t)\|^2 = 0; \qquad (7.5.10a)$$

$$\|s_i(t) - \hat{s}_k(t)\|^2 = \|s_i(t) - s_k(t)\|^2 = 2E(1 - r_{ik}), \qquad (7.5.10b)$$

where r_{ik} is the cross-correlation coefficient between the signals $s_i(t)$ and $s_k(t)$; E is energy of the signals $s_i(t)$ and $s_k(t)$.

The relationship (7.5.8b) implies that on an arbitrary signal-to-interference (signal-to-noise) ratio in the process $x(t) = s_k(t) \vee n(t)$, the correlation coefficients $\rho[s_i(t), \hat{s}_i(t)]$ and $\rho[s_k(t), \hat{s}_i(t)]$ between the signals $s_i(t)$ and $s_k(t)$ and the estimator $\hat{s}_i(t)$ of the signal $s_i(t)$ in the i-th processing channel are equal to:

$$\rho[s_i(t), \hat{s}_i(t)] = 1 - \frac{1}{4}(1 - r_{ik}); \qquad (7.5.11a)$$

$$\rho[s_k(t), \hat{s}_i(t)] = 1 - \frac{3}{4}(1 - r_{ik}), \qquad (7.5.11b)$$

and squared metrics from (7.5.8a) are determined by the following relationships:

$$\|s_i(t) - \hat{s}_i(t)\|^2 = \frac{1}{2}E(1 - r_{ik}); \qquad (7.5.12a)$$

$$\|s_k(t) - \hat{s}_i(t)\|^2 = \frac{3}{2}E(1 - r_{ik}); \qquad (7.5.12b)$$

$$\|s_i(t) - s_k(t)\|^2 = 2E(1 - r_{ik}). \qquad (7.5.12c)$$

The relationships (7.5.9a) and (7.5.11a) imply that while receiving the signal $s_k(t)$ in the process $x(t) = s_k(t) \vee n(t)$ in the k-th processing channel in the output of integrator (see Fig. 7.5.1), at the instant $t = t_0 + T$, the maximum value of the correlation integral (7.5.7) formed is equal to $E \cdot \rho[s_k(t), \hat{s}_k(t)] = E$. In the i-th processing channel ($i \neq k$) at the same time, the value of the correlation integral $\int_{t \in T_s} y_i(t) s_i(t) dt$ formed is equal to $E \cdot \rho[s_i(t), \hat{s}_i(t)] < E$.

Thus, regardless of the conditions of parametric and nonparametric prior uncertainty and the probabilistic-statistical properties of interference (noise), the optimal unit of deterministic signal classification (optimal demodulator) in signal space with lattice properties realizes error-free classification of the signals from the given set $S = \{s_i(t)\}, i = 1, \ldots, m$.

The three segments of Fig. 7.5.2 illustrate the signals $z_i(t)$ and $u_i(t)$ in the

7.5 Classification of Deterministic Signals in Metric Space with Lattice Properties

outputs of the correlation integral computation circuit and the strobing circuit, the strobing pulses in the first, second, and third processing channels obtained by statistical modeling under the condition, that the signals $s_1(t)$, $s_2(t)$, $s_3(t)$, and $s_1(t)$ were received in the input of classification unit in the mixture $x(t) = s_i(t) \vee n(t)$, $i = 1, \ldots, m$ successively in time in the intervals $[0, T]$, $[T, 2T]$, $[2T, 3T]$, $[3T, 4T]$, respectively.

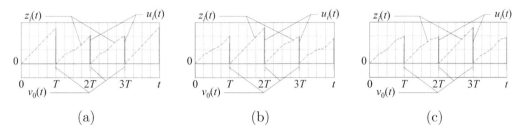

FIGURE 7.5.2 Signals in outputs of correlation integral computation circuit $z_i(t)$ and strobing circuit $u_i(t)$ within (a) first, (b) second, and (c) third processing channels

The signals $s_1(t)$, $s_2(t)$, $s_3(t)$ are orthogonal phase-shift keying with equal energies. Signal-to-interference (signal-to-noise) ratio E/N_0 is equal to the quantity $E/N_0 = 10^{-8}$, where E is an energy of the signal $s_i(t)$; N_0 is a power spectral density of interference (noise). In the case of receiving of the k-th signal $s_k(t)$ from the set of deterministic signals $S = \{s_i(t)\}$, $i = 1, \ldots, m$, in the k-th processing channel, the function $z_k(t)$ observed in the output of the correlation integral computation circuit is linear. Conversely, if in the i-th processing channel the j-th signal is received, $j \neq i$, then the function $z_i(t)$ formed in the output of correlation integral computation circuit, differs from the linear one.

As shown in Fig. 7.5.2, although the signal-to-interference (signal-to-noise) ratio is rather small, the signals $u_i(t)$ in the inputs of the decision gate (in the outputs of the strobing circuits) in each processing channel can be accurately distinguished over their amplitude. The relationships (7.5.9a) and (7.5.11a) imply that while receiving the signal $s_k(t)$ in the observed process $x(t) = s_k(t) \vee n(t)$, in the k-th processing channel in the output of the integrator (see Fig. 7.5.1), at the instant $t = t_0 + jT$, $j = 1, 2, 3, \ldots$, the maximum value of correlation integral (7.5.7) formed is equal to $E \cdot \rho[s_k(t), \hat{s}_k(t)] = E$. In the i-th processing channel ($i \neq k$), at the same time, the value of the correlation integral formed is equal to $E \cdot \rho[s_i(t), \hat{s}_i(t)] = 0.75E$.

The Shannon theorem on capacity of the communication channel with additive white Gaussian noise assumes the existence of lower bound $\inf[E_b/N_0]$ of the ratio E_b/N_0 of an energy E_b that falls at one bit of information transferred by the signal to the quantity of noise power spectral density N_0 called the ultimate Shannon limit [51], [52], [164].

This value $\inf[E_b/N_0] = \ln 2$ establishes the limit below which error-free information transmitting cannot be realized. The previous example implies that while solving the problem of deterministic signal classification in the presence of interference (noise) in signal space with lattice properties, the value $\inf[E_b/N_0]$ can be arbitrarily small as can the probability of signal receiving error while receiving a signal from the set of deterministic signals $S = \{s_i(t)\}$, $i = 1, \ldots, m$.

However, this does not mean that in such signal spaces one can achieve unbounded values of the capacity of a noisy communication channel. Sections 5.2 and 6.5 show that the capacity of communication channel, even in the absence of interference (noise), is a finite quantity. It is impossible "... to transmit all the information in the *Encyclopedia Britannica* in the absence of noise by the only signal $s_i(t)$", that, in fact, follows from Theorem 5.1.1.

The results of investigation of deterministic signal classification problem in signal space with lattice properties permit us to draw the following conclusions.

1. Formulation of the synthesis problem of a deterministic signal classification algorithm on the basis of minimization of a squared metric in Hilbert space (7.5.2b) with the simultaneous rejection of data concerning the prior probabilities of signal receiving and interference (noise) distribution provides the invariance property of the synthesized algorithm with respect to the conditions of parametric and nonparametric prior uncertainty. The obtained values of correlation coefficients (7.5.9a,b) and (7.5.11a,b) confirm this fact.

2. While solving the problem of deterministic signal classification in the presence of interference (noise) in a signal space with lattice properties, no constraints are imposed on a lower bound of the ratio E_b/N_0 of the energy E_b, that falls at one bit of transmitted information, to the quantity of power spectral density of interference (noise) N_0. This ratio can be arbitrarily small as can the probability of error while receiving a signal from the set of deterministic signals $S = \{s_i(t)\}$, $i = 1, \ldots, m$, $m \in \mathbf{N}$.

3. The essential distinctions in signal classification quality between the linear signal space and the signal space with lattice properties are elucidated by fundamental differences in the interactions of the signal $s_i(t)$ and interference (noise) $n(t)$ in these spaces: the additive $s_i(t) + n(t)$, and the interactions in the form of binary operations of join $s_i(t) \vee n(t)$ and meet $s_i(t) \wedge n(t)$, respectively.

7.6 Resolution of Radio Frequency Pulses in Metric Space with Lattice Properties

The *uncertainty function* [118] introduced by P. M. Woodward as a generalized characteristic of resolution of a signal processing unit accurately describes the properties of all the signals applied in practice. However, utilizing uncertainty function as an analytical tool for theoretical investigations of signal resolution problem in nonEuclidean spaces in general and in signal space with lattice properties in particular is impossible due to specificity of the properties of the signals and the algorithms of their processing in these signal spaces; so another approach is necessary here.

Parameter $\boldsymbol{\lambda}'$ of received signal $s'(t, \boldsymbol{\lambda}')$ is usually mismatched with respect to parameter λ of the expected useful signal $s(t, \boldsymbol{\lambda})$ matched with some filter used for signal processing. The effects of mismatching take place during signal detection, and influence signal resolution and estimation of signal parameters. Mismatching can be evaluated over the signal $w(\boldsymbol{\lambda}', \boldsymbol{\lambda})$ in the output of signal processing unit matched with the expected signal $s(t, \boldsymbol{\lambda})$. In the case of a signal $s'(t, \boldsymbol{\lambda}')$ with a mismatched value of parameter $\boldsymbol{\lambda}'$ in the input of the processing unit in the absence of interference (noise), the output response $w(\boldsymbol{\lambda}', \boldsymbol{\lambda})$ is called a mismatching function [267]. The *normalized mismatching function* $\rho(\boldsymbol{\lambda}', \boldsymbol{\lambda})$ is also introduced in [267]:

$$\rho(\boldsymbol{\lambda}', \boldsymbol{\lambda}) = \frac{w(\boldsymbol{\lambda}', \boldsymbol{\lambda})}{\sqrt{w(\boldsymbol{\lambda}', \boldsymbol{\lambda}')w(\boldsymbol{\lambda}, \boldsymbol{\lambda})}}. \qquad (7.6.1)$$

The function so determined is called a *normalized time-frequency mismatching function* of a processing unit, if the vector parameter of the expected signal includes two scalar parameters: delay time t_0' and Doppler frequency shift F_0' [267]. Vector parameter $\boldsymbol{\lambda}'$ of the received signal can be expressed by two similar scalar parameters $t_0' = t_0 + \tau$ and $F_0' = F_0 + F$, where τ and F are time delay and Doppler frequency mismatching, respectively.

This section has a twofold goal. First, it is necessary to synthesize the algorithm and unit of resolution of radio frequency (RF) pulses without intrapulse modulation in signal space with lattice properties. Second, it is necessary to determine potential resolution of this unit.

Synthesis and analysis of optimal algorithm of RF pulse resolution are fulfilled on the following assumptions. In synthesis, interference (noise) distribution is considered arbitrary, and useful signals are considered harmonic whose amplitude, time of arrival, and initial phase are unknown nonrandom. Other parameters of useful signals are considered to be known. Under the further analysis of signal processing algorithm, interference (noise) is assumed to be Gaussian.

Consider the model of interaction between two harmonic signals $s_1(t)$ and $s_2(t)$ and interference (noise) $n(t)$ in signal space $\mathcal{L}(\vee, \wedge)$ in the form of distributive lattice properties with operations of join $a(t) \vee b(t)$ and meet $a(t) \wedge b(t)$, respectively: $a(t) \vee b(t) = \sup_L(a(t), b(t))$, $a(t) \wedge b(t) = \inf_L(a(t), b(t))$; $a(t), b(t) \in \mathcal{L}(\vee, \wedge)$:

$$x(t) = s_1(t) \vee s_2(t) \vee n(t). \qquad (7.6.2)$$

Let the model of the received signals $s_1(t), s_2(t)$ be determined by the expression:

$$s_i(t) = \begin{cases} A_i \cos(\omega_0 t + \varphi), & t \in T_i; \\ 0, & t \notin T_i, \end{cases} \qquad (7.6.3)$$

where A_i is an unknown nonrandom amplitude of the useful signal $s_i(t)$; $\omega_0 = 2\pi f_0$; f_0 is known carrier frequency of the signal $s_i(t)$; φ is an unknown nonrandom initial phase of the signal $s_i(t)$, $\varphi \in [-\pi, \pi]$; T_i is a domain of definition of the signal $s_i(t)$, $T_i = [t_{0i}, t_{0i} + T]$, $i = 1, 2$; t_{0i} is an unknown time of arrival of the signal $s_i(t)$; T

is known signal duration, $T = NT_0$; N is a number of periods of harmonic signal $s_i(t)$; $N \in \mathbf{N}$, $\mathbf{N}$ is the set of natural numbers; T_0 is a period of a signal carrier.

We suppose interference (noise) $n(t)$ is characterized by such statistical properties that two neighbor independent interference (noise) samples $n(t_j)$ and $n(t_{j\pm 1})$ are distant over the interval $\Delta\tau = 1/f_0 = T_0$. The instantaneous values (samples) of the signal $\{s(t_j)\}$ and interference (noise) $\{n(t_j)\}$ are the elements of the signal space: $s(t_j), n(t_j) \in \mathcal{L}(\vee, \wedge)$.

The observation equation, according to the interaction Equation (7.6.2), takes the form:
$$x(t_j) = s_1(t_j) \vee s_2(t_j) \vee n(t_j), \quad (7.6.4)$$

where $t_j = t - jT_0$, $j = 0, 1, \ldots, J-1$, $t_j \in T^*$; T^* is a processing interval: $T^* = [\min_{i=1,2}\{t_{0i}\}, \max_{i=1,2}\{t_{0i} + T\}]$; $T^* \subset T_1 \cup T_2$; $J \in \mathbf{N}$, $\mathbf{N}$ is the set of natural numbers.

The resolution problem lies in forming in the output of a processing unit such estimators $\hat{s}_1(t)$, $\hat{s}_2(t)$ of the signals $s_1(t)$, $s_2(t)$ based on processing of the observed signal $x(t)$ (7.6.2) that allow (according to the chosen criteria) distinguishing these signals by time parameter. The resolution problem $Res[s_{1,2}(t)]$ for two harmonic signals $s_1(t)$ and $s_2(t)$ in signal space $\mathcal{L}(\vee, \wedge)$ with lattice properties is formulated and solved on the basis of step-by-step processing of statistical collection $\{x(t_j)\}$ determined by the observation equation (7.6.4):

$$Res[s_{1,2}(t)] = \begin{cases} PF[s_{1,2}(t)]; & (a) \\ IP[s_{1,2}(t)]; & (b) \\ Sm[s_{1,2}(t)], & (c) \end{cases} \quad (7.6.5)$$

where $PF[s_{1,2}(t)]$ is primary filtering; $IP[s_{1,2}(t)]$ is intermediate processing; $Sm[s_{1,2}(t)]$ is smoothing; together they form successive processing stages of the general algorithm of resolution $Res[s_{1,2}(t)]$ of the signals $s_1(t)$ and $s_2(t)$ (7.6.5).

The optimality criteria determining every signal processing stage $PF[s_{1,2}(t)]$, $IP[s_{1,2}(t)]$, $Sm[s_{1,2}(t)]$ are interrelated and involved in single systems:

$$PF[s_{1,2}(t)] = \begin{cases} y(t) = \arg\min_{y^\circ(t) \in Y; t, t_j \in T^*} |\bigwedge_{j=0}^{J-1}[x(t_j) - y^\circ(t)]|; & (a) \\ w(t) = F[y(t)]; & (b) \\ J = \arg\min_{y(t) \in Y}[\int_{t \in T^*} |y(t)|dt\big|_{x(t) = s_i(t) \vee 0}], \int_{t \in T^*} |y(t)|dt \neq 0; & (c) \\ \int_{t \in T^*} |w_{12}(t) - [w_1(t) \vee w_2(t)]|dt\big|_{|t_{01} - t_{02}| \in \Delta_{t_{0i}}} \to \min_{w(t) \in W}; & (d) \\ w_{12}(t) = w(t)\big|_{x(t) = s_1(t) \vee s_2(t)}, \ w_{1,2}(t) = w(t)\big|_{x(t) = s_{1,2}(t) \vee 0}, \end{cases} \quad (7.6.6)$$

where $y(t)$ is the solution of the problem of minimization of metric $|\bigwedge_{j=0}^{J-1}[x(t_j) - y^\circ(t)]|$ between the observed statistical collection $\{x(t_j)\}$ and optimization variable (function) $y^\circ(t)$; $w(t)$ is some deterministic function $F[*]$ of the result $y(t)$ of minimization of the function of the observed collection $\{x(t_j)\}$ (4); t_{0i} is an unknown

7.6 Resolution of Radio Frequency Pulses in Metric Space with Lattice Properties

time of arrival of the signal $s_i(t)$, $i = 1, 2$; $\Delta_{t_{0i}} = [0.75T_0, 1.25T_0]$; T^* is a processing interval; J is a number of the samples of stochastic process $x(t)$ used during signal processing, $J \in \mathbf{N}$, $\mathbf{N}$ is the set of natural numbers;

$$IP[s_{1,2}(t)] = \begin{cases} \mathbf{M}\{u^2(t)\}\big|_{s_{1,2}(t)\equiv 0} \to \min_{\mathcal{L}}; & (a) \\ \mathbf{M}\{[u(t) - s_i(t)]^2\}\big|_{x(t)=s_i(t)\vee 0} = \varepsilon; & (b) \\ u(t) = L[w(t)], & (c) \end{cases} \quad (7.6.7)$$

where $\mathbf{M}\{*\}$ is the symbol of mathematical expectation; $L[w(t)]$ is the functional transformation of the process $w(t)$ into the process $u(t)$; ε is a constant defined by some function of a power of the signal $s(t)$;

$$Sm[s_{1,2}(t)] =$$
$$= \begin{cases} v(t) = \arg\min_{v^\circ(t)\in V;\, t, t_k \in \tilde{T}} \sum_{k=0}^{M-1} |u(t_k) - v^\circ(t)|; & (a) \\ \Delta\tilde{T} : \delta_d(\Delta\tilde{T}) = \delta_{d,\text{sm}}; & (b) \\ M = \arg\max_{M' \in \mathbf{N}\cap[M^*,\infty[} [\delta_f(M')]\big|_{M^*:\delta_f(M^*)=\delta_{f,\text{sm}}}, & (c) \end{cases} \quad (7.6.8)$$

where $v(t)$ is smoothing function of the process $u(t)$ that is the solution of the problem of minimizing the metric $\sum_{k=0}^{M-1} |u(t_k) - v^\circ(t)|$ between the instantaneous values of stochastic process $u(t)$ and optimization variable $v^\circ(t)$; $t_k = t - \frac{k}{M}\Delta\tilde{T}$, $k = 0, 1, \ldots, M-1$, $t_k \in \tilde{T} =]t - \Delta\tilde{T}, t]$; $\tilde{T}$ is an interval, in which smoothing of stochastic process $u(t)$ is realized; $M \in \mathbf{N}$, $\mathbf{N}$ is the set of natural numbers; M is a number of samples of stochastic process $u(t)$ used during smoothing in the interval $\tilde{T}$; $\delta_d(\Delta\tilde{T})$ and $\delta_f(M)$ are relative dynamic and fluctuation errors of smoothing as the dependences on the quantity $\Delta\tilde{T}$ of the smoothing interval $\tilde{T}$ and the number of the samples M, respectively; $\delta_{d,\text{sm}}$ and $\delta_{f,\text{sm}}$ are the quantities of dynamic and fluctuation errors of smoothing, respectively.

We now explain the optimality criteria and the single relationships included into the systems (7.6.6), (7.6.7), (7.6.8) determining the successive processing stages $PF[s_{1,2}(t)]$, $IP[s_{1,2}(t)]$, $Sm[s_{1,2}(t)]$ of the general algorithm of resolution $Res[s_{1,2}(t)]$ of the signals $s_1(t)$ and $s_2(t)$ (7.6.5).

Equation (7.6.6a) of the system (7.6.6) determines the criterion of minimum of metric between the statistical set of the observation $\{x(t_j)\}$ and the result of primary processing $y(t)$. The choice of function of metric $|\bigwedge_{j=0}^{J-1}[x(t_j) - y^\circ(t)]|$ should take into account the metric convergence and the convergence in probability to the estimated parameter of the sequence for the interaction in the form (7.6.2a) (see Section 7.2). Equation (7.6.6b) establishes the interrelation between stochastic processes $y(t)$ and $w(t)$. The relationship (7.6.6c) determines the criterion of the choice of number of periods J of the signals $s_1(t)$, $s_2(t)$, used while processing on the basis of minimization of the norm $\int_{t\in T^*} |y(t)|dt$. The criterion (7.6.6c) is considered under three constraint conditions: (1) interference (noise) is identically equal to zero: $n(t) \equiv 0$; (2) the second signal $s_k(t)$, $k \neq i$, $k = 1, 2$ is absent; (3) the norm $\int_{t \in T^*} |y(t)|dt$ of the function $y(t)$ is not equal to zero: $\int_{t \in T^*} |y(t)|dt \neq 0$.

The relationship (7.6.6d) defines the criterion of minimum of norm of the difference $w_{12}(t) - [w_1(t) \vee w_2(t)]$ of the processing unit responses $w_{12}(t)$ and $w_{1,2}(t)$ under joint interaction between the signals $s_1(t)$ and $s_2(t)$ ($x(t) = s_1(t) \vee s_2(t)$), and also under the effect of only one of the signals $s_i(t)$, $i = 1, 2$ ($x(t) = s_{1,2}(t) \vee 0$), respectively. The criterion (7.6.6d) is considered under the constraint condition that the module of the difference $|t_{01} - t_{02}|$ of time of arrival t_{0i} of the signals $s_i(t)$, $i = 1, 2$ takes the values within the interval $0.75T_0 \leq |t_{01} - t_{02}| \leq 1.25T_0$.

Equations (7.6.7a), (7.6.7b), (7.6.7c) define the criterion of the choice of the functional transformation $L[w(t)]$. Equation (7.6.7a) determines the criterion of minimum of the second moment of the process $u(t)$ in the absence of the signals $s_i(t)$, $i = 1, 2$ in the input of processing unit. Equation (7.6.7b) establishes the quantity of the second moment of the difference between the signals $u(t)$, $s_{1,2}(t)$ under two constraint conditions: (1) interference (noise) is identically equal to zero: $n(t) \equiv 0$; (2) the second signal $s_k(t)$, $k \neq i$, $k = 1, 2$ is absent.

Equation (7.6.8a) of the system (7.6.8) determines the criterion of minimum of metric between the process $u(t)$ and the result of its smoothing $v(t)$ in the interval $\tilde{T}$. The criterion (7.6.8a) is considered under the constraint condition that both useful signals are identically equal to zero: $s_i(t) \equiv 0$, $i = 1, 2$, whereas interference (noise) $n(t)$ differs from zero: $x(t) = n(t) \vee 0$. The relationship (7.6.8b) determines the rule of choosing the quantity $\Delta \tilde{T}$ of the smoothing interval $\tilde{T}$ based on a given quantity of relative dynamic error $\delta_{d,\text{sm}}$ of smoothing. Equation (7.6.8c) determines the number of samples M of stochastic process $u(t)$ based on the relative fluctuation error $\delta_{f,\text{sm}}$ of its smoothing.

To solve the problem of resolution algorithm synthesis according to the chosen criteria, we obtain an expression for the process $v(t)$ in the output of the processing unit by successive solving the equation system (7.6.5).

To solve the problem of minimization of the function $|\bigwedge_{j=0}^{J-1}[x(t_j) - y^\circ(t)]|$ (7.6.6a) of the system (7.6.6), we find its extremum, setting the derivative with respect to $y^\circ(t)$ to zero:

$$d|\bigwedge_{j=0}^{J-1}(x(t_j) - y^\circ(t))|/dy^\circ(t) = -\text{sign}[\bigwedge_{j=0}^{J-1}(x(t_j) - y^\circ(t))] = 0. \quad (7.6.9)$$

The solution of Equation (7.6.9) is the value of the estimator $y(t)$ in the form of the meet of the observation results $\{x(t_j)\}$:

$$y(t) = \bigwedge_{j=0}^{J-1} x(t_j) = \bigwedge_{j=0}^{J-1} x(t - jT_0). \quad (7.6.10)$$

The derivative of the function $|\bigwedge_{j=0}^{J-1}[x(t_j) - y^\circ(t)]|$, according to the relationship (7.6.9), at point $y(t)$ changes its sign from minus to plus. Thus, the extremum determined by the formula (7.6.10) is the minimum point of this function and the solution of Equation (7.6.6a) that determines this criterion of estimation.

7.6 Resolution of Radio Frequency Pulses in Metric Space with Lattice Properties

The condition of the criterion (7.6.6c) of the system (7.6.6) $x(t) = s_i(t) \vee 0$, $i = 1, 2$ determines the observation Equation (7.6.2) of the following form: $x(t_j) = s_i(t_j) \vee 0$, $j = 0, 1, \ldots, J-1$; therefore, according to the relationship (7.6.10), the identities hold:

$$y(t)\big|_{x(t)=s_i(t)\vee 0} = [s_i(t) \vee 0] \wedge [s_i(t - (J-1)T_0) \vee 0]. \tag{7.6.11}$$

On the basis of the identity (7.6.11), we obtain the value of the norm $\int_{t \in T^*} |y(t)| dt$ from the criterion (7.6.6c):

$$\int_{t \in T^*} |y(t)| dt = \begin{cases} 4(N - J + 1)A_i/\pi, & J \le N; \\ 0, & J > N, \end{cases} \tag{7.6.12}$$

where N is a number of periods of harmonic signal $s_i(t)$; A_i is an unknown nonrandom amplitude of the signal $s_i(t)$.

From Equation (7.6.12), according to the criterion (7.6.6c), we obtain optimal value J of a number of signal periods used on primary processing (7.6.10) equal to N:

$$J = N. \tag{7.6.13}$$

Under joint effects of the signals $s_1(t)$ and $s_2(t)$ ($x(t) = s_1(t) \vee s_2(t)$), the response $w_{12}(t)$ of the processing unit is the function of the expression (7.6.10) on $J = N$:

$$y(t)\big|_{x(t)=s_1(t)\vee s_2(t)} = y_+(t)\big|_{x(t)=s_1(t)\vee s_2(t)} + y_-(t)\big|_{x(t)=s_1(t)\vee s_2(t)}; \tag{7.6.14}$$

$$y_+(t)\big|_{x(t)=s_1(t)\vee s_2(t)} = [s_1(t) \vee s_2(t) \vee 0] \wedge \\ \wedge [s_1(t-(N-1)T_0) \vee s_2(t-(N-1)T_0) \vee 0]; \tag{7.6.14a}$$

$$y_-(t)\big|_{x(t)=s_1(t)\vee s_2(t)} = [s_1(t) \wedge s_1(t-(N-1)T_0) \wedge 0] \vee \\ \vee [s_2(t) \wedge s_2(t-(N-1)T_0) \wedge 0], \tag{7.6.14b}$$

where $y_+(t)\big|_{x(t)=s_1(t)\vee s_2(t)}$ and $y_-(t)\big|_{x(t)=s_1(t)\vee s_2(t)}$ are the positive and negative parts of the function $y(t)\big|_{x(t)=s_1(t)\vee s_2(t)}$, respectively:

$$y_+(t)\big|_{x(t)=s_1(t)\vee s_2(t)} = y(t)\big|_{x(t)=s_1(t)\vee s_2(t)} \vee 0; \tag{7.6.15a}$$

$$y_-(t)\big|_{x(t)=s_1(t)\vee s_2(t)} = y(t)\big|_{x(t)=s_1(t)\vee s_2(t)} \wedge 0. \tag{7.6.15b}$$

Under the effect of only one of the signals $s_1(t)$ or $s_2(t)$ ($x(t) = s_{1,2}(t) \vee 0$), the response $w_{1,2}(t)$ of the processing unit is the function (7.6.6b) of the expression (7.6.10) on $J = N$:

$$y(t)\big|_{x(t)=s_{1,2}(t)\vee 0} = y_+(t)\big|_{x(t)=s_{1,2}(t)\vee 0} + y_-(t)\big|_{x(t)=s_{1,2}(t)\vee 0}; \tag{7.6.16}$$

$$y_+(t)\big|_{x(t)=s_{1,2}(t)\vee 0} = [s_{1,2}(t) \vee 0] \wedge [s_{1,2}(t-(N-1)T_0) \vee 0]; \tag{7.6.16a}$$

$$y_-(t)\big|_{x(t)=s_{1,2}(t)\vee 0} = 0, \tag{7.6.16b}$$

where $y_+(t)\big|_{x(t)=s_{1,2}(t)\vee 0}$ and $y_-(t)\big|_{x(t)=s_{1,2}(t)\vee 0}$ are the positive and negative parts of the function $y(t)\big|_{x(t)=s_{1,2}(t)\vee 0}$, respectively.

The identity (7.6.16b) is stipulated by the absence of the negative component $x_-(t)$ of the input effect $x(t) = s_{1,2}(t) \vee 0$: $x_-(t) = x(t) \wedge 0 = 0$.

If the criterion (7.6.6d) $|t_{01} - t_{02}| \in \Delta_{t_{0i}} = [0.75T_0, 1.25T_0]$ holds, then the expression (7.6.14a) can be represented in the following form:

$$y_+(t)\big|_{x(t)=s_1(t)\vee s_2(t)} = y_+(t)\big|_{x(t)=s_1(t)\vee 0} + y_+(t)\big|_{x(t)=s_2(t)\vee 0}, \tag{7.6.17}$$

where $y_+(t)\big|_{x(t)=s_{1,2}(t)\vee 0}$ is the function determined by the relationship (7.6.16a).

Analyzing the relationships (7.6.14), (7.6.16), (7.6.17), it is easy to conclude that the norm $\int_{t\in T^*} |w_{12}(t) - [w_1(t) \vee w_2(t)]|dt$ of the difference of the functions $w_{12}(t)$ and $w_1(t) \vee w_2(t)$ is minimal and equal to zero if and only if between the functions $w(t)$ and $y(t)$, the following identity holds:

$$w(t)\big|_{x(t)=s_1(t)\vee s_2(t)} = y(t)\big|_{x(t)=s_1(t)\vee s_2(t)} \vee 0. \tag{7.6.18}$$

The summand $y_-(t)\big|_{x(t)=s_1(t)\vee s_2(t)}$ determined by the expression (7.6.14b) is excluded from further processing, and the following equalities hold:

$$w_{12}(t) = w_1(t) \vee w_2(t); \tag{7.6.19a}$$

$$w(t)\big|_{x(t)=s_1(t)\vee s_2(t)} = w(t)\big|_{x(t)=s_1(t)\vee 0} + w(t)\big|_{x(t)=s_2(t)\vee 0}. \tag{7.6.19b}$$

Thus, in the result of minimization of the norm of the difference of the functions $w_{12}(t)$ and $w_1(t) \vee w_2(t)$ — $\int_{t\in T^*} |w_{12}(t) - [w_1(t) \vee w_2(t)]|dt$, according to the criterion (7.6.6d), the coupling equation (7.6.6b) between the functions $w(t)$ and $y(t)$ is determined by the relationship (7.6.18):

$$F[y(t)] = y(t) \vee 0.$$

It is obvious that the coupling equation (7.6.6b) has to be invariant with respect to the presence (absence) of interference (noise) $n(t)$, so the final variant of the coupling equation can be written on the basis of the Equation (7.6.18) in the form:

$$w(t) = F[y(t)] = y(t) \vee 0. \tag{7.6.20}$$

Hence, the identity (7.6.20) determines the form of the coupling equation (7.6.6b) obtained on the basis of the criterion (7.6.6d). According to the relationship (7.6.20), the noninformative component of the process $y(t)$, determined by its negative part $y_-(t) = y(t) \wedge 0$, must be excluded from signal processing, while the positive part $y_+(t) = y(t) \vee 0$ of the process $y(t)$ takes part in the further processing, and $y_+(t)$ is the informational component of $y(t)$. From the energetic standpoint, informational $y_+(t)$ and noninformational $y_-(t)$ components contain $1/N$ and $(N-1)/N$ parts of the norm $\int_{t\in T^*} |y(t)|dt\big|_{x(t)=s_{1,2}(t)}$, respectively, in the presence of the only signal $s_1(t)$ or $s_2(t)$ in the input of the processing unit: $x(t) = s_{1,2}(t)$.

In the absence of interference (noise) $n(t) = 0$, between the signals in the input and output of processing unit, the following relationships hold:

$$x_{12}(t) = x_1(t) \vee x_2(t); \tag{7.6.21a}$$

$$w_{12}(t) = w_1(t) \vee w_2(t). \tag{7.6.21b}$$

where $x_{12}(t) = s_1(t) \vee s_2(t) \vee 0$; $x_{1,2}(t) = s_{1,2}(t) \vee 0$; $w_{12}(t) = w(t)\big|_{x(t)=s_1(t)\vee s_2(t)}$; $w_{1,2}(t) = w(t)\big|_{x(t)=s_{1,2}(t)\vee 0}$

The relationships (7.6.21a,b) determine homomorphism between the signals in the input and output of the processing unit. If the time difference $|t_{01} - t_{02}|$ between arrivals of the signals $s_1(t)$ and $s_2(t)$ exceeds the quantity $1.5T_0$: $|t_{01} - t_{02}| \geq 1.5T_0$, then the relationship (7.6.21b) does not hold. This fact is stipulated by nonlinear interaction of the responses of the signals $s_1(t)$ and $s_2(t)$ under their simultaneous effect in the input of the processing unit.

The solution $u(t)$ of the relationships (7.6.7a), (7.6.7b), (7.6.7c) of the system (7.6.7), which establish the criterion of the choice of functional transformation of the process $w(t)$ is the function $L[w(t)]$ determining the gain characteristic of the limiter:

$$u(t) = L[w(t)] = [(w(t) \wedge a) \vee 0] + [(w(t) \vee (-a)) \wedge 0], \tag{7.6.22}$$

where a is the parameter of the limiter chosen to equal $a = \sup_A \Delta_A = A_{\max}$, $\Delta_A =]0, A_{\max}]$; $A_{\max}$ is maximal possible value of the signal amplitude.

To obtain the expression for the process $v(t)$ in the output of the processing unit, we solve the function minimization equation on the basis of the criterion (7.6.8a) of the system (7.6.8). To solve this problem, we find the extremum of the function $\int_{t \in \tilde{T} \subset T^*} |u(t) - v^\circ(t)| dt \big|_{x(t)=n(t)\vee 0}$, setting its derivative with respect to $v^\circ(t)$ to zero:

$$d\{\sum_{k=0}^{M-1} |u(t_k) - v^\circ(t)|\big|_{x(t)=n(t)\vee 0}\}/dv^\circ(t) =$$

$$= -\sum_{k=0}^{M-1} \text{sign}[u(t_k) - v^\circ(t)]\big|_{x(t)=n(t)\vee 0} = 0.$$

The solution of the last equation is the value of the estimator $v(t)$ in the form of the sample median $\text{med}\{*\}$ of the sample collection $\{u(t_k)\}$ of the stochastic process $u(t)$ in the interval $\tilde{T} =]t - \Delta\tilde{T}, t]$:

$$v(t) = \underset{t_k \in \tilde{T}}{\text{med}}\{u(t_k)\}, \ t_k \in \tilde{T}, \tag{7.6.23}$$

where $t_k = t - \frac{k}{M}\Delta\tilde{T}$, $k = 0, 1, \ldots, M-1$, and the quantities $\Delta\tilde{T}$ and M are chosen according to the criteria (7.6.8b) and (7.6.8c) of the system (7.6.8), respectively.

The derivative of the function $\sum_{k=0}^{M-1} |u(t_k) - v^\circ(t)|\big|_{x(t)=n(t)\vee 0}$ at the point $v(t)$

changes its sign from minus to plus. Hence, the extremum determined by the formula (7.6.23) is minimum of this function and the solution of the equation (7.6.8a) determining this estimation criterion.

Thus, summarizing the relationships (7.6.10), (7.6.13), (7.6.20), (7.6.22), (7.6.23), we conclude that the estimator $v(t)$ of the signals $s_1(t)$ and $s_2(t)$ received in the presence of interference (noise) $n(t)$ is the function of smoothing of a stochastic process $u(t)$ obtained by limitation of the process $w(t)$ that is the positive part $y_+(t)$ of the process $y(t)$: $w(t) = y_+(t) = y(t) \vee 0$, which is the result of primary processing of the observed statistical collection $\{x(t - jT_0)\}$, $j = 0, 1, \ldots, N - 1$: $y(t) = \bigwedge_{j=0}^{N-1} x(t - jT_0)$:

$$v(t) = \operatorname{med}\{u(t - \frac{k}{M}\Delta \tilde{T})\}; \tag{7.6.24a}$$

$$u(t) = L[w(t)]; \tag{7.6.24b}$$

$$w(t) = [\bigwedge_{j=0}^{N-1} x(t - jT_0)] \vee 0, \tag{7.6.24c}$$

where $k = 0, 1, \ldots, M - 1$; $\Delta \tilde{T}$ is a quantity of the smoothing interval $\tilde{T}$; $\tilde{T} =]t - \Delta \tilde{T}, t]$.

The block diagram of the signal resolution unit, according to the processing algorithm (7.6.24), involves transversal filter realizing primary over-period processing in the form $\bigwedge_{j=0}^{N-1} x(t - jT_0)$; the unit of formation of the positive part $w(t) = y_+(t)$ of the process $y(t)$ (7.6.20) that excludes its negative instantaneous values from further processing; the limiter $L[w(t)]$; and also median filter (MF) (see Fig. 7.6.1).

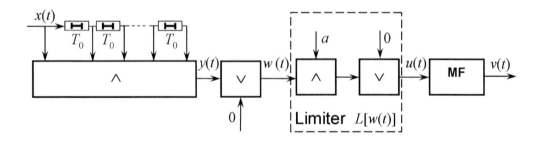

FIGURE 7.6.1 Block diagram of signal resolution unit

We first shall analyze the resolution ability of the obtained processing unit on the basis of normalized mismatching function (7.6.1), not taking into account the influence of interference (noise), then we shall do it in the presence of interference (noise) within the model (7.6.2). This analysis will be carried out by the signal $w(t)$ (7.6.24c) in the input of the limiter (see Fig. 7.6.1), that avoids inessential details while preserving the physical sense of the features of this processing.

In the absence of interference (noise) $n(t) = 0$ in the input of the resolution unit

7.6 Resolution of Radio Frequency Pulses in Metric Space with Lattice Properties

(see Fig. 7.6.1), under the signal model (7.6.3), interaction Equation (7.6.2) takes the form:

$$x(t) = s(t) \vee 0; \tag{7.6.25a}$$

$$s(t) = \begin{cases} A\cos(\omega_0 t + \varphi), & t \in T_s; \\ 0, & t \notin T_s, \end{cases} \tag{7.6.25b}$$

where all the variables have the same sense contained in the signal model (7.6.3): A is an unknown nonrandom amplitude of the useful signal $s(t)$; $\omega_0 = 2\pi f_0$; f_0 is the known carrier frequency of the signal $s(t)$; φ is an unknown nonrandom initial phase of the signal $s(t)$, $\varphi \in [-\pi, \pi]$; T_s is a domain of definition of the signal $s(t)$, $T_s = [t_0, t_0+T]$; t_0 is an unknown time of arrival of the signal $s(t)$; T is known signal duration, $T = NT_0$; N is a number of periods of harmonic signal $s(t)$; $N \in \mathbf{N}$, $\mathbf{N}$ is the set of natural numbers; T_0 is a period of carrier.

Information concerning the values of unknown nonrandom amplitude A and initial phase φ of the signal $s(t)$ is contained in its estimator $\hat{s}(t) = w(t)$ in the interval $T_{\hat{s}}$, $t \in T_{\hat{s}}$:

$$T_{\hat{s}} = [t_0 + (N-1)T_0, t_0 + NT_0], \tag{7.6.26}$$

where t_0 is an unknown time of arrival of the signal $s(t)$; N is a number of periods of harmonic signal $s(t)$; $N \in \mathbf{N}$, $\mathbf{N}$ is the set of natural numbers; T_0 is a period of carrier.

We note that this filter cannot be described adequately by the pulse-response characteristic used for linear filter definition. We can easily make sure that the filter response on δ-function is identically equal to zero. The response of the filter that realizes processing of the harmonic signal $s(t)$ (7.6.25b) in the absence of interference (noise) is the estimator $\hat{s}(t)$, which, according to the expression (7.6.24c), takes the values:

$$\hat{s}(t) = \begin{cases} s(t) = A\cos[\omega_0(t - t_0) + \varphi], & s(t) \geq 0, \; t \in T_{\hat{s}}; \\ 0, & s(t) < 0, \; t \in T_{\hat{s}} \text{ or } t \notin T_{\hat{s}}. \end{cases} \tag{7.6.27}$$

Due to this property, the filter of the signal space with lattice properties fundamentally differs from the filter of the linear signal space, matched with the same signal $s(t)$, whose response is determined by the autocorrelation function of the signal. The relationship (7.6.27) shows that the filter, realizing primary processing (7.6.24c), provides the compression of the useful signal in $N = Tf_0$ times, and as any nonlinear device, expands the spectrum of the processed signal. Using the known analogy, the result (7.6.27) can be interpreted as the potential possibilities of the filter in signal resolution in a time domain under an extremely large signal-to-noise ratio $E/N_0 \to \infty$ in the input. The expression (7.6.27) implies that filter resolution Δ_τ in time parameter is about a quarter of a carrier period T_0: $\Delta_\tau \sim 1/4f_0 = T_0/4$, where f_0 is a carrier frequency of the signal.

Figure 7.6.2 illustrates the signal $w(t)$ in the output of the unit of formation of the positive part (see Fig. 7.6.1) during the interaction of two harmonic signals $s_1(t)$ and $s_2(t)$ in the input of a synthesized unit in the absence of interference (noise) $n(t) = 0$. In the figure, 1 is the signal $s_1(t)$; 2 is the signal $s_2(t)$; 3 is the response

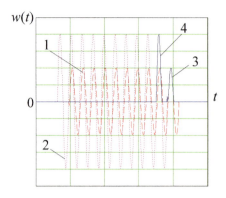

FIGURE 7.6.2 Signal $w(t)$ in input of limiter in absence of interference (noise). 1 and 2: signals $s_1(t)$ and $s_2(t)$, respectively; 3 and 4: responses $\hat{s}_1(t)$ and $\hat{s}_2(t)$ of signals $s_1(t)$ and $s_2(t)$, respectively

$\hat{s}_1(t)$ of the signal $s_1(t)$; 4 is the response $\hat{s}_2(t)$ of the signal $s_2(t)$. The responses 3 and 4 of the signals $s_1(t)$ and $s_2(t)$ are shown by the solid line.

We can determine normalized time-frequency mismatching function (7.6.1) of the filter that realizes the primary processing algorithm (7.6.24c) of harmonic signal $s(t)$ within the model (7.6.25b) in signal space with lattice properties in the absence of interference (noise), assuming for simplicity that $\varphi = -\pi/2$:

$$s(t) = A\cos[\omega_0(t - t_0) - \pi/2], \ t \in T_s = [t_0, t_0 + T]. \tag{7.6.28}$$

The received signal $s'(t)$ is transformed as a result of the Doppler effect, and its time-frequency characteristics differ from those of the initial signal $s(t)$:

$$s'(t) = A'\cos[\omega'_0(t - t'_0) - \pi/2], \ t \in T'_s = [t'_0, t'_0 + T'], \tag{7.6.28a}$$

where A' is a changed amplitude of the received signal $s'(t)$; $\omega'_0 = 2\pi f'_0$ is a changed cyclic frequency of a carrier; $f'_0 = f_0(1 + \delta F)$ is a changed carrier frequency of the received signal $s'(t)$; $\delta F = F/f_0$ is a relative quantity of Doppler frequency shift; F is an absolute quantity of Doppler frequency shift; t'_0 is time of arrival of the received signal; $T' = T/(1 + \delta F) = N \cdot T'_0$ is a changed duration of the signal; N is a number of periods of the received signal $s'(t)$; $T'_0 = T_0/(1 + \delta F)$ is a changed period of a carrier.

The response $w(t)$ of the signal $s'(t)$ in the output of the filter in the absence of interference (noise) is described by the function:

$$w(t) = \begin{cases} w_\uparrow(t) = & A'\cdot\sin[2\pi f_0(1 + \delta F)(t - t'_1)], \quad t'_1 \leq t < t'_m; \\ w_\downarrow(t) = & -A'\cdot\sin[2\pi f_0(1 + \delta F)(t - t'_2)], \quad t'_m \leq t < t'_2, \end{cases} \tag{7.6.29}$$

where t'_1 and t'_2 are times of beginning and ending of the response $w(t)$; $t'_m = (t'_1 + t'_2)/2$ is time corresponding to maximum value of the response $w(t)$; A' is an amplitude of the transformed signal $s'(t)$.

7.6 Resolution of Radio Frequency Pulses in Metric Space with Lattice Properties

The first part $w_\uparrow(t)$ of the function $w(t)$ characterizes the leading edge of the pulse in the output of the filter, and the second part $w_\downarrow(t)$ is its trailing edge. The leading edge $w_\uparrow(t)$ corresponds to the first quarter of the wave of the first period of the signal $s'(t+(N-1)T'_0)$ delayed on $N-1$ periods T'_0 of a carrier oscillation. The trailing edge $w_\downarrow(t)$ corresponds to the second quarter of wave of the last period of the received signal $s'(t)$.

In the case of the positive $\delta F > 0$ and negative $\delta F < 0$ relative Doppler shifts, the values t'_1, t'_m, and t'_2 are determined by the following relationships:

$$\begin{cases} t'_1 = t_m + (N-1)\Delta T_0 \cdot 1(-\delta F) - 0.25 T'_0; \\ t'_2 = t_m + (N-1)\Delta T_0 \cdot 1(\delta F) + 0.5 \Delta T_0 + 0.25 T'_0; \\ t'_m = (t'_1 + t'_2)/2 = t_m + 0.5[(N-1) + 0.5]\Delta T_0, \end{cases} \quad (7.6.30)$$

where t_m is time corresponding to maximum value of the response $w(t)$ in the output of the filter on zero Doppler shift $\delta F = 0$; $T'_0 = T_0/(1+\delta F)$ is a period of a changed carrier of transformed signal $s'(t)$; $\Delta T_0 = T'_0 - T_0 = T_0(-\delta F)/(1+\delta F)$ is the difference of the periods of a carrier of the transformed $s'(t)$ and the initial $s(t)$ signals; $1(x)$ is Heaviside step function.

Normalizing the function, according to the definition (7.6.1), and realizing transformation of the variable t in the formula (7.6.29) with respect to location t_m and scale T'_0 parameters:

$$\delta \tau = (t - t_m)/T'_0, \quad (7.6.31)$$

we obtain the expression for normalized time-frequency mismatching function $\rho(\delta \tau, \delta F)$ of the filter:

$$\rho(\delta \tau, \delta F) = \begin{cases} \sin[2\pi(1+\delta F)(\delta \tau - \delta \tau_1)], & \delta \tau_1 \leq \delta \tau < \delta \tau_m; \\ -\sin[2\pi(1+\delta F)(\delta \tau - \delta \tau_2)], & \delta \tau_m \leq \delta \tau < \delta \tau_2, \end{cases} \quad (7.6.32)$$

where $\delta \tau$ and δF are relative time delay and frequency shift, respectively; $\delta \tau_1$ and $\delta \tau_2$ are relative time of beginning and time of ending of mismatching function $\rho(\delta \tau, \delta F)$ on F=const, respectively; $\delta \tau_m = (\delta \tau_1 + \delta \tau_2)/2$ is relative time corresponding to maximum value of mismatching function $\rho(\delta \tau, \delta F)$ on F=const.

In the case of the positive $\delta F > 0$ and negative $\delta F < 0$ relative frequency shifts, the values τ_1, τ_m, and τ_2 are determined, according to the transformation (7.6.31) and the relationship (7.6.30), by the following expressions:

$$\begin{cases} \delta \tau_1 = (N-1)\delta T_0 \cdot 1(-\delta F) - 0.25; \\ \delta \tau_2 = (N-1)\delta T_0 \cdot 1(\delta F) + 0.5 \delta T_0 + 0.25; \\ \delta \tau_m = (\delta \tau_1 + \delta \tau_2)/2 = 0.5[(N-1) + 0.5]\delta T_0, \end{cases} \quad (7.6.33)$$

where $\delta T_0 = (T'_0 - T_0)/T_0 = (-\delta F)/(1+\delta F)$ is a relative difference of the carrier periods of transformed $s'(t)$ and initial $s(t)$ signals.

The form of normalized time-frequency mismatching function $\rho(\delta \tau, \delta F)$ of the filter on $N = 50$ is shown in Fig. 7.6.3. Cut projections of normalized time-frequency mismatching function $\rho(\delta \tau, \delta F)$ (7.6.32), made by horizontal planes $\rho(\delta \tau, \delta F)$=const that are parallel to coordinate plane $(\delta \tau, \delta F)$, on $N \gg 1$, are similar by form to a parallelogram with a center in the origin of coordinates, so that its small diagonal belongs to the axis $\delta\tau$, and two sides are perpendicular to this diagonal. They are parallel to the axis δF and belong to the second and fourth quadrants. The sizes of cut projection $\rho(\delta\tau, \delta F) = 0$ along the axes $\delta\tau$ and δF are equal to $1/2$ and $1/[2(N-1)]$, respectively. Cut projections of normalized time-frequency mismatching function $\rho(\delta\tau, \delta F)$ made by horizontal planes $\rho(\delta\tau, \delta F)$=const that are parallel to coordinate plane $(\delta\tau, \delta F)$ on $N = 50$, are shown in Fig. 7.6.4.

The curves 1 throug 5 in Fig. 7.6.4 correspond to the values $\rho(\delta\tau, \delta F) = 0$; 0.2; 0.4; 0.6; and 0.8, respectively.

Any two copies of the signal whose mutual relative mismatching in time $\delta\tau$ and in frequency δF exceeds *resolution measure* of the filter in relative time delay δ_τ and frequency shift δ_F, respectively:

$$|\delta\tau| > \delta_\tau, \quad |\delta F| > \delta_F, \tag{7.6.34}$$

are considered resolvable, so that the quantities δ_τ and δ_F are the doubled minimal values of the roots of the equations $\rho(\delta\tau, 0) = 0.5$, $\rho(0, \delta F) = 0.5$, respectively:

$$\delta_\tau = 2 \inf_{\delta\tau \in \Delta_{\delta\tau}} \{\arg[\rho(\delta\tau, 0) = 0.5]\}; \tag{7.6.35a}$$

$$\delta_F = 2 \inf_{\delta F \in \Delta_{\delta F}} \{\arg[\rho(0, \delta F) = 0.5]\}. \tag{7.6.35b}$$

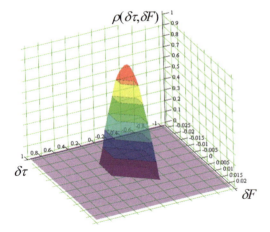

FIGURE 7.6.3 Normalized time-frequency mismatching function $\rho(\delta\tau, \delta F)$ of filter on $N = 50$

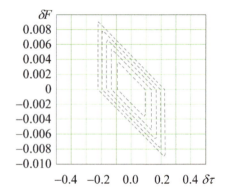

FIGURE 7.6.4 Cut projections of normalized time-frequency mismatching function. Lines 1 through 5 correspond to $\rho(\delta\tau, \delta F) = 0$; 0.2; 0.4; 0.6; and 0.8, respectively

The relationships (7.6.32) and (7.6.33) imply that *potential resolutions* of the filter, matched with harmonic signal (7.6.28a) in signal space with lattice properties, in relative time delay δ_τ and relative frequency shift δ_F are equal to:

$$\delta_\tau = 1/3; \quad \delta_F = 1/[3(N-1)]. \tag{7.6.36}$$

This means that *potential resolutions* of such a filter in time delay Δ_τ and frequency shift Δ_F are determined by the relationships:

$$\Delta_\tau = T_0/3; \quad \Delta_F = f_0/[3(N-1)], \tag{7.6.37}$$

and their product is determined by a number of periods N of harmonic signal $s(t)$:

$$\Delta_\tau \cdot \Delta_F = \delta_\tau \cdot \delta_F = 1/[9(N-1)]. \tag{7.6.38}$$

The relationship (7.6.38) implies that to provide simultaneously the desired values of resolution in time delay Δ_τ and frequency shift Δ_F, it is necessary to use the signals with sufficiently large numbers of periods N of oscillations. This implies that to provide simultaneously high resolution in both time and frequency parameters in a signal space with lattice properties, it is not necessary to use signals with large time-bandwidth products, inasmuch as this problem can be solved easily by means of harmonic signals in the form (7.6.28).

We now determine: how does the presence of interference (noise) affect the filter resolution? While receiving the realization $s^*(t)$ of the signal $s(t)$, conditional probability density function (PDF) $p_y(z;t/s^*) \equiv p_y(z/s^*)$ of instantaneous value $y(t)$ of statistics (7.6.10): $y(t) = \bigwedge_{j=0}^{N-1} x(t-jT_0)$, $t \in T_{\hat{s}}$ (see Formula (7.6.26)) is determined by the expression:

$$p_y(z/s^*) = P(C_c) \cdot \delta(z - s^*(t)) + N \cdot p_n(z)[1 - F_n(z)]^{N-1} \cdot 1(z - s^*(t)), \tag{7.6.39}$$

where $P(C_c) = 1 - P(C_e)$ is a probability of correct formation of the estimator $\hat{s}(t) = s^*(t)$:

$$P(C_c) = 1 - [1 - F_n(s^*(t))]^N; \tag{7.6.40}$$

$P(C_e)$ is a probability of error formation of the estimator $\hat{s}(t) = s^*(t)$:

$$P(C_e) = [1 - F_n(s^*(t))]^N; \tag{7.6.41}$$

$\delta(z)$, $1(z)$ are the Dirac delta and Heaviside step functions, respectively; $F_n(z)$ is the cumulative distribution function (CDF) of interference (noise) $n(t)$.

Any interval $T_{\hat{s}}$ is a partition of two sets T' and T'', respectively:

$$T_{\hat{s}} = T' \cup T'', \; T' \cap T'' = \emptyset; \; s(t') \le 0, \; t' \in T'; \; s(t'') > 0, \; t'' \in T''; \tag{7.6.42}$$

$$T' = [t_0 + (N-1)T_0 + \frac{T_0}{4} - \frac{\varphi T_0}{2\pi}; \; t_0 + (N-1)T_0 + \frac{3T_0}{4} - \frac{\varphi T_0}{2\pi}]; \; T'' = T - T',$$

and the measures of the intervals T', T'' are equal to $m(T') = m(T'') = T_0/2$.

Then, depending on whether the sample $w(t)$ belongs to one of sets T' and T'' of a partition (7.6.42) of the interval $T_{\hat{s}}$, the conditional PDF $p_w(z;t/s^*) \equiv p_w(z/s^*)$, $t \in T_{\hat{s}}$ of the instantaneous values $w(t)$ of stochastic process $w(t) = y(t) \vee 0$ in the input of the limiter is determined by the relationships:

$$p_w(z/s^*) = P \cdot \delta(z) + N \cdot p_n(z)[1 - F_n(z)]^{N-1} \cdot 1(z), \; t \in T' \subset T_{\hat{s}}, \tag{7.6.43}$$

where $P = P(C_c) + \int_{s(t)}^{0} N \cdot p_n(z)[1 - F_n(z)]^{N-1} dz$, $s(t) \leq 0$, $t \in T'$;

$$p_w(z/s^*) = P(C_c)\delta(z - s^*(t)) + \\ + N \cdot p_n(z)[1 - F_n(z)]^{N-1} 1(z - s^*(t)), \; t \in T'' \subset T_{\hat{s}}. \quad (7.6.44)$$

Obviously, when $s(t) > 0$, $t \in T'' \subset T_{\hat{s}}$, the PDF $p_w(z/s^*)$ of the stochastic process $w(t)$ in the output of the filter is identically equal to the PDF $p_y(z/s^*)$ (7.6.39):

$$p_w(z/s^*) = p_y(z/s^*), \; t \in T'' \subset T_{\hat{s}}.$$

The probability density function $p_w(z/s^*)$, $t \in T_{\hat{s}}$ is also the PDF of the estimator $\hat{s}(t)$ of the instantaneous value of the signal $s(t)$ in the input of the limiter.

Every random variable $n(t)$ is characterized by a PDF $p_n(z)$ with zero expectation, so assuming that $s^*(t) \geq 0$, the inequality holds:

$$F_n(s^*(t)) \geq 1/2. \quad (7.6.45)$$

According to the inequality (7.6.45), the estimator of upper bound of the probability $P(C_e)$ of error formation of the estimator $\hat{s}(t)$ is determined by the relationship:

$$P(C_e) \leq 2^{-N} \Rightarrow \sup[P(C_e)] = 2^{-N}. \quad (7.6.46)$$

Correspondingly, the lower bound of the probability $P(C_c)$ of the correct formation of the estimator $\hat{s}(t)$ is determined by the inequality:

$$P(C_c) \geq 1 - 2^{-N} \Rightarrow \inf[P(C_c)] = 1 - 2^{-N}. \quad (7.6.47)$$

Analysis of the relationship (7.6.44) allows us to conclude that the response of the signal $s(t)$, $t \in T_{\hat{s}}$ is observed in the filter output in the interval $T_{\hat{s}} = [t_0 + (N-1)T_0, t_0 + NT_0]$ with extremely high probability $P(C_c) \geq 1 - 2^{-N}$ regardless of signal-to-noise ratio. The relationship (7.6.44) also implies that the estimator $\hat{s}(t)$ is biased; nevertheless, it is asymptotically unbiased and consistent, inasmuch as it converges in probability and in distribution to the estimated value $s(t)$.

The expressions (7.6.43) and (7.6.44) for conditional PDF $p_w(z/s^*)$ of the instantaneous values of stochastic process $w(t)$ in the limiter output for arbitrary instants should be specified, inasmuch as they were obtained on the assumption that $t \in T_{\hat{s}}$, where the interval $T_{\hat{s}}$ (7.6.26) corresponds to the domain of definition of the signal response in the filter output (7.6.27), and, respectively, to domain of definition of the estimator $\hat{s}(t)$ of the signal $s(t)$.

Figure 7.6.5 illustrates the realization $w^*(t)$ of the signal $w(t)$ including the signal response and the residual overshoots in both sides of the signal response with amplitudes equal to the instantaneous values of the signal $s(t) \geq 0$ in the random instants $t \in T_k$ from the intervals $\{T_k\}$ corresponding to the time positions of the positive semiperiods of the autocorrelation function of the harmonic signal $s(t)$:

$$T_k = [t_0 + ((N-1) + k)T_0; \; t_0 + (N+k)T_0], \; k = 0, \pm 1, \ldots, \pm(N-1), \quad (7.6.48)$$

where N is a number of periods of harmonic signal $s(t)$, $N = Tf_0$; T_0 is a period of harmonic signal; f_0 is a carrier frequency of the signal; t_0 is time of signal arrival; $T_{k=0} \equiv T_{\hat{s}}$.

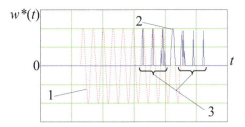

FIGURE 7.6.5 Realization $w^*(t)$ of signal $w(t)$ including signal response and residual overshoots. 1 = signal $s(t)$; 2 = signal response $\hat{s}(t)$; 3 = residual overshoots

Fig. 7.6.5 shows: 1 is the signal $s(t)$; 2 is the signal response $\hat{s}(t)$, $t \in T_{\hat{s}}$; 3 represents the residual overshoots.

Any interval T_k is a partition of two sets T'_k and T''_k, respectively:

$$T_k = T'_k \cup T''_k, \ T'_k \cap T''_k = \emptyset; \ s(t') \leq 0, \ t' \in T'_k; \ s(t'') > 0, t'' \in T''_k; \quad (7.6.49)$$

$$T'_k = [t_0+[(N-1)+k]T_0+\frac{T_0}{4}-\frac{\varphi T_0}{2\pi}; \ t_0+[(N-1)+k]T_0+\frac{3T_0}{4}-\frac{\varphi T_0}{2\pi}]; \ T''_k = T_k - T'_k,$$

and the measures of the intervals T'_k and T''_k are equal to $m(T'_k) = m(T''_k) = T_0/2$.

Formation of the signal $w(t)$ in the input of the limiter is realized according to the rule (7.6.24c):

$$w(t) = y(t) \vee 0 = [\bigwedge_{j=0}^{N-1} x(t - jT_0)] \vee 0, t \in T_k, \quad (7.6.50)$$

where T_k is determined by the formula (7.6.48), $k = 0, \pm 1, \ldots, \pm(N-1)$, $T_{k=0} \equiv T_{\hat{s}}$.

The expression (7.6.50) implies that as a result of $N - |k|$ tests from N, at the instant $t \in T_k$, the signal $w(t)$ can take the values from the set $\{0, s(t_i), n(t_l)\}$, $i = 1, \ldots, N - |k|$; $l = 1, \ldots, N - |k|$. As the result of $|k|$ tests from N, it can take the values from the set $\{0, n(t_m)\}$, $m = 1, \ldots, |k|$. Thus, in the intervals $\{T_k\}$, the signal $s(t), t \in T_k$ is used less often (at $|k|$) to form the estimator than in the interval $T_{k=0} \equiv T_{\hat{s}}$, that naturally makes the signal processing quality worse. Thus, at the instant $t \in T_k$, the signal $w(t)$ is determined by the least $y_{\min}(t) = y_1(t) \wedge y_2(t)$ of two random variables $y_1(t), y_2(t)$:

$$w(t) = y_{\min} \vee 0 = [y_1(t) \wedge y_2(t)] \vee 0, \ t \in T_k, \quad (7.6.51)$$

where $y_1(t) = \bigwedge_{j=0}^{|k|-1} x(t_j)$, $y_2(t) = \bigwedge_{j=|k|}^{N-1} x(t_j)$.

We determine the conditional PDF $p_w(z/s*)$, $t \in T_k$, based on the fact that only the samples in the form $\{0 \vee n(t_j)\}$ take part in formation of random variable $y_1(t)$, and only the samples in the form $\{s(t_j) \vee n(t_j)\}$ take part in formation of the random variable $y_2(t)$. Then, the PDF $p_y(z/s*)$, $t \in T_k$ of random variable $y_{\min}(t) = y_1(t) \wedge y_2(t)$ is determined by the relationship:

$$p_y(z/s^*) = p_{y1}(z/s^* = 0)[1 - F_{y2}(z/s^*)] + p_{y2}(z/s^*)[1 - F_{y1}(z/s^* = 0)], \quad (7.6.52)$$

where, according to (7.6.39), the PDFs $p_{y1}(z/s^* = 0)$ and $p_{y2}(z/s^*)$ of random variables $y_1(t)$ and $y_2(t)$ are determined by the expressions:

$$p_{y1}(z/s^* = 0) = P(C_c)|_{|k|, s=0} \cdot \delta(z) + |k| \cdot p_n(z)[1 - F_n(z)]^{|k|-1} \cdot 1(z);$$

$$p_{y2}(z/s^*) = P(C_c)|_{N-|k|} \cdot \delta(z - s^*(t)) + (N-|k|) \cdot p_n(z)[1 - F_n(z)]^{N-|k|-1} \cdot 1(z - s^*(t)),$$

and their CDFs $F_{y1}(z/s^* = 0)$ and $F_{y2}(z/s^*)$ are determined by the relationships:

$$F_{y1}(z/s^* = 0) = P(C_c)|_{|k|} \cdot 1(z), \quad F_{y2}(z/s^*) = P(C_c)|_{N-|k|} \cdot 1(z - s^*(t));$$

where $P(C_c)|_q = 1 - [1 - F_n(s^*(t))]^q$ and $P(C_e)|_q = [1 - F_n(s^*(t))]^q$, q=const.

Then PDF $p_{y2}(z/s^*)$ (7.6.52) can be represented in the form:

$$p_y(z/s^*) = p_{y1}(z/s^* = 0)\{1 - 1(z - s^*(t)) + [1 - F_n(z)]^{N-|k|} \cdot 1(z - s^*(t))\} +$$
$$+ p_{y2}(z/s^*)\{1 - 1(z) + [1 - F_n(z)]^{|k|} \cdot 1(z)\}. \quad (7.6.53)$$

Depending on the values taken by the signal $s(t)$ $t \in T_k$ in the interval T_k (7.6.49), $s(t') \leq 0$, $t' \in T'_k$ or $s(t'') > 0$, $t'' \in T''_k$, PDF $p_y(z/s^*)$ (7.6.53) is determined by the expressions:

$$p_y(z/s^*)|_{t \in T'_k} = P(C_c)|_{N-|k|} \cdot \delta(z - s^*(t)) +$$
$$+ (N - |k|) \cdot p_n(z)[1 - F_n(z)]^{N-|k|-1} \cdot [1(z - s^*(t)) - 1(z)] +$$
$$+ P(C_c)|_{|k|, s=0} P(C_e)|_{N-|k|, s=0} \delta(z) + N \cdot p_n(z)[1 - F_n(z)]^{N-1} \cdot 1(z); \quad (7.6.54)$$

$$p_y(z/s^*)|_{t \in T''_k} = P(C_c)|_{|k|, s=0} \cdot \delta(z) + |k| \cdot p_n(z)[1 - F_n(z)]^{|k|-1} \cdot [1(z) - 1(z - s^*(t))] +$$
$$+ P(C_c)|_{N-|k|} P(C_e)|_{|k|} \delta(z - s^*(t)) +$$
$$+ N \cdot p_n(z)[1 - F_n(z)]^{N-1} \cdot 1(z - s^*(t)). \quad (7.6.55)$$

Due to its non-negative definiteness, under the condition that $s(t) > 0$, $t \in T''_k$, the PDF $p_y(z/s^*)|_{t \in T''_k}$, $p_w(z/s^*)|_{t \in T''_k}$ in the filter output is identically equal to PDF (7.6.55):

$$p_w(z/s^*)|_{t \in T''_k} \equiv p_y(z/s^*)|_{t \in T''_k}. \quad (7.6.56)$$

If the signal $s(t)$ takes the values $s(t) \leq 0$, $t \in T'_k$ in the interval T_k, then the PDF is equal to:

$$p_w(z/s^*)|_{t \in T'_k} = [P(C_c)|_{|k|} + P + P(C_c)|_{|k|, s=0} \cdot P(C_e)|_{N-|k|, s=0}] \delta(z) +$$
$$+ N \cdot p_n(z)[1 - F_n(z)]^{N-1} \cdot 1(z), \quad (7.6.57)$$

7.6 Resolution of Radio Frequency Pulses in Metric Space with Lattice Properties

where

$$P = \int_{s^*(t)}^{0} (N - |k|) \cdot p_n(z)[1 - F_n(z)]^{N-|k|-1} dz;$$

$$P(C_c)|_{|k|,s=0} = 1 - 2^{-|k|}; \quad P(C_e)|_{N-|k|,s=0} = 2^{-N+|k|}.$$

The identity (7.6.56) implies that on small signal-to-interference (signal-to-noise) ratio in the filter input, at the instant $t \in T_k''$, the output of the filter forms: (1) the signal estimator with the probability $P[C(w = \hat{s})] \approx 2^{-|k|} - 2^{-N}$; (2) the value equal to zero $w(t) = 0$ with the probability $P[C(w = 0)] = 1 - 2^{-|k|}$, or (3) the least positive value $n_{\min}$ of interference (noise) $n(t)$ from the set $\{n(t_j)\}$ with the probability $P[C(w = n_{\min})] \approx 2^{-N}$. The expression (7.6.57) implies that on small signal-to-interference (signal-to-noise) ratio in the filter input, at the instant $t \in T_k'$ the output of the filter forms: (1) the values equal to zero $w(t) = 0$ with the probability $P[C(w = 0)] = 1 - 2^{-N}$, or (2) the values equal to the least positive value $n_{\min}$ of interference (noise) $n(t)$ from the set $\{n(t_j)\}$ with probability $P[C(w = n_{\min})] \approx 2^{-N}$.

The signal $w(t)$ in the limiter input can be represented by the linear combination of signal $w_s(t)$ and interference (noise) $w_n(t)$ components (see (7.3.25)):

$$w(t) = w_s(t) + w_n(t).$$

The signal component $w_s(t)$ is a stochastic process, whose realization instantaneous values in each interval T_k are equal to the positive instantaneous values of the signal $s(t)$ with probability $P[C(w = \hat{s})]$ or to zero with probability $P[C(w = 0)]$. The interference (noise) component $w_n(t)$ is a stochastic process, whose realization instantaneous values in each interval T_k are equal to the least positive instantaneous value of interference (noise) $n_{\min}$ with probability $P[C(w = n_{\min})]$, or to zero with probability $P[C(w = 0)]$.

The relationships (7.6.54) and (7.6.55) imply that on small signal-to-interference (signal-to-noise) ratio in the filter input, the PDF $p_{w_s}(z/s^*)$ of the signal component $w_s(t)$ in its output is equal to:

$$p_{w_s}(z/s^*) = \begin{cases} [1 - 2^{-|k|} + 2^{-N}]\delta(z) + [2^{-|k|} - 2^{-N}]\delta(z - s^*(t)), & t \in T_k''; \\ \delta(z), & t \in T_k'. \end{cases} \quad (7.6.58)$$

Hence, mathematical expectation $\mathbf{M}\{w_s(t)\}$ of the signal component $w_s(t)$ of the process $w(t)$ is equal to:

$$\mathbf{M}\{w_s(t)\} = \int_{-\infty}^{\infty} z p_{w_s}(z/s^*) dz =$$

$$= [2^{-|k|} - 2^{-N}] \cdot [s^*(t) \vee s^*(t - (N-1)T_0) \vee 0], \quad t \in T_k. \quad (7.6.59)$$

Figure 7.6.6 illustrates the realization $w^*(t)$ of stochastic process $w(t)$ in the output of the unit of formation of the positive part, and Fig. 7.6.7 shows the realization $v^*(t)$ of stochastic process $v(t)$ in the output of median filter (see Fig. 7.6.1), under the interaction between two harmonic signals $s_1(t)$, $s_2(t)$ and interference (noise) $n(t)$ in the input of synthesized unit obtained by statistical modeling of processing of the input signal $x(t)$. In the figures 1 denotes the signal $s_1(t)$; 2 is the signal $s_2(t)$; 3 is the response of the signal $s_1(t)$; 4 is the response of the signal $s_2(t)$; 5 represents residual overshoots of the signal component $w_s^*(t)$ of realization $w^*(t)$ of stochastic process $w(t)$.

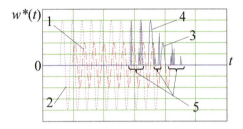
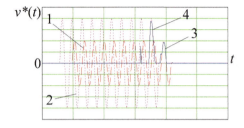

FIGURE 7.6.6 Realization $w^*(t)$ of stochastic process $w(t)$ in input of limiter and residual overshoots

FIGURE 7.6.7 Realization $v^*(t)$ of stochastic process $v(t)$ in output of median filter

The examples correspond to the following conditions. The signals $s_1(t)$ and $s_2(t)$ are narrowband RF pulses without intrapulse modulation; interference is a quasi-white Gaussian noise with the ratio of maximum frequency of interference power spectral density $f_{n,\max}$ to a carrier frequency f_0 of the signals $s_1(t)$ and $s_2(t)$: $f_{n,\max}/f_0 = 8$; the signal-to-interference (signal-to-noise) ratios for the signals $s_1(t)$ and $s_2(t)$ take the values $E_{s1}/N_0 = 8 \cdot 10^{-7}$ and $E_{s2}/N_0 = 3.2 \cdot 10^{-6}$, respectively (where E_{s1}, E_{s2} are the energies of the signals $s_1(t)$ and $s_2(t)$; N_0 is power spectral density of interference (noise)). The number of samples N of the input signal $x(t)$ used in signal processing and determined by the number of periods of a carrier of the signals $s_1(t)$ and $s_2(t)$ is equal to 10. The delay of the signal $s_1(t)$ with respect to the signal $s_2(t)$ is equal to $1.25T_0$, where T_0 is a period of a carrier oscillation: $T_0 = 1/f_0$.

The responses of the signals $s_1(t)$ and $s_2(t)$, shown in Fig. 7.6.6, are easily distinguished in the remnants of the nonlinear interaction between interference (noise) and the signals $s_1(t)$, $s_2(t)$ in the form of residual overshoots (line 5) of the signal component $w_s^*(t)$ of realization $w^*(t)$ of stochastic process $w(t)$. As can be seen from Fig. 7.6.7, median filter (see Fig. 7.6.1) removes residual overshoots (line 5) of the signal component $w_s^*(t)$ of realization $w^*(t)$ of stochastic process $w(t)$ and slightly cuts the top of the responses of the signals $s_1(t)$ and $s_2(t)$. The results of statistical modeling of processing of the input signal $x(t)$ shown in Figs. 7.6.6 and 7.6.7, confirm the high efficiency of harmonic signal resolution in the presence of strong interference provided by the synthesized processing algorithm.

Using the identity $w(t) = s^*(t)$ with respect to the formula (7.6.59), where

7.6 Resolution of Radio Frequency Pulses in Metric Space with Lattice Properties

$w(t)$ is determined by the function (7.6.29), normalizing this function along with transformation of the variable t (7.6.31), we obtain the expression for normalized time-frequency mismatching function $\rho(\delta\tau, \delta F)$ of the filter in the presence of strong interference (see Fig. 7.6.8):

$$\rho(\delta\tau, \delta F) = \begin{cases} a \cdot \sin[2\pi(1+\delta F)(\delta\tau - \delta\tau_{1,k})], & \delta\tau_{1,k} \leq \delta\tau < \delta\tau_{m,k}; \\ -a \cdot \sin[2\pi(1+\delta F)(\delta\tau - \delta\tau_{2,k})], & \delta\tau_{m,k} \leq \delta\tau < \delta\tau_{2,k}, \end{cases} \quad (7.6.60)$$

where $\delta\tau$ and δF are relative time delay and relative frequency shift; a is a multiplier equal to $(2^{-|k|} - 2^{-N})/(1 - 2^{-N})$; $\delta\tau_{1,k}$, $\delta\tau_{2,k}$ are relative times of beginning and ending of the intervals of definition of mismatching function $\rho(\delta\tau, \delta F)$ on F=const; $\delta\tau_{m,k} = (\delta\tau_{1,k} + \delta\tau_{2,k})/2$ is relative time corresponding to the maximum value of mismatching function $\rho(\delta\tau, \delta F)$ on F=const.

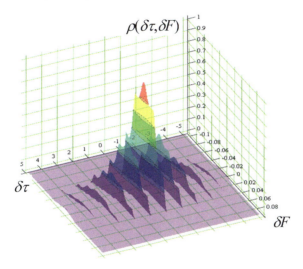

FIGURE 7.6.8 Normalized time-frequency mismatching function of filter in presence of strong interference

In cases of positive $F > 0$ and negative $F < 0$ relative frequency shifts, the values $\delta\tau_{1,k}$, $\delta\tau_{m,k}$, and $\delta\tau_{2,k}$ are determined, according to the transformation (7.6.31) and the relationship (7.6.30), by the following expressions:

$$\begin{cases} \delta\tau_{1,k} = (N-1)\delta T_0 \cdot 1(-\delta F) - 0.25 + k; \\ \delta\tau_{2,k} = (N-1)\delta T_0 \cdot 1(\delta F) + 0.5\delta T_0 + 0.25 + k; \\ \delta\tau_{m,k} = (\delta\tau_{1,k} + \delta\tau_{2,k})/2 = 0.5[(N-1) + 0.5]\delta T_0 + k, \end{cases} \quad (7.6.61)$$

where $\delta T_0 = (T_0' - T_0)/T_0 = (-\delta F)/(1 + \delta F)$ is a relative difference of the periods of a carrier of the transformed $s'(t)$ and the initial $s(t)$ signals.

The expression (7.6.60) and the relationships (7.6.61) imply that resolutions in relative time delay δ_τ and relative frequency shift δ_F of the filter, matched with the harmonic signal (7.6.28a) in signal space with lattice properties, remain invariable under interaction between the signal and interference (noise) regardless of the energetic relationships between them as determined by Equations (7.6.36).

The multipeak character of the function $\rho(\delta\tau, \delta F)$, that spreads along the axis of relative time delay $\delta\tau$ in the interval $[-0.25-(N-1),(N-1)+0.25]$ with maxima in the points $\delta\tau = k$, $k = 0, \pm 1, \ldots, \pm(N-1)$, stipulates the ambiguity of time delay measurement.

The investigation of the resolution algorithm of RF pulses without intrapulse modulation with a rectangular envelope in signal space with lattice properties allows us to draw the following conclusions.

1. While solving the problem of signal resolution in signal space with lattice properties, there exists the possibility of realization of so-called "needle-shaped" response in the output of a signal processing unit without side lobes. The independence of the responses of the transformed $s'(t)$ (7.6.28b) and the initial signal $s(t)$ (7.6.28a) in the range of time delay $\Delta\tau$ and frequency shift ΔF, that are out of the bounds of filter resolution in time Δ_τ and in frequency Δ_F parameters ($|\Delta\tau| > \Delta_\tau, |\Delta F| > \Delta_F$), is achieved by nonlinear processing in signal space with lattice properties.

2. The effect of an arbitrarily strong interference in signal space with lattice properties does not change the filter resolutions in time delay and in frequency shift. However, it does cause the ambiguity of time delay measurement.

3. The absence of the constraints imposed by uncertainty principle of Woodward [118] allows us to obtain any resolution in time delay Δ_τ and in frequency shift Δ_F in signal space with lattice properties even under harmonic signal use. The last feature is provided by the proper choice of carrier frequency f_0 and number of periods N of carrier oscillations within the signal duration T.

7.7 Methods of Mapping of Signal Spaces into Signal Space with Lattice Properties

Finally we consider some possible methods of constructing the signal space with lattice properties, which in principle differ from the direct methods of realization of signal spaces with such properties. The direct methods of realization of signal spaces with given properties assume the use of physical medium (in a general case, with nonlinear properties) in which useful and interference signals interact on the basis of operations determining algebraic properties of signal space. The methods considered in this section are based on the signal mappings from a single space (or from several spaces) into a space with given properties (in this case, lattice properties).

We first consider the method of construction of signal space with lattice properties on the base of a space with group properties (linear signal space), and then consider the construction of a signal space with lattice properties on the base of spaces with semigroup properties.

7.7.1 Method of Mapping of Linear Signal Space into Signal Space with Lattice Properties

Signal space with lattice properties $\mathcal{L}(\vee, \wedge)$ can be obtained by transformation of the signals of linear space in such a way that the results of interactions $x(t)$ and $\tilde{x}(t)$ between the signal $s(t)$ and interference (noise) $n(t)$ in signal space with lattice properties $\mathcal{L}(\vee, \wedge)$ with operations of join $\vee$ and meet $\wedge$ are realized according to the relationships:

$$x(t) = s(t) \vee n(t) = \{[s(t) + n(t)] + |s(t) - n(t)|\}/2; \quad (7.7.1a)$$

$$\tilde{x}(t) = s(t) \wedge n(t) = \{[s(t) + n(t)] - |s(t) - n(t)|\}/2, \quad (7.7.1b)$$

which are the consequences of the equations [221, Section XIII.3;(14)], [221, Section XIII.4;(22)].

The identities (7.7.1) determine the mapping T of linear signal space $\mathcal{LS}(+)$ into signal space with lattice properties $\mathcal{L}(\vee, \wedge)$: $T: \mathcal{LS}(+) \to \mathcal{L}(\vee, \wedge)$. The mapping T^{-1} inversed with respect to the initial one T: $T^{-1}: \mathcal{L}(\vee, \wedge) \to \mathcal{LS}(+)$ is determined by the known identity [221, Section XIII.3;(14)]:

$$s(t) + n(t) = s(t) \vee n(t) + s(t) \wedge n(t).$$

Based on the relationships (7.7.1), to form the results of interaction between the signal $s(t)$ and interference (noise) $n(t)$ in signal space with lattice properties $\mathcal{L}(\vee, \wedge)$, it is necessary to have two linearly independent equations with respect to $s(t)$ and $n(t)$. Let two linearly independent functions $a(t)$ and $b(t)$ of useful signal $s(t)$ and interference (noise) $n(t)$, received by a directional antenna A and an omnidirectional antenna B (see Fig. 7.7.1), arrive onto two inputs of the unit of mapping T of linear signal space into signal space with lattice properties.

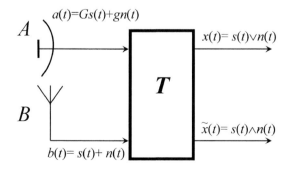

FIGURE 7.7.1 Block diagram of mapping unit

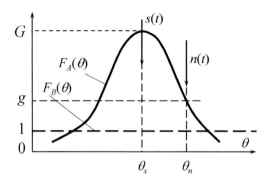

FIGURE 7.7.2 Directional field patterns $F_A(\theta)$, $F_B(\theta)$ of antennas A, B; θ_s, θ_n are directions of arrival of signal and interference (noise), respectively

There are two signals in two outputs of the mapping unit T: $x(t) = s(t) \vee n(t)$ and $\tilde{x}(t) = s(t) \wedge n(t)$ determined by the relationships (7.7.1). The useful signal $s(t)$ and interference (noise) $n(t)$ are received by the antennas of channels A and B (see Fig. 7.7.2), which are such that: (1) the antennas A and B have the same phase center and (2) the antennas A and B are characterized by the directional field patterns $F_A(\theta)$ and $F_B(\theta)$, respectively, so that $F_A(\theta_s) = G$; $F_A(\theta_n) = g$; $G > g$ and $G > 1$; $F_B(\theta) = 1$; complex gain-frequency characteristics $\dot{K}_A(\omega)$ and $\dot{K}_B(\omega)$ of the receiving channels A and B are identical: $\dot{K}_A(\omega) = \dot{K}_B(\omega)$, and there are two signals $a(t)$ and $b(t)$ in the inputs of the mapping unit T:

$$\begin{cases} a(t) = G \cdot s(t) + g \cdot n(t); & (a) \\ b(t) = s(t) + n(t), & (b) \end{cases} \qquad (7.7.2)$$

where $G = F_A(\theta_s)$ is the gain of antenna A from the direction of arrival of the signal $s(t)$; $g = F_A(\theta_n)$ is the gain of antenna A from the direction of arrival of interference (noise) $n(t)$.

We suppose that the direction of arrival of the signal θ_s is known, and the direction of arrival of interference (noise) θ_n is unknown. We also consider that interference $n(t)$ is quasi-white Gaussian noise with independent samples $\{n(t_j)\}$, $j = 0, 1, 2, \ldots$, that are distant over time interval $\Delta t = |t_j - t_{j\pm 1}| = 1/(2f_{n,\max})$, where $f_{n,\max}$ is an upper bound frequency of power spectral density of interference, and the useful signal $s(t)$ changes slightly in the interval Δt, i.e., $s(t) = s(t \pm \Delta t)$.

The equation system (7.7.2) with respect to $s(t)$ and $n(t)$ cannot be solved due to the presence of the additional unknown quantity g. To solve it, in addition to the system (7.7.2), one more equation system can be used. It is formed on the basis of the observations $a(t)$ and $b(t)$, delayed at the interval Δt, taking into account the last assumption concerning a slow (as against Δt) signal change ($s(t) = s(t \pm \Delta t)$):

$$\begin{cases} a(t') = G \cdot s(t) + g \cdot n(t'); & (a) \\ b(t') = s(t) + n(t'), & (b) \end{cases} \qquad (7.7.3)$$

where $t' = t - \Delta t$, $\Delta t = 1/(2f_{n,\max})$.

7.7 Methods of Mapping of Signal Spaces into Signal Space with Lattice Properties

We now obtain the relationships determining the algorithm of the mapping unit $T\colon a(t), b(t) \to x(t), \tilde{x}(t); a(t), b(t) \in \mathcal{LS}(+); x(t), \tilde{x}(t) \in \mathcal{L}(\vee, \wedge)$.

Joint fulfillment of both pairs of Equations (7.7.2a), (7.7.3a) and (7.7.2b), (7.7.3b) implies the system:

$$\begin{cases} a(t) - a(t') = g \cdot [n(t) - n(t')]; \\ b(t) - b(t') = n(t) - n(t'). \end{cases} \quad (7.7.4)$$

The equations of the system (7.7.4) imply the identity determining the gain g of antenna A, which affect interference (noise) $n(t)$ arriving from the direction θ_n:

$$g = F_A(\theta_n) = \frac{a(t) - a(t')}{b(t) - b(t')}. \quad (7.7.5)$$

Multiplying both parts of Equation (7.7.2b) by some coefficient k and subtracting them from the corresponding parts of Equation (7.7.2a) of the same system, multiplying the difference $a(t) - kb(t)$ by some coefficient q, we compose the auxiliary identity:

$$q[a(t) - kb(t)] = q(G - k) \cdot s(t) + q(g - k) \cdot n(t) = s(t) - n(t). \quad (7.7.6)$$

To make the identity (7.7.6) to hold, it is necessary to select values k and q, which are the roots of the system of equations:

$$\begin{cases} q(G - k) = 1; \\ q(g - k) = -1. \end{cases}$$

By solving it, we obtain the values of coefficients k and q providing the identity to hold (7.7.6):

$$k = (G + g)/2; \quad (7.7.7a)$$

$$q = 2/(G - g), \quad (7.7.7b)$$

Taking into account the identity (7.7.6), one can compose the required relationship:

$$s(t) - n(t) = q[a(t) - k \cdot b(t)], \quad (7.7.8)$$

where k, q are the coefficients determined by the relationships (7.7.7a) and (7.7.7b), respectively.

Substituting the sum $s(t) + n(t)$ from Equation (7.7.2b) and the difference $s(t) - n(t)$ from Equation (7.7.8) into the identities (7.7.1a,b), we obtain the desired relationships determining the algorithm of the mapping unit $T\colon a(t), b(t) \to x(t), \tilde{x}(t); a(t), b(t) \in \mathcal{LS}(+); x(t), \tilde{x}(t) \in \mathcal{L}(\vee, \wedge)$:

$$T = \begin{cases} x(t) = [b(t) + q|a(t) - k \cdot b(t)|]/2; & (a) \\ \tilde{x}(t) = [b(t) - q|a(t) - k \cdot b(t)|]/2; & (b) \\ k = (G + g)/2; & (c) \\ q = 2/(G - g); & (d) \\ g = [a(t) - a(t')]/[b(t) - b(t')]. & (e) \end{cases} \quad (7.7.9)$$

One possible variant of the block diagram of the signal space mapping unit is described in [268], [269].

Generally, the mapping of linear signal space $\mathcal{LS}(+)$ into the signal space with lattice properties $\mathcal{L}(\vee, \wedge)$ supposes the further use of the processing algorithms that are invariant with respect to the conditions of parametric and nonparametric prior uncertainties. If further signal processing is constrained by the use of the algorithms that are critical with respect to energetic characteristics of the useful and interference signals, the mapping unit T must form the signals in the outputs $x(t)$ and $\tilde{x}(t)$ of the following form:

$$x(t) = Gs(t) \vee n(t) = \{[Gs(t) + n(t)] + |Gs(t) - n(t)|\}/2; \tag{7.7.10a}$$

$$\tilde{x}(t) = Gs(t) \wedge n(t) = \{[Gs(t) + n(t)] - |Gs(t) - n(t)|\}/2, \tag{7.7.10b}$$

where $G = F_A(\theta_s)$ is a gain of antenna A expected from the known direction of arrival of the signal θ_s.

With an approach used while forming Equation (7.7.6), we can formulate required relationships based on the identities (7.7.10):

$$q_1[a(t) - k_1 b(t)] = q_1(G - k_1) \cdot s(t) + q_1(g - k_1) \cdot n(t) = Gs(t) - n(t); \tag{7.7.11a}$$

$$q_2[a(t) + k_2 b(t)] = q_2(G + k_2) \cdot s(t) + q_2(g + k_2) \cdot n(t) = Gs(t) + n(t). \tag{7.7.11b}$$

To ensure the identities (7.7.11) hold, it is necessary to find values $k_{1,2}$ and $q_{1,2}$, which are the roots of the following equation systems:

$$\begin{cases} q_1(G - k_1) = G; \\ q_1(g - k_1) = -1, \end{cases} \tag{7.7.12a}$$

$$\begin{cases} q_2(G + k_2) = G; \\ q_2(g + k_2) = 1. \end{cases} \tag{7.7.12b}$$

Solving the equation systems (7.7.12), we obtain the values of coefficients $k_{1,2}$ and $q_{1,2}$ that are equal, respectively, to:

$$k_1 = G(1 + g)/(G + 1); \tag{7.7.13a}$$

$$q_1 = (G + 1)/(G - g), \tag{7.7.13b}$$

$$k_2 = G(1 - g)/(G - 1); \tag{7.7.14a}$$

$$q_2 = (G - 1)/(G - g). \tag{7.7.14b}$$

Then, according to the identities (7.7.11), we can arrange the required relationships of the following form:

$$Gs(t) - n(t) = q_1[a(t) - k_1 b(t)]; \tag{7.7.15a}$$

$$Gs(t) + n(t) = q_2[a(t) + k_2 b(t)], \tag{7.7.15b}$$

where $k_{1,2}$, $q_{1,2}$ are the coefficients determined by the pairs of relationships (7.7.13a), (7.7.14a) and (7.7.13b), (7.7.14b), respectively.

Substituting the values $Gs(t) - n(t)$ and $Gs(t) + n(t)$ from Equations (7.7.15) into the identities (7.7.10), we obtain the required relationships determining the algorithm of the mapping unit $T: a(t), b(t) \to x(t), \tilde{x}(t); a(t), b(t) \in \mathcal{LS}(+); x(t), \tilde{x}(t) \in \mathcal{L}(\vee, \wedge)$:

$$T = \begin{cases} x(t) = \{q_2[a(t) + k_2 b(t)] + q_1|a(t) - k_1 \cdot b(t)|\}/2; & (a) \\ \tilde{x}(t) = \{q_2[a(t) + k_2 b(t)] - q_1|a(t) - k_1 \cdot b(t)|\}/2; & (b) \\ k_1 = G(1+g)/(G+1); & (c) \\ q_1 = (G+1)/(G-g); & (d) \\ k_2 = G(1-g)/(G-1); & (e) \\ q_2 = (G-1)/(G-g); & (f) \\ g = (a(t) - a(t'))/(b(t) - b(t')). & (g) \end{cases} \quad (7.7.16)$$

7.7.2 Method of Mapping of Signal Space with Semigroup Properties into Signal Space with Lattice Properties

The signal space with lattice properties $\mathcal{L}(\vee, \wedge)$ can be realized by transformation of the signals from two spaces with semigroup properties, namely by transformation from spaces with the properties of additive $\mathbf{SG}(+)$ and multiplicative $\mathbf{SG}(\cdot)$ semigroups (see Definition 4.1.1). To realize this method, it is necessary to synthesize the units in which the additive and multiplicative signal interactions take place.

The results of interactions $x(t)$ and $\tilde{x}(t)$ between the signal $s(t)$ and interference $n(t)$ in signal space with lattice properties $\mathcal{L}(\vee, \wedge)$ with operations of join $\vee$ and meet $\wedge$ are realized according to the relationships (7.7.1). The function $|s(t) - n(t)|$ can be formed on the basis of the known equation:

$$|s(t) - n(t)| = \sqrt{u^2(t) - 4v(t)} = w(t), \quad (7.7.17)$$

where the functions $u(t)$ and $v(t)$ are determined over operations of addition and multiplication between the signal $s(t)$ and interference $n(t)$ that take place in signal spaces with additive $\mathbf{SG}(+)$ and multiplicative $\mathbf{SG}(\cdot)$ semigroups, respectively:

$$\begin{cases} u(t) = s(t) + n(t); & (a) \\ v(t) = s(t) \cdot n(t). & (b) \end{cases} \quad (7.7.18)$$

Based on the relationships (7.7.1), to form the results of interactions $x(t)$ and $\tilde{x}(t)$ between the signal $s(t)$ and interference $n(t)$ in signal space with lattice properties $\mathcal{L}(\vee, \wedge)$, it is necessary to have two independent equations with respect to $s(t)$ and $n(t)$ (7.7.18) that form equations:

$$x(t) = s(t) \vee n(t) = [u(t) + w(t)]/2 = [u(t) + \sqrt{u^2(t) - 4v(t)}]/2; \quad (7.7.19a)$$

$$\tilde{x}(t) = s(t) \wedge n(t) = [u(t) - w(t)]/2 = [u(t) - \sqrt{u^2(t) - 4v(t)}]/2, \quad (7.7.19b)$$

where $w(t) = \sqrt{u^2(t) - 4v(t)} = |s(t) - n(t)|$ (7.7.17).

The desired relationships determining the method of mapping $T': u(t), v(t) \to$

$x(t), \tilde{x}(t)$; $u(t) \in \mathbf{SG}(+)$, $v(t) \in \mathbf{SG}(\cdot)$; and $x(t), \tilde{x}(t) \in \mathcal{L}(\vee, \wedge)$ are defined by the equations system:

$$T' = \begin{cases} x(t) = s(t) \vee n(t) = [u(t) + w(t)]/2; & (a) \\ \tilde{x}(t) = s(t) \wedge n(t) = [u(t) - w(t)]/2; & (b) \\ w(t) = \sqrt{u^2(t) - 4v(t)} = |s(t) - n(t)|; & (c) \\ u(t) = s(t) + n(t); & (d) \\ v(t) = s(t) \cdot n(t); & (e) \end{cases} \quad (7.7.20)$$

The example of signal space with mentioned properties is also a ring $\mathbf{R}(+, \cdot)$, i.e., an algebraic structure in which two binary operations are defined (addition and multiplication), which are such that: (1) $\mathbf{R}(+)$ is an additive group with neutral element 0, $0 \in \mathbf{R}(+)$: $\forall a, a \in \mathbf{R}(+)$: $\exists - a$: $a + 0 = a$; $a + (-a) = 0$; (2) $\mathbf{R}(\cdot)$ is multiplicative semigroup; (3) operations of addition and multiplication are connected through distributive laws:

$$a(b+c) = ab + ac; \quad (b+c)a = ba + ca; \quad a, b, c \in \mathbf{R}(+, \cdot)$$

Generally, the methods and algorithms of signal processing within signal spaces with lattice properties are single-channel types regardless of the number of interference sources affecting input of the processing unit. These methods can be realized in two ways: (1) utilizing the physical medium with lattice properties, where the wave interactions are determined by lattice operations (the direct methods); (2) utilizing the methods of mapping of signal spaces with group (semigroup) properties into the signal spaces with lattice properties (the indirect methods).

Unfortunately, the materials with such properties are unknown to the author; and realization of the second group methods inevitably causes various destabilizing factors exerting their negative influence upon the efficiency of signal processing, that is stipulated by inaccurate reproduction of the operations of join and meet between the signals. This circumstance causes violation of the initial properties of signal space. Signal processing methods used in signal space with lattice properties cease to be optimal; thus, it is necessary to reoptimize signal processing algorithms obtained under assumption that lattice properties hold within the signal space. That is a subject for separate consideration.

Here, however, we note, that "reoptimized" signal processing algorithms operating in signal space with lattice properties in the presence of destabilizing factors (for instance, those that decrease statistical dependence (or cross-correlation) of interference in the receiving channels), can be more efficient than the known signal processing algorithms functioning in linear signal space.

Certainly, the direct methods of realization of signal spaces with given properties are essentially more promising. They assume the use of physical medium with nonlinear properties in which useful and interference signals interact on the basis of lattice operations. The problem of synthesis of physical medium where interaction of wave processes is described by the lattice operations oversteps the bounds of the signal processing theory and requires special research.

Conclusion

The main hypothesis underlying this book can be formulated in the following way. It is impossible to construct signal processing theory without providing unity of conceptual basics with information theory (at least within its syntactical aspects), and as a consequence, without their theoretical compatibility and harmonious association. Apparently, the inverse statement is also true: it is impossible to state information theory logically (within its syntactical aspects) in isolation from signal processing theory.

There are two axiomatic statements underlying this book: the main axiom of signal processing theory and axiom of a measure of binary operation between the elements of signal space built upon generalized Boolean algebra with a measure. The first axiom establishes the relationships between information quantities contained in the signals in the input and output of processing unit. The second axiom determines qualitative and quantitative aspects of informational relationships between the signals (and their elements) in a space built upon generalized Boolean algebra with a measure. All principal results obtained in this work are the consequences from these axiomatic statements.

The main content of this work can be formulated by the following statements:

1. Information can exist only in the presence of the set of its material carriers, i.e., the signals forming the signal space.
2. Information contained in a signal exists only in the presence of the structural diversity between the signal elements.
3. Information contained in a couple of signals exists due to the identities and the distinctions between the elements of these signal structures.
4. Signal space is characterized by properties peculiar to nonEuclidean geometry.
5. Information contained in the signals of signal space becomes measurable if a measure of information quantity is introduced in such space.
6. Measure of information quantity is an invariant of a group of signal mappings in signal space.
7. Measure of information quantity induces metric in signal space.
8. From the standpoint of providing minimum losses of information contained in the signals, it is expedient to realize signal processing within the signal spaces with lattice properties.

9. Algorithms and units of signal processing in metric spaces with lattice properties are characterized by invariance property with respect to parametric and nonparametric prior uncertainty conditions.

Unlike the approaches formulated in [150], [152], [157], [166], algorithms and units of signal processing in metric spaces with lattice properties, characterized by signal processing quality indices that at qualitative and quantitative levels are unattainable for their analogues in linear spaces, have been obtained as a consequence of answering Question 1.5 in the Introduction.

In this work, the problem of qualitative increasing the signal processing efficiency under prior uncertainty conditions has been solved theoretically without overstepping the bounds of signal processing theory. The interaction between useful and interference signals is investigated on the basis of the operations of the space with lattice properties, i.e., without invasion into the subject matter of other branches of science that could answer the questions concerning the practical realization of wave process interactions with aforementioned properties.

The interrelations between information theory, signal processing theory, and algebraic structures, established in the book and shown in Fig. C.1, have the following meaning.

The notion of physical signal space has been introduced by Definition 4.1.1 on the basis of the most general (among all known algebraic structures) concept of the semigroup. Properties of signal spaces used in signal processing theory may differ, but all the signal spaces are always semigroups. In Chapters 6 and 7, the models of signal spaces have been considered on the basis of L-groups possessing both group and lattice properties. Linear spaces, widely used as signal space models are also characterized by group properties.

The main interrelation considered in this book, namely between generalized Boolean algebra with a measure and mathematical foundations of information theory, shown in Fig. C.1 has been investigated within Chapters 2 through 5. A measure introduced upon generalized Boolean algebra accomplishes two functions: (1) it is a measure of information quantity; (2) it induces metric in signal space. This statement is a cornerstone of interrelation between information theory and signal processing theory.

Informational and metric relationships between the signals interacting in signal spaces with various algebraic properties have been established on the basis of probabilistic and informational characteristics of stochastic signals that are invariants of groups of signal mappings introduced in Chapter 3. This provides indirect interrelation between probability theory and signal processing theory realized through information theory. Finally, the relationships determining the channel capacity in signal spaces with L-group properties cited in Section 6.5 establish the interrelation between information theory and L-groups.

It is impossible to consider "Fundamentals of Signal Processing..." to be accomplished, since interesting directions remain to be investigated, for instance, solving

Conclusion

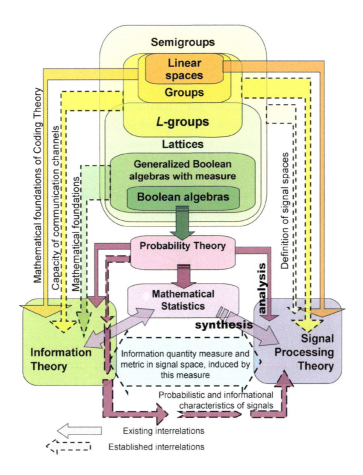

FIGURE C.1 Suggested scheme of interrelations between information theory, signal processing theory, and algebraic structures

the known signal processing problems within linear spaces on the basis of methods of signal processing in spaces with L-group properties; expanding the book's principal ideas into the signal spaces in the form of stochastic fields; development of information processing methods and units (including quantum and optic-electronic) based on generalized Boolean algebra with a measure, and others.

The solution of nonlinear electrodynamics problem of synthesis of a physical medium in which wave interaction is described by L-group operations requires special efforts from the scientific community. However, such an advance would overstep the bounds of signal processing theory subject and require a multidisciplinary approach. Practical application of suggested algorithms and units of signal processing in metric spaces with lattice properties is impossible without the breakthroughs to the aforementioned technologies and cooperative efforts noted above.

The main problem related to fundamentals of signal processing theory and information theory rely on finding three necessary compromises among: (1) mathematics and physics, (2) algebra and geometry, and (3) continuity and discreteness.

An attempt to compress the content of this book into a single paragraph would read as follows:

> Information is the property that is inherent to nonEuclidean signal space forming a set of material carriers and is also inherent to every single signal. This property is explained by identities and distinctions between signals and their structural elements. It also reveals itself in signal interactions. A measure of information quantity induces metric in signal space, and this measure is invariant of a group of signal mappings.

The paragraph above summarizes the principal idea behind this book.

A secondary idea is the corollary from the principal one. It arises from the need to increase the efficiency of signal processing by decreasing or even eliminating the inevitable losses of information that accompany the interactions of useful and interference signals. Achieving the goal of more efficient signal processing requires researching and developing new technologies based upon suggested signal spaces, in which interactions of useful and interference signals take place with essentially lesser losses of information as against linear spaces.

The author hopes the book will inform a broad research audience and respectable scientific community. We live in times when the most insane fantasies become perceptible realities with an incredible speed. The author expresses his confidence that the signal processing technologies and methods described in this book will appear sooner rather than later.

Bibliography

[1] Whittaker, E. T. On the functions which are represented by the expansions of the interpolation theory. *Proc. Royal Soc. Edinburgh*, 10(1):57–64, 1922.

[2] Carson, J.R. Notes on the theory of modulation. *Proc. IRE*, 10(1):57–64, 1922.

[3] Gabor, D. Theory of communication. *J. IEE*, 93(26):429–457, 1946.

[4] Belyaev, Yu.K. Analytical stochastic processes. *Probability Theory and Appl.*, 4(4):437–444, 1959 (in Russian).

[5] Zinoviev, V.A. and Leontiev, V.K. On perfect codes. *Prob. Inf. Trans.*, 8(1):26–35, 1975 (in Russian).

[6] Wyner, A. The common information of two dependent random variables. *IEEE Trans. Inf. Theory*, IT–21(2):163–179, 1975.

[7] Pierce, J.R. *An Introduction to Information Theory: Symbols, Signals and Noise*. Dover Publications, New York, 2nd edition, 1980.

[8] Nyquist, H. Certain factors affecting telegraph speed. *Bell Syst. Tech.J.*, 3:324–346, 1924.

[9] Nyquist, H. Certain topics in telegraph transmission theory. *Trans. AIEE*, 47:617–644, 1928.

[10] Tuller, W.G. Theoretical limitations on the rate of transmission of information. *Proc. IRE*, 37(5):468–478, 1949.

[11] Kelly, J. A new interpretation of information rate. *Bell Syst. Tech. J.*, 35:917–926, 1956.

[12] Blackwell, D., Breiman, L., and Thomasian, A.J. The capacity of a class of channels. *Ann. Math. Stat.*, 30:1209–1220, 1959.

[13] McDonald, R.A. and Schultheiss, P.M. Information rates of Gaussian signals under criteria constraining the error spectrum. *Proc. IEEE*, 52:415–416, 1964.

[14] Wyner, A.D. The capacity of the band-limited Gaussian channel. *Bell Syst. Tech. J.*, 45:359–371, 1965.

[15] Pinsker, M.S. and Sheverdyaev, A.Yu. Transmission capacity with zero error and erasure. *Probl. Inf. Trans.*, 6(1):13–17, 1970.

[16] Ahlswede, R. The capacity of a channel with arbitrary varying Gaussian channel probability function. *Trans. 6th Prague Conf. Inf. Theory*, 13–21, 1971.

[17] Blahut, R. Computation of channel capacity and rate distortion functions. *IEEE Trans. Inf. Theory*, 18:460–473, 1972.

[18] Ihara, S. On the capacity of channels with additive non-Gaussian noise. *Inf. Control*, (37(1)):34–39, 1978.

[19] El Gamal, A.A. The capacity of a class of broadcast channels. *IEEE Trans. Inf. Theory*, 25(2):166–169, 1979.

[20] Gelfand, S.I. and Pinsker, M.S. Capacity of a broadcast channel with one deterministic component. *Probl. Inf. Trans.*, 16(1):17–21, 1980.

[21] Sato, H. The capacity of the Gaussian interference channel under strong interference. *IEEE Trans. Inf. Theory*, 27(6):786–788, 1981.

[22] Carleial, A. Outer bounds on the capacity of the interference channel. *IEEE Trans. Inf. Theory*, 29:602–606, 1983.

[23] Ozarow, L.H. The capacity of the white Gaussian multiple access channel with feedback. *IEEE Trans. Inf. Theory*, 30:623–629, 1984.

[24] Teletar, E. Capacity of multiple antenna Gaussian channels. *Eur. Trans. Telecommun.*, 10(6):585–595, 1999.

[25] Hamming, R.V. Error detecting and error correcting codes. *Bell System Tech. J.*, 29:147–160, 1950.

[26] Rice, S.O. Communication in the presence of noise: probability of error for two encoding schemes. *Bell System Tech. J.*, 29:60–93, 1950.

[27] Huffman, D.A. A method for the construction of minimum redundancy codes. *Proc. IRE*, 40:1098–1101, 1952.

[28] Elias, P. Error-free coding. *IRE Trans. Inf. Theory*, 4:29–37, 1954.

[29] Shannon, C.E. Certain results in coding theory for noisy channels. *Infor. Control.*, 1:6–25, 1957.

[30] Shannon, C.E. Coding theorems for a discrete source with a fidelity criterion. *IRE Natl. Conv. Record*, 7:142–163, 1959.

[31] Bose, R.C. and Ray-Chaudhuri, D.K. On a class of error correcting binary group codes. *Inf. Control*, (3):68–79, 1960.

[32] Wozencraft, J. and Reiffen, B. *Sequential Decoding*. MIT Press, Cambridge, MA, 1961.

[33] Viterbi, A.J. On coded phase-coherent communications. *IRE Trans. Space Electron. Telem.*, 7:3–14, 1961.

[34] Gallager, R.G. *Low Density Parity Check Codes*. MIT Press, Cambridge, MA, 1963.

[35] Abramson, N.M. *Information Theory and Coding*. McGraw-Hill, New York, 1963.

[36] Viterbi, A.J. Error bounds for convolutional codes and an asymptotically optimum decoding algorithm. *IEEE Trans. Inf. Theory*, IT-13:260–269, 1969.

[37] Forney, G.D. Convolutional codes: algebraic structure. *IEEE Trans. Inf. Theory*, 16(6):720–738, 1970.

[38] Berger, T. *Rate Distortion Theory: A Mathematical Basis for Data Compression*. Prentice-Hall, Englewood Cliffs, 1971.

[39] Viterbi, A.J. Convolutional codes and their performance in communication systems. *IEEE Trans. Commun. Technol.*, 19(5):751–772, 1971.

[40] Ziv, J. Coding of sources with unknown statistics. Distortion relative to a fidelity criterion. *IEEE Trans. Inf. Theory*, 18:389–394, 1972.

[41] Slepian, D. and Wolf, J.K. A coding theorem for multiple access channels with correlated sources. *Bell Syst. Tech. J.*, 52:1037–1076, 1973.

[42] Pasco, R. Source Coding Algorithms for Fast Data Compression. PhD thesis, Stanford University, Stanford, 1976.

[43] Wolfowitz, J. *Coding Theorems of Information Theory.* Prentice-Hall, Englewood Cliffs, 1978.

[44] Viterbi, A.J. and Omura, J.K. *Principles of Digital Communication and Coding.* McGraw-Hill, New York, 1979.

[45] Blahut, R.E. *Theory and Practice of Error Control Codes.* Addison-Wesley, Reading, MA, 1983.

[46] Lin, S. and Costello, D.J. *Error Control Coding: Fundamentals and Applications.* Prentice-Hall, Englewood Cliffs, 3rd edition, 1983.

[47] Gray, R.M. *Source Coding Theory.* Kluwer, Boston, 1990.

[48] Berrou, C., Glavieux, A., and Thitimajshima, P. Near Shannon limit error-correcting codes and decoding: turbo codes. *Proc. 1993 Int. Conf. Commun.*, 1064–1070, 1993.

[49] Sayood, K. *Introduction to Data Compression.* Morgan Kaufmann, San Francisco, 1996.

[50] Hartley, R.V.L. Transmission of information. *Bell Syst. Tech. J.*, 7(3):535–563, 1928.

[51] Shannon, C.E. A mathematical theory of communication. *Bell System Tech. J.*, 27:379–423, 623–656, 1948.

[52] Shannon, C.E. Communication in the presence of noise. *Bell Syst. Tech. J.*, 37:10–21, 1949.

[53] Goldman, S. *Information Theory.* Prentice-Hall, Englewood Cliffs, 1953.

[54] Powers, K.H. A unified theory of information. Technical Report 311, R.L.E. Massachusetts Institute of Technology, 1956.

[55] Gelfand, I.M., Kolmogorov, A.N., and Yaglom, A.M. On the general definition of the quantity of information. *Dokl. Akad. Nauk SSSR*, 111(4):745–748, 1956.

[56] Feinstein, A. *A Foundation of Information Theory.* McGraw-Hill, New York, 1958.

[57] Stam, A. Some inequalities satisfied by the quantities of information of Fisher and Shannon. *Inf. Control.*, (2):101–112, 1959.

[58] Fano, R.M. *Transmission of Information: A Statistical Theory of Communication.* Wiley, New York, 1961.

[59] Kolmogorov, A.N. Three approaches to the quantitative definition of information. *Probl. Inf. Trans.*, 1(1):1–7, 1965.

[60] Jelinec, F. *Probabilistic Information Theory.* McGraw-Hill, New York, 1968.

[61] Kolmogorov, A.N. Logical basis for information theory and probability theory. *IEEE Trans. Inform. Theory*, 14:662–664, 1968.

[62] Schwartz, M. *Information, Transmission, Modulation and Noise.* McGraw-Hill, New York, 1970.

[63] Harkevich, A.A. Information Theory. Pattern Recognition. In *Selected Proceedings.* Volume 3. Nauka, 1972 (in Russian).

[64] Slepian, D. *Key Papers in the Development of Information Theory.* IEEE Press, New York, 1974.

[65] Cover, T.M. Open problems in information theory. *Proc. Moscow Inf. Theory Workshop*, 35–36, 1975.

[66] Stratonovich, R.L. *Information Theory.* Soviet Radio, Moscow, 1975 (in Russian).

[67] Blahut, R.E. *Principles and Practice of Information Theory.* Addison-Wesley, Reading, MA, 1987.

[68] Dembo, A., Cover, T.M., and Thomas, J.A. Information theoretic inequalities. *IEEE Trans. Inf. Theory*, (37 (6)):1501–1518, 1991.

[69] Ihara, S. *Information Theory for Continuous Systems*. World Scientific, Singapore, 1993.

[70] Cover, T.M. and Thomas, J.A. *Elements of Information Theory*. Wiley, Hoboken, 2nd edition, 2006.

[71] Bennett, W.R. Time-division multiplex systems. *Bell Syst. Tech. J.*, 20:199–221, 1941.

[72] Shannon, C.E. Communication theory of secrecy systems. *Bell Syst. Tech. J.*, 28:656–715, 1949.

[73] Wozencraft, J.M. and Jacobs, I.M. *Principles of Communication Engineering*. Wiley, New York, 1965.

[74] Gallager, R.G. *Information Theory and Reliable Communication*. Wiley, New York, 1968.

[75] Fink, L.M. *Discrete Messages Transmission Theory*. Soviet Radio, Moscow, 1970 (in Russian).

[76] Liao, H. Multiple Access Channels. PhD thesis, Department of Electrical Engineering, University of Hawaii, Honolulu, 1972.

[77] Penin, P.I. *Digital Information Transmission Systems*. Soviet Radio, Moscow, 1976 (in Russian).

[78] Lindsey, W.G. and Simon, M.K. *Telecommunication Systems Engineering*. Prentice-Hall, Englewood Cliffs, 1973.

[79] Thomas, C.M., Weidner, M.Y., and Durrani, S.H. Digital amplitude-phase keying with m-ary alphabets. *IEEE Trans. Commun.*, 22(2):168–180, 1974.

[80] Spilker, J. *Digital Communications by Satellite*. Prentice-Hall, Englewood Cliffs, 1977.

[81] Penin, P.I. and Filippov, L.I. *Information Transmission Electronic Systems*. Radio i svyaz, Moscow, 1984 (in Russian).

[82] Varakin, L.E. *Communication Systems with Noise-Like Signals*. Radio i svyaz, Moscow, 1985 (in Russian).

[83] Zyuko, A.G., Falko, A.I., and Panfilov, I.P. *Noise Immunity and Efficiency of Communication Systems*. Radio i svyaz, Moscow, 1985 (in Russian).

[84] Zyuko, A.G., Klovskiy, D.D., Nazarov, M.V., and Fink, L.M. *Signal Transmission Theory*. Radio i svyaz, Moscow, 1986 (in Russian).

[85] Wiener, N. *Cybernetics, or Control and Communication in the Animal and the Machine*. Wiley, New York, 1948.

[86] Bar-Hillel, Y. Semantic information and its measures. *Trans. 10th Conf. Cybernetics*, 33-48, 1952.

[87] Rashevsky, N. Life, information theory and topology. *Bull. Math. Biophysics*, 17:229–235, 1955.

[88] Cherry, C.E. *On Human Communication: A Review, a Survey, and a Criticism*. MIT Press, Cambridge, MA, 3rd edition, 1957.

[89] Kullback, S. *Information Theory and Statistic*. Wiley, New York, 1959.

[90] Brillouin, L. *Science and Information Theory*. Academic Press, New York, 1962.

[91] Shreider, Yu.A. On a model of semantic information theory. *Probl. Cybernetics*, (13):233–240, 1965 (in Russian).

[92] Mityugov, V.V. *Physical Foundations of Information Theory*. Soviet Radio, Moscow, 1976 (in Russian).

[93] Chaitin, G.J. *Algorithmic Information Theory*. Cambridge University Press, Cambridge, 1987.

[94] Yockey, H.P. *Information Theory and Molecular Biology*. Cambridge University Press, New York, 1992.

[95] Goppa, V.D. *Introduction to Algebraic Information Theory*. Nauka, Moscow, 1995 (in Russian).

[96] Bennet, C.H. and Shor, P.W. Quantum information theory. *IEEE Trans. Inf. Theory*, 44:2724–2742, 1998.

[97] Holevo, A.S. *Introduction to Quantum Information Theory*. MCCME, Moscow, 2002 (in Russian).

[98] Chernavsky, D.S. *Synergetics and Information (Dynamic Information Theory)*. Editorial URSS, Moscow, 2nd edition, 2004 (in Russian).

[99] Schottky, W. Über spontane stromschwankungen in verschiedenen elektrizitätsleitern. *Annalen der Physik*, 57:541–567, 1918.

[100] Schottky, W. Zur berechnung und beurteilung des schroteffektes. *Annalen der Physik*, 68:157–176, 1922.

[101] Nyquist, H. Thermal agitation of electric charge in conductors. *Phys. Rev.*, 32:110–113, 1928.

[102] Johnson, J.B. Thermal agitation of electricity in conductors. *Phys. Rev.*, 32:97–109, 1928.

[103] Papaleksy, N.D. *Radio Noise and Noise Reduction*. Gos. Tech. Izdat, Moscow, 1942 (in Russian).

[104] Rice, S.O. Mathematical analysis of random noise. *Bell System Tech. J.*, 23(3):282–332, 1944.

[105] Rice, S.O. Statistical properties of a sine wave plus random noise. *Bell System Tech. J.*, 27(1):109–157, 1948.

[106] Davenport, W.B. and Root, W.L. *Introduction to the Theory of Random Signals and Noise*. McGraw-Hill, New York, 1958.

[107] Bunimovich, V.I. *Fluctuating Processes in Radio Receivers*. Soviet Radio, Moscow, 1951 (in Russian).

[108] Tihonov, V.I. *Fluctuating Noise in the Elements of Receiving and Amplifying Sets*. VVIA, Moscow, 1951 (in Russian).

[109] Doob, J.L. *Stochastic Processes*. Wiley, New York, 1953.

[110] Zheleznov, N.A. On some questions of spectral-correlative theory of non-stationary signals. *Radiotekhnika i elektronika*, 4(3):359–373, 1959 (in Russian).

[111] Deutsch, R. *Nonlinear Transformations of Random Processes*. Prentice-Hall, Englewood Cliffs, 1962.

[112] Tihonov, V.I. *Statistical Radio Engineering*. Soviet Radio, Moscow, 1966 (in Russian).

[113] Levin, B.R. *Theoretical Basics of Statistical Radio Engineering*. Volume 1. Soviet Radio, Moscow, 1969 (in Russian).

[114] Malahov, A.N. *Cumulant Analysis of Stochastic non-Gaussian Processes and Their Transformations.* Soviet Radio, Moscow, 1978 (in Russian).

[115] Tihonov, V.I. *Statistical Radio Engineering.* Radio i svyaz, Moscow, 1982 (in Russian).

[116] Tihonov, V.I. *Nonlinear Transformations of Stochastic Processes.* Radio i svyaz, Moscow, 1986 (in Russian).

[117] Stark, H. and Woods, J.W. *Probability and Random Processes with Applications to Signal Processing.* Prentice-Hall, Englewood Cliffs, 2002.

[118] Woodward, P.M. *Probability and Information Theory, with Application to Radar.* Pergamon Press, Oxford, 1953.

[119] Siebert, W.M. Studies of Woodward's uncertainty function. Technical report, R.L.E. Massachusetts Institute of Technology, 1958.

[120] Wilcox, C.H. The synthesis problem for radar ambiguity functions. Technical Report 157, Mathematical Research Center, U.S. Army, University of Wisconsin, Madison, 1958.

[121] Cook, C.E. and Bernfeld, M. *Radar Signals. An Introduction to Theory and Application.* Academic Press, New York, 1967.

[122] Franks, L.E. *Signal Theory.* Prentice-Hall, Englewood Cliffs, 1969.

[123] Papoulis, A. *Signal Analysis.* McGraw-Hill, New York, 1977.

[124] Varakin, L.E. *Theory of Signal Systems.* Soviet Radio, Moscow, 1978 (in Russian).

[125] Kolmogorov, A.N. Interpolation and extrapolation of stationary random sequences. *Izv. AN SSSR. Ser. Math.*, 5(1):3–11, 1941 (in Russian).

[126] North, D.O. Analysis of the factors which determine signal/noise discrimination in pulsed carrier systems. Technical Report PTR-6C, RCA Lab., Princeton, N.J., 1943. (reprinted in *Proc. IRE*, Volume 51, 1963, 1016–1027).

[127] Kotelnikov, V.A. *Theory of Potential Noise Immunity.* Moscow Energetic Institute, Moscow, 1946 (in Russian).

[128] Wiener, N. *Extrapolation, Interpolation and Smoothing of Stationary Time Series.* MIT Press, Cambridge, MA, 1949.

[129] Slepian, D. Estimation of signal parameters in the presence of noise. *IRE Trans. Inf. Theory*, 3:68–89, 1954.

[130] Middleton, D. and Van Meter, D. Detection and extraction of signals in noise from the point of view of statistical decision theory. *J. Soc. Ind. Appl. Math.*, 3:192–253, 1955.

[131] Amiantov, I.N. *Application of Decision Making Theory to the Problems of Signal Detection and Signal Extraction in Background Noise.* VVIA, Moscow, 1958 (in Russian).

[132] Kalman, R.E. A new approach to linear filtering and prediction problem. *J. Basic Eng. (Trans. ASME)*, 82D:35–45, 1960.

[133] Blackman, R.B. *Data Smoothing and Prediction.* Addison-Wesley, Reading, MA, 1965.

[134] Rabiner, L.R. and Gold, B. *Theory and Application of Digital Signal Processing.* Prentice-Hall, Englewood Cliffs, 1975.

[135] Oppenheim, A.V. and Schafer, R.W. *Digital Signal Processing.* Prentice-Hall, Englewood Cliffs, 1975.

[136] Applebaum, S.P. Adaptive arrays. *IEEE Trans. Antennas Propag.*, 24(5):585–598, 1976.

[137] Monzingo, R. and Miller, T. *Introduction to Adaptive Arrays*. Wiley, Hoboken, NJ, 1980.

[138] Widrow, B. and Stearns, S.D. *Adaptive Signal Processing*. Prentice-Hall, Englewood Cliffs, 1985.

[139] Alexander, S.T. *Adaptive Signal Processing: Theory and Applications*. Springer Verlag, New York, 1986.

[140] Oppenheim, A.V. and Schafer, R.W. *Discrete-Time Signal Processing*. Prentice-Hall, Englewood Cliffs, 1989.

[141] Lim, J.S. *Two-dimensional Signal and Image Processing*. Prentice-Hall, Englewood Cliffs, 1990.

[142] Haykin, S. *Adaptive Filter Theory*. Prentice-Hall, Englewood Cliffs, 2002.

[143] Middleton, D. *An Introduction to Statistical Communication Theory*. McGraw-Hill, New York, 1960.

[144] Vaynshtein, L.A. and Zubakov, V.D. *Signal Extraction in Background Noise*. Soviet Radio, Moscow, 1960 (in Russian).

[145] Viterbi, A.J. *Principles of Coherent Communication*. McGraw-Hill, New York, 1966.

[146] Van Trees, H.L. *Detection, Estimation, and Modulation Theory*. Wiley, New York.

[147] Sage, A.P. and Melsa, J.L. *Estimation Theory with Applications to Communications and Control*. McGraw Hill, New York, 1971.

[148] Amiantov, I.N. *Selected Questions of Statistical Communication Theory*. Soviet Radio, Moscow, 1971 (in Russian).

[149] Levin, B.R. *Theoretical Basics of Statistical Radio Engineering*. Volume 2. Soviet Radio, Moscow.

[150] Levin, B.R. *Theoretical Basics of Statistical Radio Engineering*. Volume 3. Soviet Radio, Moscow.

[151] Gilbo, E.P. and Chelpanov, I.B. *Signal Processing on the Basis of Ordered Selection: Majority Transformation and Others That are Close to it*. Soviet Radio, Moscow, 1977 (in Russian).

[152] Repin, V.G. and Tartakovsky, G.P. *Statistical Synthesis under Prior Uncertainty and Adaptation of Information Systems*. Soviet Radio, Moscow, 1977 (in Russian).

[153] Sosulin, Yu.G. *Detection and Estimation Theory of Stochastic Signals*. Soviet Radio, Moscow, 1978 (in Russian).

[154] Kulikov, E.I. and Trifonov, A.P. *Estimation of Signal Parameters in the Background of Noise*. Soviet Radio, Moscow, 1978 (in Russian).

[155] Tihonov, V.I. *Optimal Signal Reception*. Radio i Svyaz, Moscow, 1983 (in Russian).

[156] Akimov, P.S., Bakut, P.A., Bogdanovich, V.A., et al. *Signal Detection Theory*. Radio i Svyaz, Moscow, 1984 (in Russian).

[157] Kassam, S.A. and Poor, H.V. Robust techniques for signal processing: A Survey. *Proc. IEEE*, 73(3):433–481, 1985.

[158] Trifonov, A.P. and Shinakov, Yu. S. *Joint Classification of Signals and Estimation of Their Parameters in the Noise Background*. Radio i Svyaz, Moscow, 1986 (in Russian).

[159] Levin, B.R. *Theoretical Basics of Statistical Radio Engineering*. Radio i Svyaz, Moscow, 1989 (in Russian).

[160] Kay, S.M. *Fundamentals of Statistical Signal Processing*. Prentice-Hall, Englewood Cliffs, 1993.

[161] Poor, H.V. *An Introduction to Signal Detection and Estimation*. Springer, New York, 2nd edition, 1994.

[162] Helstrom, C.W. *Elements of Signal Detection and Estimation*. Prentice-Hall, Englewood Cliffs, 1994.

[163] Middleton, D. *An Introduction to Statistical Communication Theory*. IEEE, New York, 1996.

[164] Sklar, B. *Digital Communications: Fundamentals and Applications*. Prentice-Hall, Englewood Cliffs, 2nd edition, 2001.

[165] Minkoff, J. *Signal Processing Fundamentals and Applications for Communications and Sensing Systems*. Artech House, Norwood, MA, 2002.

[166] Bogdanovich, V.A. and Vostretsov, A.G. *Theory of Robust Detection, Classification and Estimation of Signals*. Fiz. Math. Lit, Moscow, 2004 (in Russian).

[167] Vaseghi, S.V. *Advanced Digital Signal Processing and Noise Reduction*. Wiley, New York, 2006.

[168] Huber, P.J. *Robust Statistics*. Wiley, New York, 1981.

[169] Fraser, D.A. *Nonparametric Methods in Statistics*. Wiley, New York, 1957.

[170] Noether, G.E. *Elements of Nonparametric Statistics*. Wiley, New York, 1967.

[171] Hajek, J. and Sidek, Z. *Theory of Rank Tests*. Academic Press, New York, 1967.

[172] Lehmann, E.L. *Nonparametrics: Statistical Methods Based on Ranks*. Holden-Day, Oakland, CA, 1975.

[173] Kolmogorov, A.N. Complete metric Boolean algebras. *Philosophical Studies*, 77(1):57–66, 1995.

[174] Horn, A. and Tarski, A. Measures in Boolean algebras. *Trans. Amer. Math. Soc.*, 64:467–497, 1948.

[175] Vladimirov, D.A. *Boolean Algebras*. Nauka, Moscow, 1969 (in Russian).

[176] Sikorsky, R. *Boolean Algebras*. Springer, Berlin, 2nd edition, 1964.

[177] Boltyansky, V.G. and Vilenkin, N.Ya. *Symmetry in Algebra*. Nauka, Moscow, 1967 (in Russian).

[178] Weyl, H. *Symmetry*. Princeton University Press, 1952.

[179] Wigner, E.P. *Symmetries and Reflections: Scientific Essays*. Indiana University Press, Bloomington, 1967.

[180] Dorodnitsin, V.A. and Elenin, G.G. Symmetry of nonlinear phenomena. In *Computers and Nonlinear Phenomena: Informatics and Modern Nature Science*, pages 123–191. Nauka, Moscow, 1988 (in Russian).

[181] Bhatnagar, P.L. *Nonlinear Waves in One-dimensional Dispersive Systems*. Clarendon Press, Oxford, 1979.

[182] Kurdyumov, S.P., Malinetskiy, G.G., Potapov, A.B., and Samarskiy, A.A. Structures in nonlinear medium. In *Computers and Nonlinear Phenomena: Informatics and Modern Nature Science*, pages 5–43. Nauka, Moscow, 1988 (in Russian).

[183] Whitham, G.B. *Linear and Nonlinear Waves*. Wiley, New York, 1974.

[184] Dubrovin, B.A., Fomenko, A.T., and Novikov, S.P. *Modern Geometry*. Nauka, Moscow, 1986 (in Russian).

[185] Klein, F. *Non-Euclidean Geometry*. Editorial URSS, Moscow, 2004 (in Russian).

[186] Rosenfeld, B.A. *Non-Euclidean Spaces*. Nauka, Moscow, 1969 (in Russian).

[187] Dirac, P.A.M. *Principles of Quantum Mechanics*. Clarendon Press, Oxford, 1930.

[188] Shannon, C.E. The bandwagon. *IRE Trans. Inf. Theory*, 2(1):3, 1956.

[189] Ursul, A.D. *Information. Methodological Aspects*. Nauka, Moscow, 1971 (in Russian).

[190] Berg, A.I. and Biryukov, B.V. Cybernetics: the way of control problem solving. In *Future of Science*. Volume 3. Nauka, 1970 (in Russian).

[191] Steane, A.M. Quantum computing. *Rep. Progr. Phys.*, (61):117–173, 1998.

[192] Feynman, R.P. Quantum mechanical computers. *Found. Phys.*, (16):507–531, 1987.

[193] Wiener, N. *I Am a Mathematician*. Doubleday, New York, 1956.

[194] Chernin, A.D. *Physics of Time*. Nauka, Moscow, 1987 (in Russian).

[195] Leibniz, G.W. *Monadology: An Edition for Students*. University of Pittsburgh Press, 1991.

[196] Riemann, B. *On the Hypotheses Which Lie at the Foundation of Geometry*. Göttingen University, 1854.

[197] Klein, F. *Highest Geometry*. Editorial URSS, Moscow, 2004 (in Russian).

[198] Zheleznov, N.A. *On Some Questions of Informational Electric System Theory*. LKVVIA, Leningrad, 1960 (in Russian).

[199] Yaglom, A.M. and Yaglom, I.M. *Probability and Information*. Nauka, Moscow, 1973 (in Russian).

[200] Prigogine, I. and Stengers, I. *Time, Chaos and the Quantum: Towards the Resolution of the Time Paradox*. Harmony Books, New York, 1993.

[201] Shreider, Yu.A. On quantitative characteristics of semantic information. *Sci. Tech. Inform.*, (10), 1963 (in Russian).

[202] Ogasawara, T. Compact metric Boolean algebras and vector lattices. *J. Sci. Hiroshima Univ.*, 11:125—128, 1942.

[203] Mibu, Y. Relations between measures and topology in some Boolean spaces. *Proc. Imp. Acad. Tokio*, 20:454–458, 1944.

[204] Ellis, D. Autometrized Boolean algebras. *Canadian J. Math.*, 3:87–93, 1951.

[205] Tomita, M. Measure theory of complete Boolean algebras. *Mem. Fac. Sci. Kyusyu Univ.*, 7:51–60, 1952.

[206] Hewitt, E. A note on measures in Boolean algebras. *Duke Math. J.*, 20:253–256, 1953.

[207] Vulih, B.Z. On Boolean measure. *Uchen. Zap. Leningr. Ped. Inst.*, 125:95–114, 1956 (in Russian).

[208] Lamperti, J. A note on autometrized Boolean algebras. *Amer. Math. Monthly*, 64:188–189, 1957.

[209] Heider, L.J. A representation theorem for measures on Boolean algebras. *Mich. Math. J.*, 5:213–221, 1958.

[210] Kelley, J.L. Measures in Boolean algebras. *Pacific J. Math.*, 9:1165–1177, 1959.

[211] Vladimirov, D.A. On the countable additivity of a Boolean measure. *Vestnik Leningr. Univ. Mat. Mekh. Astronom*, 16(19):5–15, 1961 (in Russian).

[212] Vinokurov, V.G. Representations of Boolean algebras and measure spaces. *Math. Sb.*, 56 (98)(3):374–391, 1962 (in Russian).

[213] Vladimirov, D.A. Invariant measures on Boolean algebras. *Math. Sb.*, 67 (109)(3):440–460, 1965 (in Russian).

[214] Stone, M.H. Postulates for Boolean algebras and generalized Boolean algebras. *Amer. J. Math.*, 57:703–732, 1935.

[215] Stone, M.H. The theory of representations for Boolean algebras. *Trans. Amer. Math. Soc.*, 40:37–111, 1936.

[216] McCoy, N.H. and Montgomery, D. A representation of generalized Boolean rings. *Duke Math. J.*, 3:455–459, 1937.

[217] Grätzer, G. and Schmidt, E.T. On the generalized Boolean algebras generated by a distributive lattice. *Nederl. Akad. Wet. Proc.*, 61:547–553, 1958.

[218] Subrahmanyan, N.V. Structure theory for generalized Boolean rings. *Math. Ann.*, 141:297–310, 1960.

[219] Whitney, H. The abstract properties of linear dependence. *Amer. J. Math.*, 37:507–533, 1935.

[220] Menger, K. New foundations of projective and affine geometry. *Ann. Math.*, 37:456–482, 1936.

[221] Birkhoff, G. *Lattice Theory*. American Mathematical Society, Providence, 1967.

[222] Blumenthal, L.M. Boolean geometry. *Rend. Coirc. Math. Palermo*, 1:1–18, 1952.

[223] Artamonov, V.A., Saliy, V.N., Skornyakov, L.A., Shevrin, L.N., and Shulgeyfer, E.G. *General Algebra*. Volume 2. Nauka, Moscow, 1991 (in Russian).

[224] Hilbert, D. *The Foundations of Geometry*. Open Court Company, 2001.

[225] Marczewski, F. and Steinhaus, H. On certain distance of sets and the corresponding distance of functions. *Colloq. Math.*, 6:319–327, 1958.

[226] Zolotarev, V.M. *Modern Theory of Summation of Independent Random Variables*. Nauka, Moscow, 1986 (in Russian).

[227] Buldygin, V.V. and Kozachenko, Yu.V. *Metric Characterization of Random Variables and Random Processes*. American Mathematical Society, Providence, 2000.

[228] Samuel, E. and Bachi, R. Measure of distance of distribution functions and some applications. *Metron*, 13:83–112, 1964.

[229] Dudley, R.M. Distances of probability measures and random variables. *Ann. Math. Statist*, 39(5):1563–1572, 1968.

[230] Senatov, V.V. On some properties of metrics at the set of distribution functions. *Math. Sb.*, 31(3):379–387, 1977.

[231] Kendall, M.G. and Stuart, A. *The Advanced Theory of Statistics. Inference and Relationship*. Charles Griffin, London, 1961.

[232] Cramer, H. *Mathematical Methods of Statistics*. Princeton University Press, 1946.

[233] Melnikov, O.V., Remeslennikov, V.N., Romankov, V.A., Skornyakov, L.A., and Shestakov, I.P. *General Algebra*. Volume 1. Nauka, Moscow, 1990 (in Russian).

[234] Prudnikov, A.P., Brychkov, Yu.A., and Marichev, O.I. *Integrals and Series: Elementary Functions*. Gordon & Breach, New York, 1986.

[235] Paley, R.E. and Wiener, N. *Fourier Transforms in the Complex Domain*. American Mathematical Society, Providence, 1934.

[236] Baskakov, S.I. *Radio Circuits and Signals*. Vysshaya Shkola, Moscow, 2nd edition, 1988 (in Russian).

[237] Oxtoby, J. *Measure and Category*. Springer, New York, 2nd edition, 1980.

[238] Kotelnikov, V.A. On the transmission capacity of "ether" and wire in electrocommunications. In *Modern Sampling Theory: Mathematics and Applications*. Birkhauser, Boston, 2000. (Reprint of 1933 edition).

[239] Whittaker, J.M. Interpolatory function theory. *Cambridge Tracts on Math. and Math. Physics*, (33), 1935.

[240] Jerry, A.J. The Shannon sampling theorem: its various extensions and applications: a tutorial review. *Proc. IEEE*, (65):1565–1596, 1977.

[241] Dmitriev, V.I. *Applied Information Theory*. Vysshaya shkola, Moscow, 1989 (in Russian).

[242] Popoff, A.A. Sampling theorem for the signals of space built upon generalized Boolean algebra with a measure. *Izv. VUZov. Radioelektronika*, (1):31–39, 2010 (in Russian). Reprinted in *Radioelectronics and Communications Systems*, 53 (1): 25–32, 2010.

[243] Tihonov, V.I. and Harisov, V.N. *Statistical Analysis and Synthesis of Electronic Means and Systems*. Radio i svyaz, Moscow, 1991 (in Russian).

[244] Harkevich, A.A. *Noise Reduction*. Nauka, Moscow, 1965 (in Russian).

[245] Deza, M.M. and Laurent, M. *Geometry of Cuts and Metrics*. Springer, Berlin, 1997.

[246] Aleksandrov, P.S. *Introduction to Set Theory and General Topology*. Nauka, Moscow, 1977 (in Russian).

[247] Borisov, V.A., Kalmykov, V.V., and Kovalchuk, Ya.M. *Electronic Systems of Information Transmission*. Radio i svyaz, Moscow, 1990 (in Russian).

[248] Zyuko, A.G., Klovskiy, D.D., Korzhik, V.I., and Nazarov, M.V. *Electric Communication Theory*. Radio i svyaz, Moscow, 1999 (in Russian).

[249] Tihonov, V.I. and Mironov, M.A. *Markov Processes*. Soviet Radio, Moscow, 1977 (in Russian).

[250] Zacks, S. *The Theory of Statistical Inference*. Wiley, New York, 1971.

[251] Lehmann, E.L. *Theory of Point Estimation*. Wiley, New York, 1983.

[252] David, H.A. *Order Statistics*. Wiley, New York, 1970.

[253] Gumbel, E.J. *Statistics of Extremes*. Columbia University Press, New York, 1958.

[254] Van Trees, H. L. *Detection, Estimation, and Modulation Theory*. Wiley, New York, 1968.

[255] Grätzer, G. *General Lattice Theory*. Akademie Verlag, Berlin, 1978.

[256] Le Cam, L. On some asymptotic properties of maximum likelihood estimates and related bayes estimates. *Univ. California Publ. Statist.*, 1:277–330, 1953.

[257] Mudrov, V.I. and Kushko, V.L. *The Least Modules Method*. Znanie, Moscow, 1971 (in Russian).

[258] Kendall, M.G. and Stuart, A. Distribution theory. In *Advanced Theory of Statistics*. Volume 1. Charles Griffin, 1960.

[259] Cohn, P.M. *Universal Algebra*. Harper & Row, New York, 1965.

[260] Grätzer, G. *Universal Algebra*. Springer, Berlin, 1979.

[261] Polyà, G. Remarks on characteristic functions. *Proc. First Berkeley Symp. Math. Stat. Probabil.*, (1):115–123, 1949.

[262] Dwight, H.B. *Tables of Integrals and Other Mathematical Data*. MacMillan, New York, 1961.

[263] Tukey, J.W. *Exploratory Data Analysis*. Addison-Wesley, Reading, MA, 1977.

[264] Tihonov, V.I. *Overshoots of Random Processes*. Nauka, Moscow, 1970 (in Russian).

[265] Mallows, C.L. Some theory of nonlinear smoothers. *Ann. Stat.*, 8:695–715, 1980.

[266] Gallagher, N.C. and Wise, G.L. A theoretical analysis of the properties of median filters. *IEEE Trans. Acoustics, Speech and Signal Proc.*, ASSP-29:1136–1141, 1981.

[267] Shirman, Ya.D. and Manzhos, V.N. *Theory and Technique for Radar Data Processing in the Presence of Noise*. Radio i Svyaz, Moscow, 1981 (in Russian).

[268] Popoff, A.A. Unit of signal space mapping. Patent of Ukraine 56926, H 04 B 15/00, 2011.

[269] Popoff, A.A. Method of signal space mapping. Patent of Ukraine 60051, H 04 B 15/00, 2011.

[270] Popoff, A.A. Interrelation of signal theory and information theory. The ways of logical difficulties overcoming. *J. State Univ. Inform. Comm. Tech.*, 4(4):312–324, 2006 (in Russian).

[271] Popoff, A.A. Probabilistic-statistical and informational characteristics of stochastic processes that are invariant with respect to the group of bijective mappings. *J. State Univ. Inform. Comm. Tech.*, 5(1):52–62, 2007 (in Russian).

[272] Popoff, A.A. Informational relationships between the elements of signal space built upon generalized Boolean algebra with a measure. *J. State Univ. Inform. Comm. Tech.*, 5(2):175–184, 2007 (in Russian).

[273] Popoff, A.A. Information quantity measure in signal space built upon generalized Boolean algebra with a measure. *J. State Univ. Inform. Comm. Tech.*, 5(3):253–261, 2007 (in Russian).

[274] Popoff, A.A. Informational and physical signal interactions in signal space. The notion of ideal signal interaction. *J. State Univ. Inform. Comm. Tech.*, 5(4):19–27, 2007 (in Russian).

[275] Popoff, A.A. Invariants of one-to-one functional transformations of stochastic processes. *Izv. VUZov. Radioelektronika*, (11):35–43, 2007 (in Russian). Reprinted in *Radioelectronics and Communications Systems*, 50 (11): 609–615, 2007.

[276] Popoff, A.A. Information quantity transferred by binary signals in the signal space built upon generalized Boolean algebra with a measure. *J. State Univ. Inform. Comm. Tech.*, 6(1):27–32, 2008 (in Russian).

[277] Popoff, A.A. Homomorphic mappings in the signal space built upon generalized Boolean algebra with a measure. *J. State Univ. Inform. Comm. Tech.*, 6(3):238–247, 2008 (in Russian).

[278] Popoff, A.A. Information quantity transferred by m-ary signals in the signal space built upon generalized Boolean algebra with a measure. *J. State Univ. Inform. Comm. Tech.*, 6(4):287–295, 2008 (in Russian).

[279] Popoff, A.A. The unified axiomatics of signal theory and information theory. *Inform. Sec. Sci. Proc.*, (5):199–205, 2008 (in Russian).

[280] Popoff, A.A. Comparative analysis of estimators of unknown nonrandom signal parameter in linear space and K-space. *Izv. VUZov. Radioelektronika*, (7):29–40, 2008 (in Russian). Reprinted in *Radioelectronics and Communications Systems*, 51 (7): 368–376, 2008.

[281] Popoff, A.A. Possibilities of processing the signals with completely defined parameters under interference (noise) background in signal space with algebraic lattice properties. *Izv. VUZov. Radioelektronika*, (8):25–32, 2008 (in Russian). Reprinted in *Radioelectronics and Communications Systems*, 51 (8): 421–425, 2008.

[282] Popoff, A.A. Characteristics of processing the harmonic signals in interference (noise) background under their interaction in K-space. *Izv. VUZov. Radioelektronika*, (10):69–80, 2008 (in Russian). Reprinted in *Radioelectronics and Communications Systems*, 51 (10): 565–572, 2008.

[283] Popoff, A.A. Informational characteristics and properties of stochastic signal considered as subalgebra of generalized Boolean algebra with a measure. *Izv. VUZov. Radioelektronika*, (11):57–67, 2008 (in Russian). Reprinted in *Radioelectronics and Communications Systems*, 51 (11): 615–621, 2008.

[284] Popoff, A.A. Noiseless channel capacity in signal space built upon generalized Boolean algebra with a measure. *J. State Univ. Inform. Comm. Tech.*, 7(1):54–62, 2009 (in Russian).

[285] Popoff, A.A. Geometrical properties of signal space built upon generalized Boolean algebra with a measure. *J. State Univ. Inform. Comm. Tech.*, 7(3):27–32, 2009 (in Russian).

[286] Popoff, A.A. Characteristics and properties of signal space built upon generalized Boolean algebra with a measure. *Izv. VUZov. Radioelektronika*, (5):34–45, 2009 (in Russian). Reprinted in *Radioelectronics and Communications Systems*, 52 (5): 248–255, 2009.

[287] Popoff, A.A. Peculiarities of continuous message filtering in signal space with algebraic lattice properties. *Izv. VUZov. Radioelektronika*, (9):29–40, 2009 (in Russian). Reprinted in *Radioelectronics and Communications Systems*, 52 (9): 474–482, 2009.

[288] Popoff, A.A. Informational characteristics of scalar random fields that are invariant with respect to group of their bijective mappings. *Izv. VUZov. Radioelektronika*, (11):67–80, 2009 (in Russian). Reprinted in *Radioelectronics and Communications Systems*, 52 (11): 618–627, 2009.

[289] Popoff, A.A. Analysis of stochastic signal filtering algorithm in noise background in K-space of signals. *J. State Univ. Inform. Comm. Tech.*, 8(3):215–224, 2010 (in Russian).

[290] Popoff, A.A. Resolution of the harmonic signal filter in the space with algebraic lattice properties. *J. State Univ. Inform. Comm. Tech.*, 8(4):249–254, 2010 (in Russian).

[291] Popoff, A.A. Classification of the deterministic signals against background noise in signal space with algebraic lattice properties. *J. State Univ. Inform. Comm. Tech.*, 9(3):209–217, 2011 (in Russian).

[292] Popoff, A.A. Advanced electronic Counter-Counter-Measures technologies under extreme interference environment. *Mod. Inform. Tech. Sphere Defence*, (2):65–74, 2011.

[293] Popoff, A.A. Quality indices of APSK signal processing in signal space with L-group properties. *Mod. Special Tech.*, 25(2):61–72, 2011 (in Russian).

[294] Popoff, A.A. Invariants of groups of bijections of stochastic signals (messages) with application to statistical analysis of encryption algorithms. *Mod. Inform. Sec.*, 10(1):13–20, 2012 (in Russian).

[295] Popoff, A.A. Detection of the deterministic signal against background noise in signal space with lattice properties. *J. State Univ. Inform. Comm. Tech.*, 10(2):65–71, 2012 (in Russian).

[296] Popoff, A.A. Detection of the harmonic signal with joint estimation of time of signal arrival (ending) in signal space with L-group properties. *J. State Univ. Inform. Comm. Tech.*, 10(4):32–43, 2012 (in Russian).

[297] Popoff, A.A. Invariants of groups of mappings of stochastic signals samples in metric space with L-group properties. *J. State Univ. Inform. Comm. Tech.*, 11(1):28–38, 2013 (in Russian).

[298] Popoff, A.A. Comparative analysis of informational relationships under signal interactions in spaces with various algebraic properties. *J. State Univ. Inform. Comm. Tech.*, 11(2):53–69, 2013 (in Russian).

[299] Popoff, A.A. Unit of digital signal filtering. Patent of Ukraine 57507, G 06 F 17/18, 2011.

[300] Popoff, A.A. Method of digital signal filtering. Patent of Ukraine 57507, G 06 F 17/18, 2011.

[301] Popoff, A.A. Radiofrequency pulse resolution unit. Patent of Ukraine 59021, H 03 H 15/00, 2011.

[302] Popoff, A.A. Radiofrequency pulse resolution method. Patent of Ukraine 65236, H 03 H 15/00, 2011.

[303] Popoff, A.A. Unit of signal filtering. Patent of Ukraine 60222, H 03 H 17/00, 2011.

[304] Popoff, A.A. Method of signal filtering. Patent of Ukraine 61607, H 03 H 17/00, 2011.

[305] Popoff, A.A. Deterministic signals demodulation unit. Patent of Ukraine 60223, H 04 L 27/14, 2011.

[306] Popoff, A.A. Deterministic signals demodulation method. Patent of Ukraine 60813, H 04 L 27/14, 2011.

[307] Popoff, A.A. Transversal filter. Patent of Ukraine 71310, H 03 H 15/00, 2012.

[308] Popoff, A.A. Transversal filter. Patent of Ukraine 74846, H 03 H 15/00, 2012.

[309] Popoff, A.A. *Fundamentals of Signal Processing in Metric Spaces with Lattice Properties. Part I. Mathematical Foundations of Information Theory with Application to Signal Processing.* Central Research Institute of Armament and Defence Technologies, Kiev, 2013 (in Russian).

Index

Symbols
σ-additive measure, 28

A
Abelian group, 59
Abit (absolute unit), 63, 169, 177–178, 182, 200–201
Abit per second (abit/s) measure of channel capacity, 249
Absolute geometry, 45–46
Absolute unit (abit), 63, 169, 177–178, 182, 200–201
Active mapping, 56
Additive interaction
    channel capacity in presence of noise, 245
    features of signal interaction in signal space, 147–156
        ideal/quasi-ideal interactions, 147–156
        information quantity in a signal, 146
        optimality criteria, 146
        quality indices problems
            signal classification, 230–232
            signal detection, 238
    signal classification problem, 352
    signal detection problem, 304
    unknown nonrandom parameter estimation in sample space with lattice properties, 271
Additive measure on Boolean algebra, 28
Additive noise and information quantity, 16
Amplitude, 173
Amplitude estimation
    harmonic signal detection algorithm synthesis and analysis, 324–341
    linear frequency modulated signal detection algorithm synthesis and analysis, 342–351
Analytical stochastic processes, 104–105
Analyticity condition of Gaussian stochastic process, 104
Angle, 48
Angular measure, 52
Autocorrelation function (ACF), 175, 179–181
Axioms
    main axiom of signal processing theory, 22–23, 385
    measure of binary operations, 59–60, 109–110, 126, 385
    signal spaces with lattice properties, 269–270
    system of metric space built upon generalized Boolean algebra with a measure, 46–51
        congruence, 48–49
        connection (containment), 47
        continuity, 49
        order, 47–48
        parallels for sheet and plane, 49–50

B
Barycenter, 54, 55, 58, 318, 321, 333, 338, 348, 350
Binary signals, information quantity carried by, 174–179
Bit per second (bit/s) measure of channel capacity, 244, 247, 249
Bits as measure of information quantity, 178, 200–201

Boolean algebra, 27–28
Boolean algebra with a measure, 23, 25, 177, *See also* Generalized Boolean algebra with a measure
Boolean lattice, 27
Boolean ring, 28
Butterworth filters, 107–108

C

Cantor's axiom, 49
Capacity of a communication channel, *See* Channel capacity
Carrier signal, 173, 208
Channel capacity, *See* Channel capacity, 173
   always a finite quantity, 7, 188, 196, 199, 201, 247, 358
   bit/s and abit/s measures, 244, 247, 249
   definition, 243, 248
   generalized discrete random sequence, 183–190
   information quantity carried by binary signals, 174–179
   information quantity carried by continuous signals, 190–196
   information quantity carried by m-ary signals, 179–190
   noiseless channels, 196–206
      continuous channel capacity, 200–201
      discrete channel capacity, 199
      evaluation of, 201–206
   presence of interference (noise), 242–250
      continuous channels, 244–247
      discrete channels, 247–250
   related concepts and notation, 173
   Shannon's limit, 155–156, 357
   signal classification problem, 357–358
   Wiener's statement, 186
Classification, *See* Signal classification
Closed interval, 38

Coefficient of statistical interrelation, 78, 116
Communication channel, 196, 243, *See also* Channel capacity
Communication channel capacity, *See* Channel capacity
Complemented lattice, 27
Complements, 27
Conditional probability density functions, *See* Probability density functions
Conditional probability of false alarm, 237, 310, 314, 322, 330, 344, 351
Congruence axiom for metric space built upon generalized Boolean algebra with a measure, 48–49
Connection (containment) axiom for metric space built upon generalized Boolean algebra with a measure, 47
Continuity axiom for metric space built upon generalized Boolean algebra with a measure, 49
Continuous channel capacity, 190–196
   distinguishing continuous and discrete channels, 196
   noiseless channels, 200–201
   presence of interference (noise), 244–247
Continuous random variable entropy, 15
Correlation ratio, 76
Cosine theorem for metric space built upon generalized Boolean algebra with a measure, 51–54
Cos-invariant theorem, 52
Cramer-Rao lower bound, 271
Cryptographic coding paradox, 17
Cumulative distribution functions (CDFs)
   estimator for quality indices of unknown nonrandom parameter, 219-221
   probabilistic measure of statistical interrelationship, 98-99

Index 405

RF signal resolution algorithm, 371
signal detection quality index, 239–240
signal extraction algorithm analysis, 294–296
signal resolution-detection quality indices, 252
signal resolution-estimation quality indices, 262
stochastic process description, 76
unknown nonrandom parameter estimation in sample spaces with lattice properties, 274–275
Curvature measure, 68, 137
Cybernetics, 6

D

Decision gate (DG), 309–310, 320, 335, 337, 349, 354
Deterministic approach, 2
combining probabilistic approaches for measures of information quantity, 18–19
Deterministic signal classification, algorithm synthesis and analysis, 352–358
Deterministic signal detection, algorithm synthesis and analysis, 305–311
Difference, 27
Differential entropy, 15-16
Differential entropy noninvariance property, 17
Dirac delta function, 108, 132, 262, 295, 371
Discrete channel capacity
distinguishing continuous and discrete channels, 196
generalized discrete random sequence, 183–190
noiseless channels, 199
normalized autocorrelation function, 179–180
presence of interference (noise), 247–250

Discretization, 67–68, 134–136
Disjoint elements, 65
Distributive lattice, 26–27
Diversity, 2, 3, 5, 12, 21
Doppler shifts, 368–369
Dynamic error of signal filtering, 301
Dynamic error of signal smoothing, 302

E

Effective width, 79–80, 197, 203–206
Elliptic geometry, 51
Entropy, 7, 8
differential entropy noninvariance problem, 17
information relationship, 9, 14–18, 21
transition from discrete to continuous random variable, 15–16
Envelope computation unit (ECU), 320, 337, 348
Equivalence principle in sampling theorem, 134, 136, *See also* Equivalent representation theorems
Equivalent representation theorems, 69–71, 138–141
Estimation errors, 271
Estimation problem, 208–209
quality indices of unknown parameter, 218–229 *See also* Estimators; Quality indices
Estimator error, fluctuation component of, 301
Estimator formation unit (EFU), 348–349
Estimators, 208, 272
for known and unknown modulating functions, 302–303
quality index of estimator definitions, 212, 227, 233–234, 240, 253, 283
quality indices for metric sample space, 282–287
superefficiency of, 226, 271, 281

unknown nonrandom parameters in sample spaces with lattice properties, 271–281 *See also* Quality indices

Estimator space with metric, 261–262, 282

Euclidean geometry, 10–11

Everywhere dense set, 41

Extraction problem, *See* Signal filtering

F

False alarm conditional probability, 237, 314, 322, 330, 344, 351

Filtering (extraction) problem, *See* Signal filtering

Filtering error, fluctuation component of, 301

Filter of signal space with lattice properties, 367

Finite additive measure, 28

Fischer's measure of information quantity, 5

Fluctuation component of signal estimator error, 301

Fourier transform, 78–79

Franks, Lewis, 11

Frequency modulated signal detection, algorithm synthesis and analysis for metric space with L-group properties, 342–351

Frequency modulation, 173

G

Galilei, Galileo, 9

Gaussian communication channel, 196

Generalized Boolean algebra, 27

Generalized Boolean algebra with a measure, 23, 26, 122, 126, 386
  axiom of measure of binary operations, 109–110, 126, 385
  definition, 28–29
  geometrical properties of metric space built upon, 29–59
  homomorphic mappings in signal space built upon, 133–145

ideal/quasi-ideal signal interactions during additive interactions in signal space, 151–156

informational properties of information carrier space, 59–73, *See also* Information carrier space, informational properties

notation and identities, 26–28

subalgebra for physical and informational signal space, 124

Generalized Boolean lattice, 27

Generalized Boolean ring, 27, 59, 130

Generalized discrete random sequence, 183–190

Generator points of plane, 44

Generator points of sheet, 42

Genetic information transmission, 6

H

Half-perimeter of triangle, 52

Harmonic signal detection, algorithm synthesis and analysis, 311
  block diagrams, 319
  with joint estimation of signal arrival, 311–324
  with joint estimation of signal arrival, initial phase, and amplitude, 324–341

Hartley's logarithmic measure, 14, 184, 191

Heaviside step function, 132, 139, 182, 198, 262, 275, 294, 371

Hexahedron, 30–32

Hilbert transform, 303, 314, 330, 344

Hilbert's axiom system, 35–36

Homomorphic mappings in signal space built upon generalized Boolean algebra with a measure, 133–145

Homomorphism of stochastic process, 134–140

Hyperbolic geometry, 51

Hyperspectral density (HSD), 79–81, 202–206

Hypothesis testing, 305

I

Ideal and quasi-ideal signal interaction, 147–156
Ideal filter, 197
Identity of signals in informational sense, 150
Indexed set, 26, 39
Information, defining, 4–5, 25, 388
Informational inequality (informational paradox) in signal space, 149–150
Informational inequality of signal processing theorem, 213–214
Informational interrelations, 125
Informational paradox (informational inequality), 149–150
Informational signal space, 123–133
    features of signal interaction, *See* Signal space, features of signal interaction in space with algebraic properties
    information quantity definitions and relationships, 126–133, 149
    morphism of physical signal space, 122
    units of information quantity, 127–128 *See also* Information carrier space; Metric space built upon generalized Boolean algebra with a measure; Signal space
Informational variable, 173
Information and entropy relationship, 9, 14–18
Information carrier space
    definition, 28–29
    informational properties, 59–73
        axiom of a measure of binary operation, 59–60
        information losses of first and second genus, 68–69
        information quantity definitions, 60–64
        information quantity relationships, 65–67
        sets under discretization, 67–68
        theorems on equivalent representation, 69–71
        theorems on isomorphic mapping, 71–73
    notation and identities, 26–28 *See also* Informational signal space; Metric space built upon generalized Boolean algebra with a measure; Signal space
Information distribution density (IDD) of stochastic processes, 101–109, 112, 122, 134, 198
    features of signal interaction in signal space with algebraic properties, 148
    homomorphism of continuous stochastic process, 134, 136–140
    informational signal space definition, 125
    information quantity carried by binary signals, 176
    information quantity carried by m-ary signals, 181–182
    information quantity definitions and relationships, 110–113
    mutual IDD, 115–122
    noiseless channel capacity evaluation, 201–206
    physical signal space definition, 124 *See also* Mutual information distribution density
Information losses, 267
    of the first genus, 68, 137
    in physical signal space during additive interactions, 149–151, 267
    of the second genus, 68–69, 137
    signal processing quality indices, *See* Quality indices
Information quantity, 267
    carried by binary signals, 174–179

carried by continuous signals, 190–196
carried by m-ary signals, 179–190
entropy as measure of, 21
generalized discrete random sequence, 183–190
main axiom of signal processing theory, 22–23
measures, *See* Measure of information quantity
mutual information in a signal, 146–147
theoretical problems, 14–22 *See also* Channel capacity; Information distribution density (IDD) of stochastic processes; Normalized measure of statistical interrelationship; Probabilistic measure of statistical interrelationship; specific quantities

Information quantity, definitions and relationships
features of signal interaction in signal space with algebraic properties, 148, 157–158
informational signal space, 126–133, 149
information carrier space, 60–67
information losses of first and second genus, 68–69, 137
metric and informational relationships in signal space, 164, 167–170
physical signal space, 149
for stochastic processes, 110–113
theorems on equivalent representation, 69–71, 138–141
theorems on isomorphic mapping, 71–73, 143–145 *See also* specific quantities

Information theory
constructing using Boolean algebra with a measure, 23, *See also* Generalized Boolean algebra with a measure
distinguishing continuous and discrete channels, 196
information quantity in a signal, 146–147
measurement and, 7–8
morphism of physical signal space, 124
natural sciences and, 5–10
signal processing theory relationship, 25–26, 385, 386–387
subjective perception of message problem, 18
theoretical problems, 14–22 *See also* Information quantity; Measure of information quantity

Information transmitting and receiving channel, 196, *See also* Channel capacity

Initial phase estimation
harmonic signal detection algorithm synthesis and analysis, 324–341
linear frequency modulated signal detection algorithm synthesis and analysis, 342–351

Interference (noise) conditions, channel capacity and, 242–250, *See also* Channel capacity

Interpolation problem, *See* Signal interpolation

Invariance, 2–3, 75, 386
differential entropy noninvariance problem, 17
signal-resolution detection quality indices problem, 259
theorem of normalized function of statistical interrelationship, 80–81

Invariance principles, 2

Isomorphic mapping theorems, 71–73, 143–145

Isomorphism of stochastic process, 135

Isomorphism preserving a measure, 29, 125

J
Join, 82, 88, 92, 153–156, 164
Joint detection and estimation, 312, 314, 325, 329, 332, 343, 344, 347

K
Kepler, Johannes, 1
Keying, 173
Klein, Felix, 11

L
Lattice, 26–27, 82–83, 269
    algebraic properties, 269–271
    ideal/quasi-ideal signal interactions during additive interactions in signal space, 153–154
    signal interaction informational relationships in signal spaces with algebraic properties, 156
    See also L-groups; Metric space with lattice properties; Sample space with lattice properties
Lattice-ordered groups, *See* L-groups
Lattice with relative components, 26–27
Least modules method (LMM) based estimators, 272
Least-squares method (LSM) based estimators, 272
Leibniz, Gottfried, 10–11
L-groups (lattice-ordered groups), 83–84, 218, 270, 386
    algebraic axioms, 270
    channel capacity, *See* Channel capacity
    main signal processing problems, 208–210
    quality indices for signal processing in metric spaces, *See* Quality indices *See also* Metric space with lattice properties; Sample space with lattice properties

Likelihood ratio computation, 305, 352-358
Limit of everywhere dense linearly ordered indexed set, 41
Line, 4
    with linearly ordered elements, 40
    in metric space built upon generalized Boolean algebra with a measure, 32–41
    with partially ordered elements, 39
Linear frequency modulated (LFM) signal detection, algorithm synthesis and analysis for metric space with L-group properties, 342–351
Linearly ordered indexed set, 39
Linear sample space, 271–272
    signal detection problem, 304–305
    unknown nonrandom parameter estimation, 271–276, 286
    quality indices of estimators in metric sample space, 283
    relative estimation efficiency, 276–281 *See also* Sample space with lattice properties
Linear spaces, 11–12, 19–20, 270, 386
    mapping into signal space with lattice properties, 379–383 *See also* Linear sample space; Physical signal space
Logarithmic measure of information quantity, 14, 184, 191–193

M
Main axiom of signal processing theory, 22–23, 385
Mapping methods for metric space, 56
Mapping signal spaces into signal space with lattice properties, 378–384
    linear signal space, 379–383
    signal space with semigroup properties, 383–384
M-ary signals, information quantity carried by, 179–190
Matched filtering in signal space with

L-group properties, 312–313, 327, 331, 343, 345
Matched filtering unit (MFU), 335–336, 338
Measurement, 7–8
Measurement errors, 271, 272
Measure of binary operations axiom, 59–60, 109–110, 126
Measure of information quantity, 14, 17–18, 25, 157, 385, 386, 388
   absolute unit (abit), 63, 169, 177–178, 200–201
   bits, 178, 200–201
   building using Boolean algebra with a measure, 25
   combining probabilistic and deterministic approaches, 18–19
   entropy as, 21, *See also* Entropy
   Fischer's measure, 5
   logarithmic measure, 14, 184, 191–193
   Shannon's measure, 17–18
   suggested criteria for, 22
   units of in signal space, 127–128
     See also Generalized Boolean algebra with a measure; Information quantity; Metric space built upon generalized Boolean algebra with a measure; Normalized measure of statistical interrelationship; Probabilistic measure of statistical interrelationship
Meet, 82, 88, 92, 153–156, 164
Metric, 29, 31, 77–78, 80, 84, 101, 115, 125, 385
   interacting samples in *L*-group, 83–84
   between stochastic processes, 78, 116
   between two samples, 78, 84 *See also* Measure of information quantity
Metric and trigonometrical relationships, metric space built upon generalized Boolean algebra with a measure, 51–54
Metric function theorems for quality indices of unknown nonrandom parameter estimation, 221–225
Metric function theorems for signal interactions in physical signal space, 165, 172
Metric inequality of signal processing, 212
Metric space, 11, 29–30, 56–57, 125, 385
Metric space built upon generalized Boolean algebra with a measure, 29
   algebraic properties, 58–59
   definition, 28, 56–57
   geometrical properties, 57–58
     axiomatic system, 46–51
     line, 32–41
     main relationships between elements, 29–32
     metric and trigonometrical relationships, 51–54
     properties with normalized metric, 54–59
     sheet and plane, 41–46
   informational space definition, 125
   mapping methods, 56
   normalized measure of statistical interrelationship properties, 93–98
   pseudometric space, 166 *See also* Informational signal space; Information carrier space; Physical signal space; Sample space with lattice properties; Signal space
Metric space with lattice properties, 386
   channel capacity in presence of noise, 242–250, *See also* Channel capacity
   quality indices of unknown nonrandom parameter estimators, 282–287

signal classification algorithm and
analysis, 352–358
signal detection algorithm synthesis
and analysis, 304–351
signal processing algorithm
synthesis and analysis, *See*
Signal processing algorithms,
synthesis and analysis
signal processing quality indices,
See Quality indices
stochastic signal extraction
algorithm synthesis and
analysis, 287–304 *See also*
Sample space with lattice
properties
Mills' function, 279
Mismatching parameter for signal
resolution algorithm, 359,
368–370, 377
Mismatching function, 359
Mityugov, Vadim, 247
Modulating functions, 173, 178, 208,
210, 247, 260, 302–303
Modulation, 173
Mutual information distribution density
(mutual IDD), 115–122
physical signal space definition, 124
Mutual information quantity, *See*
Quantity of mutual information
Mutual normalized function of
statistical interrelationship
(mutual NFSI), 77–78, 115, 118

N

Natural sciences
concepts and research methodology,
1–5
information theory and, 5–10
space-related problems, 10–12
Newton, Isaac, 8, 10
Noise conditions, channel capacity and,
See Channel capacity
Noiseless channel capacity, 196–206
Non-Euclidean geometries, 3–4, 385,
388

Nonlinearity, 3, 75, 387
Normalized autocorrelation function
(ACF), 175, 179–181
Normalized function of statistical
interrelationship (NFSI),
76–82, 122, 148
definition, 77, 113
discrete random sequence with
arbitrary state transition
probabilities, 179–190
hyperspectral density, 79–81
informational signal space
definition, 125
information distribution density,
101–109
information quantity carried by
binary signals, 175–176
invariance theorem, 80–81
mutual NFSI, 77–78, 115, 118
necessary condition for carrying
information, 106–107
noiseless channel capacity
evaluation, 202, 204
quality index of signal
classification, 233
Normalized measure, 28, 58
Normalized measure of statistical
interrelationship (NMSI),
82–98, 122, 163
definition, 84–85
quality index of estimator
definitions, 212, 240
signal interaction informational
relationships in signal spaces
with algebraic properties,
157–158
signal processing inequality
theorem, 213
Normalized metric, metric space built
upon generalized Boolean
algebra with a measure, 54–59
Normalized mismatching function, 359
Normalized time-frequency
mismatching function, 359,
368–370, 377

Normalized variance function, 76
Null element, 26

O

Optimality criteria, 146, 207
　choice considerations, 268–269
　harmonic signal detection
　　algorithm, 325–329
　linear frequency modulated signal
　　detection algorithm, 343, 345
　RF signal resolution algorithm,
　　360–362 See also Quality
　　indices
Order axiom for metric space built
　upon generalized Boolean
　algebra with a measure, 47–48
Ordinate set, 58, 108, 114–115, 118,
　120–121
Orthogonality relationships, metric
　space built upon generalized
　Boolean algebra with a
　measure, 30–32
Overall quantity of information, 64–66,
　111, 130, 132, 132–133, 137,
　139, 197, See also Quantity of
　overall information
　invariance property of, 72, 73, 144,
　　145, 171

P

Paley-Wiener condition, 105, 107
Parabolic geometry, 51
Parallels for a plane, axiom, 50
Parallels for a sheet, axiom, 50–51
Parameter filtering, 208
Partially ordered set, 82–84, 100, 156,
　269
Passive mapping, 56
Phase estimation
　harmonic signal detection algorithm
　　synthesis and analysis, 324–341
　linear frequency modulated signal
　　detection algorithm synthesis
　　and analysis, 342-351
Phase modulation, 173

Physical interaction, 124, 269
Physical signal space, 123–124, 269, 386
　algebraic properties, 123–124
　ideal/quasi-ideal interactions during
　　additive interactions, 151–156
　information losses during additive
　　interactions, 149–151, 267
　information quantity definitions
　　and relationships, 149, 157–158
　metric and informational
　　relationships between
　　interacting signals, 163–172
　morphism into informational signal
　　space, 122, 124
　signal interaction informational
　　relationships in spaces with
　　algebraic properties, 156–163
　stochastic process generalization,
　　156 See also Linear spaces;
　　Metric space built upon
　　generalized Boolean algebra
　　with a measure; Signal space
Plane and sheet, notions for metric
　space built upon generalized
　Boolean algebra with a
　measure, 41–46
　axioms for parallels, 49–50
Potential resolutions of a filter, 359,
　370–371
Power spectral density, 81–82, 105, 107
Prior uncertainty, 122, 229, 237, 242,
　247, 259, 264, 268, 303–304,
　310–311, 351, 356, 358, 386
Probabilistic approach, 2, 75
　combining deterministic approaches
　　for measures of information
　　quantity, 18–19
Probabilistic measure of statistical
　interrelationship (PMSI),
　98–101, 122, 156–157, 163
Probabilities, 14
Probability density functions (PDFs)
　entropy and information
　　relationship, 14–15

generalized discrete random sequence, 183–185
information quantity carried by binary signals, 174
probabilistic measure of statistical interrelationship, 98–99
RF signal resolution algorithm, 371–376
signal detection quality index, 239
signal extraction algorithm analysis, 294–301
signal resolution-detection quality indices, 252
signal resolution-estimation quality indices, 261–263
stochastic process description, 76–77, 82
unknown nonrandom parameter estimation in sample spaces with lattice properties, 274–276

Pseudometric space, 166

Q

Quality indices, 207
estimator space with metric, 261–262, 282
main signal processing problems, 208–210
metric function theorems, 221–225
optimality criteria, 146
quality index of estimator definitions, 212, 227, 233–234, 240
resolution-detection, 250–259
resolution-estimation, 259–265
sample size and, 228
signal classification, 229–237
signal detection, 237–242, 321
signal filtering (extraction), 210–218
signal processing inequality theorems, 212–214
unknown nonrandom parameter estimation, 218–229 See also Optimality criteria

Quantity of absolute information, 60–63, 110–111, 126, 127, 148, 149–150, 164, 168–169
  invariance property of, 72, 144, 171
  unit of measure (abit), 63, 169, 177–178
Quantity of mutual information, 60–61, 110, 112, 120, 121, 126–127, 148, 149–150, 157, 164, 167, 211
  invariance property of, 72, 73, 144, 145, 171
Quantity of overall information, 60–61, 110, 121, 126, 148, 149, 150, 163–164, 164, 168, 176
  invariance property of, 72, 73, 144, 145, 171
Quantity of particular relative information, 60, 62, 110, 121, 127
  invariance property of, 72, 73, 144, 145
Quantity of relative information, 60–62, 110–111, 121, 127, 167
  invariance property of, 72, 73, 144, 145, 171
Quantum information theory, 7
Quasi-ideal signal interactions in signal space, 152–156

R

Radio frequency (RF) pulse resolution, 358–378, See also Signal resolution
Relative component, 27
Relative quantity of information, 64–66, 111, 130, 132–133, 137, 197, 200, See also Quantity of relative information
  invariance property of, 72, 144
Research methodology, 1–5
Resolution, See Signal resolution
Resolution-classification, 209
Resolution-detection, 210 quality index, 250–259 See also Signal detection; Signal resolution

Resolution-estimation, 210
    quality indices, 259–265
    resolution-detection-estimation problem, 259–260 *See also* Signal resolution
Resolution in relative frequency shift, 377
Resolution in relative time delay, 377
Resolution measure of a filter, 370
Resolution of radio frequency (RF) pulses, 358–378, *See also* Signal resolution
Riemann, Bernhard, 11
Ring, 384

S

Sample space with lattice properties, 218, 269–271
    algebraic properties, 268–269
    mapping signal space into, 378–384
        linear signal space, 379–383
        signal space with semigroup properties, 383–384
    matched filtering problem, 312–313, 327, 331, 343, 345
    unknown nonrandom parameter estimation, 271–276, 286–287
        efficiency based on quality indices in metric sample space, 282–287
        estimator variance theorem and corollary, 276–279
        indirect measurement models, 272–273
        quality indices of estimators in metric sample space, 282–287
        relative estimation efficiency with respect to linear sample space, 276–281 *See also* L-groups; Metric space with lattice properties
Sample space with L-group properties, 218, *See also* Sample space with lattice properties
Sampling, 134, 135, 141
Sampling interval, 136, 140, 141

Sampling theorem, 12–13, 19, 133–134
    equivalence principle, 134, 136, *See also* Equivalent representation theorems
    homomorphic mappings in signal space built upon generalized Boolean algebra with a measure, 134
Second Law of thermodynamics, 8
Semantic information theory, 10
Semigroup properties
    defined binary operations, 157
    mapping signal spaces into signal space with lattice properties, 383-384
    physical signal space definition, 123, 269–270
    usual interaction of signals, 158
Shannon, Claude, and information theory, 4, 6, 9, 15
    limit on channel capacity, 155–156, 357
    measure of information quantity, 17–18
Sheet and plane, 41–46
    axioms for parallels, 49–50
Signal amplitude estimation
    harmonic signal detection algorithm synthesis and analysis, 324–341
    linear frequency modulated signal detection algorithm synthesis and analysis, 342–351
Signal classification, 209
    algorithm synthesis and analysis, 352–358
    likelihood ratio computation, 352–353
    processing unit block diagram, 355
    quality indices, 229–237
    resolution-classification, 209–210
Signal detection, 209, 304
    algorithm synthesis and analysis, 304, 351
        block diagrams, 308, 323, 335, 336, 348–349

deterministic signal, 305–311
harmonic signal, 311–341
likelihood ratio computation, 305
linear frequency modulated signal, 342–351
optimality criteria, 325–329, 343, 345
quality indices, 237–242, 321
quality indices of resolution-detection, 250–259
resolution-detection-estimation problem, 259–260

Signal estimator, 208

Signal estimator error, fluctuation component of, 301

Signal extraction, *See* Signal filtering

Signal filtering (extraction), 208, 210
algorithm synthesis and analysis, 287–288, 303–304
analysis of optimal algorithm, 294–302
further processing possibilities, 302–303
optimal algorithm synthesis, 288–293
dynamic error, 301
estimators for known and unknown modulating functions, 302–303
processing unit block diagram, 293, 303
quality indices, 210–218

Signal initial phase estimation
harmonic signal detection algorithm synthesis and analysis, 324–341
linear frequency modulated signal detection algorithm synthesis and analysis, 342–351

Signal interpolation, 208, *See also* Signal smoothing

Signal parameter estimation, 208, *See also* Estimation problem

Signal parameter estimator, 208, *See also* Estimators

Signal parameter filtering, 208, *See also* Signal filtering

Signal processing algorithms, synthesis and analysis, 267–268
block diagrams
classification unit, 355
detection units, 308, 319, 323, 335, 336, 348–349
extraction unit, 293, 303, 336
matched filtering unit, 336
resolution unit, 366
choice of optimality criteria, 268–269
decision gates, 309–310, 320, 335, 337, 349, 354
deterministic signal classification, 352–358
estimators for known and unknown modulating functions, 302–303
mapping signal spaces into signal space with lattice properties, 378–384
matched filtering problem, 312–313, 327, 331, 343, 345
prior uncertainty problem, 268
signal detection, 304, 351
deterministic signal, 305–311
harmonic signal, 311-341
linear frequency modulated signal, 342–351
optimality criteria, 325–329, 343, 345
signal resolution, 358–378
block diagram, 366
conditional PDF, 371–376
Doppler shifts, 368–369
filter of signal space with lattice properties, 367
normalized time-frequency mismatching function, 359, 368–370, 377
optimality criteria, 360–362
stochastic signal extraction, 287–288, 303–304
analysis of optimal algorithm, 294–302

further processing possibilities, 302–303
optimal algorithm synthesis, 288–293
processing unit block diagram, 293
Signal processing inequality theorems, 212–214
Signal processing problems, 207–210, 267
Signal processing theory
constructing using Boolean algebra with a measure, 23, *See also* Generalized Boolean algebra with a measure
information theory and, 25–26, 385, 386–387
main axiom, 22–23, 385
main problems, general model of interaction, 267
main problems, quality indices context, 207–210, *See also* Quality indices
sampling theorem, 12–13, 133–134
Signal receiving error probability, 230
Signal representation equation, 173
Signal resolution, 358–360
mismatching parameter, 359, 368–370, 377
quality indices, 259–265
in relative frequency shift, 377
in relative time delay, 377
resolution-classification, 209–210
resolution-detection, 210
resolution-detection-estimation problem, 259–260
resolution-detection quality indices, 250–259
resolution-estimation, 210
RF pulse resolution algorithm synthesis and analysis, 358–378
block diagram, 366
conditional PDF, 371–376
Doppler shifts, 368–369
filter of signal space with lattice properties, 367
filter potential resolution, 370–371
filter resolution measure, 370
normalized time-frequency mismatching function, 368–369, 377
optimality criteria, 360–362
uncertainty function, 358–359
Signal smoothing, 208, 302
Signal space, 11–12, 19–20, 123, 385
homomorphic mappings of continuous signal, 133–143
theorems on equivalent representation, 138–141
theorems on isomorphic mapping, 143–145
information carrier space definition, 28–29
information carrier space notation and identities, 26–28
information quantity definitions and relationships, 126–133, *See also* Information quantity
mapping into signal space with lattice properties, 378–384
physical and informational, 122, 123–133, *See also* Informational signal space; Physical signal space
theoretical problems, 25
units of information quantity, 127–128 *See also* Information carrier space; Linear sample space; Metric space built upon generalized Boolean algebra with a measure; Sample space with lattice properties
Signal space, features of signal interaction in space with algebraic properties, 146–163
additive interaction and ideal/quasi-ideal interaction, 147–156

Index 417

informational paradox (informational inequality), 149–150
information quantity definitions and relationships, 148–150
optimality criteria, 146
space with various algebraic properties, 156–163
summary conclusions, 162–163
Signal space, metric and informational relationships between interacting signals, 163–172
information quantity definitions and relationships, 164, 167–170
theorems on metric functions, 165, 172
Signal spaces with lattice properties, mapping of signal spaces into, 378–384
linear signal space, 379–383
signal space with semigroup properties, 383–384 *See also* Sample space with lattice properties
Signal theory, space in, 11–12, 19–20, *See also* Signal space
Signal time of arrival (ending) estimation
harmonic signal detection algorithm synthesis and analysis, 311–341
linear frequency modulated signal detection algorithm synthesis and analysis, 342–351
Simplex, 54, 55–56
Sine theorem for metric space built upon generalized Boolean algebra with a measure, 52
Sin-invariant theorem, 53
Smoothing problem, *See* Signal smoothing
Space, 3-4
estimator space with metric, 261–262, 282
probabilistic approach, 75

in signal theory, 11–12, 19–20, *See also* Signal space
theoretical problems, 10–12 *See also* Informational signal space; Information carrier space; Linear spaces; Metric space built upon generalized Boolean algebra with a measure; Physical signal space; Sample space with lattice properties; Signal space
Statistical approaches, 5, 14
Statistical hypotheses testing, 305
Stochastic processes, 75–76, 191
analyticity, 104–105
Butterworth filters and, 107–108
classification, 78
dependence between instantaneous values of interacting signals, 82–83
discretization of continuous signal, 135–136
generalized for physical signal space as a whole, 156
homomorphism of continuous process, 134–140
hyperspectral density, 79–81
informational characteristics and properties, 108–122, 123, *See also* Information quantity
axiom of measure of binary operations, 109–110
information distribution density, 101–108, 112, *See also* Information density distribution (IDD) of stochastic processes
information quantity definitions and relationships, 110–113
mutual information density distribution, 115–122
necessary condition for carrying information, 106–107
isomorphism of, 135
metric between, 78, 116

normalized function of statistical interrelationship, 77–82, 113

normalized measure of statistical interrelationship, 82–98

probabilistic measure of statistical interrelationship, 98–101

quantity of information carried by signals, *See* Information quantity

signal extraction in metric space with lattice properties, 287–303

theorems on equivalent representation, 69–71, 143–145

theorems on isomorphic mapping, 71–73, 143–145 *See also* Normalized function of statistical interrelationship; Normalized measure of statistical interrelationship; Probabilistic measure of statistical interrelationship

Stone's duality, 28, 59, 64, 130

Straight lines, 4

Strobing circuit (SC), 355

Superefficient estimators, 226, 271, 281

Superposition principle, 3

Swift, Jonathan, 17–18

Symmetry, 2–3

System of elements, 28

T

Tetrahedron, 30, 32, 55

Theorem on cos-invariant, 52

Theorem on sin-invariant, 53

Theorems on equivalent representation, 69–71, 138–141

Theorems on isomorphic mapping, 71–73, 143–145

Thermodynamic entropy, 7, 8

Time direction, 8

Time of signal arrival (ending) estimation

harmonic signal detection algorithm synthesis and analysis, 311–341

linear frequency modulated signal detection algorithm synthesis and analysis, 342–351

Triangle, 31, 48, 51

Triangle angle sum, 53

Trigonometrical relationships, metric space built upon generalized Boolean algebra with a measure, 51–54

Tuller, W. G., 23

U

Ultimate Shannon's limit, 155–156

Uncertainty function, 358–359

Unit element, 26

Unit of measure of information quantity (abit), 63, 169, 177–178

Unity, 26

Universal algebra, 26–27, 282

Usual interaction of two signals, 158

V

Valuation, 84, 222, 233

Valuation identity, 88

W

Weak stationary stochastic process, 78

Wiener, Norbert, 6, 9, 186

Woodward, P. M., 268, 358

Z

Zero, 26

Zheleznov, Nikolai, 13

PGSTL 10/31/2017